D1241924

GAUGE FIELD THEORIES

GAUGE FIELD THEORIES

AN INTRODUCTION WITH APPLICATIONS

MIKE GUIDRY
Department of Physics
University of Tennessee
Knoxville, Tennessee

and

Oak Ridge National Laboratory
Oak Ridge, Tennessee

A Wiley-Interscience Publication
JOHN WILEY & SONS, INC.
New York / Chichester / Brisbane / Toronto / Singapore

Copyright © 1991 by John Wiley & Sons, Inc.

All rights reserved. Published simultaneously in Canada.

Reproduction or translation of any part of this work
beyond that permitted by Section 107 or 108 of the
1976 United States Copyright Act without the permission
of the copyright owner is unlawful. Requests for
permission or further information should be addressed to
the Permissions Department, John Wiley & Sons, Inc.

Library of Congress Cataloging in Publication Data:

Gúidry, M.W.
    Gauge field theories: an introduction with applications/ Mike
Guidry.
        p.   cm.
    "A Wiley-Interscience publication."
    Includes bibliographical references.
    ISBN 0-471-63117-5
    1. Gauge fields (Physics)   2. Quantum field theory. I. Title.
    QC793.3.F5G85   1991
    530.1′435 – dc20   90-12943
                                                CIP

Printed in the United States of America

10 9 8 7 6 5 4 3 2 1

*For my Mother, Michael,
and Amanda Michelle*

Preface

Over the last two decades a revolution has taken place concerning the way physicists view the fundamental processes taking place in our universe. This revolution has its basis in the belief that *all* fundamental interactions are associated with a particularly beautiful and powerful kind of quantum field theory–a theory of *local gauge fields*. This tenet of faith is termed the *gauge principle*. These theories certainly satisfy the criterion of beauty, and they have passed sufficient experimental tests to ensure that they are at least approximately correct for the weak, electromagnetic, and strong interactions. There is even some indication that superstring theory may bring gravitation into the unified gauge fold, but this is presently far from certain.

The intent of this book is to allow a person who has little background in elementary particle physics and quantum field theory to acquire sufficient expertise to speak the language and to confront the literature in this and related disciplines. In particular, I have in mind three areas of research that should present challenging opportunities in the relatively near future for particle, nuclear, and astrophysicists. The first is an elaboration of the role that gauge field theories, or effective theories derived from gauge theories, play in nuclear physics: where are the quarks and gluons in nuclei, and what is their relation to the traditional nucleon and meson degrees of freedom? There is considerable phenomenological understanding, but microscopically we have only scratched the surface of this problem. The second concerns the possibility of creating fundamental new states of matter such as an extended quark–gluon plasma in ultrarelativistic heavy ion collisions; this also is a field in its infancy. The third is the relation of gauge theories to the creation and evolution of the universe: what role do these theories play in cosmology, and how does cosmology test and illuminate our understanding of elementary particle physics?

The book is divided into three parts. Part I is an introduction to the general principles of relativistic quantum field theory. I assume only that the reader has a background in nonrelativistic quantum mechanics, and develop field theory first using the traditional method of canonical quantization, and then from the modern perspective of path integrals.

In Part II we sharpen our focus and introduce the essential ingredients of gauge fields for the weak, electromagnetic, and strong interactions. This section begins with an introduction to those principles of group theory that will be required in the subsequent discussion of gauge fields. The phenomenology of the weak interactions is then used to motivate the introduction of non-Abelian gauge fields (Yang–Mills fields) and the pivotal concept of spontaneous symmetry breaking in the Goldstone and Higgs modes. We then examine the Glashow–Salam–Weinberg electroweak gauge theory and the non-Abelian gauge theory of the strong interactions, quantum chromodynamics (QCD). Finally, Part II concludes with a short introduction to grand unified theories (GUTs), which attempt to describe all non-gravitational interactions within a single gauge theory. In Part II, as in Part I, no knowledge is assumed of elementary particle physics beyond that which a student is likely to have encountered in general introductions to modern physics.

In Part III, we shift our attention to the interface between modern elementary particle physics and "applied disciplines" such as nuclear physics, astrophysics, and cosmology. Models and approximations of non-perturbative QCD are a dominant theme of this part, because of their relevance for nuclear physics. Here the reader will find introductions to phenomenological models of quark confinement such as chiral bags and quark potential models, to lattice gauge theories, and to topological solitons such as the instanton and Skyrmion. Finally, in the last two chapters topics that are of particular concern to nuclear physicists and cosmologists are addressed. In Ch. 14 we consider the quark–gluon plasma that presumably characterized the universe microseconds after the big bang, and that may be produced soon in laboratory experiments, and in Ch. 15 we examine the part that gauge theories may have taken in producing some of the essential characteristics of our universe.

As we proceed, the relevance of experimental results will be stressed, but technical experimental details will be omitted. Of necessity, considerable mathematics will be invoked, but we will emphasize the essential conceptual points and avoid the most tedious details wherever possible (forest > trees principle). As noted above, the reader will be assumed innocent of any knowledge of elementary particle physics, but conversant with an intermediate level quantum mechanics (angular momentum theory and rudimentary second-quantization methods). A background in more advanced quantum mechanics is desirable but not mandatory, since the necessary advanced topics will be reviewed. As far as possible, the presentation is self-contained. To accomplish this without obscuring the basic thread of the story I have often resorted to placing important background material, proofs, and derivations in exercises scattered throughout the text. There are 209 such exercises (counting each independent part as a separate exercise).

Outline solutions to these exercises, or a reference to the literature where a clear discussion of the solutions may be found, are contained in the 60 pages of Appendix D. As a consequence of this device, there are two levels at which this book may be used. Without the exercises it constitutes an introduction to concepts and terminology that should be useful to nonspecialists, or as a prelude to the more advanced literature for those concentrating in related disciplines. With the exercises included, this material would be suitable as a graduate level introduction for students intending to do research in the areas discussed above. The exercises range in difficulty from rather trivial to more advanced. Each is included for one or more definite reasons: (1) to make an important point omitted from the main text, (2) to elaborate on assertions made without proof in the text, or (3) to illustrate mathematical techniques that are commonly used in this field.

This book originated in a series of lectures given to graduate students at the University of Tennessee and to research scientists at the Oak Ridge National Laboratory in the period 1983–1985. I wish to thank the students and colleagues who attended those lectures and contributed useful insights through their comments and questions. I also wish to acknowledge collectively the multitude of secretaries who typed the original lectures on which this book is based.

I thank the corresponding publishers for permission to reproduce the following illustrations: Figures 9.2 and 9.3 reproduced, by permission, from C. Quigg, *Gauge Theories of the Strong, Weak, and Electromagnetic Interactions,* (Addison–Wesley, 1983). Figure 10.20 reproduced, with permission, from the *Annual Review of Nuclear and Particle Science,* Volume **31**, ©1981 by Annual Reviews Inc. Figure 13.9 reproduced, with permission, from the *Annual Review of Nuclear and Particle Science,* Volume **35**, ©1985 by Annual Reviews Inc. Figure 14.7 reproduced, with permission, from the *Annual Review of Nuclear and Particle Science,* Volume **35**, ©1985 by Annual Reviews Inc. Figure 14.8 reproduced, with permission, from the *Annual Review of Nuclear and Particle Science,* Volume **35**, ©1985 by Annual Reviews Inc. Figure 14.9 reproduced, with permission, from the *Annual Review of Nuclear and Particle Science,* Volume **35**, ©1985 by Annual Reviews Inc. Figure 15.4 reproduced, with permission, from the *Annual Review of Nuclear and Particle Science,* Volume **33**, ©1983 by Annual Reviews Inc.

A number of colleagues took time from their busy schedules to read portions of the manuscript and to make valuable suggestions: J. Q. Chen, K. T. R. Davies, R. Donangelo, D. H. Feng, A. H. Guth, E. G. Harris (who read it all), R. W. Kincaid, A. Klein, Z. P. Li, P. Ring, S. P. Sorensen, C. W. Wong, A. Q. Wu, C. L. Wu, and W. M. Zhang. In addition to ferreting out error, they have contributed significantly to the clarity of the presentation.

Several people played vital roles in the production of the book. Wayne Kincaid was diligent in a careful reading of the manuscript and was responsible for many grammatical and style suggestions that have improved both the readability and the accuracy. The final typesetting was performed by Ming-Bo Liu, with assistance from Wayne Kincaid, using $\mathcal{A}\mathcal{M}\mathcal{S}$-TeX 2.0 on an IBM PS/2 Model 50Z. Phillip Bingham typeset portions of the preliminary manuscript. All drawings were done by Xiao-Ling Han on a Macintosh SE and Macintosh IIsi using Superpaint and Aldus Freehand graphics. The TeX files were printed at 2000 dpi on a Chelgraph IBX2000 printer; the Macintosh drawings were converted to Postscript format and printed at 2400 dpi on a Verityper 5300 printer. Final page makeup was the responsibility of production editor Bob Hilbert and his staff at Wiley Interscience. I wish to express my gratitude to chief editor Maria Taylor, her predecessor Bea Shube, and to production editor Hilbert for a pleasant working relationship.

Finally, I would like to extend special thanks to my wife Jo Ann, who was forbearing while I neglected various other responsibilities in the course of preparing the original lectures and the subsequent manuscript. These tasks would have been considerably more difficult without her patience and understanding.

MIKE GUIDRY

Knoxville
April, 1991

Contents

GAUGE FIELD THEORIES

Part I

QUANTUM FIELD THEORY

CHAPTER 1

Relativistic Wave Equations

The phenomena that we will be investigating take place over localized regions of spacetime. To probe spacetime at high resolution requires high energies and momenta, as suggested by simple uncertainty principle arguments. Therefore, we require covariant generalizations of the Schrödinger equation. In this chapter we review the principles of special relativity and Lorentz transformations, and apply these principles to the construction of special relativistic wave equations for fermions and bosons—the Dirac and Klein–Gordon equations. In the process, considerable nomenclature and convention will be introduced. It will be found that the interpretation of these equations as single-particle wave equations is fraught with difficulty. In Chs. 2–4 we will introduce the techniques of quantum field theory, which correspond to a "second quantization" of the wave equations constructed in Ch. 1. Only then will we have a mathematical apparatus potent enough to handle the theories that are of interest to us.

1.1 Special Relativity and Spacetime

The 4-dimensional spacetime (ct, x, y, z) in which we find ourselves is called a *Minkowski space*, and a point in such a space is termed an *event*. The equation

$$s^2 \equiv c^2 t^2 - x^2 - y^2 - z^2 = 0 \qquad (1.1)$$

defines the *light cone* (Fig. 1.1). Events that have $s^2 > 0$ are called *timelike*; events that have $s^2 < 0$ are called *spacelike*; events on the light cone $(s^2 = 0)$ are connected by signals propagating at the speed of light from the origin and are called *lightlike*. Timelike events can be causally connected to the origin with signals for which $v < c$, but spacelike events cannot be connected to the origin except by signals for which $v > c$.

The *special theory of relativity* is based on two postulates: (1) the velocity of light in a vacuum is constant in all inertial systems (an inertial system is a system of reference in which Newton's first law of motion holds), and (2) the laws of physics are invariant under transformations between inertial systems (*covariance*). The transformations that respect these postulates are called the *Lorentz transformations*. Before discussing them we take a short notational detour.

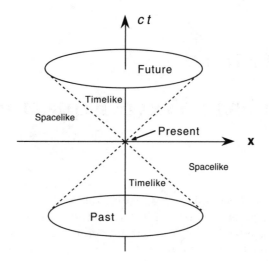

Fig. 1.1   The light cone.

1.2 Contravariant and Covariant Vectors

It is customary to introduce added notation and nomenclature that are not actually required in special relativity, but that considerably simplify general relativity. Even though we will normally require only fidelity to special relativity in our equations, we will maintain the usual practice and introduce the distinction between covariant and contravariant vectors. Our convention will conform rather closely with that of Jackson (1975).

We will most commonly work in the previously introduced non-Euclidian vector space (Minkowski space) for which the spacetime continuum is associated with the coordinates $(t, x, y, z) \equiv (x^0, x^1, x^2, x^3)$, where $\hbar = c = 1$ units have been used (see Appendix A). It will be assumed that there exists some transformation law that yields new coordinates (x'^0, x'^1, x'^2, x'^3):

$$x'^\mu = x'^\mu\left(x^0, x^1, x^2, x^3\right) \qquad \mu = 0, 1, 2, 3, \tag{1.2}$$

and that the transformation may be inverted. We may define tensors of rank k associated with the spacetime point $x \equiv (x^0, x^1, x^2, x^3)$ by their properties under the transformation $x \to x'$.

A *scalar* is unchanged by the transformation. Two tensors of rank 1, that is *vectors*, may be distinguished. A *contravariant vector* $A \equiv (A^0, A^1, A^2, A^3)$ transforms as

$$A'^\mu = \sum_\nu \frac{\partial x'^\mu}{\partial x^\nu} A^\nu \equiv \frac{\partial x'^\mu}{\partial x^\nu} A^\nu. \tag{1.3}$$

Here, the Einstein summation convention is introduced: *an index appearing twice on one side of an equation (once as superscript, once as subscript) implies a summation over that index.*[‡] We will usually reserve Greek letters for indices ranging over 0, 1, 2, and 3, but Latin letters $(i, j, k, \ldots)$ for indices ranging only over 1, 2, and 3. Explicitly then, for a contravariant vector

$$A'^{\mu} = \frac{\partial x'^{\mu}}{\partial x^0} A^0 + \frac{\partial x'^{\mu}}{\partial x^1} A^1 + \frac{\partial x'^{\mu}}{\partial x^2} A^2 + \frac{\partial x'^{\mu}}{\partial x^3} A^3. \tag{1.4}$$

Likewise a *covariant vector* $A_{\mu} \equiv (A_0, A_1, A_2, A_3)$ obeys the transformation law

$$A'_{\mu} = \frac{\partial x^{\nu}}{\partial x'^{\mu}} A_{\nu}. \tag{1.5}$$

Three kinds of rank-2 tensors may be defined. A *contravariant tensor of rank 2* is composed of 16 quantities transforming as

$$F'^{\mu\nu} = \frac{\partial x'^{\mu}}{\partial x^{\gamma}} \frac{\partial x'^{\nu}}{\partial x^{\delta}} F^{\gamma\delta}, \tag{1.6}$$

while the corresponding *covariant tensor of rank 2* transforms as

$$F'_{\mu\nu} = \frac{\partial x^{\gamma}}{\partial x'^{\mu}} \frac{\partial x^{\delta}}{\partial x'^{\nu}} F_{\gamma\delta}, \tag{1.7}$$

and the *mixed second-rank tensor* as

$$F'^{\mu}_{\nu} = \frac{\partial x'^{\mu}}{\partial x^{\gamma}} \frac{\partial x^{\delta}}{\partial x'^{\nu}} F^{\gamma}_{\delta}. \tag{1.8}$$

In an analogous fashion, higher order tensors may be defined.

These definitions of covariant, contravariant, and mixed tensors are required in general relativity where the transformations may be spacetime dependent. For the flat spacetime of special relativity the transformations between inertial systems are linear and independent of spacetime coordinates. The transformation of a contravariant coordinate vector x^{μ} is then given by

$$x'^{\mu} = \alpha^{\mu}_{\nu} x^{\nu}, \tag{1.9}$$

[‡]An index appearing as a superscript in a denominator counts as a subscript as far as the summation convention is concerned. Likewise, a subscript in a denominator counts as a superscript for summation purposes. In later chapters we will use a summation convention on internal (not Lorentz) indices in which any repeated index, whether superscript or subscript, implies a summation. A comparison of indices on the two sides of an equation usually indicates the summation convention in use and gives a quick consistency check: indices that are not repeated should balance on the two sides of an equation.

with α^μ_ν independent of the coordinates, so

$$\frac{\partial x'^\mu}{\partial x^\nu} = \alpha^\mu_\nu. \tag{1.10}$$

Therefore, the derivatives appearing in the previous definitions for various kinds of tensors are just constants, and we may simply define for flat spacetime the transformation laws

Scalars

$$A' = A, \tag{1.11}$$

Contravariant vectors

$$A'^\mu = \alpha^\mu_\nu A^\nu, \tag{1.12}$$

Covariant vectors

$$A'_\mu = \alpha^\nu_\mu A_\nu, \tag{1.13}$$

Contravariant tensor of rank 2

$$F'^{\mu\nu} = \alpha^\mu_\gamma \alpha^\nu_\delta F^{\gamma\delta}, \tag{1.14}$$

Covariant tensor of rank 2

$$F'_{\mu\nu} = \alpha^\gamma_\mu \alpha^\delta_\nu F_{\gamma\delta}, \tag{1.15}$$

Mixed tensor of rank 2

$$F'^\mu_\nu = \alpha^\mu_\gamma \alpha^\delta_\nu F^\gamma_\delta, \tag{1.16}$$

and so on for higher order tensors. The coefficients α^ν_μ are elements of a Lorentz transformation and are discussed below.

The *scalar product* of vectors is defined by

$$A \cdot B \equiv A^\mu B_\mu. \tag{1.17}$$

As the name implies, (1.17) defines an object that is an invariant. In an analogous way, the *contractions* of higher order tensors may be defined by employing a repeated index (one upper, one lower). For example,

$$A^\nu_\mu = B^{\nu\lambda} C_{\lambda\mu}. \tag{1.18}$$

In a contraction the implied summation over any repeated index removes it as an index. The vector scalar product is just a special case of this, and tensor contraction is a generalization of taking the vector scalar product.

The differential length element in a space is given by

$$(ds)^2 = g_{\mu\nu} dx^\mu dx^\nu, \tag{1.19}$$

where $g_{\mu\nu}$ is the *metric tensor* for the space. For the spacetime of special relativity the geometry may be defined in terms of the invariant interval

$$(ds)^2 \equiv (dx^0)^2 - (dx^1)^2 - (dx^2)^2 - (dx^3)^2 \qquad (1.20)$$

[see (1.1) and the discussion of Lorentz transformations below].

Comparing (1.19) and (1.20), we see that the metric tensor of special relativity is diagonal:

$$g_{\mu\nu} = g_{\nu\mu} = \begin{pmatrix} 1 & 0 & 0 & 0 \\ 0 & -1 & 0 & 0 \\ 0 & 0 & -1 & 0 \\ 0 & 0 & 0 & -1 \end{pmatrix} \qquad (1.21)$$

($g_{\mu\nu}$ is not so simple in the curved spacetime of general relativity; there the metric tensor is not fixed in advance, but is determined by the Einstein equations). For flat spacetime the covariant and contravariant metric tensors are identical,

$$g_{\mu\nu} = g^{\mu\nu}, \qquad (1.22)$$

and their contraction on one index yields the four-dimensional Kronecker delta

$$g_{\mu\lambda}g^{\lambda\nu} = g_\mu^\nu = \delta_\mu^\nu, \qquad (1.23)$$

where $\delta_\mu^\nu = 1$ for $\mu = \nu = 0, 1, 2, 3$ and $\delta_\mu^\nu = 0$ if $\mu \neq \nu$.

The general procedure for converting an index on a tensor from upper to lower, or vice versa, is to contract it with the metric tensor. For example, to transform between covariant and contravariant vector components,

$$x_\mu = g_{\mu\nu}x^\nu \qquad (1.24)$$

$$x^\mu = g^{\mu\nu}x_\nu. \qquad (1.25)$$

Since $g_{00} = 1$ and $g_{11} = g_{22} = g_{33} = -1$, if a contravariant 4-vector has components $A^\mu = (A^0, A^1, A^2, A^3)$, its covariant partner has components $A_0 = A^0$, $A_1 = -A^1$, $A_2 = -A^2$, and $A_3 = -A^3$. That is,

$$A^\mu = (A^0, \mathbf{A}), \qquad (1.26)$$

but the corresponding covariant vector is

$$A_\mu = (A^0, -\mathbf{A}). \qquad (1.27)$$

The 3-vector $\mathbf{A}$ has components (A^1, A^2, A^3), and we use a notation where boldface denotes 3-vectors, while 4-vectors are set in a normal font. In this notation the scalar product (1.17) of two 4-vectors is

$$A \cdot B = A^0 B^0 - \mathbf{A} \cdot \mathbf{B} = g_{\mu\nu} A^\mu B^\nu$$
$$= A^0 B^0 - A^1 B^1 - A^2 B^2 - A^3 B^3$$
$$= A^0 B^0 + A_1 B^1 + A_2 B^2 + A_3 B^3. \tag{1.28}$$

For a general tensor, index raising or lowering is accomplished by a contraction of the form

$$F^{..\mu..}_{..} = g^{\mu\nu} F^{...}_{.\nu} \qquad F^{..}_{...\mu} = g_{\mu\nu} F^{..\nu}_{....} \tag{1.29}$$

The derivatives with respect to the spacetime coordinates obey

$$\frac{\partial}{\partial x'^\mu} = \frac{\partial x^\nu}{\partial x'^\mu} \frac{\partial}{\partial x^\nu}, \tag{1.30}$$

which is just the transformation law for covariant vectors [eq. (1.5)]. It follows that the operator implying differentiation with respect to a contravariant (covariant) coordinate vector component transforms as a component of a covariant (contravariant) vector, suggesting the notation

$$\partial^\mu \equiv \frac{\partial}{\partial x_\mu} = (\partial^0, \partial^1, \partial^2, \partial^3) = \left(\frac{\partial}{\partial x^0}, -\mathbf{\nabla} \right)$$
$$\partial_\mu \equiv \frac{\partial}{\partial x^\mu} = (\partial_0, \partial_1, \partial_2, \partial_3) = \left(\frac{\partial}{\partial x^0}, \mathbf{\nabla} \right). \tag{1.31}$$

The final steps are dictated by (1.26) and (1.27), and

$$\mathbf{\nabla} = \left(\frac{\partial}{\partial x^1}, \frac{\partial}{\partial x^2}, \frac{\partial}{\partial x^3} \right) = (\partial_1, \partial_2, \partial_3) = \left(-\partial^1, -\partial^2, -\partial^3 \right)$$

is the 3-divergence (thus, $\mathbf{\nabla} \cdot \mathbf{A} = \partial_k A^k$). We may then define a *4-divergence* of a 4-vector A^μ by

$$\partial_\mu A^\mu = \partial^\mu A_\mu = \frac{\partial A^0}{\partial x^0} + \mathbf{\nabla} \cdot \mathbf{A} = \frac{\partial A^0}{\partial x^0} + \nabla_i A^i. \tag{1.32}$$

In this notation the 4-dimensional d'Alembertian operator is the contraction

$$\Box \equiv \partial_\mu \partial^\mu = \frac{\partial^2}{\partial x^0 \partial x_0} - \mathbf{\nabla}^2 = \frac{\partial^2}{\partial t^2} - \mathbf{\nabla}^2, \tag{1.33}$$

and is a scalar under Lorentz transformations. Another differential operator that we will find useful is expressed in this notation by

$$A \overset{\leftrightarrow}{\partial_\mu} B \equiv A(\partial_\mu B) - (\partial_\mu A)B, \tag{1.34}$$

where the derivatives act only inside the parentheses.

These are the essential notational results that will be required for subsequent discussion. For a more extensive discussion of covariant and contravariant vectors, tensors, and metric spaces in special and general relativity, see Landau and Lifshitz (1971), Jackson (1975), and Harris (1975). Fortified with this elegant mathematical machinery, we now return to our discussion of Lorentz transformations and special relativity.

1.3 Lorentz Transformations

Consider two inertial systems S and S', moving with constant velocity $\mathbf{v}$ with respect to each other. For simplicity, we orient the axes as in Fig. 1.2, and let the origins coincide at $t = t' = 0$. A light pulse is emitted from the coincident origins and it propagates as a spherical wave. In coordinate system S its wavefront is defined by $(c = 1)$

$$t^2 - x_1^2 - x_2^2 - x_3^2 = 0, \tag{1.35}$$

while in coordinate system S'

$$t'^2 - x_1'^2 - x_2'^2 - x_3'^2 = 0, \tag{1.36}$$

where the invariance of c has been used.

Introducing 4-vectors and the summation convention, we may express these relations in a highly compact notation

$$x_\mu x^\mu = 0 \qquad x'_\mu x'^\mu = 0, \tag{1.37}$$

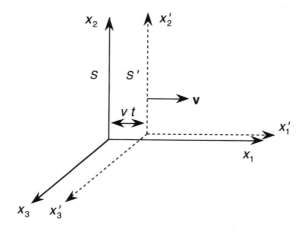

Fig. 1.2   Two inertial systems.

and by combining these equations the invariance of c is embodied in the concise algebraic expression,

$$x_\mu x^\mu = x'_\mu x'^\mu. \tag{1.38}$$

A particular physical event will be described in the two inertial systems by the respective sets of numbers (x_0, x_1, x_2, x_3) and (x'_0, x'_1, x'_2, x'_3). The coordinates in the two systems are connected by Lorentz transformations

$$x'_\mu = \alpha^\nu_\mu(v)x_\nu \qquad (\mu = 0,1,2,3), \tag{1.39}$$

where the $\alpha^\nu_\mu(v)$ are functions of the relative velocity v (and of the relative spatial orientation of the two systems), and are determined by imposing the constancy of c; that is, through eq. (1.38). We may also formulate Lorentz transformations among contravariant vectors

$$x'^\mu = \alpha^\mu_\nu(v)x^\nu, \tag{1.40}$$

and employ contractions with the metric tensor to raise or lower indices.

The definition (1.39) corresponds to what is more precisely termed the *homogeneous Lorentz transformation*. A more general *inhomogeneous Lorentz transformation (Poincaré transformation)* is defined by

$$x'_\mu = \alpha^\nu_\mu x_\nu + b_\mu. \tag{1.41}$$

This corresponds to a displacement of the origin by b_μ, as well as a rotation of the spacetime coordinate system. By Lorentz transformations we will normally mean homogeneous Lorentz transformations.

The analysis of rotations in 3-dimensional space begins with the definition of a rotation as an operation that leaves invariant the quantity $x^2 + y^2 + z^2 \equiv r^2$. From this, the machinery of tensor analysis may be constructed. The Lorentz transformation leaves invariant the quantity $s^2 = t^2 - x_1^2 - x_2^2 - x_3^2$; therefore, it corresponds to a generalized rotation in 4-dimensional Minkowski space, and many well-known results from the 3-dimensional coordinate system rotation may be immediately generalized to the 4-dimensional case. For example, the Lorentz transformation may be written in matrix form

$$x' = Ax, \tag{1.42}$$

where x and x' denote 4-component vectors and the α^ν_μ constitute a 4×4 matrix A

$$\begin{pmatrix} x'_0 \\ x'_1 \\ x'_2 \\ x'_3 \end{pmatrix} = \begin{pmatrix} \alpha^0_0 & \alpha^1_0 & \alpha^2_0 & \alpha^3_0 \\ \alpha^0_1 & \alpha^1_1 & \alpha^2_1 & \alpha^3_1 \\ \alpha^0_2 & \alpha^1_2 & \alpha^2_2 & \alpha^3_2 \\ \alpha^0_3 & \alpha^1_3 & \alpha^2_3 & \alpha^3_3 \end{pmatrix} \begin{pmatrix} x_0 \\ x_1 \\ x_2 \\ x_3 \end{pmatrix}$$

and the coefficients satisfy

$$\alpha_\lambda^\nu \alpha_\mu^\lambda = \delta_\mu^\nu. \tag{1.43}$$

The continuous Lorentz transformations correspond to rotations of the coordinate system within a particular frame, and *Lorentz boosts* between frames. The explicit forms of the matrices for Lorentz transformations are given in many places. Most of our discussion will require only their formal properties.

EXERCISE 1.1[‡] Prove the orthogonality condition (1.43). Show that the determinant of a Lorentz transformation matrix is equal to ±1.

Quantities that are not changed by a Lorentz transformation [eq. (1.11)] are Lorentz tensors of rank zero (scalars). Examples include

- the 4-dimensional volume element d^4x,
- the 4-vector scalar product $A \cdot B$,
- the d'Alembertian operator $\Box = \partial_\mu \partial^\mu$.

A quantity transforming in the same way as the coordinates under the Lorentz transformations is termed a *4-vector* (Lorentz tensor of rank 1). We have already had an example in eq. (1.30): the derivatives transform in flat spacetime as

$$\frac{\partial}{\partial x'_\mu} = \alpha_\nu^\mu \frac{\partial}{\partial x_\nu} \equiv \alpha_\nu^\mu \partial^\nu. \tag{1.44}$$

Hence, the 4-gradient $(\partial^0, \partial^1, \partial^2, \partial^3)$ transforms as a 4-vector. Other examples of 4-vectors include

- the 4-momentum $p^\mu = (E, \mathbf{p})$, with $E = \sqrt{\mathbf{p}^2 + m^2}$;
- the 4-current $j^\mu = (\rho, \mathbf{j})$, where ρ is the charge density and $\mathbf{j}$ the ordinary spatial 3-current from the Schrödinger equation;
- the electromagnetic 4-potential $A^\mu = (A^0, \mathbf{A})$, where A^0 is the scalar potential and $\mathbf{A}$ is the vector potential.

Second-rank tensors transform as (1.14)–(1.16). Examples that we will employ include the antisymmetric electromagnetic field tensor $F^{\mu\nu}$, which is encountered in Ch. 2, and the metric tensor $g_{\mu\nu}$.

The importance of formulating equations in terms of such scalars, 4-vectors, and 4-tensors is clear; by doing so we ensure that the two basic postulates of special relativity are observed: (1) the Lorentz transformation is constructed explicitly to secure the constancy of light velocity in all inertial frames, and (2) quantities with a tensor structure lead to equations that are form-invariant *(covariant)* under coordinate system transformations.

[‡]The solutions of this and subsequent exercises may be found in Appendix D.

1.3.1 Improper and Proper Rotations

The parity transformation corresponds to inverting the coordinates but not the time, while the time-reversal transformation reverses the time but not the spatial coordinates. The matrices that accomplish this are

$$\pi = \begin{pmatrix} 1 & 0 & 0 & 0 \\ 0 & -1 & 0 & 0 \\ 0 & 0 & -1 & 0 \\ 0 & 0 & 0 & -1 \end{pmatrix} \tag{1.45}$$

and

$$\tau = \begin{pmatrix} -1 & 0 & 0 & 0 \\ 0 & 1 & 0 & 0 \\ 0 & 0 & 1 & 0 \\ 0 & 0 & 0 & 1 \end{pmatrix}. \tag{1.46}$$

These matrices do not correspond to generalized rotations. For example, the parity operation is equivalent to a rotation by $180°$ plus a reflection through a plane. Operations such as parity or time reversal are called *improper transformations*. Their transformation matrices M may be characterized by the condition[‡]

$$\text{Det } M = -1, \tag{1.47}$$

while the *proper* Lorentz transformation matrices R satisfy

$$\text{Det } R = +1. \tag{1.48}$$

The proper Lorentz transformations can be built from a succession of infinitesimal transformations; in contrast, the improper Lorentz transformations are discrete and cannot be obtained by compounding infinitesimal transformations.

1.3.2 Particle Symmetries

We will find it useful to classify particles, fields, and currents with respect to their properties under proper and improper Lorentz transformations, and subgroups of these transformations. It is normal to label them according to their angular momentum and parity quantum numbers J^P. The standard terminology for integer spins is shown in Table 1.1. For example, the π-meson has $J^P = 0^-$ and is termed a *pseudoscalar particle*. In quantum field theory the pion is associated with a *pseudoscalar field*.

[‡]Eq. (1.47) is sufficient, but not necessary, to classify a transformation as improper. For example, space–time inversion is an improper transformation for which det $M = +1$. Eq. (1.48) is a necessary condition for proper transformations.

Table 1.1

Particle Symmetries

Name	J^P	Independent Components for Massive Structureless Particle	
Scalar	0^+	1	
Pseudoscalar	0^-	1	
Vector	1^-	3	(also called a polar vector)
Axial vector	1^+	3	(also called a pseudovector)
Rank-2 tensor	$2^\pm$	5	

The table gives the number of independent components for structureless particles; the particles may also possess internal degrees of freedom. For example, the π-meson is a pseudoscalar in spacetime, but it has three components in the internal isospin space (it is an *isovector*). Massless spin-1 or spin-2 particles have only two states of polarization.

Generally, a massive particle of integer spin J is associated with an irreducible Lorentz tensor of rank J which has $2J + 1$ independent components (see Harris, 1975, §20.4; Ramond, 1981, §I.3). This is modified if the particle is massless. For example, the free photon is a massless vector particle having two, rather than three, states of polarization. We will discuss this in Ch. 2.

Fermions constitute a class of objects (of half-integer spin) that transform with well-defined angular momentum properties, but that do not have a Lorentz tensor structure, and hence do not have tensor transformation properties under the Lorentz group (§1.5.5). They are called *spinors* (or bispinors). Although the fermion operators are not Lorentz tensors, certain bilinear combinations have a tensor structure, as we will discuss in §1.5.6.

EXERCISE 1.2 Demonstrate that the 4-vector scalar product $A \cdot B$ and the differential volume element d^4x are invariant under Lorentz transformations.

EXERCISE 1.3 (a) Show that the d'Alembertian operator $\Box \equiv \partial_\mu \partial^\mu$ is a scalar under Lorentz transformations. (b) Prove that if A_μ is a 4-vector, then $F_{\mu\nu} = \partial_\mu A_\nu - \partial_\nu A_\mu$ transforms as an antisymmetric rank-2 tensor.

1.3.3 Group Structure of Lorentz Transformations

The homogeneous Lorentz transformations form a group, called the *homogeneous Lorentz group* (often just termed the Lorentz group), which leaves invariant the quantity $s^2 = t^2 - x^2 - y^2 - z^2$. This group is a 6-parameter Lie group in four dimensions (Lie groups are reviewed in Ch. 5). The 3-dimensional orthogonal group of rotations $O(3)$ is a subgroup of the homogeneous Lorentz group. If the group of translations is added to the homogeneous Lorentz group we obtain the 10-parameter inhomogeneous Lorentz group or *Poincaré group*. The *proper orthochronous Poincaré group* excludes improper Lorentz transformations such as time reversal and space reflection.[‡]   The laws of physics are thought to be absolutely invariant under this group. This is an elegant way of stating that nature seems indifferent to the choice of coordinate system origin or time origin, spatial orientation of the system, or state of uniform rectilinear motion. As is well known, fundamental conservation laws such as those for angular momentum and energy follow from this. For a more extensive discussion of the group structure associated with the Lorentz transformations see Ramond (1981), Messiah (1958), and Jackson (1975).

1.4 Klein–Gordon Equation

Having reviewed the theory of special relativity and Lorentz transformations, we now seek a relativistic analog to the single-particle Schrödinger equation. The simplest approach is to begin with the relativistic energy expression

$$E^2 = \mathbf{p}^2 c^2 + m^2 c^4, \tag{1.49}$$

which may be written, using $E = \hbar\omega$ and $\mathbf{p} = \hbar\mathbf{k}$,

$$\hbar^2\omega^2 = \hbar^2 c^2 \mathbf{k}^2 + m^2 c^4, \tag{1.50}$$

where ω is a frequency and $\mathbf{k}$ a wave vector. We may also express these equations in covariant notation by introducing the 4-momentum and 4-wave vector

$$p^\mu = (E, \mathbf{p}) \tag{1.51}$$

$$k^\mu = (\omega, \mathbf{k}). \tag{1.52}$$

Then (1.49) and (1.50) become (in $\hbar = c = 1$ units)

$$p_\mu p^\mu = m^2 \tag{1.53}$$

$$k_\mu k^\mu = m^2. \tag{1.54}$$

Now we make the substitution familiar from ordinary quantum mechanics

$$E = \hbar\omega \rightarrow -\frac{\hbar}{i}\frac{\partial}{\partial t} \tag{1.55}$$

[‡]Generally, for the Poincaré group one finds Det $\alpha = +1$, and either $\alpha_0^0 \geqslant +1$ or $\alpha_0^0 \leqslant -1$. The group is noncompact and not connected (see Ch. 5), with four pieces. The choice Det $\alpha = +1$ and $\alpha_0^0 \geqslant +1$ identifies the proper orthochronous part of the group, which is the portion continuously connected to the group identity.

$$\mathbf{p} = \hbar \mathbf{k} \rightarrow \frac{\hbar}{i} \boldsymbol{\nabla}, \tag{1.56}$$

and set $\hbar = c = 1$ to obtain

$$\boldsymbol{\nabla}^2 \phi - \frac{\partial^2 \phi}{\partial t^2} = m^2 \phi, \tag{1.57}$$

where ϕ is the wavefunction on which the operators are assumed to act. Finally, using the definition (1.33) we obtain the Klein–Gordon (KG) equation

$$\left(\Box + m^2 \right) \phi = 0. \tag{1.58}$$

We have already shown in Exercise 1.3 that the operator $\Box$ is a Lorentz scalar. Therefore, the covariance of the wave equation will be ensured if ϕ is a scalar, which implies quantum numbers $J^P = 0^+$ for the particle that we wish to describe with the Klein–Gordon equation (a scalar boson).

Although the KG equation is covariant, there are serious problems associated with its use as a relativistic single-particle wave equation. Let us seek a plane-wave solution of the free-particle KG equation in the form

$$\phi(\mathbf{x}, t) \simeq e^{-iEt + i\mathbf{p} \cdot \mathbf{x}} = e^{-ip \cdot x}, \tag{1.59}$$

where the exponent is written as a 4-vector scalar product

$$p \cdot x \equiv p_\mu x^\mu = Et - \mathbf{p} \cdot \mathbf{x}. \tag{1.60}$$

As can be verified by direct substitution, this solves the KG equation only if $E^2 = \mathbf{p}^2 + m^2$. Thus, for a given 3-momentum $\mathbf{p}$ there are two possible solutions for the energy,

$$E = \pm\sqrt{\mathbf{p}^2 + m^2}. \tag{1.61}$$

The negative-energy solutions cannot simply be ignored because this leads quickly to inconsistencies for realistic systems. In retrospect, we know that the negative-energy solutions are associated with antiparticles, so what first appeared to be a disaster will prove to be a highly desirable aspect of the theory. However, there are more serious problems.

By analogy with the procedure for the Schrödinger equation, we may derive a conservation law for a probability current of the free-particle Klein–Gordon equation. From Exercise 1.4,

$$\frac{\partial \rho}{\partial t} + \boldsymbol{\nabla} \cdot \mathbf{j} = 0, \tag{1.62}$$

where

$$\rho = i\phi^* \overleftrightarrow{\partial}_t \phi \equiv i \left[\phi^* \frac{\partial \phi}{\partial t} - \left(\frac{\partial \phi^*}{\partial t} \right) \phi \right] \tag{1.63}$$

$$\mathbf{j} = -i\phi^* \overleftrightarrow{\boldsymbol{\nabla}} \phi \equiv -i \left[\phi^* \boldsymbol{\nabla} \phi - (\boldsymbol{\nabla}\phi^*)\phi \right]. \tag{1.64}$$

Introducing a 4-current

$$j^\mu = (\rho, \mathbf{j}) = i\phi^* \overset{\leftrightarrow}{\partial^\mu} \phi = i(\phi^* \partial^\mu \phi - \phi \partial^\mu \phi^*), \tag{1.65}$$

this conservation condition may be written in the compact form

$$\partial_\mu j^\mu = 0. \tag{1.66}$$

The spatial current $\mathbf{j}$ of (1.65) is of the same form as the Schrödinger current. However, the component ρ contains time derivatives of ϕ [eq. (1.63)], because the KG equation is second order in $\partial/\partial t$. As a consequence, ρ need not be positive definite, in contrast to the Schrödinger result where $\rho = \psi^* \psi$. This can be seen explicitly for the free-particle solution: from (1.59), (1.61), and (1.63) we find (Exercise 1.4)

$$\rho \simeq E = \pm\sqrt{\mathbf{p}^2 + m^2}. \tag{1.67}$$

The presence of negative-energy solutions and negative probability densities, and the discovery that the KG equation predicted the wrong spectrum for hydrogen, caused it to be abandoned shortly after its inception. As we will see in Ch. 2, it can be rehabilitated if interpreted as a meson *field equation* (rather than as a single-particle equation), in which case ρ becomes a charge density with the positive and negative signs corresponding to particles and antiparticles.

EXERCISE 1.4 (a) Prove that the probability current associated with the Klein–Gordon equation obeys equations (1.62)–(1.66). (b) Show that for the plane wave solution (1.59) the probability density ρ can be negative.

First, we proceed historically and see how Dirac managed to obtain a covariant wave equation that evaded these problems. The preceding difficulties with the KG equation stem from two sources: (1) the squared energy–momentum relation $E^2 = \mathbf{p}^2 + m^2$ allows negative-energy solutions, and (2) the KG equation is second-order in the time derivative, leading to terms containing $\partial/\partial t$ in the probability density and hence to possible negative values for ρ. The first problem will be converted to an asset when the negative-energy solutions are interpreted. To circumvent the second problem we need a wave equation linear in the time derivative (Dirac equation).

Before considering the Dirac equation, we will dispose of one other variant of the preceding discussion that would give us a wave equation linear in time, but with some rather unpleasant properties (Baym, 1969, Ch. 22). Starting from $i\frac{\partial}{\partial t}\psi = E\psi$ and (1.49), we could write the momentum-space equation

$$i\frac{\partial}{\partial t}\psi_p(t) = \sqrt{\mathbf{p}^2 + m^2}\,\psi_p(t). \tag{1.68}$$

However, if we Fourier transform both sides back to coordinate space a non-local equation results

$$i\frac{\partial}{\partial t}\psi(\mathbf{x},t) = \int d^3x'\, K(\mathbf{x}-\mathbf{x}')\psi(\mathbf{x}',t), \tag{1.69}$$

where the kernel

$$K(\mathbf{x}-\mathbf{x}') = \int \frac{d^3p}{8\pi^3}\, e^{i\mathbf{p}\cdot(\mathbf{x}-\mathbf{x}')}\sqrt{\mathbf{p}^2+m^2} \tag{1.70}$$

is significant as long as $\mathbf{x}'$ is within a Compton wavelength $\hbar/mc$ of $\mathbf{x}$; therefore, $\partial\psi(\mathbf{x},t)/\partial t$ can depend on values of ψ at $(\mathbf{x}',t)$ *outside* the light cone with its origin at $(\mathbf{x},t)$, and causality will be violated if this equation is used to describe particles localized to less than a Compton wavelength.

1.5 Dirac Equation

Since space and time enter relativity on the same footing their derivatives should appear in equivalent orders for a relativistic wave equation. We have seen that an equation of first order in the time derivatives is required to avoid difficulties such as negative probabilities.

1.5.1 Linear Equation in ∇ and $\partial/\partial t$

A wave equation *first order in both time and space derivatives* may be obtained by making the replacement

$$E = \sqrt{\mathbf{p}^2+m^2} \rightarrow \boldsymbol{\alpha}\cdot\mathbf{p} + \beta m, \tag{1.71}$$

with $\boldsymbol{\alpha}$ and β to be determined. Then, upon making the usual substitutions $E \rightarrow i\partial/\partial t$ and $\mathbf{p} \rightarrow -i\nabla$ we obtain the Dirac wave equation

$$i\frac{\partial\psi(\mathbf{x},t)}{\partial t} = H\psi(\mathbf{x},t) = (-i\,\boldsymbol{\alpha}\cdot\nabla + \beta m)\,\psi(\mathbf{x},t)$$

$$= (\boldsymbol{\alpha}\cdot\mathbf{p} + \beta m)\,\psi(\mathbf{x},t), \tag{1.72}$$

where $\boldsymbol{\alpha}\cdot\nabla = \alpha^k\partial_k$ with $k = 1, 2, 3$. If this is to be a valid wave equation it must (1) be Lorentz covariant and consistent with the correct energy relationship $E^2 = \mathbf{p}^2+m^2$, and (2) allow a non-contradictory continuity equation and probability interpretation for ψ. As shown in many places, these requirements are incompatible if $\boldsymbol{\alpha}$ and β are numbers, but they can be fulfilled for 4×4

noncommuting matrices satisfying

$$\alpha_1^2 = \alpha_2^2 = \alpha_3^2 = \beta^2 = 1 \tag{1.73}$$

$$\{\alpha_i, \alpha_j\} \equiv \alpha_i \alpha_j + \alpha_j \alpha_i = 2\delta_{ij} \tag{1.74}$$

$$\{\alpha_i, \beta\} \equiv \alpha_i \beta + \beta \alpha_i = 0, \tag{1.75}$$

where

$$1 \equiv \begin{pmatrix} 1 & 0 & 0 & 0 \\ 0 & 1 & 0 & 0 \\ 0 & 0 & 1 & 0 \\ 0 & 0 & 0 & 1 \end{pmatrix}. \tag{1.76}$$

A conventional choice (not the only one in use) is

$$\boldsymbol{\alpha} = \begin{pmatrix} 0 & \boldsymbol{\sigma} \\ \boldsymbol{\sigma} & 0 \end{pmatrix} \qquad \beta = \begin{pmatrix} 1 & 0 \\ 0 & -1 \end{pmatrix}, \tag{1.77}$$

where $\boldsymbol{\sigma}$ denotes the Pauli matrices

$$\sigma_1 = \begin{pmatrix} 0 & 1 \\ 1 & 0 \end{pmatrix} \qquad \sigma_2 = \begin{pmatrix} 0 & -i \\ i & 0 \end{pmatrix} \qquad \sigma_3 = \begin{pmatrix} 1 & 0 \\ 0 & -1 \end{pmatrix}, \tag{1.78}$$

and where we employ an obvious shorthand notation where each entry of the matrix in (1.77) is itself a 2×2 matrix. Thus, in explicit 4×4 matrix notation,

$$\alpha_1 = \begin{pmatrix} 0 & 0 & 0 & 1 \\ 0 & 0 & 1 & 0 \\ 0 & 1 & 0 & 0 \\ 1 & 0 & 0 & 0 \end{pmatrix} \qquad \alpha_2 = \begin{pmatrix} 0 & 0 & 0 & -i \\ 0 & 0 & i & 0 \\ 0 & -i & 0 & 0 \\ i & 0 & 0 & 0 \end{pmatrix}$$

$$\tag{1.79}$$

$$\alpha_3 = \begin{pmatrix} 0 & 0 & 1 & 0 \\ 0 & 0 & 0 & -1 \\ 1 & 0 & 0 & 0 \\ 0 & -1 & 0 & 0 \end{pmatrix} \qquad \beta = \begin{pmatrix} 1 & 0 & 0 & 0 \\ 0 & 1 & 0 & 0 \\ 0 & 0 & -1 & 0 \\ 0 & 0 & 0 & -1 \end{pmatrix}.$$

Since the operators appearing in the Dirac equation are 4×4 matrices the wavefunctions that satisfy it are 4-component column or row vectors

$$\psi = \begin{pmatrix} \psi_1 \\ \psi_2 \\ \psi_3 \\ \psi_4 \end{pmatrix} \qquad \psi^\dagger = (\psi_1^* \ \psi_2^* \ \psi_3^* \ \psi_4^*), \tag{1.80}$$

where we use a dagger to represent Hermitian conjugation: transpose the matrix and complex conjugate its elements. These wavefunctions do not transform as Lorentz tensors; instead they obey transformation relations appropriate to 4-component spinors (bispinors), as discussed in §1.5.5.

Thus the Dirac wavefunction is a 4-component column matrix in the space where β and $\boldsymbol{\alpha}$ operate, and each of the four components ψ_λ is in turn an element of Hilbert space. The corresponding Dirac equation is a matrix equation representing four simultaneous differential equations.

1.5.2 The Dirac Current

The free Dirac equation is linear in the time derivative, so it is not surprising to find that a positive definite probability density exists. Let us multiply the Dirac equation (1.72) from the left by the Hermitian conjugate spinor $\psi^\dagger$:

$$i\psi^\dagger \frac{\partial}{\partial t}\psi = -i\psi^\dagger \alpha^k \partial_k \psi + m\psi^\dagger \beta\psi.$$

Now form the Hermitian conjugate of the Dirac equation [complex conjugate everything in (1.72) and transpose all matrices], and multiply from the right by ψ:

$$-i\frac{\partial \psi^\dagger}{\partial t}\psi = i(\partial^k \psi^\dagger)\alpha_k \psi + m\psi^\dagger \beta\psi.$$

In this step we have used $\boldsymbol{\alpha} = \boldsymbol{\alpha}^\dagger$ and $\beta = \beta^\dagger$, and rearrangements such as $(\beta\psi)^\dagger = \psi^\dagger \beta^\dagger = \psi^\dagger \beta$; see Exercises 1.5 and 1.6. Subtracting these equations gives a continuity equation of the form

$$\frac{\partial \rho}{\partial t} + \boldsymbol{\nabla}\cdot\mathbf{j} = 0, \tag{1.81a}$$

with a probability density

$$\rho = \psi^\dagger \psi = |\psi_1|^2 + |\psi_2|^2 + |\psi_3|^2 + |\psi_4|^2 > 0, \tag{1.81b}$$

and a probability current

$$j^k = \psi^\dagger \alpha^k \psi. \tag{1.81c}$$

Unlike the corresponding Klein–Gordon result, this is an acceptable continuity equation because ρ is now a positive definite quantity; observe also that $\mathbf{j}$ does not contain $\boldsymbol{\nabla}$, as it does in the Schrödinger or Klein–Gordon case. These are consequences of the linearity of the Dirac equation in the derivatives.

One further implication of this linear structure concerns boundary conditions. The requirement that ρ and $\mathbf{j}$ be continuous for solutions to the Schrödinger equation demands that both ψ and $\boldsymbol{\nabla}\psi$ be continuous, but in the Dirac theory the absence of $\boldsymbol{\nabla}$ in the probability current means that only ψ need be continuous.

The continuity equation can be expressed in covariant form if we introduce the current 4-vector

$$j = (\rho, \mathbf{j}), \tag{1.82a}$$

which allows (1.81a) to be written

$$\partial_\mu j^\mu = 0. \tag{1.82b}$$

1.5.3 Interaction with an Electromagnetic Field

The interaction of a Dirac particle with an electromagnetic field may be incorporated by the standard prescription from nonrelativistic quantum mechanics (*minimal coupling*—see §2.7).

$$p_\mu \rightarrow p_\mu - \frac{q}{c}A_\mu \qquad \text{(classical)}$$
$$\rightarrow i(\partial_\mu + iqA_\mu) \qquad \text{(quantum-mechanical, } c = 1\text{)}, \tag{1.83}$$

where $A_\mu = (A^0, \mathbf{A})$ is the 4-potential, $p_\mu = (E, \mathbf{p})$ is the 4-momentum, and q is the charge. Then the Dirac equation becomes

$$i\frac{\partial \psi}{\partial t} = \left[\boldsymbol{\alpha} \cdot (-i\boldsymbol{\nabla} - q\mathbf{A}) + qA^0 + \beta m\right]\psi. \tag{1.84}$$

The same prescription may be employed to incorporate electromagnetic interactions in the Klein–Gordon equation. As demonstrated in Ch. 2, the form of this substitution is dictated by the requirement of local gauge invariance for the electromagnetic field.

1.5.4 Covariant Notation

To investigate the covariance of the Dirac equation it is useful to introduce a form that is more symmetric in space and time. The γ-*matrices* are defined by

$$\gamma^0 = \beta \qquad \gamma^i = \beta\alpha^i = \gamma^0\alpha^i. \tag{1.85}$$

They satisfy the anticommutation relation

$$\{\gamma^\mu, \gamma^\nu\} \equiv \gamma^\mu\gamma^\nu + \gamma^\nu\gamma^\mu = 2g^{\mu\nu}, \tag{1.86}$$

with metric tensor (1.21). This can also be written as

$$\{\gamma_\mu, \gamma^\nu\} = 2\delta_\mu^\nu, \tag{1.87}$$

because of (1.23). Since the γ-matrices are related to the α- and β-matrices by (1.85) their explicit form in the representation we are using is

$$\gamma^0 = \begin{pmatrix} 1 & 0 \\ 0 & -1 \end{pmatrix} \qquad \gamma^i = \begin{pmatrix} 0 & \sigma_i \\ -\sigma_i & 0 \end{pmatrix}, \tag{1.88}$$

where the σ_i are Pauli matrices (1.78).[‡]   Two combinations of γ-matrices occur frequently and merit a special notation:

$$\sigma^{\mu\nu} \equiv \tfrac{i}{2}[\gamma^\mu, \gamma^\nu] \qquad (1.89)$$

(we will use [ ] to denote commutators, and { } to denote anticommutators), and

$$\gamma^5 = \gamma_5 \equiv i\gamma^0\gamma^1\gamma^2\gamma^3. \qquad (1.90)$$

In this representation we have explicitly

$$\sigma^{ij} = \epsilon_{ijk} \begin{pmatrix} \sigma_k & 0 \\ 0 & \sigma_k \end{pmatrix} \qquad (1.91)$$

$$\sigma^{0k} = i\alpha^k = i \begin{pmatrix} 0 & \sigma_k \\ \sigma_k & 0 \end{pmatrix} \qquad (1.92)$$

$$\gamma^5 = \gamma_5 = \begin{pmatrix} 0 & 1 \\ 1 & 0 \end{pmatrix}. \qquad (1.93)$$

The scalar product (inner product) of γ-matrices and 4-vectors occurs often enough to justify a special notation, the *Feynman slash convention:*

$$\slashed{A} \equiv \gamma_\mu A^\mu = \gamma^0 A^0 - \boldsymbol{\gamma} \cdot \mathbf{A} \qquad (1.94)$$

$$\boldsymbol{\gamma} \cdot \mathbf{A} \equiv \gamma^1 A^1 + \gamma^2 A^2 + \gamma^3 A^3 \qquad (1.95)$$

$$\slashed{p} \equiv \gamma_\mu p^\mu = E\gamma^0 - \boldsymbol{\gamma} \cdot \mathbf{p} \qquad (1.96)$$

$$i\slashed{\partial} \equiv i\slashed{\nabla} = i\gamma^\mu \frac{\partial}{\partial x^\mu} = i\gamma^\mu \partial_\mu = i\gamma_0 \frac{\partial}{\partial t} + i\boldsymbol{\gamma} \cdot \boldsymbol{\nabla}. \qquad (1.97)$$

Taking the trace of a matrix (Tr M = sum of diagonal elements in the matrix M) is a frequent operation, and some useful trace theorems and γ-matrix identities are listed in Table 1.2. A more extensive listing of Dirac matrix properties may be found in Appendix A2 of Itzykson and Zuber (1980). With the notation just introduced the Dirac equation can be written in several

---

[‡]The representation (1.88) or (1.79) is called the *Pauli–Dirac representation.* Other representations of the γ-matrices can be obtained from this one by matrix transformation (Itzykson and Zuber, 1980, Appendix A2). Two useful ones are the *Majorana representation,* with

$$\gamma^\mu_{\text{Majorana}} = U\gamma^\mu_{\text{Dirac}}U^\dagger \qquad U = U^\dagger = \frac{1}{\sqrt{2}} \begin{pmatrix} 1 & \sigma_2 \\ \sigma_2 & -1 \end{pmatrix},$$

and the *chiral representation,* with

$$\gamma^\mu_{\text{chiral}} = U\gamma^\mu_{\text{Dirac}}U^\dagger \qquad U = \frac{1}{\sqrt{2}} \begin{pmatrix} 1 & 1 \\ -1 & 1 \end{pmatrix}.$$

where $\gamma^\mu_{\text{Dirac}}$ denotes a matrix in the Pauli–Dirac representation.

Table 1.2
Properties of Dirac γ-Matrices

USEFUL RELATIONS:

$$\gamma_0 = \gamma^0 \equiv \beta \qquad \gamma^i \equiv \gamma^0 \alpha^i \qquad \gamma_5 = \gamma^5 \equiv i\gamma^0\gamma^1\gamma^2\gamma^3 \qquad \sigma^{\mu\nu} \equiv \tfrac{i}{2}[\gamma^\mu, \gamma^\nu]$$

$$\{\gamma^\mu, \gamma^\nu\} = 2g^{\mu\nu} \qquad \gamma_0\gamma_\mu^\dagger\gamma_0 = \gamma_\mu \qquad \{\gamma_5, \gamma^\mu\} = 0$$

$$(\gamma^i)^2 = -1 \qquad (\gamma^0)^2 = 1 \qquad (\gamma^5)^2 = 1$$

$$(\gamma_0)^\dagger = \gamma_0 \qquad (\gamma^i)^\dagger = -\gamma^i \qquad (\gamma^5)^\dagger = \gamma^5$$

TRACE THEOREMS AND IDENTITIES‡

(A) $\slashed{A}\slashed{B} = A \cdot B - i\sigma_{\mu\nu}A^\mu B^\nu$

(B) Trace of the product of an odd number of γ_μ vanishes

(C) $\operatorname{Tr}\gamma^5 = 0$

(D) $\operatorname{Tr}1 = 4$

(E) $\operatorname{Tr}\slashed{A}\slashed{B} = 4A \cdot B$

(F) $\operatorname{Tr}\slashed{A_1}\slashed{A_2}\slashed{A_3}\slashed{A_4} = 4\left(A_1 \cdot A_2 A_3 \cdot A_4 - A_1 \cdot A_3 A_2 \cdot A_4 + A_1 \cdot A_4 A_2 \cdot A_3\right)$

(G) $\operatorname{Tr}\gamma_5\slashed{A}\slashed{B} = 0$

(H) $\operatorname{Tr}\gamma_5\slashed{A}\slashed{B}\slashed{C}\slashed{D} = 4i\epsilon_{\mu\nu\gamma\delta}A^\mu B^\nu C^\gamma D^\delta$

(I) $\gamma_\mu\slashed{A}\gamma^\mu = -2\slashed{A}$

(J) $\gamma_\mu\slashed{A}\slashed{B}\gamma^\mu = 4A \cdot B$

(K) $\gamma_\mu\slashed{A}\slashed{B}\slashed{C}\gamma^\mu = -2\slashed{C}\slashed{B}\slashed{A}$

‡Bjorken and Drell (1964). The antisymmetric tensor $\epsilon_{\alpha\beta\gamma\delta}$ is defined in (2.130).

compact forms:

$$i\left(\gamma^0\frac{\partial}{\partial x^0} + \gamma^1\frac{\partial}{\partial x^1} + \gamma^2\frac{\partial}{\partial x^2} + \gamma^3\frac{\partial}{\partial x^3}\right)\psi - m\psi = 0 \qquad (1.98)$$

$$(i\slashed{\partial} - m)\psi = (i\slashed{\nabla} - m)\psi = 0 \qquad (1.99)$$

$$(\slashed{p} - m)\psi = 0 \qquad (1.100)$$

where m is understood to be multiplied by the unit 4×4 matrix and p is the 4-momentum operator. The minimal substitution (1.83) yields (1.84), which in this notation is

$$(i\slashed{\partial} - q\slashed{A} - m)\psi = 0, \qquad (1.101)$$

for the Dirac particle interacting with an electromagnetic field.

EXERCISE 1.5 Verify (1.73)–(1.75) and (1.86); verify the expressions (1.91) and (1.92); show that for the matrices important in the Dirac problem σ_i, α_i, β, γ^0, γ^5, and σ^{ij} are Hermitian, while γ^i and σ^{0i} are antihermitian ($A = -A^\dagger$).

EXERCISE 1.6 (a) If you are not already convinced, verify the following properties of matrices:

1. $(ABC\ldots)^* = A^*B^*C^*\ldots$
2. $(ABC\ldots)^\dagger = \ldots C^\dagger B^\dagger A^\dagger$
3. $(ABC\ldots)^\mathrm{T} = \ldots C^\mathrm{T} B^\mathrm{T} A^\mathrm{T}$
4. $(ABC\ldots)^{-1} = \ldots C^{-1}B^{-1}A^{-1}$
5. The trace of a product of matrices is invariant under cyclic permutations of the product: e.g., $\mathrm{Tr}\,(ABC) = \mathrm{Tr}\,(CAB) = \mathrm{Tr}\,(BCA)$ [but generally, $\mathrm{Tr}\,(ABC) \neq \mathrm{Tr}\,(ACB)$].

(b) Prove that the trace of the product of an odd number of γ-matrices vanishes (Theorem B of Table 1.2). *Hint:* consider the cyclic property of the trace applied to $\gamma^5 \Gamma \gamma^5$, where Γ is a product of an odd number of γ-matrices. (c) Prove that $\mathrm{Tr}\,A\!\!\!/\,B\!\!\!/ = 4A \cdot B$ if A_μ and B_ν commute (Theorem E of Table 1.2). *Hint:* use the cyclic property of the trace to write $\mathrm{Tr}\,A\!\!\!/\,B\!\!\!/ = \frac{1}{2}\,\mathrm{Tr}\,(A\!\!\!/\,B\!\!\!/ + B\!\!\!/\,A\!\!\!/)$.

1.5.5 Covariance of the Dirac Equation

The Dirac equation leads to a sensible probability current and incorporates the correct relativistic energy expression. It remains for us to demonstrate that it is form invariant under Lorentz transformations (covariant), and to deduce the meaning of the 4-component wavefunction; first we address covariance (Bjorken and Drell, 1964; Sakurai, 1967).

Because p^μ and A^μ are 4-vectors, the addition of minimal electromagnetic coupling by the replacement (1.83) has no effect on the question of covariance and we may work with the simple equation

$$(i\partial\!\!\!/ - m)\,\psi(x) = \left(i\gamma^\mu \frac{\partial}{\partial x^\mu} - m\right)\psi(x) = 0. \tag{1.102}$$

The principle of relativity requires that under a Lorentz transformation to a primed coordinate system the Dirac equation retain the same form:

$$\left(i\gamma'^\mu \frac{\partial}{\partial x'^\mu} - m\right)\psi'(x') = 0, \tag{1.103}$$

where $\psi(x)$ and $\psi'(x')$ describe the same physical state and x and x' are related through (1.39) and (1.40). It can be shown that the γ matrices in the two systems are equivalent up to a unitary transformation, so we drop the distinction between γ and γ' and write for the transformed Dirac equation

$$\left(i\gamma^\mu \frac{\partial}{\partial x'^\mu} - m\right)\psi'(x') = 0. \tag{1.104}$$

We assume the transformation between ψ and ψ' to be linear,

$$\psi'(x') = S\psi(x), \tag{1.105}$$

where the transformation matrix S is 4×4 and depends on the relative velocities and spatial orientations of the systems, but not on the coordinates. Neither system is privileged, so the inverse transformation must also exist:

$$\psi(x) = S^{-1}\psi'(x'). \tag{1.106}$$

Using the inverse transformation, and multiplying from the left by S, the original Dirac equation may be written

$$\left(iS\gamma^{\mu}S^{-1}\frac{\partial}{\partial x^{\mu}} - m \right)\psi'(x') = 0. \tag{1.107}$$

But using the Lorentz transformation (1.40) and (1.30)

$$\frac{\partial}{\partial x^{\mu}} = \alpha^{\nu}_{\mu}\frac{\partial}{\partial x'^{\nu}}, \tag{1.108}$$

and (1.107) becomes

$$\left(iS\gamma^{\mu}S^{-1}\alpha^{\nu}_{\mu}\frac{\partial}{\partial x'^{\nu}} - m \right)\psi'(x') = 0. \tag{1.109}$$

Comparing with (1.104) we obtain an identical form if

$$S\gamma^{\mu}S^{-1}\alpha^{\nu}_{\mu} = \gamma^{\nu}. \tag{1.110}$$

This may be taken as the defining equation for the matrix operator S. Since we originally introduced S as the transformation operator for Dirac wavefunctions, the solution of (1.110) determines the Lorentz transformation properties of the wavefunction, and the existence of a solution establishes the covariance of the Dirac equation. The solution is discussed in many places (see Bjorken and Drell, 1964). For the proper Lorentz transformations the finite transformation may be built by compounding infinitesimal ones. As an example, under rotation by an angle θ about the z-axis,

$$\psi'(x') = e^{\frac{i}{2}\theta\sigma^{12}}\psi(x), \tag{1.111}$$

where σ^{12} is given by (1.91)

$$\sigma^{12} = \begin{pmatrix} \sigma_3 & 0 \\ 0 & \sigma_3 \end{pmatrix}. \tag{1.112}$$

For improper transformations this technique fails, but direct solutions of the defining equation can be found (see Exercise 1.8).

Quantities obeying the law of transformation (1.105) with S defined by eq. (1.110) are called Dirac spinors, bispinors, or Lorentz spinors.[‡] By construc-

---

[‡]The two-component wavefunctions associated with spin in the Schrödinger equation may be termed *Pauli spinors* to distinguish them from 4-component Dirac spinors.

tion, the solutions of the Dirac equation must have a bispinor structure if the equation is to satisfy at once the requirements of quantum mechanics and special relativity.

Because the exponent in (1.111) involves $\theta/2$, a rotation of 4π (not 2π!) is required to return $\psi(x)$ to its original value (see §13.2.2). This implies that physical observables in the Dirac theory must be constructed from even powers of $\psi(x)$ and $\psi^\dagger(x)$.

The combination $\psi^\dagger\gamma^0$ occurs frequently and it is convenient to define the *adjoint spinor* $\overline{\psi}(x)$

$$\overline{\psi}(x) \equiv \psi^\dagger\gamma^0 = (\psi_1^*, \psi_2^*, -\psi_3^*, -\psi_4^*), \tag{1.113}$$

which has the behavior

$$\overline{\psi}'(x') = \overline{\psi}(x)S^{-1} \tag{1.114}$$

under Lorentz transformations. From (1.81), (1.85), and (1.113)

$$\rho = \psi^\dagger\psi = \psi^\dagger\gamma^0\gamma^0\psi = \overline{\psi}\gamma^0\psi \tag{1.115}$$

$$j^i = \psi^\dagger\alpha^i\psi = \psi^\dagger\gamma^0\gamma^i\psi = \overline{\psi}\gamma^i\psi, \tag{1.116}$$

where

$$(\gamma^0)^2 = 1 \tag{1.117}$$

has also been used. Then, from (1.115) and (1.116) the current 4-vector (1.82a) can be written

$$j^\mu = \overline{\psi}\gamma^\mu\psi. \tag{1.118}$$

1.5.6 Bilinear Covariants

Bilinear forms in $\psi^\dagger$ and ψ transforming as Lorentz tensors are called *bilinear covariants*. As an example, let us consider the Lorentz properties of the Dirac current j^μ, which transforms as $j'^\mu = \overline{\psi}'\gamma^\mu\psi'$. From (1.105) and (1.114), $j'^\mu = \overline{\psi}S^{-1}\gamma^\mu S\psi$, and utilizing (1.110)

$$j'^\mu = \alpha_\nu^\mu \overline{\psi}\gamma^\nu\psi = \alpha_\nu^\mu j^\nu. \tag{1.119}$$

Therefore, the Dirac current j^μ behaves as a vector under proper Lorentz transformations [see eq. (1.12)]. To completely specify its transformation properties we must also examine the behavior of j^μ under the improper Lorentz transformations. Under space reflection (1.45) (for which $S = \gamma^0$; see Exercise 1.8), $j^0 \to j^0$ and $j^k \to -j^k$, so j^μ is a true vector rather than an axial vector.

Sixteen linearly independent 4×4 matrices may be constructed from products of γ-matrices. These independent matrices may be used to construct an algebra, the *Clifford algebra*, which was known long before the Dirac equation. Any 4×4 matrix can be expressed in terms of this 16-component basis.

Table 1.3
Bilinear Dirac Covariants

Bilinear Form	Transforms as Lorentz	Components
$\overline{\psi}\psi$	Scalar	1
$\overline{\psi}\gamma_5\psi$	Pseudoscalar	1
$\overline{\psi}\gamma^\mu\psi$	Vector	4
$\overline{\psi}\gamma_5\gamma^\mu\psi$	Axial vector	4
$\overline{\psi}\sigma^{\mu\nu}\psi$	Antisymmetric tensor	6

Sandwiching these matrices between $\overline{\psi}$ and ψ generates the bilinear covariants listed in Table 1.3.

EXERCISE 1.7 Prove that $\overline{\psi}\psi$ transforms as a scalar, $\overline{\psi}\gamma_5\psi$ a pseudoscalar, and $\overline{\psi}\sigma^{\mu\nu}\psi$ a rank-2 tensor under Lorentz transformations. *Hint*: for proper Lorentz transformations $[S, \gamma_5] = 0$.

EXERCISE 1.8 (a) Show that for parity transformations an acceptable solution for (1.105) is $S = \eta\gamma^0$, where η is an irrelevant phase. (b) Show that the result of part (a) implies that fermions and antifermions have opposite intrinsic parities. (c) Find an appropriate charge conjugation operator (an operator that relates particles to antiparticles; see §1.6) for the Dirac wavefunctions. Investigate its action on the spinors (1.123). *Hint*: find a sequence of operations that converts (1.101) into the corresponding equation for a particle with the same m, but with $q \to -q$. (d) Show that for the special case of particles at rest the Dirac spinors can be labeled by eigenvalues of the spin projection operator

$$\frac{\hbar}{2}\Sigma_3 = \frac{\hbar}{2}\begin{pmatrix} \sigma_3 & 0 \\ 0 & \sigma_3 \end{pmatrix}$$

(but see Exercise 12.10 for particles in motion). Show that the charge conjugation operation defined in part (c) changes the sign of the Σ_3 eigenvalue for a rest spinor.

1.5.7 Solutions of the Dirac Equation for Free Particles

We now examine explicit solutions of the Dirac equation. First, consider a free fermion at rest; the wavefunction must then be uniform over all space, so that spatial differentials in the Dirac equation may be ignored and (1.72) becomes

$$H\psi = i\frac{\partial\psi}{\partial t} \simeq \beta m\psi = \begin{pmatrix} m & 0 & 0 & 0 \\ 0 & m & 0 & 0 \\ 0 & 0 & -m & 0 \\ 0 & 0 & 0 & -m \end{pmatrix}\psi, \qquad (1.120)$$

in the Pauli–Dirac representation. There are four solutions, two of positive energy

$$u^{(1)}(\mathbf{p}=0) = \begin{pmatrix} 1 \\ 0 \\ 0 \\ 0 \end{pmatrix} \qquad u^{(2)}(\mathbf{p}=0) = \begin{pmatrix} 0 \\ 1 \\ 0 \\ 0 \end{pmatrix}, \tag{1.121a}$$

and two of negative energy

$$u^{(3)}(\mathbf{p}=0) = \begin{pmatrix} 0 \\ 0 \\ 1 \\ 0 \end{pmatrix} \qquad u^{(4)}(\mathbf{p}=0) = \begin{pmatrix} 0 \\ 0 \\ 0 \\ 1 \end{pmatrix}, \tag{1.121b}$$

with the corresponding plane waves given by ($\alpha = 1, 2$)

$$\begin{aligned} \psi_+^{(\alpha)}(x) &= u^{(\alpha)}(\mathbf{p}=0)e^{-imt} \qquad \text{(positive energy)} \\ \psi_-^{(\alpha)}(x) &= u^{(\alpha+2)}(\mathbf{p}=0)e^{imt} \qquad \text{(negative energy)}. \end{aligned} \tag{1.122}$$

The negative energies originally encountered in the Klein–Gordon equation are still with us. By constructing a wave equation first order in $\partial/\partial t$ and ∇ we solved the negative probability difficulty of the KG equation, but we are still using the energy expression (1.49); this necessarily leads to positive and negative energies. In the nonrelativistic limit the first two solutions correspond to the spin degrees of freedom for a Pauli–Schrödinger electron. However, for relativistic problems both positive-energy and negative-energy solutions are required to avoid inconsistencies. We will interpret the negative-energy solutions in terms of antiparticles in the next section.

For fermions in uniform translational motion the momentum-dependent eigenspinors are

$$u^{(1)}(\mathbf{p}) = N \begin{pmatrix} \chi^{(1)} \\ \dfrac{\boldsymbol{\sigma} \cdot \mathbf{p}}{E+m} \chi^{(1)} \end{pmatrix} = N \begin{pmatrix} 1 \\ 0 \\ \dfrac{p_3}{E+m} \\ \dfrac{p_+}{E+m} \end{pmatrix}$$

$$\tag{1.123a}$$

$$u^{(2)}(\mathbf{p}) = N \begin{pmatrix} \chi^{(2)} \\ \dfrac{\boldsymbol{\sigma} \cdot \mathbf{p}}{E+m} \chi^{(2)} \end{pmatrix} = N \begin{pmatrix} 0 \\ 1 \\ \dfrac{p_-}{E+m} \\ \dfrac{-p_3}{E+m} \end{pmatrix}$$

$$v^{(1)}(\mathbf{p}) = N \begin{pmatrix} \dfrac{\boldsymbol{\sigma} \cdot \mathbf{p}}{E + m} \chi^{(2)} \\ \chi^{(2)} \end{pmatrix} = N \begin{pmatrix} \dfrac{p_-}{E + m} \\ \dfrac{-p_3}{E + m} \\ 0 \\ 1 \end{pmatrix}$$

$$(1.123a')$$

$$v^{(2)}(\mathbf{p}) = N \begin{pmatrix} \dfrac{\boldsymbol{\sigma} \cdot \mathbf{p}}{E + m} \chi^{(1)} \\ \chi^{(1)} \end{pmatrix} = N \begin{pmatrix} \dfrac{p_3}{E + m} \\ \dfrac{p_+}{E + m} \\ 1 \\ 0 \end{pmatrix}$$

Corresponding to the plane wave solutions

$$\psi_+^{(\alpha)}(x) = u^{(\alpha)}(\mathbf{p})e^{-ip \cdot x} \qquad \psi_-^{(\alpha)}(x) = v^{(\alpha)}(\mathbf{p})e^{ip \cdot x}, \tag{1.123b}$$

where we introduce the two-component spinors

$$\chi^{(1)} = \begin{pmatrix} 1 \\ 0 \end{pmatrix} \qquad \chi^{(2)} = \begin{pmatrix} 0 \\ 1 \end{pmatrix} \tag{1.123c}$$

and choose a normalization

$$N = \sqrt{\frac{E + m}{2m}}. \tag{1.123d}$$

In anticipation of the identification of negative-energy solutions with antiparticles (§1.6), positive-energy antiparticle spinors $v^{(1,2)}(\mathbf{p})$ have been defined in terms of negative-energy particle spinors $u^{(3,4)}(\mathbf{p})$ through (Sakurai, 1967, §3-10 and Appendix B; Halzen and Martin, 1984, §5.4)

$$\begin{aligned} v^{(1)}(\mathbf{p})e^{ip \cdot x} &= u^{(4)}(-\mathbf{p})e^{-i(-p) \cdot x} \\ v^{(2)}(\mathbf{p})e^{ip \cdot x} &= u^{(3)}(-\mathbf{p})e^{-i(-p) \cdot x}. \end{aligned} \tag{1.123e}$$

[The reason for the association $(1, 2) \leftrightarrow (4, 3)$ of the spinor indices in (1.123e) is given in the solution to Exercise 1.8d.] In these and the following expressions the quantity E is always positive, and refers to the physical particle or antiparticle. The second form for each spinor in (1.123a) results from the observation that (Exercise 1.9c)

$$\boldsymbol{\sigma} \cdot \mathbf{p} = \begin{pmatrix} p_3 & p_1 - ip_2 \\ p_1 + ip_2 & -p_3 \end{pmatrix} \equiv \begin{pmatrix} p_3 & p_- \\ p_+ & -p_3 \end{pmatrix}.$$

These spinors satisfy the following relations:

$$(\not{p} - m)u^{(\alpha)}(\mathbf{p}) = 0 \qquad (\not{p} + m)v^{(\alpha)}(\mathbf{p}) = 0 \qquad (1.124a)$$

$$\bar{u}^{(\alpha)}(\mathbf{p})(\not{p} - m) = 0 \qquad \bar{v}^{(\alpha)}(\mathbf{p})(\not{p} + m) = 0 \qquad (1.124b)$$

$$\bar{u}^{(\alpha)}(\mathbf{p})u^{(\beta)}(\mathbf{p}) = \delta_{\alpha\beta} \qquad \bar{v}^{(\alpha)}(\mathbf{p})v^{(\beta)}(\mathbf{p}) = -\delta_{\alpha\beta} \qquad (1.125a)$$

$$\bar{u}^{(\alpha)}(\mathbf{p})v^{(\beta)}(\mathbf{p}) = \bar{v}^{(\alpha)}(\mathbf{p})u^{(\beta)}(\mathbf{p}) = 0 \qquad (1.125b)$$

$$u^{(\alpha)\dagger}(\mathbf{p})u^{(\beta)}(\mathbf{p}) = \frac{E}{m}\delta_{\alpha\beta} \qquad v^{(\alpha)\dagger}(\mathbf{p})v^{(\beta)}(\mathbf{p}) = \frac{E}{m}\delta_{\alpha\beta} \qquad (1.125c)$$

$$\sum_{\alpha=1,2} u^{(\alpha)}(\mathbf{p})\bar{u}^{(\alpha)}(\mathbf{p}) = \frac{\not{p} + m}{2m} \qquad \sum_{\alpha=1,2} v^{(\alpha)}(\mathbf{p})\bar{v}^{(\alpha)}(\mathbf{p}) = \frac{\not{p} - m}{2m} \qquad (1.126)$$

where $\bar{q} \equiv q^{\dagger}\gamma^{0}$, and α, β are spinor indices taking the values 1 or 2. The relations (1.124) are Dirac equations; (1.125) defines the orthogonality properties and normalization convention, and (1.126) is a statement of completeness. Note that (1.126) defines *matrix equations*—see Exercise 2.8c. The probability density is normalized so that

$$\psi_{+}^{(\alpha)\dagger}(x)\psi_{+}^{(\beta)}(x) = \psi_{-}^{(\alpha)\dagger}(x)\psi_{-}^{(\beta)}(x) = \frac{E}{m}\delta_{\alpha\beta},$$

with the factor E/m compensating for Lorentz contraction of the volume element to preserve the normalization under Lorentz transformations (see Exercise 1.9b).

If we choose the z-axis parallel to the momentum, $\mathbf{p} \equiv (0, 0, p)$, the spinors (1.123a) are also eigenstates of the *helicity operator*

$$\frac{1}{2}\mathbf{\Sigma}\cdot\hat{\mathbf{p}} \equiv \frac{1}{2}\begin{pmatrix} \boldsymbol{\sigma} & 0 \\ 0 & \boldsymbol{\sigma} \end{pmatrix}\cdot\hat{\mathbf{p}} = \frac{1}{2}\begin{pmatrix} \boldsymbol{\sigma}\cdot\hat{\mathbf{p}} & 0 \\ 0 & \boldsymbol{\sigma}\cdot\hat{\mathbf{p}} \end{pmatrix},$$

where $\hat{\mathbf{p}} = \mathbf{p}/|\mathbf{p}|$. In particular,

$$\tfrac{1}{2}\mathbf{\Sigma}\cdot\hat{\mathbf{p}}u^{(\alpha)}(\mathbf{p}) = s_{\alpha}u^{(\alpha)}(\mathbf{p}) \qquad \tfrac{1}{2}\mathbf{\Sigma}\cdot\hat{\mathbf{p}}v^{(\alpha)}(\mathbf{p}) = -s_{\alpha}v^{(\alpha)}(\mathbf{p}), \qquad (1.127)$$

where the helicity quantum numbers are $s_1 = +\frac{1}{2}$ and $s_2 = -\frac{1}{2}$. Physically, helicity is the projection of the fermion spin along the momentum direction, which is a constant of motion for a free Dirac particle. Helicity is invariant under spatial rotations, but it is Lorentz invariant only in the limit of massless fermions (see §6.2.2).

EXERCISE 1.9 (a) Prove the orthogonality relation (1.125b). (b) Show that the spinor normalization (1.125) is Lorentz invariant. (c) Show that

$$\boldsymbol{\sigma}\cdot\mathbf{p} = \begin{pmatrix} p_3 & p_1 - ip_2 \\ p_1 + ip_2 & -p_3 \end{pmatrix} \equiv \begin{pmatrix} p_3 & p_- \\ p_+ & -p_3 \end{pmatrix},$$

where $p_\pm = p_1 \pm ip_2$. Show that $(\boldsymbol{\sigma} \cdot \mathbf{p})^2 = \mathbf{p}^2$.   (d) Write the Dirac spinors in two-component form, $\psi = \begin{pmatrix} u_A \\ u_B \end{pmatrix}$. Show that for positive-energy particle solutions in the nonrelativistic limit the upper ("large") components u_A dominate the lower ("small") components u_B.

1.6   Prescriptions for Negative-Energy States

We have seen how wave equations that are Lorentz covariant may be constructed. These equations incorporate the correct relativistic energy expression and, for the Dirac case, a sensible interpretation of the probabilities and currents emerges. We still must understand the negative-energy solutions, since they will not go away and cannot be ignored. The process of interpreting these solutions will reveal the beginnings of a way to rescue the Klein–Gordon equation, which was in dire straits when last encountered. We will also see in these considerations the reasons that eventually require us to go beyond the simple Dirac and KG equations to a relativistic quantum field theory.

1.6.1   Dirac Hole Theory and Negative-Energy Solutions

The negative-energy solutions of the Dirac equation appear to constitute a serious problem. It is easy to compute that interactions with the radiation field would cause all positive-energy electrons to tumble to oblivion in the negative-energy states in a very short period. This eventuality clearly is not in agreement with the observed stability of atomic matter.

Dirac proposed solving this problem by assuming that the negative-energy states are all filled. Then, since electrons are fermions the Pauli principle prevents positive-energy electrons from making transitions to the negative-energy states. This set of filled negative-energy states is called the *Fermi sea*. If the sea were quiescent it would not be difficult to believe that such a uniform background would have no observable consequences (except possibly for general relativity, where absolute mass is significant because it determines the curvature of space).

However, this sea cannot be quiescent because of quantum-mechanical fluctuations of the vacuum. From the free Dirac equation the gap between the filled negative-energy states and the positive-energy states is twice the fermion rest mass $2mc^2$, as illustrated in Fig. 1.3. Therefore, energy supplied to the vacuum either by virtual fluctuations, or by real physical processes such as incident radiation, may cause a fermion to be promoted from the negative-energy sea to an unoccupied positive-energy state, leaving a hole in the sea (Fig. 1.3). Such a particle–hole excitation of the vacuum produces a positive-energy fermion and a hole in the sea that will behave, because of the negative sign on the energy, as a fermion of opposite charge—an *antifermion*.

The application of this idea to the electron by Dirac led to the prediction

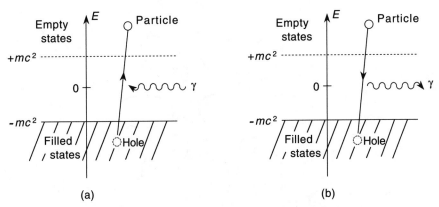

Fig. 1.3 Pair production and pair annihilation.

of the positron, and its subsequent discovery changed the tenacious negative-energy solutions from embarrassing liabilities to one of the greatest triumphs of the theory. Thus, Fig. 1.3a describes pair creation and Fig. 1.3b describes pair annihilation. As noted above, these processes can create virtual particle–hole fluctuations existing for fleeting instants allowed by the uncertainty principle, or real particle–hole pairs if at least twice the rest mass energy is supplied externally. Both processes have observable consequences.

Dirac's hole theory may be formulated in terms of the charge conjugate wavefunction ψ_c defined in Exercise 1.8c,d. An extensive discussion may be found in Sakurai (1967).

1.6.2 Vacuum Polarization

We have come full circle! Having set out to find a covariant analog to the single-particle Schrödinger equation we have succeeded in obtaining the Dirac equation, with which an impressive body of data has been correlated (for example, electron spin and magnetic moments, atomic spin–orbit coupling, and the existence and properties of positrons). However, the hole-theory interpretation shows explicitly that the Dirac equation is generally not a single-particle equation—because of the particle–hole excitations of the vacuum it is inherently a many-body equation.

As an example, consider the phenomenon of *vacuum polarization* in quantum electrodynamics. Suppose we toss a bare positive-energy electron into the Fermi sea, and try to measure its charge. Because of zero-point energy the Fermi sea is fluctuating with virtual electron–positron excitations, but a test charge immersed in any dielectric medium will polarize the medium, as illustrated in Fig. 1.4. Thus the electron will preferentially attract the virtual positrons, which shield its charge when viewed from large distances. A macroscopic probe will see an effective charge, and only at very short distances (if at all) could the bare electronic charge become visible: the vacuum has been

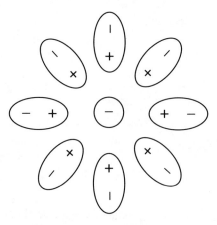

Fig. 1.4   Polarization of a dielectric by a test charge.

polarized. Such vacuum polarization effects are small, but clearly seen (see §6.4). Observe that since test charges are always immersed in the Fermi sea, an effective rather than a bare charge is measured in all experiments. This is important in later discussions of divergences in quantum field theory.

We may legitimately use the Klein–Gordon or Dirac equations as single-particle equations only for weak, slowly varying fields for which a broad gap $\simeq 2mc^2$ remains between positive-energy and negative-energy states. Otherwise, phenomena such as *zitterbewegung* or the *Klein paradox* assert themselves; see Sakurai (1967), or Baym (1969) for examples. In elementary particle physics one typically deals with energies that are large compared with $2mc^2$ for electrons, or even heavier particles. Then the Dirac equation must be handled as a many-body equation. The nonrelativistic many-body problem is most elegantly approached using the formalism of second quantization, which treats the Schrödinger equation as a field equation to be quantized. An analogous situation occurs in relativistic quantum mechanics: for relativistic many-body physics the most powerful microscopic techniques involve second quantization of the corresponding one-body wave equations. Thus we are led in Ch. 2 to the construction of relativistic quantum field theories.

1.6.3   Partial Rescue of the Klein–Gordon Equation

Having trod the historical path, we have seen that the Dirac equation overcomes the most important objection to the original implementation of the Klein–Gordon equation—the interpretation of the probability density. In this section we demonstrate that the particle–antiparticle correspondence introduced in the Dirac hole formalism can be used in the Klein–Gordon theory to provide a sensible reinterpretation of the KG probability density. With this interpretation the Klein–Gordon equation will be as viable a candidate for a relativistic wave equation as the Dirac equation. The KG equation will be the appropriate single-particle equation for bosons, whereas the Dirac equation will be the corresponding equation for fermions; *each* will be seriously deficient when interpreted as a single-particle equation in high-energy processes.

The Klein–Gordon equation (1.58) in the presence of the minimal electro-magnetic substitution (1.83) is

$$[(i\partial^\mu - qA^\mu)^2 - m^2)]\phi = 0. \tag{1.128}$$

The current continuity equations [the analog of (1.66)] in the presence of an electromagnetic field then become $\partial_\mu j^\mu = 0$, with

$$j^\mu = \phi^*(i\partial^\mu - qA^\mu)\phi - \phi(i\partial^\mu + qA^\mu)\phi^*. \tag{1.129}$$

Now we take the complex conjugate of (1.128). All operators are real, so only the explicit i's change and the wavefunction gets an asterisk:

$$[(i\partial^\mu + qA^\mu)^2 - m^2]\phi^* = 0. \tag{1.130}$$

Comparing (1.128) with (1.130), and observing that a change in sign of the 4-vector potential $A \to -A = (-A^0, -\mathbf{A})$ is equivalent to changing the sign of the particle's charge, we find an important symmetry of the KG equation: *if ϕ is a solution of the Klein–Gordon equation with a charge q, then ϕ^* is a solution with the same mass, but charge $-q$.* This symmetry is called *charge conjugation* (C), and the corresponding symmetry operation relates particle and antiparticle (see Exercise 1.8c,d). The wavefunction ϕ^* is called the *charge–conjugate* wavefunction. The density ρ_c and current $\mathbf{j}_c$ associated with the charge-conjugate solution are

$$\rho_c = -\rho \qquad \mathbf{j}_c = -\mathbf{j}, \tag{1.131}$$

where ρ and $\mathbf{j}$ are the corresponding quantities calculated for (1.128).

Here then is the key to a consistent interpretation of the KG continuity equation. The KG equation, and the Dirac equation, inherently contain particle as well as antiparticle solutions. The Klein–Gordon density ρ should be interpreted as a *charge density* rather than a probability density, and it is natural (required!) that it can be positive or negative; in a similar way the currents are to be interpreted as *charge currents*.

Notice also that complex conjugation reverses the sign of all frequencies and momenta in ϕ. Therefore, the complex conjugate of a negative-energy solution to the Klein–Gordon equation is a positive-energy solution of the charge-conjugated equation with the sign of the charge reversed: to interpret a negative-energy solution, complex conjugate ϕ and interpret it as a positive-energy solution, but with opposite charge. These ideas may be applied to the Dirac equation as well as the KG equation (see Exercise 1.8c,d), and form the basis for the "backward in time" prescription for handling negative-energy solutions that will be discussed in the following section.

As demonstrated in many places (see Bjorken and Drell, 1964, §9.7; Baym, 1969, Ch. 22), the KG equation can be rewritten as a first-order equation in the time derivative with a two-component column vector wavefunction. The two components are required to accommodate particle–antiparticle symmetry, and the internal degree of freedom represented by the wavefunction components is not spin; instead, it represents charge. The Dirac equation is also first order in time, but it requires a four-component wavefunction, roughly because we must allow for particles and antiparticles, and for two spin degrees of freedom for the spin-$\frac{1}{2}$ particles.

What about particles that are uncharged? We must distinguish two classes for such objects: (1) neutral particles with a neutral antiparticle that may be distinguished by an interaction other than electromagnetic, and (2) neutral particles that are their own antiparticle. An example in the first class is the neutral K^0 meson, which has as antiparticle the neutral $\overline{K}^0$ meson. Although they have the same electromagnetic charge, these particles carry different strangeness: $S = 1$ for K^0, but $S = -1$ for $\overline{K}^0$. They may be distinguished through strangeness selection rules in decays and reactions (see Fig. 5.7 and Perkins, 1987, §7.14).

In the second class we have objects such as the π^0 and the photon; they are their own antiparticles, and the above arguments do not apply. However, one generally finds that charged particles require a complex wavefunction [for uncharged particles the wavefunction may be chosen real $(\psi = \psi^*)$]. This should not come as a surprise, considering the preceding discussion: charged particles have distinguishable antiparticles and the corresponding relativistic wavefunction must accommodate more degrees of freedom than for a neutral particle which is its own antiparticle. The required doubling comes either from a two-component wavefunction if we separate the KG equation into two equations linear in $\partial/\partial t$, or a complex wavefunction for the normal KG equation (1.58). But for a real wavefunction the density and current vanish identically, as is obvious from (1.63) and (1.64). This is as it should be for neutral particles, given our interpretation of ρ as a charge density and $\mathbf{j}$ as an electrical current. Thus a sensible continuity equation exists for both charged and uncharged bosons in the Klein–Gordon equation, and the original reason for rejecting it is no longer relevant.

1.6.4 Backward in Time Prescription

Historically, the interpretation of the negative-energy solutions and the prediction of the positron issued from the Dirac hole theory. However, this interpretation will not work for bosons since they are not subject to the exclusion principle.

There is a second prescription for interpreting the negative-energy states,

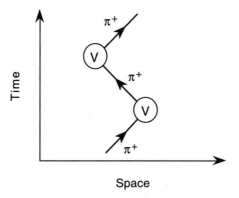

Fig. 1.5 Second-order scattering in nonrelativistic quantum mechanics. In this and subsequent diagrams in this chapter time is assumed to increase vertically.

associated with the names of Stückelberg and Feynman, that exploits particle–antiparticle symmetry but does not rely on the filled Fermi sea and the exclusion principle (Stückelberg, 1941; Feynman, 1949). This approach allows the correct probability amplitudes to be obtained for fermions *and* bosons from a single interpretation for the negative-energy solutions. The basic ideas underlying the prescription have been introduced in the preceding section, and the Stückelberg–Feynman recipe for handling the negative-energy solutions is

Negative-energy particle solutions propagating backward in time may be replaced by positive-energy antiparticle solutions propagating forward in time. Positive-energy particles propagate only forward in time; negative-energy particles propagate only backward.

This prescription may be applied to either fermions or bosons,[‡] and is a consequence of the observation that $\exp[-i(-E)(-t)] = \exp(-iEt)$. In the remainder of this section we give some simple examples suggesting the correctness of this prescription. In the next section we expand the Stückelberg–Feynman recipe into a systematic method of calculating probability amplitudes for physical processes.

In nonrelativistic quantum mechanics a second-order scattering of (say) a π^+ can be pictured as free-particle propagation forward in time between scatterings; Fig. 1.5 illustrates in a spacetime diagram. [The subsequent discussion follows Aitchison (1972) and Aitchison and Hey (1982) rather closely.] In the Stückelberg–Feynman version of relativistic quantum mechanics we must also allow for a spacetime diagram of the form shown in Fig. 1.6. In this diagram

---

[‡]To be sure, for fermions there are some signs that must be put in by hand when using this approach; their justification is provided by field theory (see Ch. 3).

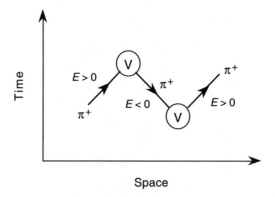

Fig. 1.6   A backward in time process that contributes in addition to the process in Fig. 1.5 for relativistic quantum mechanics.

the π^+ can be scattered *backward in time*; by our prescription the particle scattering backward in time must be assigned a negative energy, and Fig. 1.6 may be reinterpreted as Fig. 1.7: the particle propagating backward in time with negative energy is equivalent to an antiparticle propagating forward in time with positive energy.

For such diagrams the emission of an antiparticle of 4-momentum p^μ is physically indistinguishable from the absorption of a particle of 4-momentum $-p^\mu$. Likewise, the absorption of an antiparticle of 4-momentum p^μ is equivalent to the emission of a particle with 4-momentum $-p^\mu$. This prescription allows processes involving antiparticles to be converted to equivalent ones involving *only particles* by changing the signs for all antiparticle 4-momenta and reversing the role of entry and exit states for the antiparticle lines; that is, by

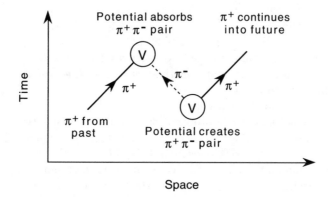

Fig. 1.7   Feynman reinterpretation of the process depicted in Fig. 1.6. Dashed lines indicate antiparticles.

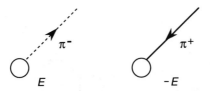

Fig. 1.8 Equivalent processes by Feynman prescription.

changing the direction of the arrows. For example, this implies that the two processes depicted in Fig. 1.8 are physically indistinguishable.

In the next section we will use these ideas to calculate probability amplitudes in a Lorentz-covariant perturbation theory. Before doing that, it is useful pedagogically to see the Stückelberg–Feynman prescription in action for nonrelativistic, first-order, time-dependent perturbation theory (Feynman, 1961; Aitchison and Hey, 1982). The first-order amplitude is

$$a_{\mathrm{fi}} = -i \int d^3x \int dt\, \phi_{\mathrm{f}}^*(\mathbf{x}) e^{iE_{\mathrm{f}}t} V(\mathbf{x},t) \phi_{\mathrm{i}}(\mathbf{x}) e^{-iE_{\mathrm{i}}t}, \tag{1.132}$$

and we consider a simple periodic potential

$$V(\mathbf{x},t) \simeq V_0(\mathbf{x}) e^{\pm i\omega t}. \tag{1.133}$$

Then, defining

$$V_{\mathrm{fi}} = -i \int d^3x\, \phi_{\mathrm{f}}^*(\mathbf{x}) V_0(\mathbf{x}) \phi_{\mathrm{i}}(\mathbf{x}), \tag{1.134}$$

and recalling the integral representation of the Dirac delta function

$$\int dy\, e^{ixy} = 2\pi\delta(x), \tag{1.135}$$

we obtain the first-order transition amplitude

$$a_{\mathrm{fi}} = V_{\mathrm{fi}} \int dt\, e^{i(E_{\mathrm{f}} - E_{\mathrm{i}} \pm \omega)t} = 2\pi V_{\mathrm{fi}} \delta(E_{\mathrm{f}} - E_{\mathrm{i}} \pm \omega). \tag{1.136}$$

Thus the transition vanishes unless $E_{\mathrm{f}} = E_{\mathrm{i}} \mp \omega$, implying that an external potential $e^{-i\omega t}$ gives up an energy $\hbar\omega$ to the system, while a potential $e^{+i\omega t}$ absorbs an energy $\hbar\omega$ from the system.

Let us see how this works in some simple pion scattering examples. For simplicity we set $V_{\mathrm{fi}} = 1$ and take

$$V \simeq e^{-i\omega t} \tag{1.137}$$

which, by the above discussion, should transfer an energy $\hbar\omega$ in a first-order interaction with the system.

(1) π^+ scattering:

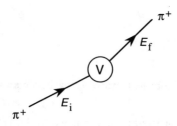

From (1.59) we have an initial plane-wave state corresponding to a positive-energy solution $\psi_i = N_i e^{-i p_i \cdot x}$, where N_i denotes a normalization and the 4-momentum is $p_i^\mu = (E_i, \mathbf{p}_i)$. The final state is $\psi_f = N_f e^{-i p_f \cdot x}$, with $p_f^\mu = (E_f, \mathbf{p}_f)$. Since we are concerned only with the time dependence at the moment, the first-order amplitude is

$$a_{fi} \simeq \int dt\, \psi_f^* e^{-i\omega t} \psi_i = \int dt\, e^{i(E_f - E_i - \omega)t} = 2\pi\delta(E_f - E_i - \omega).$$

Thus an energy $\hbar\omega$ is transferred to the pion, as expected: $E_f = E_i + \omega$.

(2) π^- scattering:

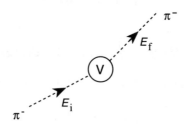

(dashed lines for antiparticles) may be redrawn as

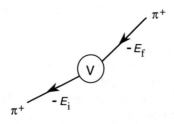

where we have invoked the recipe of reversing the 4-momenta and the role of entrance and exit states to convert an antiparticle graph to a particle graph. Thus

$$p_i^\mu = (-E_f, -\mathbf{p}_f) \qquad p_f^\mu = (-E_i, -\mathbf{p}_i),$$

and the amplitude is

$$a_{\rm fi} \simeq \int e^{i(-E_{\rm i}-\omega+E_{\rm f})t}\,dt = 2\pi\delta(-E_{\rm i}+E_{\rm f}-\omega).$$

Again we find that $E_{\rm f} = E_{\rm i} + \omega$, as expected. The prescription seems to work, allowing us to obtain sensible results from processes involving particles or antiparticles by using only particle wavefunctions.

Observe carefully the rule for determining which is the entry and which is the exit state in such processes: *redraw according to Feynman and follow the arrows.* Thus the terms "entry" and "exit" in the Feynman recipe have nothing directly to do with time; instead, they specify whether an arrow points into (entry or ingoing) or out of (exit or outgoing) an interaction. That is, exit and entry refer to the order in which wavefunctions appear in the matrix elements: in $\int \psi_2^* V \psi_1\,dx$, the entry state is ψ_1 and the exit state is ψ_2. For particles the entry state in the matrix element is the initial (i.e., past) state and the exit state is the final (i.e., future) state; for antiparticles the entry state is the final state and the exit state is the initial state (Feynman, 1961).

(3) As a third example, consider pair creation by the potential $V = e^{-i\omega t}$:

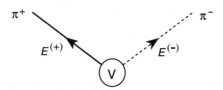

By steps analogous to those of previous diagrams this may be converted to

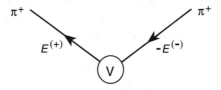

and for this process $\omega = E^{(+)} + E^{(-)}$, as expected (Exercise 1.10).

EXERCISE 1.10 Carry out the steps in the above demonstration that for a pair creation process $\hbar\omega = E^{(+)} + E^{(-)}$.

1.7 Feynman Diagrams

In subsequent chapters we will examine physical processes within the context of quantum field theory. Before rolling out the heavy artillery it is useful to consider a few examples that rely primarily on intuition and the Feynman spacetime pictures just discussed. This approach is termed the *propagator method,* and it often yields correct and intuitively satisfying results without the full rigors of field theory. A detailed exposition of such methods can be found in Bjorken and Drell (1964), and a more heuristic treatment is contained in Aitchison and Hey (1982). The former is more rigorous, but the latter is more in keeping with our intentions in this section and we rely heavily on Ch. 2 of Aitchison and Hey (1982) for the examples that follow.

1.7.1 Electromagnetic $\pi^+ K^+$ Scattering

Let us outline the construction of an amplitude for the electromagnetic scattering of a π^+ from a K^+ (both spinless bosons—the example is rather pedagogical as the electromagnetic interaction would typically be overwhelmed by the strong interaction for low energy scattering in the $\pi^+ - K^+$ system). To construct the amplitude we use the spacetime (propagator) approach, with bosons deliberately chosen to avoid complications from spinor algebra in this simple example.

The parameter $e^2/4\pi$ is the obvious expansion parameter in treating the electromagnetic interaction by perturbation theory. In natural units its value is (Appendix A)

$$\alpha = \frac{e^2}{4\pi} \simeq \frac{1}{137}, \tag{1.138}$$

so the perturbation expansion should converge rapidly and we consider only the lowest order in the fine structure constant α.

We may proceed in two steps: (1) calculate the vector potential A^μ produced by the K^+ particles, and (2) calculate the interaction of the π^+ with the K^+ vector potential. First, let us construct the general equation for a π^+ in a 4-vector potential. Schematically, the interaction looks like

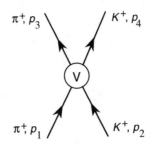

These are spinless bosons so we use the Klein–Gordon equation (1.58) with the minimal substitution (1.83). This gives (1.128), which may be written in the form

$$(\Box + m^2)\phi = -V\phi, \tag{1.139}$$

where (Exercise 1.11a)

$$V \equiv ie\left(\partial_\mu A^\mu + A^\mu \partial_\mu\right) - e^2 A^2. \qquad (1.140)$$

We will calculate to lowest order, which justifies keeping only the first term in (1.140). The first-order scattering amplitude in this potential is

$$\mathcal{A} = -i \int d^4x\, \psi_3^* V \psi_1,$$

where the plane wave states are

$$\psi_1 = N_1 e^{-ip_1 \cdot x} \qquad \psi_3 = N_3 e^{-ip_3 \cdot x},$$

and N_1 and N_3 are normalizations. A corresponding diagram is

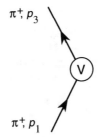

Therefore

$$\mathcal{A} = N_1 N_3 e \int d^4x\, e^{ip_3 \cdot x}\left(\partial_\mu A^\mu + A^\mu \partial_\mu\right) e^{-ip_1 \cdot x}, \qquad (1.141)$$

and this can be written (Exercise 1.11c)

$$\mathcal{A} = -i \int d^4x\, j_\mu(\pi^+) A^\mu, \qquad (1.142)$$

where the *EM transition current* is defined by

$$j_\mu(\pi^+) = ie\left[\psi_3^*(\partial_\mu \psi_1) - (\partial_\mu \psi_3^*)\psi_1\right]. \qquad (1.143)$$

[The similarity of this to the current (1.65) motivates the appellation transition current]. For plane waves (1.143) is easily evaluated, and (Exercise 1.11b)

$$\mathcal{A} = -ieN_1 N_3 (p_1 + p_3)_\mu \int d^4x\, A^\mu e^{-iq \cdot x}, \qquad (1.144)$$

where the 4-momentum transfer is

$$q \equiv p_1 - p_3. \qquad (1.145)$$

EXERCISE 1.11 (a) Show that the Klein–Gordon equation with minimal coupling gives equations (1.139) and (1.140). (b) (Aitchison and Hey, 1982) Assuming that $A_0 \to 0$ as $t \to \pm\infty$, and that $|\mathbf{A}| \to 0$ as $|\mathbf{x}| \to \infty$, use integration by parts to show that

$$\int d^4x\, e^{ip_f\cdot x}(\partial_\mu A^\mu + A^\mu \partial_\mu)e^{-ip_i\cdot x} = -i(p_f + p_i)_\mu \int d^4x\, e^{ip_f\cdot x} A^\mu e^{-ip_i\cdot x}.$$

(c) Show that eq. (1.142) follows from eq. (1.141) for plane waves.

Now we must determine the vector potential A^μ generated by the K^+ and insert it in (1.144). In the Lorentz gauge (see the discussion in §2.7.1 of classical gauge invariance) the Maxwell equations are [eq. (2.124)]

$$\Box A^\mu = j^\mu(K^+) \qquad \text{(Lorentz gauge)}, \tag{1.146}$$

where $j^\mu(K^+)$ is the current associated with the K^+. Since this analysis could as well proceed by calculating the scattering of a K^+ in the vector potential generated by the π^+, consistency suggests that the current j^μ associated with the K^+ is just the analog of the transition current (1.143) for the pions:

$$j^\mu(K^+) = ie\left[\psi_4^*(\partial^\mu \psi_2) - (\partial^\mu \psi_4^*)\psi_2\right]. \tag{1.147}$$

Inserting plane-wave states ψ_2 and ψ_4 for initial and final kaons

$$j^\mu(K^+) = eN_2 N_4 (p_2 + p_4)^\mu e^{i(p_4 - p_2)\cdot x}. \tag{1.148}$$

Hence, (1.146) must be solved for A^μ with j^μ given by (1.148). If we introduce the 4-momentum transfer (1.145)

$$q^\mu \equiv (p_1 - p_3)^\mu = (p_4 - p_2)^\mu, \tag{1.149}$$

it is easily seen that the required solution is

$$A^\mu = \frac{-j^\mu(K^+)}{q^2}. \tag{1.150}$$

Therefore, on inserting (1.150) in (1.142) the matrix element to first order in α becomes

$$\begin{aligned}
\mathcal{A}_{\pi^+ K^+} &= i\int d^4x\, j_\mu(\pi^+)\frac{1}{q^2}j^\mu(K^+) \\
&= ie^2 N_1 N_2 N_3 N_4 (p_1 + p_3)_\mu (p_2 + p_4)^\mu \\
&\quad \frac{1}{q^2}\int d^4x\, e^{i(p_3 - p_1)\cdot x} e^{i(p_4 - p_2)\cdot x}.
\end{aligned} \tag{1.151}$$

The integral defines the Dirac δ-function in four dimensions

$$\int d^4x\, e^{ix\cdot y} = (2\pi)^4 \delta^{(4)}(y), \tag{1.152}$$

and imposes 4-momentum conservation. After lowering the index on $(p_2+p_4)^\mu$ by contraction with the metric tensor we may finally write

$$\mathcal{A}_{\pi^+K^+} = -iN_1N_2N_3N_4(2\pi)^4\delta^{(4)}(p_3 + p_4 - p_1 - p_2)$$

$$e(p_1 + p_3)_\mu \left(\frac{-g^{\mu\nu}}{q^2}\right) e(p_2 + p_4)_\nu$$

$$= -i(2\pi)^4 N_1N_2N_3N_4\delta^{(4)}(p_3 + p_4 - p_1 - p_2)\mathcal{M}_{\pi^+K^+}, \tag{1.153a}$$

where the *invariant Feynman amplitude*

$$\mathcal{M}_{\pi^+K^+} \equiv e^2(p_1 + p_3)_\mu \left(\frac{-g^{\mu\nu}}{q^2}\right)(p_2 + p_4)_\nu \tag{1.153b}$$

has been separated. Such an amplitude is conveniently represented by a *Feynman diagram* (Fig. 1.9), and we may formulate a compact set of rules for associating the diagram with the matrix element:

- There is a normalization factor N_i for each external line.
- The interaction is mediated by virtual photon exchange (wavy line), associated with the *photon propagator* $-ig^{\mu\nu}/q^2$ (Lorentz gauge).
- Two vertices appear in the diagram, corresponding to factors of the form $-ie(p_1 + p_3)_\mu$. This is because the coupling of the photon to the particles at the vertex is proportional to e in this order, and because some momentum factors are required to construct a 4-vector at each vertex to couple to the photon.
- The distribution of factors $\pm i$ is chosen for consistency with the rules for higher order diagrams to be discussed later.

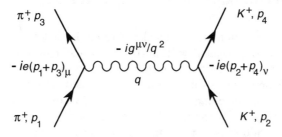

Fig. 1.9 Feynman diagram for electromagnetic π^+K^+ scattering to first order in the fine structure constant α.

- 4-momentum conservation is assumed implicitly, corresponding to the factor $(2\pi)^4\delta^{(4)}(p_3+p_4-p_1-p_2)$, but it usually is not displayed explicitly.

Two comments are in order concerning the photon propagator. The first is that the form (1.150) is valid for the Lorentz gauge; the form in other gauges will be discussed later, but physical quantities will be independent of gauge choice. The second concerns some terminology. The photon in the Feynman diagram is called a *virtual photon*, and q^2 (the square of the 4-momentum transfer) is termed the *mass squared of the virtual photon*. To see why, notice that a real photon satisfies the free-field equation (see §2.7)

$$\Box A^\mu = 0, \tag{1.154}$$

and for plane wave solutions $e^{iq\cdot x}$ we have

$$q^2 = 0, \tag{1.155}$$

meaning that a real photon is massless [see eq. (1.53)]. But the interacting photon satisfies

$$\Box A^\mu = j^\mu; \tag{1.156}$$

then $q^2 \neq 0$, and we say that the interacting photon is *virtual* or *off mass shell*, with a "mass squared" q^2. Some other common terminology arises from comparing the energy momentum 4-vector $(E,\mathbf{p})$ with the prototypical 4-vector, the spacetime coordinate $(t,\mathbf{x})$. By analogy with the discussion in §1.1 we call a virtual photon that has $q^2 > 0$ *timelike*, and one that has $q^2 < 0$ *spacelike*; of course, real photons satisfy $q^2 = 0$. For example, in e^+e^- colliding beam experiments a head-on collision produces a virtual photon with large energy but no momentum, so that the photon is timelike. Conversely, in deep inelastic electron scattering the kinematic conditions are such that the virtual photon carries more 3-momentum than energy, and it is spacelike (see Exercise 10.2).

1.7.2 Electromagnetic $\pi^+\pi^+$ Scattering

The power of the Feynman graphical technique is that the rules associating factors in the matrix element with portions of the diagrams are quite general, and that we can work backward: from a diagram it is possible to construct the matrix element. Let us illustrate this by considering $\pi^+\pi^+$ scattering, to first order in the electromagnetic interaction. Two diagrams may contribute (Fig. 1.10), and since these are identical bosons the amplitudes for the two graphs must be added with the same sign to ensure exchange symmetry:

$$\mathcal{A} = \mathcal{A}_{\text{direct}} + \mathcal{A}_{\text{exchange}}. \tag{1.157}$$

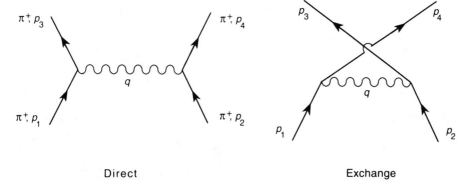

Direct   Exchange

Fig. 1.10   Feynman diagrams for electromagnetic $\pi^+\pi^+$ scattering.

Applying the Feynman rules by analogy with the previous example of $\pi^+ K^+$ scattering in the Lorentz gauge we obtain

$$\mathcal{A}_{\pi^+\pi^+} = -i(2\pi)^4 N_1 N_2 N_3 N_4 \, \delta^{(4)}(p_3 + p_4 - p_1 - p_2)\mathcal{M}_{\pi^+\pi^+}, \qquad (1.158a)$$

where the invariant Feynman amplitude is

$$\mathcal{M}_{\pi^+\pi^+} \equiv \left(\frac{-e^2(p_1 + p_3)_\mu (p_2 + p_4)^\mu}{(p_2 - p_4)^2} + \frac{-e^2(p_1 + p_4)_\mu (p_2 + p_3)^\mu}{(p_2 - p_3)^2} \right).$$
$$(1.158b)$$

1.7.3 Electromagnetic $\pi^+\pi^-$ Scattering

Now let us apply the Feynman prescription to the analogous $\pi^+\pi^-$ scattering. First we reinterpret as follows

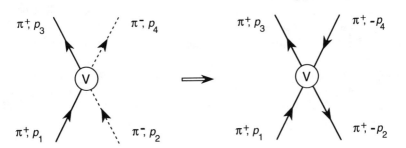

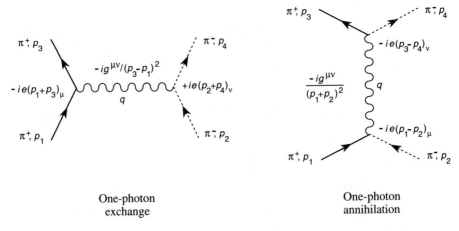

Fig. 1.11   Feynman graphs contributing to $\pi^+\pi^-$ electromagnetic scattering.

where the usual Stückelberg–Feynman recipe has been invoked to convert the π^- to a π^+ moving backward in time. We could now proceed as before, but there is no need. The reinterpreted process is just a $\pi^+\pi^+$ scattering, so we can immediately write the amplitude $\mathcal{A}_{\pi^+\pi^-}$ by switching some momenta in (1.158):

$$\mathcal{M}_{\pi^+\pi^-}(p_1, p_2, p_3, p_4) = \mathcal{M}_{\pi^+\pi^+}(p_1, -p_4, p_3, -p_2)$$

$$= \left(\frac{-e^2(p_1+p_3)_\mu(-p_4-p_2)^\mu}{(-p_4+p_2)^2} + \frac{-e^2(p_1-p_2)_\mu(-p_4+p_3)^\mu}{(p_4+p_3)^2} \right), \quad (1.159)$$

with the other factors in (1.158a) unchanged. This method of obtaining a particle–antiparticle amplitude from the corresponding particle–particle amplitude goes by the rubric of *crossing symmetry;* it can be of considerable utility in practical calculations (see Exercise 3.5e for an example). By inspection of the vertex factors in (1.158) and (1.159), and comparison with Fig. 1.10, we see that the two terms in (1.159) correspond to the graphs in Fig. 1.11. Notice that the relative sign of the vertices has changed between the π^+K^+ diagram in Fig. 1.9 and the $\pi^+\pi^-$ exchange diagram in Fig. 1.11. This reflects the commonplace that unlike charges attract and like charges repel. (Remember, we are only considering the electromagnetic portion of the interaction in these examples; the strong force would dominate for low energy meson scattering in the real world.)

To relate our theory to physical observables the amplitudes we have obtained must be used to construct quantities that can be measured, such as differential cross sections. This is straightforward, but not particularly instructive in the context of the current pedagogical discussion, and is deferred until Ch. 3.

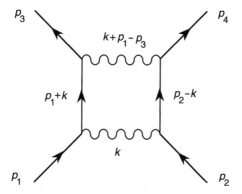

Fig. 1.12   A loop diagram.

1.8 Loops, Trees, and Infinities

The preceding examples have been restricted to order α in the fine structure constant. Higher order diagrams may be constructed by combining propagators and vertices of the type considered in the diagrams that are first order in α. For example, a two-photon exchange process is proportional to α^2 (or e^4), and contributes a diagram of the form shown in Fig. 1.12. Higher order diagrams containing closed loops like Fig. 1.12 are called *loop diagrams*, whereas those without loops, such as in Fig. 1.11, are called *tree diagrams*. It is typical of weakly interacting theories that the tree diagrams give rather accurate results, but for stronger interactions, or for very precise results, higher order diagrams with their loops must be included.

The Feynman rules set out for the first-order diagrams generalize to the higher order diagrams (for example, in Fig. 1.12 there are four vertices and two photon propagators, and also propagators for the internal particle lines), but a new feature enters: overall 4-momentum conservation in the external lines is not sufficient to prescribe the momentum in the loop. By considering momentum conservation at each vertex of Fig. 1.12 we see that there is a free *loop momentum k* entering the problem. An obvious guess extends the Feynman rules for the lower order diagrams to include an integration $\int d^4k$ over any free loop momentum. Derivation of the Feynman rules from field theory indicates that this is the correct procedure; however, a seemingly catastrophic result follows from this prescription: *most of the loop integrals are divergent*, and the theory that gives finite and sensible answers in first-order perturbation theory finds itself inundated by a torrent of infinities in higher order!

The process by which these infinities can be domesticated is termed *renormalization*; it is highly technical, and will be discussed qualitatively in Chs.

3 and 6. For now we are content with remaining at tree level. However, it is worth emphasizing at this point that the most important field theories for which the divergences can be tamed to yield sensible results are precisely the local gauge theories that will occupy our attention in subsequent chapters.

1.9 Background and Further Reading

The material on Lorentz transformations is standard in many textbooks; I have found Jackson (1975), Goldstein (1981), Landau and Lifshitz (1971), and Harris (1975) to be particularly useful. For the treatment of Dirac and Klein–Gordon equations Bjorken and Drell (1964), Baym (1969), Ziman (1969), Sakurai (1967), Messiah (1958), and Aitchison (1972) may be consulted. The propagator approach to relativistic quantum mechanics is developed in Feynman (1949, 1961). Most of our discussion of propagator methods has been borrowed from the lucid presentation in Aitchison and Hey (1982) and Aitchison (1972). Other discussion in the same vein may be found in Halzen and Martin (1984). A more technical description of these methods is given in Bjorken and Drell (1964).

CHAPTER 2

Canonical Quantization of Local Field Theories

In the processes that we wish to discuss it is no longer adequate to employ theories that treat quantum-mechanical particles interacting with classical fields. In addition, because of processes such as vacuum polarization a single-particle theory loses its relevance: the quantum fluctuations of the fields create virtual particle–hole pairs and generate an inherently many-body system. To remedy the first problem we must consider quantization of classical fields such as the electromagnetic field. In the case of the Maxwell (electromagnetic) field the quantization procedure leads to the fruitful picture of a vector quantal field and a massless field particle, the photon. Such a theory is extremely well suited to the description of processes in which photons are emitted or absorbed, because it employs field operators that create or destroy photons.

In the case of other fermions and bosons we know how to find quantum mechanical descriptions both nonrelativistically (Schrödinger equation) and in a Lorentz covariant formulation (Dirac or Klein–Gordon equations). For these cases, however, it turns out to be highly advantageous to quantize the wave equation as if it were a classical field. This procedure is called *second quantization*, the first quantization corresponding to the introduction of the wave equation with appropriate boundary conditions, the second corresponding to quantization of the field that would result from interpreting the wave equation as a classical field equation.

It is easily demonstrated in second quantization of the Schödinger equation that the second quantization adds no new physics—the first- and second-quantized theories have equivalent physical content. However, for particular applications such as many-body theory a second-quantized formalism with creation and annihilation operators automatically satisfying the appropriate quantum statistics is far easier to use than a first-quantized theory.

A similar situation prevails in the domain of relativistic wave mechanics. By analogy with the Maxwell case, the second quantization of the Dirac or Klein–Gordon equations leads to a picture of fields that create and destroy particles (the quanta of the fields), with the interactions in physical systems resulting from the field couplings. As we shall see, this is a far more natural

49

approach to the physics of elementary particles than one based on single-particle (first-quantized) wave equations.

2.1   Quantization in Discrete Mechanics

We begin our discussion of field quantization by examining a simpler problem. Consider a 1-dimensional classical system with a Lagrangian $L(q, \dot{q})$, where q is the generalized coordinate, the velocity is $\dot{q} \equiv dq/dt$, and the Lagrangian is constructed from the definition $L = T - V$, where T is the total kinetic energy and V is the total potential energy of the system.

2.1.1   Lagrange Equations of Motion

By Hamilton's principle the classical dynamical content of a system is prescribed by the requirement that the time integral of the Lagrangian (the *classical action S*) be an extremum,

$$\delta S = \delta \int_{t_1}^{t_2} L(q, \dot{q}) dt = 0, \tag{2.1}$$

where δ means an *arbitrary variation of a path with fixed endpoints* (Fig. 2.1): if $q(t)$ is a classical path, small variations $q(t) \to q(t) + \delta q(t)$ do not alter the classical action to first order in the variation. Defining the variation $\delta q(t)$ in terms of a parameter α

$$\delta q(t) = \alpha g(t), \tag{2.2}$$

where $g(t)$ is a function of t vanishing at the endpoints, the path variation is

$$q(t) \to q(t) + \alpha g(t). \tag{2.3}$$

As shown in Exercise 2.1 and Goldstein (1981, Ch. 2), the meaning of the

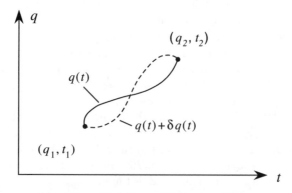

Fig. 2.1   Variation with fixed endpoints.

action variational condition is that

$$\delta S = \int_{t_1}^{t_2} \left(\frac{\partial L}{\partial q} - \frac{d}{dt} \frac{\partial L}{\partial \dot{q}} \right) \delta q \, dt = 0, \tag{2.4}$$

$$\delta S \equiv \left(\frac{\partial S}{\partial \alpha} \right)_{\alpha=0} d\alpha \qquad \delta q \equiv \left(\frac{\partial q}{\partial \alpha} \right)_{\alpha=0} d\alpha.$$

The requirement that this variational equation be satisfied for arbitrary δq leads to the *Euler–Lagrange equations of motion*

$$\frac{d}{dt} \left(\frac{\partial L}{\partial \dot{q}} \right) - \frac{\partial L}{\partial q} = 0. \tag{2.5}$$

For a particle of mass m in a one-dimensional potential $V(x)$, a Lagrangian $L = \frac{1}{2}m\dot{x}^2 - V(x)$ inserted in (2.5) gives Newton's equation, $m\ddot{x} = -dV/dt$.

EXERCISE 2.1 Derive the Euler–Lagrange equation (2.5) from eq. (2.1).

2.1.2 Quantization of the Classical Hamiltonian

To quantize this system we first introduce the Hamiltonian, which is related to the Lagrangian by the *Legendre transformation*

$$H(p,q) = p\dot{q} - L(q,\dot{q}), \tag{2.6}$$

where the generalized momentum p conjugate to q is $p = \partial L/\partial \dot{q}$. Differentiation of (2.6) leads to Hamilton's equations of motion,

$$\dot{q} = \frac{\partial H}{\partial p} \qquad \dot{p} = -\frac{\partial H}{\partial q}. \tag{2.7}$$

The canonical quantization procedure constructs a quantum mechanical Hamiltonian from the classical one through the replacement $(p,q) \to (\hat{p}, \hat{q})$, with the operators $\hat{p}$ and $\hat{q}$ satisfying[‡]

$$[\hat{p}, \hat{q}] = -i. \tag{2.8a}$$

An explicit coordinate-space replacement consistent with this requirement is

$$q \to \hat{q} \qquad p \to \hat{p} \equiv -i\frac{\partial}{\partial q}, \tag{2.8b}$$

[‡]In much of this section and the next we follow Bjorken and Drell (1965), Ch. 11. Unless confusion between classical quantities and quantum operators is likely we will omit the hats on the operators in the subsequent discussion.

where the quantities on the left are classical dynamical variables, and the ones on the right are operators acting in a linear vector space (Hilbert space).

There are many "pictures" or "representations" in which wave mechanics may be formulated. Probably the most familiar is the *Schrödinger representation*, with the particle dynamics contained in a wavefunction ψ satisfying

$$i\frac{\partial\psi(t)}{\partial t} = H\psi(t). \tag{2.9}$$

In the Schrödinger formulation the wavefunctions are time dependent, but the operators are not. Alternatively, we may implement quantum mechanics in a representation where the time dependence is lodged in the operators instead of the wavefunctions. This is termed the *Heisenberg representation*, and the Schrödinger state vector $\psi_S(t)$ is related to the Heisenberg state vector ψ_H by

$$\psi_S(t) = e^{-iHt}\psi_S(t=0) \equiv e^{-iHt}\psi_H \tag{2.10}$$

[hence $\psi_H = \psi_S(t=0)$], while the Schrödinger operators O_S are related to the Heisenberg operators $O_H(t)$ by

$$O_H(t) = e^{iHt}O_S e^{-iHt}. \tag{2.11}$$

These transformations are unitary if H is Hermitian, and it is trivial to verify that an arbitrary matrix element in the Heisenberg representation $\langle\psi_H'|O_H|\psi_H\rangle$ is equivalent to the corresponding matrix element $\langle\psi_S'|O_S|\psi_S\rangle$ in the Schrödinger representation. In the Schrödinger picture we solve for the time development of the wavefunctions using eq. (2.9), but in the Heisenberg picture we must solve for the time development of the operators. Differentiation of (2.11) leads to the *Heisenberg equation of motion*[‡]

$$\frac{dO_H(t)}{dt} = i[H, O_H(t)]. \tag{2.12}$$

The matrix elements are equivalent in the two representations, and which to use in nonrelativistic quantum mechanics is a matter of taste and convenience. However, in relativistic quantum field theory the Heisenberg representation[‡‡] is often preferable to the Schrödinger representation. There are basically

---

[‡]If the operator is an explicit function of time this is modified to

$$\frac{dO_H(t)}{dt} = i[H, O_H(t)] + \frac{\partial O_H}{\partial t}.$$

We will not consider such cases.

[‡‡]Or a representation intermediate between these termed the *interaction representation*, which will be introduced in §3.1. In the interaction representation part of the time dependence resides in the operators and part in the wavefunctions.

two reasons for this: (1) In relativistic field theory the wavefunction is more complicated than in nonrelativistic wave mechanics, and the dynamics of the operators may be easier to describe than that of ψ. (2) In the Heisenberg representation both time and coordinates appear in the field operators; this makes it easier to display Lorentz covariance.

In the Heisenberg picture the equal-time commutators of p and q retain their form for an arbitrary time,

$$[p(t), q(t)] = -i, \tag{2.13}$$

and the operator equations of motion (2.12) for p and q take the same form as the classical Hamilton's equations of motion for the dynamical variables

$$\frac{dp(t)}{dt} = -\frac{\partial H}{\partial q(t)} \tag{2.14}$$

$$\frac{dq(t)}{dt} = \frac{\partial H}{\partial p(t)}. \tag{2.15}$$

This similarity follows from the parallel roles of Poisson brackets in classical mechanics and commutators in quantum mechanics.

The time evolution of the Hermitian operators $p(t)$ and $q(t)$ is now described by differential equations first-order in time, and the only remaining task is to define initial conditions. In specifying $p(0)$ and $q(0)$ we must take care that the commutation relation (2.13) is obeyed at $t = 0$ for all physical states.

2.1.3 Harmonic Oscillator in the Heisenberg Picture

To illustrate the use of the Heisenberg representation in canonical quantization we consider a 1-dimensional harmonic oscillator. The Hamiltonian is

$$H = \tfrac{1}{2}(p^2 + \omega q^2), \tag{2.16}$$

where ω is the oscillator frequency and we take unit mass for simplicity, and the system is quantized by the requirement (2.13). The Heisenberg equations of motion are

$$\frac{dp(t)}{dt} = -\frac{\partial H}{\partial q} = -\omega^2 q \qquad \dot{q} \equiv \frac{dq(t)}{dt} = \frac{\partial H}{\partial p} = p, \tag{2.17}$$

but from the second equation, $\ddot{q} \equiv d^2 q/dt^2 = dp/dt$. Inserting this in the first equation the equations of motion become

$$\ddot{q} + \omega^2 q = 0 \qquad \dot{q} = p. \tag{2.18}$$

Introducing

$$a = \sqrt{\tfrac{1}{2\omega}}(\omega q + ip) \qquad a^\dagger = \sqrt{\tfrac{1}{2\omega}}(\omega q - ip), \tag{2.19}$$

the equations of motion may be written

$$\dot{a}(t) = -i\omega a(t) \qquad \dot{a}^\dagger(t) = i\omega a^\dagger(t). \tag{2.20}$$

These have solutions

$$a(t) = a_0 e^{-i\omega t} \qquad a^\dagger(t) = a_0^\dagger e^{i\omega t}, \tag{2.21}$$

where from the commutator (2.13) and the definition (2.19) we see that

$$[a(t), a^\dagger(t)] = [a_0, a_0^\dagger] = 1$$
$$[a(t), a(t)] = [a_0, a_0] = [a^\dagger(t), a^\dagger(t)] = [a_0^\dagger, a_0^\dagger] = 0. \tag{2.22}$$

In terms of these operators the Hamiltonian is

$$H = \tfrac{1}{2}\omega(a^\dagger a + aa^\dagger) = \tfrac{1}{2}\omega(a_0^\dagger a_0 + a_0 a_0^\dagger)$$
$$= \omega(a^\dagger a + \tfrac{1}{2}) = \omega(a_0^\dagger a_0 + \tfrac{1}{2}). \tag{2.23}$$

By steps familiar from elementary quantum mechanics we find that the operators $a_0^\dagger$ and a_0 create and annihilate quanta of energy $\hbar\omega$, that a number operator $N = a_0^\dagger a_0$ counts the number of oscillator quanta, and that the annihilation operator a_0 gives zero when applied to the ground state (the *vacuum state*, corresponding to no oscillator quanta). An n-quantum state, with n denoting the eigenvalue of the operator N, has a wavefunction

$$\psi_n = \frac{1}{\sqrt{n!}}\left(a_0^\dagger\right)^n \psi_0, \tag{2.24}$$

with an energy

$$\omega_n = \left(n + \tfrac{1}{2}\right)\omega. \tag{2.25}$$

These wavefunctions are mutually orthogonal, and the matrix elements of $a_0^\dagger$ and a_0 in this representation are easily shown to be

$$\langle\psi_{n+1}| a_0^\dagger |\psi_n\rangle = \langle\psi_n| a_0 |\psi_{n+1}\rangle = \sqrt{n+1}, \tag{2.26}$$

with a selection rule $\Delta n = \pm 1$. More generally,

$$\langle\psi_{n+1}| a^\dagger(t) |\psi_n\rangle = e^{i\omega t} \langle\psi_{n+1}| a_0^\dagger |\psi_n\rangle. \tag{2.27}$$

These equations represent a complete solution to the problem since all matrix elements may be given in terms of the operators $a^\dagger(t)$ and $a(t)$, and the basis states ψ_n.

The preceding ideas may be generalized in an obvious way to a system with n independent degrees of freedom. In the Heisenberg representation we introduce n Hermitian coordinate operators $q_i(t)$, and n conjugate momentum operators $p_i(t)$, leading to $2n$ equations of motion for the operators,

$$\dot{p}_i(t) = i[H, p_i(t)] \qquad \dot{q}_i(t) = i[H, q_i(t)]. \tag{2.28}$$

The n oscillators are taken to be independent, and the initial commutator conditions are

$$[p_i(0), q_j(0)] = -i\delta_{ij} \qquad [p_i(0), p_j(0)] = [q_i(0), q_j(0)] = 0. \tag{2.29}$$

This completely specifies the dynamical problem because the Heisenberg equations of motion preserve the equal-time commutation rules for all times.

We are now ready to extend the concept of canonical quantization from discrete to field mechanics. The approach is rather prosaic: we expect that in the limit $n \to \infty$ the previous example goes smoothly over to a canonically quantized continuum mechanics. This prescription makes intuitive sense, but it is not trivial to provide a rigorous formal justification within the context of relativistic quantum field theory. As a practical matter it seems to work, and that for us will be taken as sufficient justification.

The prototype of such a problem in classical physics is the 1-dimensional vibrating string. In this, as in more complicated field theories, each point of space (an infinitesimal segment of string in this example) may be identified with an independent coordinate. If we divide the string into n segments, solve n coupled oscillator equations for the motion, and take the limit $n \to \infty$ a continuous string results described by a field $\phi(\mathbf{x}, t)$. The discrete index i labeling a finite number of independent oscillators becomes a continuous coordinate $\mathbf{x}$, the value of $\phi(\mathbf{x}, t)$ measures the displacement of the string at $(\mathbf{x}, t)$, and the velocity at that point is given by $\partial\phi(\mathbf{x}, t)/\partial t$. Then in the quantized analog of the vibrating string we expect $\phi(\mathbf{x}, t)$ to play the role of a coordinate operator and $\partial\phi(\mathbf{x}, t)/\partial t$ to play the role of a velocity operator.

2.2 General Properties of the Action

In a formal sense the fundamental quantity we must construct in a field theory is the action

$$S = \int d^4x \, \mathcal{L} = \int_{t_1}^{t_2} L \, dt, \tag{2.30}$$

where the Lagrangian L is related to the *Lagrangian density* $\mathcal{L}$ by spatial integration

$$L = \int d^3x \, \mathcal{L}. \tag{2.31}$$

(In the interest of brevity, many authors also use the term Lagrangian for the Lagrangian density $\mathcal{L}$.) In the method of canonical quantization it is the variation of the action (2.1) that determines the classical equations of motion to be quantized. In Feynman path integral quantization (Ch. 4) the action plays an even more obvious role since the starting point is the hypothesis that the amplitude for a process is a sum over paths weighted by e^{iS} evaluated on each path.

It is useful to enumerate some general constraints that the fields are expected to obey, since this has the desirable effect of restricting the classes of field theories that must be considered. These constraints are most elegantly formulated in terms of conditions that the action or the Lagrangian density are expected to fulfill, and typically reflect some mixture of seasoned empiricism and informed theoretical prejudice. For our purposes we will demand the following (Ramond, 1981):

1. The Lagrangian density is *local*, depending on fields and their first derivatives at one spacetime point x^μ only.
2. The action S must be real to ensure total probability conservation.
3. The action should lead to classical equations of motion having no higher than second-order derivatives. Thus $\mathcal{L}$ can contain at most two ∂_μ operations in a term.
4. The action is required to be invariant under the Poincaré group (Lorentz rotations and spacetime translations), to ensure conservation of basic quantities such as energy or angular momentum. For example, the Lagrangian density does not depend explicitly on the coordinates x^μ unless there is an external source.
5. In specific cases S will be required to be invariant under additional *internal symmetries* such as $U(1)_{\text{charge}}$, weak hypercharge, $SU(3)_{\text{color}}$, $\ldots$

We should also note as a general aid in constructing field theories that S comes packaged in units of $\hbar$, so it is dimensionless in the natural system of units we employ. Since S is the 4-dimensional integral of the Lagrangian density, it follows that $\mathcal{L}$ has dimension $[\text{length}]^{-4} = [M]^4$, and L has dimension $[M]$, where M denotes mass. From these observations and the field equations we may determine the natural-unit dimensions of various field operators and coupling constants, as illustrated in Appendix A.

2.3 Lagrangian Densities for Free Fields

By analogy with discrete mechanics, we may canonically quantize a field $\phi(\mathbf{x}, t)$ through the following prescription:

1. Construct a classical Lagrangian density $\mathcal{L}(\phi, \partial_\mu \phi)$ depending on generalized coordinates $\phi(\mathbf{x}, t)$ and derivatives; the corresponding Lagrangian

is given by

$$L = \int \mathcal{L}(\phi, \partial_\mu \phi) d^3x \, .$$

2. Define the generalized momenta

$$\pi(\mathbf{x}, t) = \frac{\partial \mathcal{L}}{\partial \dot{\phi}},$$

where $\dot{\phi} \equiv \partial_0 \phi$.

3. Construct the classical Hamiltonian density $\mathcal{H}\big(\pi(\mathbf{x}, t), \phi(\mathbf{x}, t)\big)$ by a Legendre transformation

$$\mathcal{H} = \pi \dot{\phi} - \mathcal{L},$$

and integrate the Hamiltonian density to obtain the classical Hamiltonian

$$H = \int d^3x \, \mathcal{H}\big(\pi(\mathbf{x}, t), \phi(\mathbf{x}, t)\big).$$

4. Quantize the classical Hamiltonian by the standard operator replacements. For example, if the field is a boson field ϕ and π become operators in a linear vector space obeying the commutation relations

$$[\phi(\mathbf{x}, t), \phi(\mathbf{x}', t)] = 0 \qquad [\pi(\mathbf{x}, t), \pi(\mathbf{x}', t)] = 0$$
$$[\pi(\mathbf{x}, t), \phi(\mathbf{x}', t)] = -i\delta^{(3)}(\mathbf{x} - \mathbf{x}'), \tag{2.32}$$

where the Dirac delta function $\delta^{(3)}(\mathbf{x} - \mathbf{x}')$ satisfies

$$\int d^3x' \, \delta^{(3)}(\mathbf{x} - \mathbf{x}') f(\mathbf{x}') = f(\mathbf{x}), \tag{2.33}$$

for an arbitrary function $f(\mathbf{x}')$, with $\mathbf{x}$ lying within the limits of integration. If the field is a fermion field, it will be required to satisfy a corresponding set of anticommutation relations (§2.6).

Hence canonical quantization begins with the construction of a Lagrangian density that yields the field equation to be quantized upon insertion of the Lagrangian density into the field theory generalization of the Euler–Lagrange equations (2.5). We illustrate the procedure by finding the Lagrangian density corresponding to the scalar Klein–Gordon field. The wave equation is (1.58)

$$(\partial_\mu \partial^\mu + m^2)\phi(x) = 0, \tag{2.34}$$

where ϕ is assumed to be a scalar function. Define an infinitesimal variation $\delta\phi$ in the field amplitude at x

$$\phi(x) \rightarrow \phi(x) + \delta\phi(x). \tag{2.35}$$

Hamilton's principle requires [see (2.1)]

$$\delta S = \delta \int_V d^4x \, \mathcal{L}(\phi, \partial_\mu \phi) = \int_V d^4x \left(\frac{\partial \mathcal{L}}{\partial \phi} \delta \phi + \frac{\partial \mathcal{L}}{\partial(\partial_\mu \phi)} \delta(\partial_\mu \phi) \right) = 0, \quad (2.36)$$

where $\mathcal{L}$ is the Lagrangian density and V is the 4-dimensional spacetime volume. As shown in Exercise 2.2, if surface contributions to the integral at infinity can be ignored this implies for the scalar field a classical Lagrangian density

$$\mathcal{L}(\phi, \partial_\mu \phi) = \tfrac{1}{2}(\partial_\mu \phi)(\partial^\mu \phi) - \tfrac{1}{2}m^2\phi^2, \quad (2.37)$$

and a Lagrangian

$$L = \int d^3x \, \mathcal{L}(\phi, \partial_\mu \phi), \quad (2.38)$$

with the Lagrangian density satisfying the *Euler–Lagrange field equation*

$$\partial_\mu \frac{\partial \mathcal{L}}{\partial(\partial_\mu \phi)} - \frac{\partial \mathcal{L}}{\partial \phi} = 0, \quad (2.39)$$

which is the field-theoretic generalization of (2.5). From (2.37), recalling that $(\partial_\mu \phi)(\partial^\mu \phi) = g^{\mu\nu}(\partial_\mu \phi)(\partial_\nu \phi)$, and that μ and ν are dummy indices, we find

$$\frac{\partial \mathcal{L}}{\partial(\partial_\mu \phi)} = \partial^\mu \phi \qquad \frac{\partial \mathcal{L}}{\partial \phi} = -m^2\phi,$$

so (2.39) becomes $(\partial_\mu \partial^\mu + m^2)\phi = 0$, and this is indeed the Klein–Gordon equation, as advertised.

Before continuing, we stop to note several things about the field equations (2.39). The first is that in these equations $\partial_\mu \phi$ plays the same role as a velocity in discrete mechanics, while ϕ behaves as a coordinate [compare (2.5) and (2.39)]; this was anticipated by the discussion of the quantized string at the end of §2.1. The second is that these are covariant equations provided the Lagrangian is constructed as a Lorentz scalar; notice that in our example (2.37) satisfies this requirement. The third is that the equations are normally unchanged by adding a total divergence $\partial_\mu \Lambda^\mu$ to the Lagrangian density. That is, if $\mathcal{L}$ satisfies (2.39), then $\mathcal{L}' = \mathcal{L} + \partial_\mu \Lambda^\mu$ also satisfies (2.39) for arbitrary Λ^μ, because by the 4-dimensional equivalent of Gauss's theorem

$$\int_V d^4x \, \partial_\mu \Lambda^\mu = \oint_S d\sigma_\mu \Lambda^\mu, \quad (2.40)$$

where the right side is an integral over the surface S that bounds the volume V, and the usual boundary conditions require the surface integral at infinity to vanish (see Exercise 2.2; see §13.7 for an example where the usual boundary conditions may not apply). Classically, $\mathcal{L}$ and $\mathcal{L}'$ are related by a canonical transformation. Finally we note that adding a constant to $\mathcal{L}$ has no effect on the classical system unless we are concerned with gravitation, where the total mass is significant.

By similar exercises in the calculus of variations the Lagrangians for other free fields of interest may be constructed. We will refrain from such indulgences and be content with listing the most important ones (Quigg, 1983, Ch. 2). Their correctness may be demonstrated (Exercise 2.3) by showing that they yield the expected field equation when inserted in (2.39), which has been constructed to incorporate the variational principle (2.1).

2.3.1 Real Scalar or Pseudoscalar Field

The Lagrangian density for a real spin-0 field is

$$\mathcal{L} = \tfrac{1}{2}(\partial^\mu \phi)(\partial_\mu \phi) - \tfrac{1}{2}m^2 \phi^2, \tag{2.41}$$

corresponding to a Klein–Gordon equation (1.58) for the motion of the field. Such fields will be appropriate for uncharged scalar or pseudoscalar particles.

2.3.2 Complex Scalar Field

A complex scalar field may be written

$$\phi(x) = \tfrac{1}{\sqrt{2}}\big(\phi_1(x) + i\phi_2(x)\big) \tag{2.42}$$

$$\phi^\dagger(x) = \tfrac{1}{\sqrt{2}}\big(\phi_1(x) - i\phi_2(x)\big). \tag{2.43}$$

The corresponding Lagrangian density is

$$\mathcal{L} = |\partial^\mu \phi|^2 - m^2 |\phi|^2, \tag{2.44}$$

where $|\partial^\mu \phi|^2 = (\partial^\mu \phi)^\dagger(\partial_\mu \phi)$ and $|\phi|^2 = \phi^\dagger \phi$; independent variation of ϕ and $\phi^\dagger$ yields two Klein–Gordon equations of motion for the fields:

$$\left(\Box + m^2\right)\phi(x) = 0 \qquad \left(\Box + m^2\right)\phi^\dagger(x) = 0. \tag{2.45}$$

Such a field is appropriate for *charged scalars or pseudoscalars*, and ϕ and $\phi^\dagger$ will turn out to be associated with particles of opposite charge that can be related through the operation of charge conjugation (§§1.6.3 and 2.5). This can be seen explicitly if we introduce a vector potential in the Lagrangian to distinguish charges.

2.3.3 Dirac Spinor Field

Defining the Dirac conjugate field $\overline{\psi}(x)$ by (1.113), the Lagrangian density is

$$\mathcal{L} = \overline{\psi}(x)(i\partial\!\!\!/ - m)\psi(x). \tag{2.46}$$

This yields the Dirac equation (1.99) when inserted in (2.39).

2.3.4 Massless Vector Field

For the electromagnetic field we may define the field-strength tensor $F^{\mu\nu}$ (see the discussion in §2.7)

$$F^{\mu\nu} \equiv \partial^\mu A^\nu - \partial^\nu A^\mu, \tag{2.47}$$

where A^μ is the 4-vector potential. Then the free field Lagrangian density is

$$\mathcal{L} = -\tfrac{1}{4} F_{\mu\nu} F^{\mu\nu}, \tag{2.48}$$

leading by Hamilton's principle to the free field Maxwell equations. This field is appropriate for massless vector particles such as the photon.

2.3.5 Massive Vector Field

The Lagrangian density is

$$\mathcal{L} = -\tfrac{1}{4} F_{\mu\nu} F^{\mu\nu} + \tfrac{1}{2} m^2 A^\mu A_\mu. \tag{2.49}$$

This yields the *Proca equation* for vector particles of mass m

$$\left(\Box + m^2\right) A^\mu = 0. \tag{2.50}$$

EXERCISE 2.2 Fill in the missing steps between (2.34) and (2.39) in the derivation of the Euler–Lagrange field equation for a Klein–Gordon field.

EXERCISE 2.3 Demonstrate that the Lagrangian densities (2.44), (2.46), (2.48), and (2.49) lead to the appropriate field equations for the complex Klein–Gordon, Dirac, Maxwell, and Proca fields, respectively.

2.4 Quantization of the Real Scalar Field

We now consider the quantization of some explicit field theories, first by quantizing the free fields, and then by including the interaction of fields. Let us begin with the Hermitian (real) spin-0 field, which satisfies a Klein–Gordon equation and is appropriate to the description of uncharged scalar or pseudoscalar particles (Lee, 1981, Ch. 2).

2.4.1 Operators and Equations of Motion

The Lagrangian density may be written in the form [see eq. (2.41)]

$$\mathcal{L} = \tfrac{1}{2}(\partial^\mu \phi)(\partial_\mu \phi) - V(\phi), \qquad (2.51)$$

where

$$V(\phi) \equiv \tfrac{1}{2}m^2\phi^2 + \mathcal{H}_{\text{int}}(\phi). \qquad (2.52)$$

If $\mathcal{H}_{\text{int}} = 0$ we recover the Lagrangian of a free spin-0 field; otherwise, $\mathcal{H}_{\text{int}}$ describes interactions of the field with itself, or with other fields. It will be assumed that $\mathcal{H}_{\text{int}}$ does not depend on ∂^μ. The Lagrangian is given by (2.31) and the action by (2.30)

$$L = \int_\Omega d^3x\, \mathcal{L}\big(\phi(\mathbf{x},t), \partial^\mu \phi\big) \qquad S = \int d^4x\, \mathcal{L} = \int L\, dt, \qquad (2.53)$$

where we display explicitly the dependence of the spin-0 field ϕ on $\mathbf{x}$ and t. The standard procedure has been adopted of enclosing the whole system in a rectangular box of volume Ω. This approach permits us to work with a denumerable set of wavefunctions, which will allow Fourier integrals to be replaced by sums; the limit $\Omega \to \infty$ taken at the end will restore Poincaré invariance.

To allow a direct transcription of results from discrete mechanics to field mechanics we divide the volume Ω into n small cubes of volume τ, as illustrated in Fig. 2.2. By imposing this lattice, the field $\phi(\mathbf{x},t)$ may be represented in each cube by a discrete value at some point $\mathbf{x}_i$ lying in the cube

$$\phi(\mathbf{x}_i, t) \equiv \phi_i(t). \qquad (2.54)$$

The generalized coordinate $q_i(t)$ associated with the ith cube is taken to be ϕ_i scaled by the volume τ

$$q_i(t) = \tau \phi_i(t), \qquad (2.55)$$

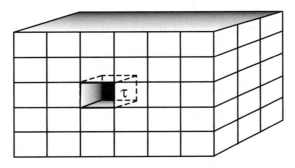

Fig. 2.2　Discrete lattice used for field quantization.

and the Lagrangian (2.53) becomes a sum. Using (1.31) in (2.51), and discretizing the integral (2.53) using (2.55) and $\Omega = n\tau$,

$$
L = \frac{1}{2} \int_{\Omega} d^3x \left[\dot{\phi}^2 - (\boldsymbol{\nabla}\phi)^2 - 2V(\phi) \right]
$$

$$
= \frac{1}{2\tau} \sum_{i=1}^{n} \dot{q}_i^2 - (\text{terms independent of } \dot{q}_i). \tag{2.56}
$$

The generalized momentum conjugate to q_i is

$$
p_i(t) \equiv \Pi(\mathbf{x}_i, t) = \frac{\partial L}{\partial \dot{q}_i} = \frac{\dot{q}_i}{\tau} = \dot{\phi}_i(t), \tag{2.57a}
$$

which corresponds in continuum notation to

$$
\Pi(\mathbf{x}, t) = \frac{\partial \mathcal{L}(\phi, \partial^\mu \phi)}{\partial \dot{\phi}(\mathbf{x}, t)}. \tag{2.57b}
$$

The Hamiltonian may be constructed using the Legendre transformation (2.6) summed over the discrete cubes,

$$
H = \sum_i p_i \dot{q}_i - L. \tag{2.58}
$$

But from (2.57a), $\dot{q}_i = \tau \, \Pi(\mathbf{x}_i, t)$ and $p_i = \Pi(\mathbf{x}_i, t)$, and in the limit $\tau \to 0$ we recover the Hamiltonian

$$
H = \int d^3x \, \mathcal{H}, \tag{2.59}
$$

with the Hamiltonian density $\mathcal{H}$ given by

$$
\begin{aligned}
\mathcal{H} = \mathcal{H}_0 + \mathcal{H}_{\text{int}} &= \Pi\dot{\phi} - \mathcal{L} \\
&= \tfrac{1}{2}\Pi^2 + \tfrac{1}{2}(\boldsymbol{\nabla}\phi)^2 + V(\phi) \\
&= \tfrac{1}{2}\Pi^2 + \tfrac{1}{2}(\boldsymbol{\nabla}\phi)^2 + \tfrac{1}{2}m^2\phi^2 + \mathcal{H}_{\text{int}},
\end{aligned} \tag{2.60}
$$

where (2.52) has been employed in the last step.

Quantization of the classical Hamiltonian is implemented by replacing Π and ϕ with field operators obeying the equal-time commutation relations

$$
[\Pi(\mathbf{x}, t), \phi(\mathbf{x}', t)] = -i\delta^{(3)}(\mathbf{x} - \mathbf{x}') \tag{2.61a}
$$

$$
[\Pi(\mathbf{x}, t), \Pi(\mathbf{x}', t)] = [\phi(\mathbf{x}, t), \phi(\mathbf{x}', t)] = 0 \tag{2.61b}
$$

[compare eq. (2.29) for discrete mechanics]. The equations of motion for the field are given by (2.12):

$$
\dot{\phi} = i[H, \phi] \qquad \dot{\Pi} = i[H, \Pi]. \tag{2.62}
$$

EXERCISE 2.4 (a) Assuming Lorentz invariance, prove that for free scalar fields the commutator between $\phi(x)$ and $\phi(y)$ vanishes outside the light cone. Therefore, measurements at spacelike separated points do not interfere, in agreement with principles of locality and microscopic causality. Show that microcausality is violated if we attempt to quantize the Klein–Gordon equation with anticommutators. The generalization of this result, and a corresponding one for the Dirac equation, is the basis of the famous *Spin–Statistics Theorem.* (b) Demonstrate that equations (2.62) are equivalent to the free Klein–Gordon equation provided $\mathcal{H}_{\text{int}} = 0$. *Hint:* use $[H, \Pi] = i\delta H/\delta\phi$, eq. (4.42), and integration by parts in (2.59) to change $(\nabla\phi)^2$ to $-\phi\nabla^2\phi$. (c) Show that for the electrodynamics of charged spin-0 particles (scalar electrodynamics) the interaction Lagrangian density contains derivatives, and the canonical momenta are of the form

$$\Pi = \partial^0\phi^\dagger - iqA^0\phi^\dagger \qquad \Pi^\dagger = \partial^0\phi + iqA^0\phi,$$

where A is the vector potential and q is the charge. Show that the interaction Hamiltonian density is

$$\mathcal{H}_{\text{int}} = -\mathcal{L}_{\text{int}}(A, \phi) - q^2(A^0)^2\phi^\dagger\phi,$$

where the interaction Lagrangian density is

$$\mathcal{L}_{\text{int}} = -iqA^\mu(\phi^\dagger \overset{\leftrightarrow}{\partial_\mu}\phi) + q^2A^2\phi^\dagger\phi.$$

Hint: the interaction is given by the minimal substitution.

2.4.2 Fourier Expansion of Field Operators

The field operators ϕ and Π operate in a Hilbert space with a structure determined by the algebra (2.61). To elucidate this structure it is useful to expand the operators in a Fourier series at fixed time (Lee, 1981, Ch. 2):

$$\phi(\mathbf{x}, t) = \Omega^{-1/2}\sum_{\mathbf{k}} e^{i\mathbf{k}\cdot\mathbf{x}} q_{\mathbf{k}}(t) \tag{2.63a}$$

$$\Pi(\mathbf{x}, t) = \Omega^{-1/2}\sum_{\mathbf{k}} e^{i\mathbf{k}\cdot\mathbf{x}} p_{-\mathbf{k}}(t). \tag{2.63b}$$

As previously noted, development of the field equations in a box of finite volume Ω leads to a discrete Fourier sum. This sum becomes a Fourier integral according to the correspondence

$$\frac{1}{\Omega}\sum_{\mathbf{k}} \xrightarrow[\Omega\to\infty]{} \frac{1}{(2\pi)^3}\int d^3k\,.$$

Since this is a real field, ϕ and Π are Hermitian operators: $\phi = \phi^\dagger$ and $\Pi = \Pi^\dagger$. Therefore, from (2.63)

$$q_{\mathbf{k}}(t) = q^\dagger_{-\mathbf{k}}(t) \qquad p_{\mathbf{k}}(t) = p^\dagger_{-\mathbf{k}}(t). \tag{2.64}$$

Let

$$\omega \equiv \omega_{\mathbf{k}} = \sqrt{\mathbf{k}^2 + \tilde{m}^2} \tag{2.65}$$

($\tilde{m}$ is not yet specified, for free fields we take $\tilde{m} = m$, where m is the mass of the particle) and define an operator [compare (2.19)]

$$a_{\mathbf{k}}(t) = \sqrt{\frac{\omega}{2}} \left(q_{\mathbf{k}}(t) + \frac{i}{\omega} p_{-\mathbf{k}}(t) \right) \tag{2.66a}$$

which, by virtue of (2.64), has the Hermitian conjugate

$$a^\dagger_{\mathbf{k}}(t) = \sqrt{\frac{\omega}{2}} \left(q_{-\mathbf{k}}(t) - \frac{i}{\omega} p_{\mathbf{k}}(t) \right). \tag{2.66b}$$

In terms of these operators

$$q_{\mathbf{k}}(t) = \sqrt{\frac{1}{2\omega}} \left(a_{\mathbf{k}}(t) + a^\dagger_{-\mathbf{k}}(t) \right) \tag{2.67a}$$

$$p_{-\mathbf{k}}(t) = -i\sqrt{\frac{\omega}{2}} \left(a_{\mathbf{k}}(t) - a^\dagger_{-\mathbf{k}}(t) \right), \tag{2.67b}$$

and the expansion (2.63) may be written in terms of the operators $a^\dagger$ and a:

$$\phi(\mathbf{x}, t) = \sum_{\mathbf{k}} (2\omega\Omega)^{-1/2} \left(a_{\mathbf{k}}(t) e^{i\mathbf{k}\cdot\mathbf{x}} + a^\dagger_{\mathbf{k}}(t) e^{-i\mathbf{k}\cdot\mathbf{x}} \right) \tag{2.68a}$$

$$\Pi(\mathbf{x}, t) = \sum_{\mathbf{k}} -i \left(\frac{\omega}{2\Omega} \right)^{1/2} \left(a_{\mathbf{k}}(t) e^{i\mathbf{k}\cdot\mathbf{x}} - a^\dagger_{\mathbf{k}}(t) e^{-i\mathbf{k}\cdot\mathbf{x}} \right). \tag{2.68b}$$

These momentum-space expansions are valid for free or interacting fields since they depend only on the completeness of the Fourier series (of course, the time dependence of the expansion coefficients will be quite different for these two cases).

We must now determine the algebra of the operators $a_{\mathbf{k}}(t)$. This may be accomplished by using (2.61) and (2.63) to construct the algebra of $p_{\mathbf{k}}(t)$ and $q_{\mathbf{k}}(t)$, and then (2.66) to determine the corresponding algebra of the operators $a_{\mathbf{k}}(t)$ and $a^\dagger_{\mathbf{k}}(t)$. First we write

$$q_{\mathbf{k}}(t) = \Omega^{-1/2} \int e^{-i\mathbf{k}\cdot\mathbf{x}} \phi(\mathbf{x}, t) d^3x \tag{2.69a}$$

$$p_{\mathbf{k}}(t) = \Omega^{-1/2} \int e^{i\mathbf{k}\cdot\mathbf{x}} \Pi(\mathbf{x}, t) d^3x , \tag{2.69b}$$

where

$$\int e^{i(\mathbf{k}-\mathbf{k}')\cdot\mathbf{x}} d^3x = \Omega \delta_{\mathbf{k}\mathbf{k}'} \tag{2.70}$$

has been used to invert (2.63). Then

$$
\begin{aligned}
[p_{\mathbf{k}}(t), q_{\mathbf{k}'}(t)] &= \frac{1}{\Omega} \int e^{i\mathbf{k}\cdot\mathbf{x}-i\mathbf{k}'\cdot\mathbf{x}'} [\Pi(\mathbf{x}, t), \phi(\mathbf{x}', t)] d^3x \, d^3x' \\
&= \frac{-i}{\Omega} \int e^{i\mathbf{k}\cdot\mathbf{x}-i\mathbf{k}'\cdot\mathbf{x}'} \delta^{(3)}(\mathbf{x} - \mathbf{x}') d^3x \, d^3x' \\
&= \frac{-i}{\Omega} \int e^{i(\mathbf{k}-\mathbf{k}')\cdot\mathbf{x}} d^3x ,
\end{aligned}
$$

where (2.61) has been employed. Therefore, using (2.70)

$$[p_{\mathbf{k}}(t), q_{\mathbf{k}'}(t)] = -i\delta_{\mathbf{k}\mathbf{k}'}. \tag{2.71a}$$

By an analogous procedure

$$[p_{\mathbf{k}}(t), p_{\mathbf{k}'}(t)] = [q_{\mathbf{k}}(t), q_{\mathbf{k}'}(t)] = 0, \tag{2.71b}$$

and from (2.66) and (2.71) we conclude that

$$[a_{\mathbf{k}}(t), a_{\mathbf{k}'}^\dagger(t)] = \delta_{\mathbf{k}\mathbf{k}'} \tag{2.72a}$$

$$[a_{\mathbf{k}}(t), a_{\mathbf{k}'}(t)] = [a_{\mathbf{k}}^\dagger(t), a_{\mathbf{k}'}^\dagger(t)] = 0. \tag{2.72b}$$

Now we see the utility of the Fourier expansion (2.63). It can be rewritten in terms of the expansion (2.68), with operators obeying the algebra (2.72). But these operators are old friends; because of the commutators (2.72) they are *harmonic oscillator creation and annihilation operators* (§2.1.3), which are completely understood—nothing excites a physicist like seeing something he already knows! We have succeeded in expressing the Klein–Gordon field as a quantum mechanical normal-mode expansion, and the field has been replaced by a set of quantized oscillators.

2.4.3 Wavefunctions

The structure of the Hilbert space for the real scalar field is now easy to derive since the detailed properties of the oscillator operators are known. The Hamiltonian may be separated into a free-field part and an interaction

$$H = H_0 + H_{\text{int}} = \int d^3x \, \mathcal{H}_0 + \int d^3x \, \mathcal{H}_{\text{int}}, \tag{2.73}$$

where the script variables denote Hamiltonian densities [see (2.59)–(2.60)]. From our experience with oscillators we may guess that

$$H_0 = \sum_{\mathbf{k}} \omega_{\mathbf{k}} \left(N_{\mathbf{k}} + \tfrac{1}{2} \right), \tag{2.74}$$

where the number operator for a given $\mathbf{k}$ is

$$N_{\mathbf{k}} = a_{\mathbf{k}}^{\dagger} a_{\mathbf{k}}. \tag{2.75}$$

This is indeed correct, as explicit substitution in (2.60) of (2.68) and some algebra will verify (Exercise 2.5b). For free fields we have $a_{\mathbf{k}}(t) = a_{\mathbf{k}} e^{-i\omega t}$, where $a_{\mathbf{k}}$ is independent of time (Exercise 2.5a). For interacting fields the expansion (2.68) remains valid, but the time dependence of the operators is more complicated than for the free field.

EXERCISE 2.5 (a) Show that the Heisenberg equations of motion (2.62) imply $a_{\mathbf{k}}(t) = a_{\mathbf{k}} e^{-i\omega t}$ for the free scalar field. (b) Start from (2.60) and verify explicitly that the free scalar field Hamiltonian is given by (2.74).

EXERCISE 2.6 Calculate the commutator of $a_{\mathbf{k}}^{\dagger}$ with the scalar field momentum operator

$$P^{\mu} = \sum_{k} k^{\mu} a_{\mathbf{k}}^{\dagger} a_{\mathbf{k}}$$

and use this to show that $a_{\mathbf{k}}^{\dagger}$ acting on a state adds a 4-momentum k^{μ}.

The eigenvectors for the system are

$$
\begin{array}{lll}
\psi_0 \simeq |0\rangle & \text{(vacuum)} & \\
\psi_1 \simeq a_{\mathbf{k}}^{\dagger} |0\rangle & \text{(one-particle state)} & (2.76) \\
\psi_2 \simeq a_{\mathbf{k}}^{\dagger} a_{\mathbf{k}'}^{\dagger} |0\rangle & \text{(two-particle state)}. &
\end{array}
$$

These states obey the eigenvalue equation

$$N_{\mathbf{k}} |\psi\rangle = n_{\mathbf{k}} |\psi\rangle \qquad (n_{\mathbf{k}} = 0, 1, 2, \dots), \tag{2.77}$$

have energy

$$E = \sum_{\mathbf{k}} \omega_{\mathbf{k}} \left(n_{\mathbf{k}} + \tfrac{1}{2} \right), \tag{2.78}$$

and form a complete basis spanning the Hilbert space. The vacuum state $|0\rangle$ contains no oscillator quanta and is the lowest energy state; it is defined by the requirement

$$a_{\mathbf{k}} |0\rangle = 0, \tag{2.79}$$

for all $\mathbf{k}$. For the two-particle states

$$a_{\mathbf{k}}^{\dagger} a_{\mathbf{k}'}^{\dagger} |0\rangle = a_{\mathbf{k}'}^{\dagger} a_{\mathbf{k}}^{\dagger} |0\rangle,$$

by virtue of the commutation relations (2.72b). Therefore the two-particle state is symmetric with respect to identical particle interchange, and the quanta of the Klein–Gordon field obey *Bose–Einstein statistics*. This will be a general feature of canonical field quantization: the creation and annihilation operators for boson fields will obey commutation relations of the form (2.72). On the other hand, we will see that fermion fields require anticommutation relations for their creation and annihilation operators [see (2.89) and Exercise 2.4a].

2.4.4 Removal of Zero-Point Energy

In quantum field theory infinities are a bothersome fact of life. Already we observe a problem in the quantization of the real Klein–Gordon field: from (2.78) the vacuum energy of the field is unbounded

$$E_0 = \tfrac{1}{2} \sum_{\mathbf{k}} \omega_{\mathbf{k}} \to \infty, \tag{2.80}$$

since it corresponds to the zero-point energy of an infinite number of harmonic oscillators. The removal of this divergence is relatively easy: only energy differences have meaning in the theories to be discussed, and we may subtract an infinite constant from H to cancel the contribution (2.80). This should have no observable consequences if gravitation is ignored, and can be accomplished by subtracting the vacuum expectation value from the energy formula (2.78), corresponding to measuring the energy relative to the vacuum state. This procedure can be cast in more systematic terms by introducing the idea of a *normal order*. Let Q be a product of creation and annihilation operators. A normal ordering of Q, denoted by $:Q:$, means that Q is rearranged so that all particle creation operators stand to the left of the particle destruction operators.[‡] As an example, for boson operators

$$:aa^{\dagger}: = a^{\dagger}a \qquad :a^{\dagger}a^{\dagger}: = a^{\dagger}a^{\dagger} \qquad :a^{\dagger}a: = a^{\dagger}a.$$

The normal-order product has the important property that its vacuum expectation value vanishes:

$$\langle 0| :Q: |0\rangle = 0. \tag{2.81}$$

[‡]If the operators are fermion operators some -1's associated with permutations are also involved. See the discussion of Wick contractions in §3.4.2.

Proof: Such a product either has a creation or destruction operator in the rightmost position. If it is a destruction operator it annihilates the vacuum (2.79) and the matrix element (2.81) vanishes. If it is a creation operator then any operator to its left is also a creation operator (definition of normal order), and (2.81) vanishes because a string consisting only of creation operators can never connect the state $\langle 0 |$ to the state $| 0 \rangle$.

Now consider an oscillator Hamiltonian such as eq. (2.23)

$$H = \omega \left(a^\dagger a + \tfrac{1}{2}\right) = \tfrac{1}{2}\omega(a^\dagger a + aa^\dagger),$$

where (2.22) has been used. Imposing normal ordering on this Hamiltonian,

$$:H: = \tfrac{1}{2}\omega:(a^\dagger a + aa^\dagger): = \omega a^\dagger a. \tag{2.82}$$

Thus normal ordering of the Hamiltonian or Lagrangian for the oscillator eliminates the zero-point energy. In field theory we often start with a Lagrangian that is implicitly or explicitly normally-ordered; the sole purpose of such a maneuver is to remove the zero-point energy of the vacuum at the beginning. Normal ordering will be encountered again when we consider Dyson–Wick contractions and the evaluation of S-matrix elements in Ch. 3.

2.5 Quantization of a Complex Scalar Field

The preceding discussion is readily extended to a system of n real fields ϕ_1, ϕ_2, ... ϕ_n. The corresponding Lagrangian density is

$$\mathcal{L} = \sum_{i=1}^{n} \left[\tfrac{1}{2}(\partial^\mu \phi^i)(\partial_\mu \phi^i) - \tfrac{1}{2}m_i^2\phi_i^2\right] + \mathcal{L}_{\text{int}}(\phi_1, \phi_2, \dots). \tag{2.83}$$

For the special case $n = 2$ and $m_1 = m_2$ the Lagrangian density (2.83) may be expressed in terms of a complex field and its Hermitian conjugate [see (2.42)–(2.43) and Exercise 2.15]

$$\phi = \tfrac{1}{\sqrt{2}}(\phi_1 + i\phi_2) \qquad \phi^\dagger = \tfrac{1}{\sqrt{2}}(\phi_1 - i\phi_2).$$

This leads to the Lagrangian density (2.44), plus an interaction term

$$\mathcal{L} = |\partial^\mu \phi|^2 - m^2|\phi|^2 + \mathcal{L}_{\text{int}}(\phi^\dagger, \phi), \tag{2.84}$$

and the quantization proceeds as in the case of the real scalar field. This Lagrangian is appropriate for a charged scalar or pseudoscalar field. In the absence of the interaction term, variation of (2.84) leads to two independent Klein–Gordon equations (2.45). Quantizing these fields and Fourier expanding yields two independent sets of oscillator operators; these may be identified with the creation and destruction operators for particles and antiparticles [see, e.g., Bjorken and Drell (1965), §12.5].

2.6 Quantization of the Dirac Field

The Lagrangian density appropriate to the spin-$\frac{1}{2}$ field is obtained by adding an interaction term to (2.46)

$$\mathcal{L} = \overline{\psi}(x)(i\not{\partial} - m)\psi(x) - \mathcal{H}_{\text{int}}, \tag{2.85}$$

where $\overline{\psi}(x) \equiv \psi^\dagger(x)\gamma^0$ and $\psi(x)$ each has four spinor components, with each component a Hilbert-space operator in the quantized theory, and $(i\not{\partial} - m)$ is a 4×4 matrix.

2.6.1 Operators and Equations of Motion

The momentum conjugate to the αth component of the field variable ψ is, using the analog of (2.57b), eq. (1.97), and $(\gamma^0)^2 = 1$,

$$\Pi_\alpha(\mathbf{x}, t) = \frac{\partial \mathcal{L}}{\partial \dot{\psi}_\alpha} = i\psi^\dagger_\alpha(\mathbf{x}, t) \tag{2.86}$$

(we assume that $\mathcal{H}_{\text{int}}$ does not depend on $\partial^\mu \psi$), but the momentum variables conjugate to $\overline{\psi}_\alpha$ are identically zero. The Hamiltonian density is, by virtue of the continuum version of (2.58),

$$\mathcal{H} = i\sum_\alpha \psi^\dagger_\alpha(\mathbf{x}, t)\dot{\psi}_\alpha(\mathbf{x}, t) - \mathcal{L}. \tag{2.87}$$

But $\dot{\psi}$ is given by the Dirac equations (1.72), or (1.98), and reverting to matrix notation (Exercise 2.7),

$$\begin{aligned}
\mathcal{H} &= \overline{\psi}(-i\gamma_j \partial^j + m)\psi + \mathcal{H}_{\text{int}} \\
&= \psi^\dagger(-i\boldsymbol{\alpha} \cdot \boldsymbol{\nabla} + \beta m)\psi + \mathcal{H}_{\text{int}} \\
&= i\psi^\dagger \frac{\partial}{\partial t}\psi + \mathcal{H}_{\text{int}} \\
&= \mathcal{H}_{\text{free}} + \mathcal{H}_{\text{int}},
\end{aligned} \tag{2.88}$$

where $(i\not{\partial} - m)\,\psi = 0$ has been used.

The quantization of a Fermi field differs from the quantization of a Bose field in that the equal time commutators (2.61) for the field operators are replaced by *anticommutation relations*

$$\{\psi_\alpha(\mathbf{x}, t), \Pi_\beta(\mathbf{x}', t)\} = i\delta^{(3)}(\mathbf{x} - \mathbf{x}')\delta_{\alpha\beta} \tag{2.89a}$$

$$\{\psi_\alpha(\mathbf{x}, t), \psi_\beta(\mathbf{x}', t)\} = \{\Pi_\alpha(\mathbf{x}, t), \Pi_\beta(\mathbf{x}', t)\} = 0. \tag{2.89b}$$

(If commutators are used instead of anticommutators the resulting theory does unacceptable things such as violate microcausality; see Exercises 2.4a and 2.9b.) Because of eq. (2.86), these anticommutation relations may be restated with the conjugate momentum eliminated in favor of $\psi^\dagger$:

$$\{\psi_\alpha(\mathbf{x},t),\psi_\beta^\dagger(\mathbf{x}',t)\} = \delta^{(3)}(\mathbf{x}-\mathbf{x}')\delta_{\alpha\beta} \qquad (2.89c)$$

$$\{\psi_\alpha(\mathbf{x},t),\psi_\beta(\mathbf{x}',t)\} = \{\psi_\alpha^\dagger(\mathbf{x},t),\psi_\beta^\dagger(\mathbf{x}',t)\} = 0. \qquad (2.89d)$$

The equations of motion are given by the Heisenberg equation (2.12),

$$\dot{\psi} = i[H,\psi] \qquad \dot{\psi}^\dagger = i[H,\psi^\dagger], \qquad (2.90)$$

where the Hamiltonian is

$$H = \int d^3x\, \mathcal{H}. \qquad (2.91)$$

For the case $\mathcal{H}_{\text{int}} = 0$, eq. (2.90) gives the free-particle Dirac equation as the equation of motion for the field.

EXERCISE 2.7 Verify that (2.88) is the Hamiltonian density appropriate for the Lagrangian density (2.85).

2.6.2 Fourier Expansion of Field Operators

As for the scalar field considered previously, it is useful to expand the Dirac field operators in a Fourier series. In the bispinor space we introduce the basis vectors $u_{\mathbf{p}s}$ and $v_{\mathbf{p}s}$ satisfying

$$(\not{p} - m)u_{\mathbf{p}s} = 0 \qquad (\not{p} + m)v_{\mathbf{p}s} = 0 \qquad (2.92)$$

$$\tfrac{1}{2}\boldsymbol{\Sigma}\cdot\hat{\mathbf{p}}\,u_{\mathbf{p}s} = s u_{\mathbf{p}s} \qquad \tfrac{1}{2}\boldsymbol{\Sigma}\cdot\hat{\mathbf{p}}\,v_{\mathbf{p}s} = s v_{\mathbf{p}s} \qquad (2.93)$$

[see (1.124)–(1.127), and note the change in notation $u^{(\alpha)}(\mathbf{p}) \to u_{\mathbf{p}s}$] where $s = \pm\tfrac{1}{2}$ is the helicity quantum number and

$$E_p = \sqrt{\mathbf{p}^2 + m^2} > 0.$$

For a given value of $\mathbf{p}$ the four eigenvectors $u_{\mathbf{p},\pm 1/2}$ and $v_{\mathbf{p},\pm 1/2}$ form a complete orthonormal basis in the bispinor space [see (1.125)–(1.126)], and we expand $\psi(\mathbf{x},t)$ on them (Lee, 1981, Ch. 3; Sakurai, 1967, §3–10)

$$\psi(\mathbf{x},t) = \Omega^{-1/2}\sum_{\mathbf{p}s}\sqrt{\frac{m}{E_p}}\left(a_{\mathbf{p}s}(t)u_{\mathbf{p}s}e^{i\mathbf{P}\cdot\mathbf{x}} + b_{\mathbf{p}s}^\dagger(t)v_{\mathbf{p}s}e^{-i\mathbf{P}\cdot\mathbf{x}}\right). \qquad (2.94)$$

In this expression u and v are c-numbers, but a and $b^\dagger$ are Hilbert-space operators. By virtue of (2.86) we may use $i\psi^\dagger$ as the generalized field momentum operator; complex conjugating (2.94) gives

$$\psi^\dagger(\mathbf{x},t) = \Omega^{-1/2} \sum_{\mathbf{p}s} \sqrt{\frac{m}{E_p}} \left(a_{\mathbf{p}s}^\dagger(t) u_{\mathbf{p}s}^\dagger e^{-i\mathbf{p}\cdot\mathbf{x}} + b_{\mathbf{p}s}(t) v_{\mathbf{p}s}^\dagger e^{i\mathbf{p}\cdot\mathbf{x}}\right). \tag{2.95}$$

From the fundamental anticommutators (2.89) these new operators satisfy the algebra

$$\{a_{\mathbf{p}s}(t), a_{\mathbf{p}'s'}^\dagger(t)\} = \{b_{\mathbf{p}s}(t), b_{\mathbf{p}'s'}^\dagger(t)\} = \delta_{\mathbf{p}\mathbf{p}'}\delta_{ss'}, \tag{2.96}$$

while other equal-time anticommutators among a, $a^\dagger$ and b, $b^\dagger$ vanish.

This momentum-space expansion depends only on the completeness property and is valid for free or interacting fields. For the special case of a free field we have (Exercise 2.9)

$$\begin{aligned} H_{\text{free}} &= \sum_{\mathbf{p}s}\left(a_{\mathbf{p}s}^\dagger a_{\mathbf{p}s} - b_{\mathbf{p}s}b_{\mathbf{p}s}^\dagger\right)E_p \\ &= \sum_{\mathbf{p}s}\left(a_{\mathbf{p}s}^\dagger a_{\mathbf{p}s} + b_{\mathbf{p}s}^\dagger b_{\mathbf{p}s}\right)E_p, \end{aligned} \tag{2.97}$$

where the anticommutator (2.96) has been used, and a constant term of -1 within the parentheses has been discarded (see §2.4.4). By similar calculations the 3-momentum operator of the free field is

$$\begin{aligned} \mathbf{P} &= -i\int d^3x\,\psi^\dagger \boldsymbol{\nabla}\psi \\ &= \sum_{\mathbf{p}s}\left(a_{\mathbf{p}s}^\dagger a_{\mathbf{p}s} - b_{\mathbf{p}s}b_{\mathbf{p}s}^\dagger\right)\mathbf{p} \\ &= \sum_{\mathbf{p}s}\left(a_{\mathbf{p}s}^\dagger a_{\mathbf{p}s} + b_{\mathbf{p}s}^\dagger b_{\mathbf{p}s}\right)\mathbf{p}, \end{aligned} \tag{2.98}$$

where the last step follows from the anticommutator and the symmetry requirement $\sum \mathbf{p} = 0$. The free-field charge is

$$\begin{aligned} Q &= \int d^3x\,q\psi^\dagger\psi \\ &= \sum_{\mathbf{p}s}\left(a_{\mathbf{p}s}^\dagger a_{\mathbf{p}s} + b_{\mathbf{p}s}b_{\mathbf{p}s}^\dagger\right)q \\ &= \sum_{\mathbf{p}s}\left(a_{\mathbf{p}s}^\dagger a_{\mathbf{p}s} - b_{\mathbf{p}s}^\dagger b_{\mathbf{p}s}\right)q, \end{aligned} \tag{2.99}$$

where again an infinite constant has been discarded.

The Heisenberg equations of motion for the free field are (see Exercise 2.9)

$$\dot{a}_{\mathbf{p}s}(t) = i[H, a_{\mathbf{p}s}(t)] = -iE_p a_{\mathbf{p}s}(t)$$
$$\dot{b}_{\mathbf{p}s}(t) = i[H, b_{\mathbf{p}s}(t)] = -iE_p b_{\mathbf{p}s}(t), \tag{2.100}$$

so that

$$a_{\mathbf{p}s}(t) = a_{\mathbf{p}s} e^{-iE_p t} \qquad b_{\mathbf{p}s}(t) = b_{\mathbf{p}s} e^{-iE_p t}. \tag{2.101}$$

with analogous expressions for their Hermitian conjugates. For interacting fields the expansions (2.94) and (2.95) remain valid, but the time behavior of the field operators will be more complicated than (2.101). From the expressions (2.94)–(2.99) we conclude that

1. The operator $a_{\mathbf{p}s}^\dagger$ creates a positive-energy particle with helicity s and 4-momentum $p^\mu = (E_p, \mathbf{p})$.
2. The operator $a_{\mathbf{p}s}$ annihilates a positive-energy particle with helicity s and 4-momentum $p^\mu = (E_p, \mathbf{p})$.
3. The operator $b_{\mathbf{p}s}^\dagger$ creates a positive-energy antiparticle with helicity s and 4-momentum $p^\mu = (E_p, \mathbf{p})$.
4. The operator $b_{\mathbf{p}s}$ annihilates a positive-energy antiparticle with helicity s and 4-momentum $p^\mu = (E_p, \mathbf{p})$.

EXERCISE 2.8 (a) Prove that $\displaystyle{\not{p}\gamma_0 + \gamma_0\not{p} = 2E = 2p_0}$, where p is the 4-momentum. (b) Show that

$$(\not{p} + m)\gamma^0(\not{p} + m) = 2E(\not{p} + m),$$

when $p^2 = m^2$. (c) Prove that the matrices

$$\Lambda_+(p) \equiv \sum_s u_{\mathbf{p}s}\bar{u}_{\mathbf{p}s} \qquad \Lambda_-(p) \equiv -\sum_s v_{\mathbf{p}s}\bar{v}_{\mathbf{p}s}$$

can be written

$$(\Lambda_+)_{\alpha\beta} = \left(\frac{\not{p} + m}{2m}\right)_{\alpha\beta} \qquad (\Lambda_-)_{\alpha\beta} = \left(\frac{-\not{p} + m}{2m}\right)_{\alpha\beta},$$

where α and β are matrix indices; show that they are projectors for positive and negative energy states. *Hint:* in part (c) use the results of parts (a) and (b), and that the spinors may be written

$$u_{p\alpha} = \frac{\not{p} + m}{\sqrt{2m(m + E)}} u_{0\alpha}(m) \qquad \bar{u}_{p\alpha} = \bar{u}_{0\alpha}(m)\frac{\not{p} + m}{\sqrt{2m(m + E)}},$$

where $u_{0\alpha}(m)$ and $\bar{u}_{0\alpha}(m)$ are rest spinors (see Messiah, 1958, §XX.24; Itzykson and Zuber, 1980, §2–2–1).

EXERCISE 2.9 (a) Verify equations (2.97)–(2.99).   (b) Show that the energy expression (2.97) leads to inconsistencies if the Dirac field is quantized with commutators. *Hint*: look at the spectrum.   (c) Show that for the free Dirac field the Heisenberg equations of motion give (2.100) and (2.101).

2.6.3 Wavefunctions

The operators $a^\dagger$ and a create and destroy fermions, while $b^\dagger$ and b create and destroy antifermions. The Hilbert space may be constructed by repeated applications of the creation operators on the vacuum, as in the case of boson fields. Introducing a number operator for positive-energy fermions of momentum $\mathbf{p}$ and helicity s

$$N_+(\mathbf{p}, s) = a^\dagger_{\mathbf{p}s} a_{\mathbf{p}s}, \qquad (2.102a)$$

and an analogous number operator for positive-energy antifermions

$$N_-(\mathbf{p}, s) = b^\dagger_{\mathbf{p}s} b_{\mathbf{p}s}, \qquad (2.102b)$$

the energy (2.97) for a free field may be written

$$H = \sum_{\mathbf{p}s} E_p \big[N_+(\mathbf{p}, s) + N_-(\mathbf{p}, s) \big]. \qquad (2.103)$$

The states are of the form

$$
\begin{array}{ccc}
|0\rangle & a^\dagger_{\mathbf{p}s}|0\rangle & b^\dagger_{\mathbf{p}s}|0\rangle \\[2mm]
a^\dagger_{\mathbf{p}s} a^\dagger_{\mathbf{p}'s'}|0\rangle & a^\dagger_{\mathbf{p}s} b^\dagger_{\mathbf{p}'s'}|0\rangle & b^\dagger_{\mathbf{p}s} b^\dagger_{\mathbf{p}'s'}|0\rangle \quad \cdots
\end{array}
\qquad (2.104)
$$

where the vacuum state $|0\rangle$ is defined by the requirement that for all $\mathbf{p}$ and s

$$a_{\mathbf{p}s}|0\rangle = b_{\mathbf{p}s}|0\rangle = 0. \qquad (2.105)$$

For two-particle states with identical particles the anticommutation conditions $\{a^\dagger_1, a^\dagger_2\} = 0$ imply exchange antisymmetry

$$a^\dagger_{\mathbf{p}s} a^\dagger_{\mathbf{p}'s'}|0\rangle = -a^\dagger_{\mathbf{p}'s'} a^\dagger_{\mathbf{p}s}|0\rangle, \qquad (2.106)$$

and the exclusion principle

$$a^\dagger_{\mathbf{p}s} a^\dagger_{\mathbf{p}s}|0\rangle = 0, \qquad (2.107)$$

with analogous expressions for the two-antiparticle states. This demonstrates explicitly that the quanta of the Dirac field are fermions, and from (2.107) we conclude that the eigenvalues of the number operators $N_+(\mathbf{p}, s)$ and $N_-(\mathbf{p}, s)$ are restricted to the values 0 or 1.

2.7 Quantization of the Electromagnetic Field

Since the classical Maxwell field has clearly observable consequences one might expect this to be the most straightforward field to quantize. However, it is not so; the reasons will occupy us in one guise or another for a major portion of this book. The problem is that the Maxwell field is associated with a 4-vector potential $A^\mu(\mu = 0, 1, 2, 3)$, but we know from classical electrodynamics that only the transverse fields are dynamical variables that deserve to be quantized. The complication of reducing four components to the two transverse components that correctly should be quantized is the reason that consideration of the Maxwell field has been deferred until now. These difficulties are intimately bound to the masslessness of the photon and to the *gauge invariance* of the classical electromagnetic field. Conversely, in the gauge invariance of the electromagnetic field we will find some extremely suggestive hints concerning the general form of legitimate quantum field theories. Accordingly, let us begin the discussion of quantizing the Maxwell field by reviewing gauge invariance in classical electrodynamics (Jackson, 1975, Chs. 6 and 11; Aitchison and Hey, 1982, Ch. 8).

2.7.1 Gauge Invariance in Classical Electrodynamics

The laws of electricity and magnetism before Maxwell were (in free space with Heaviside–Lorentz, $\hbar = c = 1$ units; see Appendix A)

$$
\begin{aligned}
\boldsymbol{\nabla}\cdot\mathbf{E} &= \rho && \text{(Gauss's Law)} \\
\boldsymbol{\nabla}\times\mathbf{E} &= -\frac{\partial\mathbf{B}}{\partial t} && \text{(Faraday's Law)} \\
\boldsymbol{\nabla}\cdot\mathbf{B} &= 0 && \text{(No free magnetic poles)} \\
\boldsymbol{\nabla}\times\mathbf{B} &= \mathbf{j} && \text{(Ampère's Law)}
\end{aligned}
\tag{2.108}
$$

where $\mathbf{E}$ and $\mathbf{B}$ are the electric and magnetic fields, ρ is the charge density, and $\mathbf{j}$ is the current vector. The continuity equation guaranteeing conservation of electrical charge is

$$
\frac{\partial\rho}{\partial t} + \boldsymbol{\nabla}\cdot\mathbf{j} = 0,
\tag{2.109}
$$

which requires that the charge variation in some arbitrary volume is caused by the flow of current through the surface of the volume.

Maxwell Equations. Maxwell noted that there was a fundamental difficulty with these equations; since the divergence of a curl vanishes by a vector calculus identity, $\boldsymbol{\nabla}\cdot(\boldsymbol{\nabla}\times\mathbf{B}) = 0$, it follows from Ampère's Law for steady currents that

$$\nabla \cdot \mathbf{j} = \nabla \cdot (\nabla \times \mathbf{B}) = 0. \tag{2.110}$$

But from the continuity equation, $\nabla \cdot \mathbf{j} = 0$ holds only if the charge density is constant in time. Maxwell changed Ampère's Law to accommodate a time dependence by introducing a term called the *displacement current*

$$\nabla \times \mathbf{B} = \mathbf{j} \quad \longrightarrow \quad \nabla \times \mathbf{B} = \mathbf{j} + \frac{\partial \mathbf{E}}{\partial t}. \tag{2.111}$$

Thus, *Maxwell's equations* are

$$\nabla \cdot \mathbf{E} = \rho \tag{2.112a}$$

$$\frac{\partial \mathbf{B}}{\partial t} + \nabla \times \mathbf{E} = 0 \tag{2.112b}$$

$$\nabla \cdot \mathbf{B} = 0 \tag{2.112c}$$

$$\nabla \times \mathbf{B} - \frac{\partial \mathbf{E}}{\partial t} = \mathbf{j}. \tag{2.112d}$$

These equations are now consistent with the continuity equation. Taking the divergence of the fourth equation,

$$\nabla \cdot (\nabla \times \mathbf{B}) = \nabla \cdot \mathbf{j} + \frac{\partial}{\partial t} \nabla \cdot \mathbf{E}.$$

Using the first equation (2.112a) and the identity $\nabla \cdot (\nabla \times \mathbf{B}) = 0$ yields the charge conservation equation (2.109).

Notice that the continuity equation (2.109) implies conservation of charge within a volume of space that can be made arbitrarily small, which means that charge is locally conserved; processes that destroy charge at one point and create it at another are forbidden, even though they may conserve the total charge. This is because global charge conservation would require the propagation of instantaneous signals between distant points, which is inconsistent with the principles of relativity. This concept of a local conservation law will play a central role in the material to follow.

It is often useful to eliminate the fields $\mathbf{E}$ and $\mathbf{B}$ appearing in the Maxwell equations in favor of a *vector potential* $\mathbf{A}$ and a *scalar potential* ϕ through the relations

$$\mathbf{B} \equiv \nabla \times \mathbf{A}$$

$$\mathbf{E} \equiv -\nabla\phi - \frac{\partial \mathbf{A}}{\partial t}. \tag{2.113}$$

These definitions automatically satisfy the second and third Maxwell equations, as can be verified with the identities $\nabla \cdot (\nabla \times \mathbf{B}) = 0$ and $\nabla \times \nabla\phi = 0$. Thus the four coupled first-order partial differential equations (2.112) are converted to two homogeneous equations that are satisfied identically, and two inhomogeneous, coupled, second-order equations:

$$\nabla^2 \phi + \frac{\partial}{\partial t}(\nabla \cdot \mathbf{A}) = -\rho$$

$$\nabla^2 \mathbf{A} - \frac{\partial^2 \mathbf{A}}{\partial t^2} - \nabla\left(\nabla \cdot \mathbf{A} + \frac{\partial \phi}{\partial t}\right) = -\mathbf{j}, \tag{2.114}$$

where the identity $\nabla \times (\nabla \times \mathbf{A}) = \nabla(\nabla \cdot \mathbf{A}) - \nabla^2 \mathbf{A}$ has been used.

The solution of the Maxwell equations is reduced to solving the coupled equations (2.114) for $\mathbf{A}$ and ϕ. We now exploit a fundamental property of classical electrodynamics, *gauge invariance*, to decouple these equations.

Gauge Transformations. The potentials $\mathbf{A}$ and ϕ as defined above are not unique. The transformations of $\mathbf{A}$ and ϕ that preserve the Maxwell equations are called *gauge transformations*. Because of the identity $\nabla \times \nabla \phi = 0$, the transformation

$$\mathbf{A} \to \mathbf{A}' \equiv \mathbf{A} + \nabla \chi, \tag{2.115a}$$

where χ is an arbitrary scalar function, will have no effect on $\mathbf{B}$; to preserve $\mathbf{E}$ the scalar potential ϕ must simultaneously be changed by

$$\phi \to \phi' \equiv \phi - \frac{\partial \chi}{\partial t}. \tag{2.115b}$$

These gauge transforms, by construction, yield the same $\mathbf{E}$ and $\mathbf{B}$ fields as the potentials before transformation, so they leave the Maxwell equations invariant.

We may use this invariance to choose a set of potentials $(\mathbf{A}, \phi)$ that satisfy

$$\nabla \cdot \mathbf{A} + \frac{\partial \phi}{\partial t} = 0. \tag{2.116}$$

[See Jackson (1975), §6.5, for a proof that such a set can always be found.] Then the second-order equations (2.114) decouple to

$$\nabla^2 \phi - \frac{\partial^2 \phi}{\partial t^2} = -\rho$$

$$\nabla^2 \mathbf{A} - \frac{\partial^2 \mathbf{A}}{\partial t^2} = -\mathbf{j}, \tag{2.117}$$

and these may be solved independently for ϕ and $\mathbf{A}$.

A procedure such as this is termed *fixing the gauge*. The particular choice of gauge defined by (2.116) is called the *Lorentz gauge* [more precisely, the Lorentz family of gauges; see Jackson (1975), §6.5], and (2.116) is termed the *Lorentz condition*. The Lorentz gauge is widely used because it leads to the two decoupled wave equations derived above, and because equations in the Lorentz gauge are easily written in covariant form.

However, other gauges are useful for certain purposes. The *Coulomb gauge*, also termed the *radiation gauge* or *transverse gauge*, is specified by the gauge-fixing condition

$$\mathbf{\nabla} \cdot \mathbf{A} = 0, \tag{2.118}$$

which is called the transversality requirement. Consulting the coupled equations we see that this also leads to a decoupling, with the scalar potential satisfying the Poisson equation

$$\nabla^2 \phi = -\rho, \tag{2.119a}$$

which has a solution

$$\phi(\mathbf{x}, t) = \int \frac{\rho(\mathbf{x}', t)}{|\mathbf{x} - \mathbf{x}'|} d^3 x', \tag{2.119b}$$

while the vector potential satisfies

$$\nabla^2 \mathbf{A} - \frac{\partial^2 \mathbf{A}}{\partial t^2} = \mathbf{\nabla} \frac{\partial \phi}{\partial t} - \mathbf{j}. \tag{2.119c}$$

The scalar potential is the instantaneous Coulomb potential associated with the charge density ρ, hence the appellation *Coulomb gauge*. As demonstrated in Jackson (1975), the current $\mathbf{j}$ appearing in the equation for the vector potential $\mathbf{A}$ can be separated into a *longitudinal* part satisfying $\mathbf{\nabla} \times \mathbf{j}_{\parallel} = 0$, and a *transverse* part satisfying $\mathbf{\nabla} \cdot \mathbf{j}_{\perp} = 0$, and the right side of (2.119c) reduces to $-\mathbf{j}_{\perp}$; whence the name *transverse gauge*. The term *radiation gauge* is a consequence of the fact that the transverse radiation fields are given by the vector potential alone. The Coulomb gauge is particularly useful in quantum electrodynamics, where only the vector potential of the photon field need be quantized.

A compact notation results if we define the *4-vector potential* $A^{\mu} \equiv (\phi, \mathbf{A})$. Then, utilizing (1.31) a gauge transformation takes the form

$$A^{\mu} \rightarrow A^{\mu} - \partial^{\mu} \chi \equiv A'^{\mu}. \tag{2.120}$$

In this notation the previous gauge-fixing conditions are

$$\partial_{\mu} A^{\mu} = 0 \quad \text{(Lorentz gauge)} \tag{2.121a}$$

$$\mathbf{\nabla} \cdot \mathbf{A} = 0 \quad \text{(Coulomb gauge)}. \tag{2.121b}$$

Notice that the Lorentz condition (2.121a) is covariant, but the Coulomb gauge condition (2.121b) is not.

Covariance of the Maxwell Equations. Introducing the d'Alembertian operator (1.33), the wave equations for ϕ and $\mathbf{A}$ in the Lorentz gauge are

$$\Box \mathbf{A} = \mathbf{j} \qquad \Box \phi = \rho, \tag{2.122}$$

where (2.117) was used. The operator $\Box$ is Lorentz invariant (Exercise 1.3). Therefore, if we define the 4-current and 4-vector potential by

$$j^{\mu} \equiv (\rho, \mathbf{J}) \qquad A^{\mu} \equiv (\phi, \mathbf{A}) = (A^0, \mathbf{A}), \tag{2.123}$$

the manifestly covariant Lorentz-gauge wave equations and gauge fixing condition are

$$\Box A^\mu = j^\mu \qquad \partial_\mu A^\mu = 0, \tag{2.124}$$

and the continuity equation is

$$\partial_\mu j^\mu = 0, \tag{2.125}$$

with a gauge transformation in this notation given by (2.120). Since the Maxwell equations are by construction invariant under gauge transformations, the covariance of the wave equation for A^μ in the Lorentz gauge implies that the Maxwell equations are covariant in all gauges. However, in a particular gauge covariance may not be immediately obvious (not *manifest*).

We may find a manifestly covariant set of Maxwell equations by appealing to the defining equations (2.113) relating the fields and the potentials to construct explicitly the six components of **B** and **E**. For example, using (2.113) and (1.31),

$$E^1 = \partial^1 A^0 - \partial^0 A^1 \qquad B^1 = \partial^3 A^2 - \partial^2 A^3. \tag{2.126}$$

In general one finds that the six independent components of **E** and **B** are elements of an antisymmetric second-rank field strength tensor $F^{\mu\nu}$:

$$F^{\mu\nu} = -F^{\nu\mu} = \partial^\mu A^\nu - \partial^\nu A^\mu. \tag{2.127}$$

In matrix form,

$$F^{\mu\nu} = \begin{pmatrix} 0 & -E^1 & -E^2 & -E^3 \\ E^1 & 0 & -B^3 & B^2 \\ E^2 & B^3 & 0 & -B^1 \\ E^3 & -B^2 & B^1 & 0 \end{pmatrix}. \tag{2.128a}$$

Contraction of $F^{\mu\nu}$ with the metric tensor yields the covariant tensor

$$F_{\mu\nu} = g_{\mu\gamma} g_{\delta\nu} F^{\gamma\delta} = \begin{pmatrix} 0 & E^1 & E^2 & E^3 \\ -E^1 & 0 & -B^3 & B^2 \\ -E^2 & B^3 & 0 & -B^1 \\ -E^3 & -B^2 & B^1 & 0 \end{pmatrix} \tag{2.128b}$$

which is related to $F^{\mu\nu}$ by $\mathbf{E} \to -\mathbf{E}$. If we also define the *dual field tensor*

$$\mathcal{F}^{\mu\nu} = \tfrac{1}{2} \epsilon^{\mu\nu\gamma\delta} F_{\gamma\delta} = \begin{pmatrix} 0 & -B^1 & -B^2 & -B^3 \\ B^1 & 0 & E^3 & -E^2 \\ B^2 & -E^3 & 0 & E^1 \\ B^3 & E^2 & -E^1 & 0 \end{pmatrix} \tag{2.129}$$

where $\epsilon^{\mu\nu\gamma\delta}$ is the completely antisymmetric fourth-rank tensor:

$$\epsilon^{\mu\nu\gamma\delta} = \begin{cases} +1 & \text{for } \mu\nu\gamma\delta = 0123 \text{ and even permutations,} \\ -1 & \text{for any odd permutation,} \\ 0 & \text{if any two indices are equal,} \end{cases} \tag{2.130}$$

$$\epsilon_{\mu\nu\gamma\delta} = -\epsilon^{\mu\nu\gamma\delta}, \tag{2.131}$$

the inhomogeneous Maxwell equations (2.112a) and (2.112d) may be written

$$\partial_\mu F^{\mu\nu} = j^\nu, \tag{2.132a}$$

and the homogeneous Maxwell equations (2.112b) and (2.112c) are

$$\partial_\mu \mathcal{F}^{\mu\nu} = 0. \tag{2.132b}$$

The quantities ∂_μ, $F^{\mu\nu}$, $\mathcal{F}^{\mu\nu}$, and j^ν all transform with well-defined properties under Lorentz transformations, and the Maxwell equations in this form are manifestly covariant. Finally, let us note for future reference that the field tensor $F^{\mu\nu}$ is a gauge-invariant construction:

$$\begin{aligned} F'^{\mu\nu} &= \partial^\mu(A^\nu - \partial^\nu\chi) - \partial^\nu(A^\mu - \partial^\mu\chi) \\ &= \partial^\mu A^\nu - \partial^\nu A^\mu = F^{\mu\nu}. \end{aligned} \tag{2.133}$$

EXERCISE 2.10 Verify the entries in the matrix (2.128) starting from the definition (2.127). Show that eqns. (2.132) are equivalent to the Maxwell equations (2.112).

2.7.2 Gauge Invariance in Quantum Mechanics

Our discussion of gauge invariance to this point has been classical. Let us now consider the meaning of gauge invariance in quantum mechanics. A particle of charge q moving with a velocity $\mathbf{v}$ in an electromagnetic field is subject to a *Lorentz force*

$$\mathbf{F} = q(\mathbf{E} + \mathbf{v} \times \mathbf{B}), \tag{2.134}$$

provided the classical Hamiltonian function is

$$H = \frac{1}{2m}(\mathbf{p} - q\mathbf{A})^2 + q\phi, \tag{2.135}$$

where $\mathbf{p}$ is the momentum, $\mathbf{A}$ is the vector potential, and ϕ is the scalar potential.

The Schrödinger equation obtained by quantizing this Hamiltonian is

$$\left[\frac{1}{2m}(-i\boldsymbol{\nabla} - q\mathbf{A})^2 + q\phi\right]\psi(\mathbf{x},t) = i\frac{\partial\psi(\mathbf{x},t)}{\partial t}. \tag{2.136}$$

Hence, for the quantum mechanics of charged particles in electromagnetic fields the Schrödinger equation may be constructed from the corresponding free-particle equation by making the operator replacements

$$\boldsymbol{\nabla} \to \mathbf{D} \equiv \boldsymbol{\nabla} - iq\mathbf{A} \qquad \frac{\partial}{\partial t} \to D^0 \equiv \frac{\partial}{\partial t} + iq\phi, \tag{2.137}$$

or, in covariant notation [see eq. (1.83)]

$$\partial^\mu \to D^\mu \equiv \partial^\mu + iqA^\mu, \tag{2.138}$$

with $A^\mu = (\phi, \mathbf{A})$ and $D^\mu = (D^0, \mathbf{D})$. The operator D^μ is of fundamental importance in our subsequent discussion of gauge field theories; it is called the *covariant derivative*. In this notation the Schrödinger equation is

$$\frac{1}{2m}(-i\mathbf{D})^2\psi(\mathbf{x},t) = iD^0\psi(\mathbf{x},t). \tag{2.139}$$

The correspondence principle suggests that (2.139) should be invariant under gauge transformations. In the notation (2.137) a gauge transformation (2.115) on (2.139) is of the form

$$\psi(\mathbf{x},t) \to \psi'(\mathbf{x},t)$$

$$\mathbf{D} \to \mathbf{D}' = \boldsymbol{\nabla} - iq\mathbf{A}'$$

$$D^0 \to D'^0 = \frac{\partial}{\partial t} + iq\phi',$$

and the gauge-transformed Schrödinger equation is (Aitchison and Hey, 1982, §8.2)

$$\frac{1}{2m}(-i\mathbf{D}')^2\psi'(\mathbf{x},t) = iD'^0\psi'(\mathbf{x},t). \tag{2.140}$$

We wish to find a relation between $\psi(\mathbf{x},t)$ and $\psi'(\mathbf{x},t)$ that will leave (2.140) in the same form as the Schrödinger equation (2.139) before the gauge transformation. It is not difficult to show that the required transformation is (see Exercise 2.11)

$$\psi'(\mathbf{x},t) = e^{iq\chi(\mathbf{x},t)}\psi(\mathbf{x},t), \tag{2.141}$$

where we note explicitly that the scalar $\chi(\mathbf{x},t)$ is a function of the spacetime coordinates. Thus, the Schrödinger equation is invariant under the set of transformations

$$\mathbf{A} \to \mathbf{A} + \boldsymbol{\nabla}\chi(\mathbf{x},t) \tag{2.142}$$

$$\phi \to \phi - \frac{\partial}{\partial t}\chi(\mathbf{x},t) \tag{2.143}$$

$$\psi(\mathbf{x}, t) \to e^{iq\chi(\mathbf{x},t)}\psi(\mathbf{x}, t), \tag{2.144}$$

and the origin of the standard prescription (1.83) for incorporating electromagnetic forces in quantum mechanics is seen to lie in the requirement of local gauge invariance, insuring harmony with the Maxwell equations in the correspondence limit.[‡]

EXERCISE 2.11 Prove that the Schrödinger equation is invariant under the gauge transformation (2.142)–(2.144).

The same argument may be extended to covariant wave equations. In terms of the 4-vector $D^\mu = \partial^\mu + iqA^\mu$, the Klein–Gordon equation

$$(D^\mu D_\mu + m^2)\psi(x) = 0, \tag{2.145}$$

and the Dirac equation

$$(i\gamma_\mu D^\mu - m)\psi(x) \equiv (i\slashed{D} - m)\psi(x) = 0, \tag{2.146}$$

in the presence of an electromagnetic field are (gauge) invariant under the combined transformation

$$A^\mu \to A^\mu - \partial^\mu\chi(x) \qquad \psi \to e^{iq\chi(x)}\psi. \tag{2.147}$$

Therefore, we have a simple prescription for obtaining the wave equation for a particle in an electromagnetic field from the corresponding relativistic or nonrelativistic free-particle equation: *replace the derivatives by the covariant derivatives*, $\partial^\mu \to D^\mu \equiv \partial^\mu + iqA^\mu$.

2.7.3 Gauge Invariance and the Photon Mass

Stringent upper limits on the photon mass have been obtained from measurements of planetary magnetic fields. From *Pioneer 10* measurements of Jupiter's field it can be concluded that (Davis, Goldhaber, and Nieto, 1975)

$$M_\gamma < 6 \times 10^{-22} \text{ MeV}/c^2. \tag{2.148}$$

Even stricter limits can be obtained from the galactic magnetic field. From (2.48) the Lagrangian density appropriate for a massless vector field is proportional to the complete contraction of $F_{\mu\nu}$ with itself, but the massive vector field has a Lagrangian density [eq. (2.49)]

$$\mathcal{L} = -\tfrac{1}{4}F_{\mu\nu}F^{\mu\nu} + \tfrac{1}{2}m^2 A^\mu A_\mu.$$

[‡]Gauge invariance does not demand the minimal substitution; rather, the minimal substitution is the simplest prescription that is gauge invariant. The data do not seem to require more complicated constructions. This is the origin of the terminology "minimal substitution".

We have just seen (2.133) that $F^{\mu\nu}$ is gauge invariant, so the Lagrangian density for the Maxwell field is unaffected by the gauge transformation $A^\mu \rightarrow A^\mu - \partial^\mu \chi$. This is not so for the massive vector field because the mass term transforms as

$$A^\mu A_\mu \rightarrow (A^\mu - \partial^\mu \chi)(A_\mu - \partial_\mu \chi) \neq A^\mu A_\mu, \qquad (2.149)$$

implying that the gauge invariance of the photon field and its masslessness are inextricably linked.

The incompatibility of gauge invariance with a mass term for such Lagrangians represents a formidable problem for any attempt to generalize the gauge symmetry of electrodynamics to the other fundamental interactions. The difficulty is that the photon is the only free massless boson observed in nature. For example, if the weak interactions are associated with gauge fields the corresponding Lagrangian density cannot have mass terms, and the weak analog of the photons (the intermediate vector bosons) should be massless. This is clearly not the case, and the way that a gauge field can acquire an effective mass without breaking gauge invariance forms one of the primary themes of this book. Essentially, this mechanism rests on the observations that a field theory Lagrangian and the vacuum state of that field theory need not have the same invariances (which can lead to *spontaneous symmetry breaking* and the appearance of massless *Goldstone bosons*), and that for a particular class of field theories a loophole called the *Higgs mechanism* can eliminate the Goldstone particles and bestow a mass on the vector particles, even though the theory remains gauge invariant. These topics will be elaborated in due course (Chs. 8–9).

2.7.4 States of Polarization and the Coulomb Gauge

The Maxwell field in its manifestly covariant incarnation is described by a real 4-vector potential, so there are four orthogonal polarization unit vectors. They may be chosen as

$$\hat{\epsilon}_0 = \begin{pmatrix} 1 \\ 0 \\ 0 \\ 0 \end{pmatrix} \qquad \hat{\epsilon}_1 = \begin{pmatrix} 0 \\ 1 \\ 0 \\ 0 \end{pmatrix} \qquad \hat{\epsilon}_2 = \begin{pmatrix} 0 \\ 0 \\ 1 \\ 0 \end{pmatrix} \qquad \hat{\epsilon}_3 = \begin{pmatrix} 0 \\ 0 \\ 0 \\ 1 \end{pmatrix} \qquad (2.150)$$

where we select the z-axis as the direction of photon propagation (Fig. 2.3):

$$k^\mu = (k, 0, 0, k) \qquad k \cdot \hat{\epsilon}_1 = k \cdot \hat{\epsilon}_2 = 0.$$

We may say that $\hat{\epsilon}_1$ and $\hat{\epsilon}_2$ are *spacelike* and perpendicular to the direction of photon propagation (they are *transverse*), $\hat{\epsilon}_3$ is spacelike and *longitudinal*, and $\hat{\epsilon}_0$ is a *timelike* (or *scalar*) polarization. In a manifestly covariant quantization of the electromagnetic field one should include all these states of polarization.

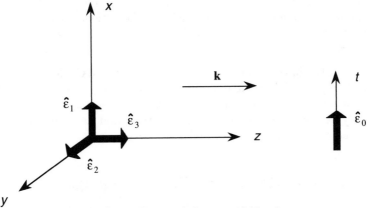

Fig. 2.3   States of photon polarization.

However, we know that (1) the classical fields are transverse, suggesting by the correspondence principle that only two independent states of polarization are required for physical observables in the quantum theory, and (2) a vector field normally has three independent states of polarization (not two or four!).

The complete resolution of these difficulties would take us further afield than we wish to go, considering our pragmatic approach to field theories. Let us be content with the following observations and assertions.

1. Our starting point is the Maxwell equations; these are Lorentz covariant.
2. Physical results, and questions of covariance, should not depend on the gauge in which we choose to work.
3. It is always possible to make a gauge transformation on the classical Maxwell field such that the transversality condition $\boldsymbol{\nabla} \cdot \mathbf{A} = 0$ may be imposed; that is, we are free to work in the Coulomb gauge. This reduces to two the number of independent polarizations for a massless vector field [see, for example, Sakurai (1967), Appendix A].
4. The massive vector field has three (rather than four) states of polarization because the momentum that would be conjugate to the scalar potential A^0 vanishes. The momentum conjugate to A^0 also vanishes for the Maxwell field, and we generally find that A^0 should be treated as a dependent rather than independent variable for vector fields. The freedom to make the gauge transformation required in point (3) is intimately associated with the masslessness of the free photon field, because the inclusion of a mass term in the Lagrangian density of the Maxwell field would break the gauge invariance (see §2.7.3). This is the basic reason that a massive vector field has three independent states of polarization, while the massless vector field has only two.
5. The photon field may be viewed as always having four states of polariza-

tion (covariant photons). However, it can be demonstrated that for *free* photons the longitudinal and timelike polarizations conspire to exactly cancel each other in their contribution to physical observables [see, e.g., Sakurai (1967), §4.6]. Thus, only the transverse fields need be quantized.

Therefore, the Maxwell field will be quantized in the Coulomb gauge, even though doing so sacrifices manifest Lorentz and gauge invariance. Despite these difficulties in the quantization procedure, calculated observables will eventually be found to respect these symmetries, and they will be in remarkable agreement with precise experimental measurements.

2.7.5   Quantization in the Coulomb Gauge

As seen in the preceding discussion, by a suitable choice of the function χ in the local gauge transformations

$$A^\mu \to A^\mu - \partial^\mu \chi \qquad \psi \to e^{iq\chi},$$

the transversality condition $\nabla \cdot \mathbf{A} = 0$ may be imposed on the electromagnetic field. The corresponding gauge is called the Coulomb gauge; we now quantize the electromagnetic field in this gauge (Bjorken and Drell, 1965, Ch. 14; Lee, 1981, Ch. 6).

The Lagrangian density for the Maxwell field is [eq. (2.48)]

$$\mathcal{L} = -\tfrac{1}{4} F_{\mu\nu} F^{\mu\nu} - \mathcal{H}_{\text{int}}$$

$$= -\tfrac{1}{4}(\partial_\mu A_\nu - \partial_\nu A_\mu)(\partial^\mu A^\nu - \partial^\nu A^\mu) - \mathcal{H}_{\text{int}}, \qquad (2.151)$$

where $\mathcal{H}_{\text{int}}$ is an interaction term to be specified later; we assume for now that it contains no derivatives (but see Exercise 2.4c). From (2.128) the components F_{00} and F^{00} vanish, and the Lagrangian density for the Maxwell field will not contain a term in $\dot{A}_0$. It follows that the momentum conjugate to the scalar potential also vanishes: $\Pi^0 = \partial \mathcal{L}/\partial \dot{A}_0 = 0$, and in quantizing the electromagnetic field we will treat the scalar potential A^0 as a c-number dependent variable that commutes with the field operators. Notice that we have now sacrificed manifest gauge invariance by fixing a particular gauge, and manifest Lorentz invariance by selecting a particular component of the vector potential for preferential treatment.

From (2.119), in the Coulomb gauge the scalar potential is given by the static Coulomb interaction

$$A^0(\mathbf{x}, t) = \int \frac{\rho(\mathbf{x}', t)}{|\mathbf{x} - \mathbf{x}'|} \, d^3 x'. \qquad (2.152)$$

For free fields it may be set to zero; in the corresponding gauge with the constraints $A^0 = 0$ and $\nabla \cdot \mathbf{A} = 0$ there will only be *two* independent components of A^μ. From (2.151) we identify the conjugate momenta

$$\Pi^0 = \frac{\partial \mathcal{L}}{\partial \dot{A}_0} = 0 \tag{2.153a}$$

$$\Pi^k = \frac{\partial \mathcal{L}}{\partial \dot{A}_k} = -\dot{A}^k + \partial^k A^0 = E^k \tag{2.153b}$$

(see Exercise 2.12). The Lagrangian density may be written

$$\mathcal{L} = -\tfrac{1}{4} F_{\mu\nu} F^{\mu\nu} - \mathcal{H}_{\text{int}} = \tfrac{1}{2}(\mathbf{E}^2 - \mathbf{B}^2) - \mathcal{H}_{\text{int}}. \tag{2.154}$$

From (2.153) and (2.154) a Legendre transformation gives the Hamiltonian density

$$\mathcal{H} = \Pi^k \dot{A}_k - \mathcal{L} = \tfrac{1}{2}(\mathbf{E}^2 + \mathbf{B}^2) + \mathbf{E} \cdot \boldsymbol{\nabla} A_0 + \mathcal{H}_{\text{int}}, \tag{2.155}$$

and the corresponding Hamiltonian is

$$H = \int d^3x \, \mathcal{H} = \tfrac{1}{2} \int d^3x \, (\mathbf{E}^2 + \mathbf{B}^2) + \int d^3x \, \mathcal{H}_{\text{int}}, \tag{2.156}$$

where the middle term of (2.155) has been eliminated using an integration by parts (Exercise 2.12).

EXERCISE 2.12 Verify eqns. (2.153)–(2.156) for the generalized momenta, Lagrangian density, and Hamiltonian density of the massless vector field.

To quantize the field we first assume that the independent components of the fields and momenta commute at equal times, and that the scalar potential A^0 commutes with everything since it is not a dynamical variable,

$$[A^\mu(\mathbf{x}, t), A^\nu(\mathbf{x}', t)] = 0 \tag{2.157a}$$

$$[\Pi^i(\mathbf{x}, t), \Pi^j(\mathbf{x}', t)] = 0 \tag{2.157b}$$

$$[\Pi^i(\mathbf{x}, t), A^0(\mathbf{x}', t)] = 0. \tag{2.157c}$$

For the commutator between Π^i and A^j we must modify the standard boson prescription (2.61) slightly to

$$[\Pi^i(\mathbf{x}, t), A^j(\mathbf{x}', t)] = [E^i(\mathbf{x}, t), A^j(\mathbf{x}', t)] = i\delta^{\text{tr}}_{ij}(\mathbf{x} - \mathbf{x}'), \tag{2.157d}$$

where the "transverse δ-function" is defined by

$$\delta^{\text{tr}}_{ij}(\mathbf{x} - \mathbf{x}') \equiv \int \frac{d^3k}{(2\pi)^3} \left(\delta_{ij} - \frac{k_i k_j}{\mathbf{k}^2} \right) e^{i\mathbf{k} \cdot (\mathbf{x} - \mathbf{x}')}. \tag{2.158}$$

The origin of this subtlety is the requirement that the commutators be consistent with (2.118), implying that both sides of (2.157d) must be divergenceless. It is demonstrated in Bjorken and Drell (1965), §14.2, that (2.158) satisfies this condition. It will be assumed that the commutation relations (2.157) are valid for both free and interacting fields.

Proceeding as in earlier cases, the field operators at time t are expanded in a Fourier series (Lee, 1981, Ch. 6). Since this is a spin-1 field, the expansion coefficients will be vectors to accommodate polarizations of that spin

$$\mathbf{A}(\mathbf{x}, t) = \sum_{\mathbf{k}} (2\omega\Omega)^{-1/2} \left(\mathbf{a}_{\mathbf{k}}(t) e^{i\mathbf{k}\cdot\mathbf{x}} + \mathbf{a}_{\mathbf{k}}^{\dagger}(t) e^{-i\mathbf{k}\cdot\mathbf{x}} \right). \qquad (2.159)$$

(Writing the field as a term plus its Hermitian conjugate ensures that the vector potential is real, as befits an uncharged field). A corresponding expansion may be written for the generalized momentum. In this expression $\omega \equiv \omega_k = |\mathbf{k}|$, $k^2 = 0$, and Ω is the volume of the box in which we quantize. The transversality condition $\nabla_{\cdot}\mathbf{A} = 0$ applied to the expansion (2.159) requires that $\mathbf{a}_{\mathbf{k}}(t)$ be perpendicular to the wave vector $\mathbf{k}$:

$$\mathbf{a}_{\mathbf{k}}(t) \cdot \mathbf{k} = 0 \qquad \mathbf{a}_{\mathbf{k}}^{\dagger}(t) \cdot \mathbf{k} = 0, \qquad (2.160)$$

as may be verified by taking the divergence of each side of (2.159) using

$$\nabla e^{i\mathbf{k}\cdot\mathbf{x}} = i k e^{i\mathbf{k}\cdot\mathbf{x}}. \qquad (2.161)$$

Thus the expansion (2.159) involves only polarizations transverse to the wave vector $\mathbf{k}$. It is convenient to choose from Fig. 2.3 the real polarization vectors $\hat{\epsilon}_1$ and $\hat{\epsilon}_2$ and the wave vector $\mathbf{k}$ to define the coordinate system of Fig. 2.4, for which

$$\hat{\epsilon}_1 \cdot \hat{\mathbf{k}} = \hat{\epsilon}_2 \cdot \hat{\mathbf{k}} = 0 \qquad (2.162)$$

$$\hat{\epsilon}_i \cdot \hat{\epsilon}_j = \delta_{ij} \qquad (2.163)$$

$$\hat{\mathbf{k}} = \frac{\mathbf{k}}{|\mathbf{k}|}. \qquad (2.164)$$

It is sometimes convenient to define the alternative set of polarization vectors

$$\hat{\mathbf{e}}_{\pm} = \frac{1}{\sqrt{2}} \left(\hat{\epsilon}_1 \pm i\hat{\epsilon}_2 \right). \qquad (2.165)$$

The set $\hat{\epsilon}_1$ and $\hat{\epsilon}_2$ describes states that are *linearly polarized*; the equivalent set $\hat{\mathbf{e}}_+$ and $\hat{\mathbf{e}}_-$ corresponds to states of *circular polarization*, with $\hat{\mathbf{e}}_+$ conventionally associated with right circular polarization and $\hat{\mathbf{e}}_-$ with left circular polarization. The states of circular polarization are helicity eigenstates, with $\hat{\mathbf{e}}_{\pm}$ corresponding to a projection ± 1 of the photon angular momentum on the propagation axis; it follows that the linearly polarized states are equal mixtures of the two helicity states.

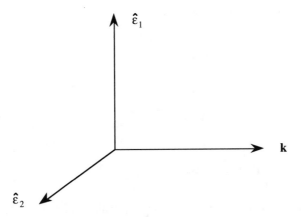

Fig. 2.4   Orthogonal polarization vectors for a transverse electromagnetic field.

The projection-zero state normally expected for a spin-1 particle is missing for the photon field because of the transversality requirement, $\nabla \cdot \mathbf{A} = 0$. The freedom to impose this condition rests on gauge invariance, and this in turn requires a massless field. Therefore, massless vector fields will normally have only two states of polarization instead of the three expected for a massive vector field.[‡]   Let us expand $\mathbf{a_k}(t)$ in terms of the polarization vectors (2.162)–(2.164) at fixed $\mathbf{k}$:

$$\mathbf{a_k}(t) = \sum_{\lambda=1,2} a_{\mathbf{k}\lambda}(t)\hat{\epsilon}_{\mathbf{k}\lambda}. \tag{2.166}$$

Then (2.159) may be written

$$\mathbf{A}(\mathbf{x},t) = \sum_{\lambda=1,2} \sum_{\mathbf{k}} (2\omega\Omega)^{-1/2} \left(a_{\mathbf{k}\lambda}(t)\hat{\epsilon}_{\mathbf{k}\lambda} e^{i\mathbf{k}\cdot\mathbf{x}} + a_{\mathbf{k}\lambda}^{\dagger}(t)\hat{\epsilon}_{\mathbf{k}\lambda} e^{-i\mathbf{k}\cdot\mathbf{x}} \right), \tag{2.167}$$

with a similar expression for the generalized momentum. As for the scalar field, we may invert to solve for the $a_{\mathbf{k}\lambda}^{\dagger}(t)$ and $a_{\mathbf{k}\lambda}(t)$, and then use (2.157) to construct the commutators. The results are

$$[a_{\mathbf{k}\lambda}(t), a_{\mathbf{k}'\lambda'}^{\dagger}(t)] = \delta_{\mathbf{k}\mathbf{k}'}\delta_{\lambda\lambda'} \tag{2.168a}$$

$$[a_{\mathbf{k}\lambda}(t), a_{\mathbf{k}'\lambda'}(t)] = [a_{\mathbf{k}\lambda}^{\dagger}(t), a_{\mathbf{k}'\lambda'}^{\dagger}(t)] = 0. \tag{2.168b}$$

[‡]In the absence of the Higgs mechanism, to be discussed in Ch. 8, which can provide massless vector fields with an effective mass and a third state of polarization.

The energy and momentum operators for free fields are

$$H = \tfrac{1}{2} \int d^3x : \left(\mathbf{E}^2 + \mathbf{B}^2\right):$$

$$= \tfrac{1}{2} \int d^3x : \left(\dot{\mathbf{A}}^2 + (\boldsymbol{\nabla} \times \mathbf{A})^2\right):$$

$$= \sum_{\mathbf{k}} \omega_k \sum_{\lambda=1,2} a^\dagger_{\mathbf{k}\lambda} a_{\mathbf{k}\lambda} \tag{2.169}$$

$$\mathbf{P} = \int d^3x : \mathbf{E} \times \mathbf{B}: = \sum_{\mathbf{k}} \mathbf{k} \sum_{\lambda=1,2} a^\dagger_{\mathbf{k}\lambda} a_{\mathbf{k}\lambda}. \tag{2.170}$$

In (2.170) we have used the quantized equivalent of the classical Poynting vector to construct the momentum operator, and in both (2.169) and (2.170) normal ordering has been imposed to remove infinities associated with the vacuum state, as discussed in §2.4.4.

In a similar way operators may be constructed that are associated with helicity eigenstates of the photon field. Defining the Hermitian conjugate operators (Lee, 1981, Ch. 6)

$$a^\dagger_{\mathbf{k},s=\pm}(t) = \mathbf{a}^\dagger_{\mathbf{k}}(t) \cdot \hat{\mathbf{e}}_\pm \qquad a_{\mathbf{k},s=\pm}(t) = \mathbf{a}_{\mathbf{k}}(t) \cdot \hat{\mathbf{e}}_\mp, \tag{2.171}$$

we can show using (2.157) and (2.159) that they obey the commutators

$$[a_{\mathbf{k}s}(t), a^\dagger_{\mathbf{k}'s'}(t)] = \delta_{\mathbf{k}\mathbf{k}'}\delta_{ss'}, \tag{2.172a}$$

$$[a_{\mathbf{k}s}(t), a_{\mathbf{k}'s'}(t)] = [a^\dagger_{\mathbf{k}s}(t), a^\dagger_{\mathbf{k}'s'}(t)] = 0, \tag{2.172b}$$

where the subscript s labels the helicity of the corresponding photon state.

From eqns. (2.167)–(2.172), it can be seen that the transverse components of the Maxwell field have been quantized in the same manner as two independent, massless, Klein–Gordon fields. Since the corresponding harmonic oscillator algebra has already been discussed, we may appropriate the results found in §2.4 for the real scalar field. For example, the operators $a^\dagger_{\mathbf{k}s}(t)$ and $a_{\mathbf{k}s}(t)$ create and destroy circularly polarized photons with helicity quantum number $s = \pm 1$. The states obviously obey boson statistics, and many-photon states may be constructed by repeated application of the creation operators to the vacuum state, which is defined by the requirement that it be annihilated by all destruction operators. The matrix elements associated with the quantized Maxwell field are those of the harmonic oscillator and require no further elaboration.

Manifest Lorentz and gauge invariance have been lost along this route of quantizing the electromagnetic field, but these invariances can be restored in the S-matrix elements (Ch. 3). Because physical observables are determined by the S-matrix elements, we adopt the point of view that such a theory is adequate. It is also possible to quantize the electromagnetic field using a covariant formalism in which all four states of polarization enter. The price that must be paid for maintaining manifest covariance is that the Hilbert space acquires an indefinite metric: there are states in the theory with negative norm (Exercise 2.13). Thus the interpretation is complicated because physical states must involve a conspiracy whereby the contributions of longitudinal and scalar polarizations compensate each other, leaving effectively only transverse states for free photons. A compressed account of such a quantization procedure in the covariant Lorentz gauge may be found in Ryder (1985), §4.4.

EXERCISE 2.13 Show that if photons are treated covariantly there are four states of polarization and there exist states in the Hilbert space of negative norm. However, for real (not virtual) photons the longitudinal and timelike polarizations exactly cancel in their contributions to a physical observable.

2.8 Noether's Theorem

Symmetry properties play a central role in the elaboration of any field theory. Some general symmetry restrictions have already been invoked in our discussions, and symmetries and group theory will be considered more systematically in Ch. 5. Here we call attention to an important property of field theories that is commonly termed Noether's Theorem: *to every continuous symmetry of a field theory Lagrangian there corresponds a conserved quantity.*[‡] Two examples taken from Quigg (1983) will be discussed.

2.8.1 Example: Conservation of 4-Momentum

Consider a spacetime transformation $x_\mu \to x'_\mu = x_\mu + a_\mu$, where the infinitesimal displacement a_μ is independent of x_μ. Poincaré invariance requires that the action be unchanged by this transformation. The corresponding change in a scalar field Lagrangian density is

$$\delta\mathcal{L} = \mathcal{L}(x') - \mathcal{L}(x) = a^\mu \partial_\mu \mathcal{L}. \tag{2.173}$$

[‡]Three points of caution before proceeding: (1) The symmetry properties of a field theory follow from the symmetry of the Lagrangian *and* the symmetry of the vacuum state; these need not be the same (*spontaneous symmetry breaking*). (2) The symmetries of a classical field are not always retained when the field is quantized (*anomalies*). (3) In some field theories conservation laws are encountered that are not a consequence of a continuous Lagrangian symmetry, but follow instead from constraints imposed by topology (*conserved topological charges*). Here we address only Noether's theorem; spontaneous symmetry breaking, anomalies, and topological charge come later, in Chs. 8, 9 and 13, respectively.

Assuming the Lagrangian density to depend on the field ϕ and its derivatives $\partial_\mu \phi$, but not explicitly on the coordinates x (translational invariance), we may also write

$$\delta\mathcal{L} = \frac{\partial\mathcal{L}}{\partial\phi}\delta\phi + \frac{\partial\mathcal{L}}{\partial(\partial_\mu\phi)}\delta(\partial_\mu\phi), \tag{2.174}$$

$$\delta\phi \equiv \phi(x') - \phi(x) = a^\mu\partial_\mu\phi \tag{2.175a}$$

$$\delta(\partial_\mu\phi) \equiv \partial_\mu\phi(x') - \partial_\mu\phi(x) = a^\nu\partial_\nu(\partial_\mu\phi). \tag{2.175b}$$

But from the Euler–Lagrange equations (2.39), $\partial\mathcal{L}/\partial\phi = \partial_\mu\big(\partial\mathcal{L}/\partial(\partial_\mu\phi)\big)$. Therefore, from (2.174) and (2.175)

$$\delta\mathcal{L} = \left(\partial_\nu\frac{\partial\mathcal{L}}{\partial(\partial_\nu\phi)}\right)a^\mu\partial_\mu\phi + \frac{\partial\mathcal{L}}{\partial(\partial_\nu\phi)}a^\mu\partial_\mu(\partial_\nu\phi)$$

$$= \partial_\nu\frac{\partial\mathcal{L}}{\partial(\partial_\nu\phi)}a^\mu\partial_\mu\phi, \tag{2.176}$$

where the validity of the last step may be checked by taking the derivative of the product, remembering that a^μ does not depend on the coordinates. Equating the expressions (2.173) and (2.176), and doing some rearranging (recall that repeated indices are dummy indices)

$$a_\nu\partial_\mu\left(\frac{\partial\mathcal{L}}{\partial(\partial_\mu\phi)}\partial^\nu\phi - g^{\mu\nu}\mathcal{L}\right) = 0,$$

where $g^{\mu\nu}$ is the metric tensor. This relation must be satisfied for arbitrary displacements a_ν. Defining the stress–energy–momentum tensor

$$\Theta^{\mu\nu} \equiv \frac{\partial\mathcal{L}}{\partial(\partial_\mu\phi)}\partial^\nu\phi - g^{\mu\nu}\mathcal{L}, \tag{2.177}$$

we are led to the local conservation law (see §2.8.3)

$$\partial_\mu\Theta^{\mu\nu} = 0. \tag{2.178}$$

Explicitly, from (2.177), (2.57b), and (2.60)

$$\Theta^{00} = \frac{\partial\mathcal{L}}{\partial(\partial_0\phi)}\partial^0\phi - g^{00}\mathcal{L}$$

$$= \left(\frac{\partial\mathcal{L}}{\partial\dot\phi}\right)\dot\phi - \mathcal{L} = \mathcal{H}, \tag{2.179}$$

and the total energy,

$$H = \int d^3x\,\mathcal{H} = \int d^3x\,\Theta^{00}, \tag{2.180}$$

is a constant of the motion. In a similar way we may show that the components Θ^{0k} are associated with momentum densities. The quantity $\Theta^{\mu\nu}$ is an example of a *conserved Noether tensor;* in this instance (2.178) links translational invariance with 4-momentum conservation.

By similar procedures, other Noether tensors may be constructed; these may relate conserved quantities either to continuous spacetime symmetries, or to · continuous internal symmetries. Let us consider an example of a conservation law associated with a continuous internal symmetry, global isospin rotations.

2.8.2 Example: Conservation of Isospin Current

Assume that the proton and neutron correspond to the same particle, but with different projections in an abstract space that we call the isospin space (see §5.3.2). In an obvious notation we may define the composite isospinor[‡]

$$\psi = \begin{pmatrix} p \\ n \end{pmatrix}. \tag{2.181}$$

Because of the isospin symmetry the neutron and proton masses are equal, and the Lagrangian density for free nucleons is [see (2.46)]

$$\mathcal{L} = \overline{p}(i\slashed{\partial} - m)p + \overline{n}(i\slashed{\partial} - m)n, \tag{2.182}$$

or more compactly

$$\mathcal{L} = \overline{\psi}(i\slashed{\partial} - m)\psi. \tag{2.183}$$

By analogy with normal rotations a global isospin rotation is generated by

$$\psi \to e^{\frac{i}{2}\tau^a \alpha_a} \psi, \tag{2.184}$$

where τ^a denotes the Pauli matrices (1.78) (it is common to represent the Pauli matrices by τ^a instead of σ^a if they operate in the isospin space rather than the spin space). The parameters α_a are constants, independent of the coordinates, so this isospin rotation is the same at all spacetime points, which is what is meant by a global transformation.[‡‡]

Because α_a and τ^a are independent of the spacetime and bispinor coordinates $\mathcal{L}$ is invariant under the global isospin rotation. By the Noether theorem there should be a conserved quantity associated with this continuous symmetry, and since continuous transformations may be built from a succession of infinitesimal ones we may isolate the conserved quantity by studying the effect

---

[‡]Do not confuse the isospinor space with the bispinor space; they are independent degrees of freedom. Each of the components of the two-component (in the isospace) spinor ψ is itself a four-component bispinor function, and there are eight degrees of freedom for ψ (and eight for $\overline{\psi}$) in the combined isospace and bispinor space.

[‡‡]We might also imagine a *local* isospin rotation, with $\alpha_a(x)$ a function of the spacetime coordinates; this will lead to the concept of a Yang–Mills field (Ch. 7).

of an infinitesimal isospin rotation on the Lagrangian (see Ch. 5). Expanding the exponential in (2.184),

$$\psi(x) \rightarrow \psi(x) + \tfrac{i}{2}\tau^a \alpha_a \psi(x) = \psi + \delta\psi. \qquad (2.185)$$

Invariance of the Lagrangian density under this transformation implies

$$\delta\mathcal{L} = 0, \qquad (2.186)$$

where

$$\delta\mathcal{L} = \frac{\partial\mathcal{L}}{\partial\psi}\delta\psi + \frac{\partial\mathcal{L}}{\partial(\partial_\mu\psi)}\delta(\partial_\mu\psi) + \frac{\partial\mathcal{L}}{\partial\overline{\psi}}\delta\overline{\psi} + \frac{\partial\mathcal{L}}{\partial(\partial_\mu\overline{\psi})}\delta(\partial_\mu\overline{\psi}). \qquad (2.187)$$

From (2.185) we identify

$$\delta\psi = \tfrac{i}{2}\tau^a \alpha_a \psi \qquad \delta(\partial_\mu\psi) = \tfrac{i}{2}\tau^a \alpha_a (\partial_\mu\psi). \qquad (2.188)$$

Furthermore, from the Lagrangian density (2.183) and the free-particle Dirac equation (1.99)

$$\frac{\partial\mathcal{L}}{\partial\overline{\psi}} = (i\slashed{\partial} - m)\psi = 0 \qquad \frac{\partial\mathcal{L}}{\partial(\partial_\mu\overline{\psi})} = 0. \qquad (2.189)$$

Whence, using (2.186)–(2.189) and (2.39),

$$\delta\mathcal{L} = \left[\partial_\mu \frac{\partial\mathcal{L}}{\partial(\partial_\mu\psi)}\right] \frac{i}{2}\tau^a \alpha_a \psi + \frac{\partial\mathcal{L}}{\partial(\partial_\mu\psi)} \frac{i}{2}\tau^a \alpha_a (\partial_\mu\psi) = 0. \qquad (2.190)$$

This is equivalent to

$$\partial_\mu \alpha_a \left[\frac{i}{2} \frac{\partial\mathcal{L}}{\partial(\partial_\mu\psi)} \tau^a \psi\right] = 0,$$

and since this must be satisfied for an arbitrary constant α_a we have

$$\partial^\mu J_\mu^a = 0, \qquad (2.191a)$$

where[‡]

$$J_\mu^a \equiv -\frac{i}{2} \frac{\partial\mathcal{L}}{\partial(\partial_\mu\psi)} \tau^a \psi = \frac{1}{2}\overline{\psi}\gamma^\mu \tau^a \psi \qquad (2.191b)$$

is a conserved current that will be called the *isospin current*; it is seen to differ from the Dirac current (1.118) only in the presence of the isospin operator $\tau^a/2$ sandwiched between $\overline{\psi}$ and ψ.

Generally, for an internal symmetry described by a set of matrix symmetry

---

[‡]The notation J_μ^a means a Lorentz 4-vector (the index μ), that is also a vector in the isospace (the index a).

generators t^a (see §5.2), the Noether currents are given by

$$J_\mu^a = -i\frac{\partial \mathcal{L}}{\partial(\partial^\mu \phi_i)}t_{ij}^a\phi_j, \tag{2.192}$$

where t_{ij}^a are the elements of the matrix generator t^a, and ϕ_i are the fields transformed by the matrices (see Exercises 2.14, 2.15 and 12.4). In the example just discussed the isospin generators are half the Pauli matrices, $t^a = \tau^a/2$, and they satisfy the $SU(2)$ Lie algebra (5.30).

2.8.3 Conserved Charges

We may formally define a charge as the spatial integral of the timelike component of a current 4-vector. That is, if $J \equiv (J^0, J^1, J^2, J^3)$ is a 4-current, the associated charge $Q(t)$ is

$$Q(t) \equiv \int d^3x\, J^0(x). \tag{2.193}$$

A familiar example is the electromagnetic 4-current $j \equiv (\rho, \mathbf{j})$, which obeys $\partial_\mu j^\mu = 0$, with the electrical charge given by

$$Q_e \equiv \int d^3x\, \rho(x) = \int d^3x\, j^0. \tag{2.194}$$

For a current that has vanishing 4-divergence (such as the electromagnetic current) the associated charge is a *conserved quantity*. Let us prove this assertion. The Heisenberg equation of motion for the charge is

$$\dot{Q} \equiv \frac{dQ(t)}{dt} = i[H, Q(t)], \tag{2.195}$$

where H is the Hamiltonian. From (2.195) and (2.193), and $\partial_\mu J^\mu = 0$,

$$\dot{Q} = \partial_0 \int d^3x\, J^0 = \int d^3x\, \partial_0 J^0 = -\int d^3x\, \boldsymbol{\nabla}\cdot\mathbf{J}.$$

But by the divergence theorem

$$\int_V d^3x\, \boldsymbol{\nabla}\cdot\mathbf{J} = \int_S \mathbf{J}\cdot\mathbf{n}\, ds, \tag{2.196}$$

where S is a closed 2-dimensional surface bounding the volume V, and $\mathbf{n}$ is an outward normal to that surface. Since it is assumed that no currents flow at infinity we obtain

$$\dot{Q} = -\int_{S_\infty} \mathbf{J}\cdot\mathbf{n}\, ds = 0, \tag{2.197}$$

and from (2.195),

$$[H, Q] = 0. \tag{2.198}$$

Thus the charge Q is a constant of motion and is associated with a symmetry of the system if the 4-divergence of the associated current vanishes. This derivation establishes the connection between conserved Noether currents in a Lagrangian formalism and the constants of motion appearing in the Hamiltonian formulation of the same problem. In a formal sense at least, it may also be useful to associate a *partially conserved current* with an approximate symmetry of the system. In the preceding derivation, if

$$\partial_\mu j^\mu = \epsilon \simeq 0 \qquad (2.199)$$

we obtain

$$[H, Q] \simeq \mathcal{O}(\epsilon), \qquad (2.200)$$

implying that Q is an approximate constant of motion.

EXERCISE 2.14 Show that the Dirac Lagrangian density (2.46) is invariant under the global $U(1)$ rotation $\psi(x) \to e^{i\alpha}\psi(x)$ (see §5.3.1). Find the corresponding Noether current and show that this leads to global conservation of electronic charge.

EXERCISE 2.15 (a) Show that the Lagrangian density (2.83) with $n = 2$ and with an interaction

$$\mathcal{H}_{\text{int}} = \tfrac{1}{4}\lambda(\phi_1^2 + \phi_2^2)^2$$

is invariant under infinitesimal rotations by the matrices

$$U = \begin{pmatrix} \cos\alpha & -\sin\alpha \\ \sin\alpha & \cos\alpha \end{pmatrix}$$

if $m_1 = m_2 \equiv m$ [$SO(2)$ invariance; see Exercise 5.4]. Find the corresponding Noether current. (b) Rewrite the Lagrangian density of part (a) in terms of the complex scalar fields

$$\phi_\pm = \tfrac{1}{\sqrt{2}}(\phi_1 \pm i\phi_2).$$

Show that this Lagrangian density is invariant under global $U(1)$ rotations of the field (see Exercise 2.14), and find the conserved current.

2.9 Interactions between Fields

For the several fields we have quantized thus far an interaction Hamiltonian density $\mathcal{H}_{\text{int}}$ has been included in the equations. Little has been said about it except that for simplicity it is assumed to be independent of the field derivatives so that the momentum variables remain those of the free field. Of course, free fields are not very interesting; it is the interactions between fields that contain the important physics, and this is the topic to which we turn in subsequent chapters. As particular cases are addressed the form of $\mathcal{H}_{\text{int}}$ will be specified; in doing so, we will be strongly guided by symmetry principles. Some of these principles will provide trivial overall conservation laws, but some of the symmetry requirements have dynamical implications that will be used to specify the form of the interaction.

As an example we consider the quantized interaction between the Maxwell field and the Dirac field (*quantum electrodynamics*). For no coupling between the fields the Lagrangian density is [see (2.46) and (2.48)]

$$\mathcal{L} = \overline{\psi}(i\partial\!\!\!/ - m)\psi - \tfrac{1}{4}F_{\mu\nu}F^{\mu\nu}. \tag{2.201}$$

A field coupling that is gauge and Lorentz invariant may be introduced by the minimal substitution, which is accomplished by replacing derivatives with covariant derivatives (2.138), $\partial_\mu \to \partial_\mu + ie_0 A_\mu$. The corresponding Lagrangian density is

$$\mathcal{L} = \overline{\psi}(i\partial\!\!\!/ - m_0)\psi - \tfrac{1}{4}F_{\mu\nu}F^{\mu\nu} - e_0\overline{\psi}A\!\!\!/\psi, \tag{2.202}$$

from which we identify an interaction term

$$\mathcal{L}_{\text{int}} = -\mathcal{H}_{\text{int}} = -e_0\overline{\psi}A\!\!\!/\psi = -e_0\overline{\psi}\gamma_\mu A^\mu\psi. \tag{2.203}$$

In these equations subscripts are attached to the mass and charge to signify that, on account of the interactions, the mass and charge parameters appearing in the Lagrangian will not generally be the physical mass and charge. This implies a process of renormalization between the bare and physical quantities that will be discussed in Chs. 3 and 6. In the interest of compact notation we will usually omit these subscripts, with the implicit understanding that physical parameters are always renormalized parameters.

The equations of motion for the coupled electron and photon fields follow from independent variation of the Lagrangian density (2.202) with respect to these fields (Exercise 2.16):

$$(i\partial\!\!\!/ - m_0)\psi(x) = e_0 A\!\!\!/(x)\psi(x) \tag{2.204a}$$

$$\partial_\nu F^{\mu\nu}(x) = e_0\overline{\psi}(x)\gamma^\mu\psi(x). \tag{2.204b}$$

These are the nonlinear field equations that must be solved for the electromagnetic interaction of the electron and photon fields.

EXERCISE 2.16 Prove that the Lagrangian density (2.202) leads to the coupled field equations (2.204).

2.10 Background and Further Reading

The material on the quantization of classical fields follows the discussion in Goldstein (1981), Ch. 2; Bjorken and Drell (1965), Chs. 11–15; Lee (1981), Chs. 1–3 and 6; Mandl (1959), Chs. 1–3; and Muirhead (1965), Chs. 3–4. Itzykson and Zuber (1980), Chs. 1, 3, and 4 may be consulted also for a systematic treatment of canonical quantization. The discussion of gauge invariance follows that in Jackson (1975), Chs. 6 and 11, and Aitchison and Hey (1982). Noether currents are standard fare in all books on quantum field theory; Quigg (1983), Cheng and Li (1984), or Campbell (1978) give clear introductions to Lagrangian symmetries and the associated currents.

CHAPTER 3

Perturbation Theory and Evaluation of the S-Matrix

In the preceding chapter we have seen how to quantize some elementary fields using the canonical prescription. However, it is necessary to do more than quantize the fields: we must learn to calculate physical observables such as differential cross sections. These observables are constructed from the amplitude for a system in some initial state to make a transition to some final state; in a field theory such transitions occur through the interactions of the fields, and the transition amplitudes are usually termed the S-matrix elements. The calculation of the S-matrix elements in a relativistic quantum field theory is a nontrivial task. The field couplings normally lead to nonlinear equations of motion for the field operators, as shown in the previous chapter for the specific example of quantum electrodynamics. Exact solutions are not possible for realistic theories, and most calculations have of necessity been done by the methods of perturbation theory. Therefore, in this chapter we consider the perturbative evaluation of S-matrix elements. The recently burgeoning industry of non-perturbative field theories will be examined when lattice gauge methods and solitons are discussed in Ch. 13.

3.1 Interaction Representation

In discrete-particle perturbation theory the starting point is usually an idealized problem that (1) can be solved exactly (either analytically or numerically), and (2) is hoped to be similar to the actual problem to be solved. Approximations to the actual problem are then developed by expansion methods. A similar approach may be employed in field theories. The starting point is typically one of non-interacting fields that obey linear wave equations. Since the equations are linear, solutions for the non-interacting fields are usually available. If the interactions among the fields are weak enough, solutions for the realistic case may be developed by expansions about the free-field ones.

We have previously worked in either the Schrödinger representation, where all time dependence is ascribed to the wavefunctions, or the Heisenberg representation, where time dependence is confined to the operators. There are many other representations that could be used, each connected to the others by unitary transformations; which to choose is purely a matter of convenience and clarity. In constructing S-matrix elements by perturbation theory it is often useful to work in the *interaction representation*, where the time dependence is allocated between the wavefunctions and operators in a particularly advantageous way.

To implement this idea quantitatively let us split the Hamiltonian for a system into a free-field part H_0 and an interaction term H_{int}

$$H = \int d^3x \, \mathcal{H}_0 + \int d^3x \, \mathcal{H}_{int} = H_0 + H_{int}, \tag{3.1}$$

a decomposition that was anticipated in the previous equations. Consider the Schrödinger time dependence for the field state amplitude Ψ_S:

$$i \frac{\partial \Psi_S(t)}{\partial t} = H^S \Psi_S(t), \tag{3.2}$$

where the subscript and superscript S denote quantities in the Schrödinger representation. For an operator F_S and state vector $\Psi_S(t)$ in the Schrödinger picture a unitary transformation to a new representation may be defined by

$$\Psi_I(t) = e^{iH_0^S t} \Psi_S(t) \qquad F_I = e^{iH_0^S t} F_S e^{-iH_0^S t}. \tag{3.3}$$

This new representation for the state vectors and operators will be termed the *interaction representation*. Notice that the transforming exponentials involve only the free-field portion of the Hamiltonian H_0, in contrast to the transformations (2.10) and (2.11) between the Schrödinger and Heisenberg representations that contain the total Hamiltonian H. The interaction representation Hamiltonian is a special case of (3.3),

$$\begin{aligned} H_I &= H_0^I + H_{int}^I \\ &= e^{iH_0^S t}(H_0^S + H_{int}^S)e^{-iH_0^S t} \\ &= H_0^S + e^{iH_0^S t}H_{int}^S e^{-iH_0^S t}, \end{aligned} \tag{3.4}$$

and from (3.1)–(3.4)

$$i \frac{\partial \Psi_I(t)}{\partial t} = H_{int}^I \Psi_I(t) \qquad \frac{dF_I}{dt} = i[H_0^I, F_I]. \tag{3.5}$$

Thus the interaction representation operators satisfy Heisenberg equations of motion, but with the commutator for the free-field Hamiltonian H_0^I, and the interaction representation state vector Ψ_I satisfies a Schrödinger equation, but for the interaction portion H_{int}^I of the Hamiltonian rather than the total Hamiltonian.

To summarize, using S, H, and I to denote Schrödinger, Heisenberg, and interaction representations respectively, and defining $H_{\mathrm{S}} = H_{\mathrm{H}} \equiv H$ and $H_0^{\mathrm{S}} = H_0^{\mathrm{I}} \equiv H_0$, the fundamental equations of quantum mechanics in these three representations are

$$i\frac{\partial \Psi_{\mathrm{S}}}{\partial t} = H\Psi_{\mathrm{S}} \qquad \dot{F}_{\mathrm{S}} = 0 \qquad \text{(Schrödinger picture)} \qquad (3.6)$$

$$\frac{\partial \Psi_{\mathrm{H}}}{\partial t} = 0 \qquad \dot{F}_{\mathrm{H}} = i[H, F_{\mathrm{H}}] \qquad \text{(Heisenberg picture)} \qquad (3.7)$$

$$i\frac{\partial \Psi_{\mathrm{I}}}{\partial t} = H_{\mathrm{int}}^{\mathrm{I}}\Psi_{\mathrm{I}} \qquad \dot{F}_{\mathrm{I}} = i[H_0, F_{\mathrm{I}}] \qquad \text{(Interaction picture)}, \qquad (3.8)$$

where F is an operator. The transformations among the three representations are summarized by the relations

$$\Psi_{\mathrm{S}}(t) = e^{-iHt}\Psi_{\mathrm{H}} \qquad (3.9)$$

$$\Psi_{\mathrm{I}}(t) = e^{iH_0 t}\Psi_{\mathrm{S}}(t) \qquad (3.10)$$

$$F_{\mathrm{S}} = e^{-iHt}F_{\mathrm{H}}(t)e^{iHt} \qquad (3.11)$$

$$F_{\mathrm{I}} = e^{iH_0 t}F_{\mathrm{S}}e^{-iH_0 t}. \qquad (3.12)$$

The interaction representation separation of the trivial free-field time dependence from the essential time-dependent interaction causing the transition between states makes it particularly useful in perturbation theory.

3.2 Definition of the S-Matrix

Let us consider a process for which we have particles that may be considered free at the beginning and end, but that interact over some limited region of space and time in the interim. We denote the state vector in the interaction representation by $\Phi(-\infty)$ for the remote past and $\Phi(\infty)$ for the remote future, and seek an operator S such that

$$\Phi(\infty) = S\Phi(-\infty). \qquad (3.13)$$

The operator S is the limit of the time evolution operator $U(t, t_0)$

$$S = \lim_{\substack{t \to +\infty \\ t_0 \to -\infty}} U(t, t_0), \qquad (3.14)$$

where the unitary operator $U(t, t_0)$ propagates the system from t_0 to t. Since in the interaction representation

$$i \frac{\partial \Phi(t)}{\partial t} = V(t)\Phi(t), \tag{3.15}$$

where

$$V(t) \equiv H_{int}^I(t) = e^{iH_0 t} \left(\int \mathcal{H}_{int}^S d^3 x \right) e^{-iH_0 t} \tag{3.16}$$

is the interaction Hamiltonian, we have by formal integration of (3.15)

$$\Phi(t) = U(t, t_0)\Phi(t_0) = \exp\left(-i \int_{t_0}^{t} V(t) dt \right) \Phi(t_0). \tag{3.17}$$

(As we will work in the interaction representation for the remainder of this chapter, the indices "I" will be omitted in this and subsequent equations.) This is only a formal solution to the problem and cannot be used directly to construct a perturbation series because generally

$$[V(t), V(t')] \neq 0, \tag{3.18}$$

and if two operators A and B fail to commute,

$$e^{A+B} \neq e^A e^B. \tag{3.19}$$

The solution is to introduce a *chronological* or *time-ordering operator* T that has the property

$$T\big(V(t_1), \ldots, V(t_n)\big) = \delta_p V(t_\alpha)V(t_\beta) \ldots V(t_\delta)V(t_\sigma), \tag{3.20}$$

with $t_\alpha \geqslant t_\beta \geqslant \ldots t_\delta \geqslant t_\sigma$. [If two times are equal the relative order of the corresponding operators is taken to be the same on the two sides of (3.20)]. The factor δ_p is 1 if only boson operators are involved. If fermion operators must be permuted $\delta_p = \pm 1$, depending on whether the number of permutations of fermion operators required for chronological ordering is even or odd. That is, T applied to a product of operators orders them according to their time arguments, with the earliest times standing to the right.[‡] This leads to the definition of S in terms of the *time-ordered exponential*

$$S = \lim_{\substack{t \to +\infty \\ t_0 \to -\infty}} U(t, t_0),$$

[‡]The operator T is called the Wick chronological operator. The definition (3.20) is often give in terms of the Dyson chronological operator P, which omits the phase δ_p. Because fermion operators will always occur in pairs in $V(t)$ and $\delta_p = 1$ for bosons, the distinction is immaterial in our discussion.

with

$$U(t, t_0) = T\left[\exp\left(-i\int_{t_0}^t V(t)\,dt\right)\right]$$

$$= T\left[\exp\left(-i\int_{t_0}^t dt\int d^3x\, \mathcal{H}_{\text{int}}(\mathbf{x}, t)\right)\right]. \tag{3.21}$$

The formal expression (3.21) is to be interpreted as the expansion

$$S = \sum_{n=0}^\infty S_n = \lim_{\substack{t\to+\infty\\ t_0\to-\infty}} T\left[1 + \frac{1}{i}\int_{t_0}^t V(t)\,dt + \dots\right.$$

$$\left. + \frac{1}{i^n n!}\left(\int_{t_0}^t V(t)\,dt\right)^n + \dots\right], \tag{3.22}$$

which will be the starting point for a perturbation series calculation of the transition probabilities. We note that for theories in which the interaction Lagrangian density $\mathcal{L}_{\text{int}}$ contains no derivative couplings the interaction Hamiltonian density differs from $\mathcal{L}_{\text{int}}$ by only a sign so that [see (2.58)]

$$V(t) = H_{\text{int}}(t) = -L_{\text{int}}(t) = -\int d^3x\, \mathcal{L}_{\text{int}}, \tag{3.23}$$

and (3.21) may be written

$$S = Te^{iA}, \tag{3.24}$$

where A is the part of the action coming from the interaction Lagrangian density

$$A = \int d^4x\, \mathcal{L}_{\text{int}}. \tag{3.25}$$

The states Φ may be expanded in a complete basis $\Phi = \sum_\alpha |\alpha\rangle\langle\alpha \mid \Phi\rangle$, and the amplitude for the transition from $\Phi(-\infty)$ to some basis state $|\beta\rangle$ at $t = +\infty$ is

$$\mathcal{A} = \langle\beta \mid \Phi(+\infty)\rangle = \langle\beta| S |\Phi(-\infty)\rangle$$

$$= \sum_\alpha \langle\beta| S |\alpha\rangle\langle\alpha \mid \Phi(-\infty)\rangle. \tag{3.26}$$

Thus the operator S can be specified by its matrix elements

$$S_{\beta\alpha} \equiv \langle\beta| S |\alpha\rangle, \tag{3.27}$$

and is termed the S-*matrix* or scattering matrix.

As is commonly the case in field theories, there are some problems with construction of the S-matrix elements that are more severe than in discrete quantum mechanics. A major source of these problems is the difficulty of finding the correct asymptotic states in a given situation. These difficulties do not play a major practical role at our level of discussion, so we will largely ignore them. The serious reader should consult one of the standard advanced field theory texts, such as Itzykson and Zuber (1980).

3.3 Interaction Picture Fourier Expansions

Operators in the interaction representation satisfy free-field equations of motion [eq. (3.8)], and their momentum space expansions take a particularly simple form. From (2.68a) and the generalization of (2.21) found in Exercise 2.5a, the spin-0 field may be expanded in the interaction representation as (Lee, 1981, Ch. 5; Mandl, 1959, Ch. 5)

$$\phi(x) = \alpha(x) + \alpha^\dagger(x) \tag{3.28a}$$

$$\alpha(x) = \sum_{\mathbf{k}} (2\omega\Omega)^{-1/2} a_{\mathbf{k}} e^{i\mathbf{k}\cdot\mathbf{x} - i\omega t}$$

$$= \sum_{\mathbf{k}} (2\omega\Omega)^{-1/2} a_{\mathbf{k}} e^{-ik\cdot x}, \tag{3.28b}$$

$$x = (t, \mathbf{x}) \qquad k = (\omega, \mathbf{k}) \qquad \omega \equiv \omega_{\mathbf{k}} = \sqrt{\mathbf{k}^2 + m^2} > 0$$

$$a_{\mathbf{k}} \equiv a_{\mathbf{k}}(t = t_0) \qquad a_{\mathbf{k}}(t) = a_{\mathbf{k}} e^{-i\omega t}, \tag{3.29}$$

with m the physical mass of the scalar particle. The operators $a_{\mathbf{k}}$ are independent of time, and the time-dependent operators $a_{\mathbf{k}}(t)$ are those of Exercise 2.5a. In this decomposition $\alpha(x)$ is termed the positive-frequency part and $\alpha^\dagger(x)$ the negative-frequency part of the operator $\phi(x)$. A similar expansion may be made for the canonical momentum $\Pi(x)$ starting from (2.68b).

For the spin-$\frac{1}{2}$ field in the interaction representation (2.94) and (2.101) give (Lee, 1981)

$$\psi(x) = u(x) + v(x), \tag{3.30a}$$

where

$$u(x) = \Omega^{-1/2} \sum_{\mathbf{p}s} \sqrt{\frac{m}{\omega}} a_{\mathbf{p}s} u_{\mathbf{p}s} e^{i\mathbf{p}\cdot\mathbf{x} - i\omega t}$$

$$= \Omega^{-1/2} \sum_{\mathbf{p}s} \sqrt{\frac{m}{\omega}} a_{\mathbf{p}s} u_{\mathbf{p}s} e^{-ip\cdot x} \tag{3.30b}$$

$$v(x) = \Omega^{-1/2} \sum_{\mathbf{p}s} \sqrt{\frac{m}{\omega}} b_{\mathbf{p}s}^\dagger v_{\mathbf{p}s} e^{ip\cdot x}. \tag{3.30c}$$

We will also need

$$\overline{\psi}(x) \equiv \psi^\dagger(x)\gamma_0 = \overline{u}(x) + \overline{v}(x) \tag{3.31a}$$

where

$$\bar{u}(x) \equiv \Omega^{-1/2} \sum_{\mathbf{p}s} \sqrt{\frac{m}{\omega}}\, a^{\dagger}_{\mathbf{p}s} u^{\dagger}_{\mathbf{p}s} \gamma_0 e^{ip \cdot x} \tag{3.31b}$$

$$\bar{v}(x) \equiv \Omega^{-1/2} \sum_{\mathbf{p}s} \sqrt{\frac{m}{\omega}}\, b_{\mathbf{p}s} v^{\dagger}_{\mathbf{p}s} \gamma_0 e^{-ip \cdot x}. \tag{3.31c}$$

By this definition the operators $\bar{u}$ and v contain the positive energy particle or antiparticle creation operators, and u and $\bar{v}$ contain the corresponding destruction operators.

In these expressions

$$\begin{aligned} a_{\mathbf{p}s}(t) &= a_{\mathbf{p}s} e^{-i\omega t} & a_{\mathbf{p}s} &\equiv a_{\mathbf{p}s}(t = t_0) \\ b^{\dagger}_{\mathbf{p}s}(t) &= b^{\dagger}_{\mathbf{p}s} e^{i\omega t} & b^{\dagger}_{\mathbf{p}s} &\equiv b^{\dagger}_{\mathbf{p}s}(t = t_0) \end{aligned} \tag{3.32}$$

and so on, while $\omega = E_p = \sqrt{\mathbf{p}^2 + m^2} > 0$, with m the physical mass of the Fermi field. The operators $a_{\mathbf{p}s}$ and $b_{\mathbf{p}s}$ and their Hermitian conjugates are independent of time, while operators such as $a_{\mathbf{p}s}(t)$ are those appearing in (2.94) and (2.95). In a manner analogous to the preceding example the first term in (3.30a) is called the positive-frequency part and the second term the negative-frequency part. The interaction representation expansion (2.95) of the generalized momentum operator $\psi^{\dagger}$ may be procured by Hermitian conjugation of (3.30).

For the photon field in the interaction representation we obtain from (2.167)

$$\mathbf{A}(x) = \mathbf{\Lambda}(x) + \mathbf{\Lambda}^{\dagger}(x) \tag{3.33a}$$

$$\mathbf{\Lambda}(x) = \sum_{\lambda=1,2} \sum_{\mathbf{k}} (2\omega\Omega)^{-1/2} a_{\mathbf{k}\lambda} \hat{\boldsymbol{\epsilon}}_{\mathbf{k}\lambda} e^{-ik \cdot x} \tag{3.33b}$$

$$a_{\mathbf{k}\lambda}(t) = a_{\mathbf{k}\lambda} e^{-i\omega t} \qquad a_{\mathbf{k}\lambda} \equiv a_{\mathbf{k}\lambda}(t = t_0), \tag{3.34}$$

and $\omega = |\mathbf{k}|$. An analogous interaction representation expansion may be written for the generalized momentum of the photon field. These are expansions in terms of the linear polarization vectors (2.162) and (2.163); the corresponding expressions for circularly polarized photons (2.165) are easily constructed.

In (3.33), as well as in (3.30) and (3.28), we observe that the positive-frequency term always contains the destruction operators, while the negative-frequency term contains the creation operators. This suggests an alternative definition of the normally ordered operator introduced in §2.4.4: if the factors in an operator are decomposed into positive- and negative-frequency Fourier components, a normally ordered operator has the positive-frequency components standing to the right of the negative-frequency components in each of its terms. As noted below in the discussion of Wick's theorem, the normal ordering of fermion fields will also entail some factors of -1 if odd numbers of operator permutations are necessary to rearrange in normal order.

3.4 Reduction by Wick's Theorem

The matrix elements of the operator (3.22) contain the physical information that we seek. Let us now consider a systematic procedure for evaluating these matrix elements.

3.4.1 General Structure of the Matrix Elements

From (3.22), (3.27), (2.76), and (2.104) the form of the nth-order contribution to the S-matrix element is

$$\langle \beta | S_n | \alpha \rangle \simeq \int dt_1 \int dt_2 \ldots \int dt_n$$
$$\cdot \langle 0 | \beta_1 \beta_2 \ldots \beta_\ell T \big(V(t_1) V(t_2) \ldots V(t_n) \big) \alpha_1^\dagger \alpha_2^\dagger \ldots \alpha_k^\dagger | 0 \rangle , \quad (3.35)$$

where

$$|\psi_i\rangle = \alpha_1^\dagger \alpha_2^\dagger \ldots \alpha_k^\dagger |0\rangle \qquad |\psi_f\rangle = \beta_1^\dagger \beta_2^\dagger \ldots \beta_\ell^\dagger |0\rangle \quad (3.36)$$

represent initial and final k- and ℓ-particle states generated by successive applications of appropriate creation operators to the vacuum state $|0\rangle$, T is the chronological operator (3.20), and $V(t)$ is the interaction portion of the Hamiltonian in the interaction representation (3.16).

The operator $V(t)$ will contain products of creation and destruction operators for the various fields interacting through each term. For example, we saw in the previous chapter that for spinor electrodynamics the interaction is of the form (2.203)

$$\mathcal{H}_{\text{int}} \simeq \overline{\psi} \gamma_\mu A^\mu \psi = \psi^\dagger \gamma_0 \gamma_\mu A^\mu \psi. \quad (3.37)$$

From (3.30)–(3.34) the operators $\psi^\dagger$ and ψ contain particle creation and destruction operators, while A^μ contains the corresponding operators for photons. Therefore $V(t)$ will involve strings of photon and particle operators that have been chronologically ordered, and the construction of the S-matrix elements (3.35) requires evaluation of products of these operators sandwiched between vacuum states. There are two keys to the systematic calculation of such matrix elements (Exercise 3.1b): (1) the destruction operators annihilate the vacuum state, and (2) the commutation or anticommutation rules allow us to bring creation operators to the left and destruction operators to the right by successive permutations of adjacent operators. Therefore, a systematic method exists to evaluate the matrix elements, but it can be tedious; we should think a little before plunging ahead with this method—perhaps there is a better way. The better way is *Wick's theorem*, which gives a systematic procedure to convert operator products into normally ordered operator products without doing all of the commutation explicitly.

3.4.2 Time-Ordered and Normally Ordered Products

Before unveiling Wick's theorem we need a few definitions. The *time-ordered product* has already been introduced in (3.20); the *normally ordered product* (which will be called the *normal product* for brevity) has been encountered in §2.4.4 for boson operators, but now we give a more general definition. Consider an operator Q, written as a product of creation and annihilation operators:

$$Q = \prod_{i=1}^{n} Q_i.$$

The normal product of Q is designated by the symbol $:Q:$ and corresponds to a permutation of the operator product $\prod Q_i$ such that the creation operators stand to the left and the annihilation operators to the right in each term. The permuted sequence is multiplied by a factor δ_p which is $+1$ unless an odd number of *fermion* operator permutations has been required, in which case it is -1. By virtue of the discussion in §3.3, an operator product is in normal order if in a Fourier decomposition its positive-frequency parts stand to the right of its negative-frequency parts. From the definitions of the time-ordered and normal products we also have the properties

$$T(A+B) = T(A) + T(B) \qquad :A+B: \, = \, :A: + :B: \quad . \tag{3.38}$$

This is all best illustrated by a few examples. First, consider a scalar boson field $\phi(x)$ with $x = (t, \mathbf{x})$:

$$\begin{aligned} T\big(\phi(x_1)\phi(x_2)\big) &= \phi(x_1)\phi(x_2) & (t_1 \geqslant t_2) \\ &= \phi(x_2)\phi(x_1) & (t_1 < t_2) \end{aligned} \tag{3.39}$$

$$\begin{aligned} :\phi(x_1)\phi(x_2): \, = \, &:\big(\alpha(x_1) + \alpha^\dagger(x_1)\big)\big(\alpha(x_2) + \alpha^\dagger(x_2)\big): \\ = \, &\alpha(x_1)\alpha(x_2) + \alpha^\dagger(x_2)\alpha(x_1) \\ &+ \alpha^\dagger(x_1)\alpha(x_2) + \alpha^\dagger(x_1)\alpha^\dagger(x_2), \end{aligned} \tag{3.40}$$

where (3.28) has been used. For a fermion field, the decomposition (3.30) gives

$$\begin{aligned} T\big(\psi_\alpha(x_1)\psi_\beta(x_2)\big) &= \psi_\alpha(x_1)\psi_\beta(x_2) & (t_1 \geqslant t_2) \\ &= -\psi_\beta(x_2)\psi_\alpha(x_1) & (t_1 < t_2) \end{aligned} \tag{3.41}$$

$$\begin{aligned} :\psi_\alpha(x_1)\overline{\psi}_\beta(x_2): \, = \, &:\big(u_\alpha(x_1) + v_\alpha(x_1)\big)\big(\overline{u}_\beta(x_2) + \overline{v}_\beta(x_2)\big): \\ = \, &- \overline{u}_\beta(x_2)u_\alpha(x_1) + u_\alpha(x_1)\overline{v}_\beta(x_2) \\ &+ v_\alpha(x_1)\overline{u}_\beta(x_2) + v_\alpha(x_1)\overline{v}_\beta(x_2), \end{aligned} \tag{3.42}$$

where α and β are bispinor indices, and (3.31)

$$\overline{\psi} = \psi^\dagger \gamma_0 \equiv \overline{u}(x) + \overline{v}(x) = u^\dagger(x)\gamma_0 + v^\dagger(x)\gamma_0.$$

3.4.3 Dyson–Wick Contraction of Operators

The Dyson–Wick *contraction* $\overline{AB}$ between two operators A and B is defined as the difference between their product and their normal product. In particular, for a time-ordered product

$$\overline{A(x_1)B(x_2)} = A(x_1)B(x_2) \equiv T\big(A(x_1)B(x_2)\big) - :A(x_1)B(x_2): \quad . \quad (3.43)$$

For example, consider a real scalar field. The Wick contraction of two field operators at different spacetime points is (see Exercise 3.1)

$$
\begin{aligned}
\overline{\phi(x_1)\phi(x_2)} &= \alpha(x_1)\alpha^\dagger(x_2) - \alpha^\dagger(x_2)\alpha(x_1) \\
&= [\alpha(x_1), \alpha^\dagger(x_2)] \qquad (t_1 \geqslant t_2), \qquad (3.44)
\end{aligned}
$$

$$\overline{\phi(x_1)\phi(x_2)} = [\alpha(x_2), \alpha^\dagger(x_1)] \qquad (t_1 < t_2).$$

As another example, from (3.30)–(3.31) and (3.41)–(3.43) for fermion fields

$$
\begin{aligned}
\overline{\psi_\alpha(x_1)\overline{\psi}_\beta(x_2)} &= -\overline{\overline{\psi}_\beta(x_2)\psi_\alpha(x_1)} \\
&= \begin{cases} \{u_\alpha(x_1), \overline{u}_\beta(x_2)\} & (t_1 \geqslant t_2) \\ -\{\overline{v}_\beta(x_2), v_\alpha(x_1)\} & (t_1 < t_2) \end{cases} \qquad (3.45)
\end{aligned}
$$

$$\overline{\psi_\alpha(x_1)\psi_\beta(x_2)} = \overline{\overline{\psi}_\alpha(x_1)\overline{\psi}_\beta(x_2)} = 0. \qquad (3.46)$$

In contrast to time-ordered products and normal products, which are Hilbert-space operators, *the interaction representation Dyson–Wick contractions are c-numbers*. This is because the free-field commutators or anticommutators are c-numbers, as may be seen by explicit construction of the contractions: utilizing the interaction representation Fourier expansion (3.28) and the equal-time commutator (2.72) in (3.44) we find that

$$
\begin{aligned}
\overline{\phi(x_1)\phi(x_2)} &= [\alpha(x_1), \alpha^\dagger(x_2)] \\
&= \sum_{\mathbf{k}} (2\omega\Omega)^{-1} e^{i[\mathbf{k}\cdot(\mathbf{x}_1 - \mathbf{x}_2) - \omega(t_1 - t_2)]} \qquad (t_1 \geqslant t_2)
\end{aligned}
$$

$$(3.47\text{a})$$

$$\overline{\phi(x_1)\phi(x_2)} = [\alpha(x_2), \alpha^\dagger(x_1)]$$

$$= \sum_{\mathbf{k}} (2\omega\Omega)^{-1} e^{i[\mathbf{k}\cdot(\mathbf{x}_2-\mathbf{x}_1)-\omega(t_2-t_1)]} \qquad (t_1 < t_2). \qquad (3.47b)$$

Since the Hilbert-space operators $a_{\mathbf{k}}$ and $a_{\mathbf{k}}^\dagger$ do not appear in these expressions (in a second-quantized field theory the coordinates x and momenta k are parameters, not Hilbert space operators), we see explicitly that the Wick contraction of the interaction representation field operators for the real scalar field is a c-number. Similar constructions can be made for other contractions, and they will also be c-numbers. Furthermore, we will see below that the momentum-space expansions of the contractions may be used to construct integral representations that are Green's functions for the corresponding wave equations. Thus from a relativistic field theory will emerge the propagators that establish the connection with the Feynman graphs. Indeed, we will find in Exercise 3.2d that the contraction in (3.47) is just the Feynman propagator for a scalar field [see (3.55)].

Utilizing the observation that a Wick contraction is a c-number, the assumed normalization of the vacuum state, eq. (3.43), and the demonstration in §2.4.4 that the vacuum expectation value of a normal product vanishes, we conclude that

$$\langle 0| \overline{AB} |0\rangle = \overline{AB} \langle 0 | 0\rangle = \overline{AB},$$

and that for time-ordered operators

$$\langle 0| \overline{AB} |0\rangle = \langle 0| T(AB) |0\rangle - \langle 0| {:}AB{:} |0\rangle = \langle 0| T(AB) |0\rangle .$$

Therefore, for time-ordered operators the Dyson–Wick contraction of an operator pair is a c-number that may be set equal to the vacuum expectation value for the operator pair:

$$\overline{AB} = \langle 0| T(AB) |0\rangle . \qquad (3.48)$$

An analogous result holds for an ordinary operator product, except that the contraction in that case is equal to the vacuum expectation value of the ordinary product.

3.4.4 Wick's Theorem

We are now ready to state Wick's Theorem: *a product of operators may be written as a sum of normal products, with all possible contractions between pairs of operators (including no contractions) executed.* This applies to an ordinary product and to a time-ordered product. Symbolically, for a time-ordered product

$$T(X_1 X_2 \ldots X_n) =$$

$$:X_1 X_2 \ldots X_n: \qquad \text{(no contractions)}$$

$$+ :\overbrace{X_1 X_2} \ldots X_n: + :\overbrace{X_1 X_2 X_3} \ldots X_n: + \ldots \qquad \text{(one contraction)}$$

$$+ :\overbrace{X_1 X_2} \overbrace{X_3 X_4} \ldots X_n: + \ldots$$

$$+ :\overbrace{X_1 X_2 X_3 X_4} \ldots X_n: + \ldots \qquad \text{(two contractions)}$$

$$+ \text{ higher contractions.} \tag{3.49}$$

For example, suppose $ABCD$ constitutes a product of operators. Then by Wick's theorem we may write

$$ABCD =:ABCD:$$

$$+ :\overbrace{AB}CD: + :\overbrace{ABC}D: + :\overbrace{ABCD}:$$

$$+ :A\overbrace{BC}D: + :A\overbrace{BCD}: + :ABC\overbrace{D}:$$

$$+ :\overbrace{AB}\overbrace{CD}: + :\overbrace{A\overbrace{BC}D}: + :A\overbrace{BCD}: \tag{3.50}$$

Since the contractions are *c*-numbers (vacuum expectation values), they may be brought outside the normal product. In doing so, each term must be multiplied by a factor $\delta_p = (-1)^p$, where p is the number of permutations of *fermion operators* required to bring the contracted operators adjacent to each other in the product. If the operators of (3.50) are fermion operators,

$$ABCD =:ABCD: + \overbrace{AB}:CD:$$

$$- \overbrace{AC}:BD: + \overbrace{AD}:BC: + \overbrace{BC}:AD:$$

$$- \overbrace{BD}:AC: + \overbrace{CD}:AB: + \overbrace{ABCD}$$

$$- \overbrace{AC}\overbrace{BD} + \overbrace{AD}\overbrace{BC}. \tag{3.51}$$

The power of Wick's theorem lies in the observation that the vacuum expectation value of a normal product vanishes. Suppose we calculate $\langle 0| ABCD |0\rangle$ from the preceding example. Only the *completely contracted terms*—those that have all operators paired by contraction—do not contain operators in normal order, so they make the only nonvanishing contribution to the vacuum expectation value. For the illustration (3.51),

$$\langle 0|\,ABCD\,|0\rangle = \overline{AB}\,\overline{CD} - \overline{AC}\,\overline{BD} + \overline{AD}\,\overline{BC}. \qquad (3.52)$$

The importance of this result is obvious when we recall that many-body matrix elements can be expressed in terms of strings of operators acting between vacuum states [see eqns. (3.35)–(3.36)]. Notice that for a contraction to be nonvanishing it must have a prescribed structure: from (3.48) one operator must create, and one must destroy, particles with the same quantum numbers. In particular, a Dyson–Wick contraction of two destruction or two creation operators is immediately seen to vanish.

> **EXERCISE 3.1** (a) Verify eqns. (3.44)–(3.47). (b) Evaluate the matrix element
> $$\mathcal{M} = \langle 0|\, a_i\, a_j\, a_k^\dagger\, a_l^\dagger\, |0\rangle\,,$$
> where $a^\dagger$ and a are time-independent creation and destruction operators obeying the anticommutator $\{a_i, a_j^\dagger\} = \delta_{ij}$, and where all destruction operators annihilate the vacuum, $a_k\,|0\rangle = 0$. Do this first by explicit anticommutation, and then by using Wick's theorem.

3.4.5   Representation of Contractions

Beginning from expressions such as (3.47) for field contractions, we may construct integral representations that can be shown to be propagators (Green's functions) associated with the appropriate wave equations for the fields. Here we reproduce the relevant formulas for several fields of interest (Bjorken and Drell, 1965, §17.5). For time-ordered fermion operators

$$\overline{\psi_\alpha(x)\,\bar\psi_\beta(y)} = \langle 0|\, T\big(\psi_\alpha(x)\bar\psi_\beta(y)\big)\,|0\rangle = iS_{\mathrm{F}}(x - y, m)_{\alpha\beta}, \qquad (3.53)$$

where the Feynman electron–positron propagator $S_{\mathrm{F}}(x - y, m)$ is given by

$$S_{\mathrm{F}}(x - y, m) = \int \frac{d^4k}{(2\pi)^4}\, \frac{e^{-ik\cdot(x-y)}\,(\slashed{k} + m)}{k^2 - m^2 + i\epsilon}. \qquad (3.54)$$

The matrix indices α and β label bispinor components ($\slashed{k} + m$ is a 4×4 matrix; see Exercise 2.8c), m is the mass of the particle, and ϵ is a positive infinitesimal specifying how the poles in (3.54) are to be handled (Exercise 3.2). For a time-ordered real or complex Klein–Gordon field

$$\overline{\phi(x)\,\phi^*(y)} = \langle 0|\, T\big(\phi(x)\phi^*(y)\big)\,|0\rangle = i\Delta_{\mathrm{F}}(x - y, m^2), \qquad (3.55)$$

with a Feynman propagator Δ_F defined by

$$\Delta_F(x - y, m^2) = \int \frac{d^4k}{(2\pi)^4} \frac{e^{-ik \cdot (x-y)}}{k^2 - m^2 + i\epsilon}. \tag{3.56}$$

By contour integration $i\Delta_F$ can be shown to be an integral representation of (3.47a,b).

Although we quantized the electromagnetic field in the Coulomb gauge, it can be shown that the corresponding transverse propagator may be replaced by the simpler *covariant photon propagator* $D_F^{\mu\nu}$, defined by

$$D_F^{\mu\nu}(x - y) = -g^{\mu\nu} \int \frac{d^4k}{(2\pi)^4} \frac{e^{-ik \cdot (x-y)}}{k^2 + i\epsilon} \tag{3.57}$$

in the "Feynman gauge" (see §3.9 and Exercise 3.9c). The proof may be found in Bjorken and Drell (1965), §§12.6, 14.6, 17.9, and Appendix C. Basically, the transverse propagator = covariant propagator + gauge terms + Coulomb term. For physical matrix elements the transverse propagator is always evaluated between conserved currents. This causes the gauge terms to vanish by current conservation. The Coulomb term is canceled by the Coulomb interaction that is independent of the transverse fields (see Sakurai, 1967, Fig. 4–20). Thus the physical matrix elements of the covariant and transverse propagators are equal. In terms of the covariant propagator, we find for chronologically-ordered photon operators that

$$\overline{A^\mu(x)A^\nu}(y) \equiv \langle 0| \, T\big(A^\mu(x)A^\nu(y)\big) \, |0\rangle = iD_F^{\mu\nu}(x - y). \tag{3.58}$$

These propagators satisfy the wave equations

$$(\Box_{x'} + m^2)\Delta_F(x' - x) = -\delta^{(4)}(x' - x) \tag{3.59}$$

$$(i\slashed{\partial}_{x'} - m)S_F(x' - x) = \delta^{(4)}(x' - x) \tag{3.60}$$

$$\Box_{x'} D_F(x' - x) = \delta^{(4)}(x' - x) \tag{3.61}$$

(where $D_F^{\mu\nu} = g^{\mu\nu}D_F$), demonstrating explicitly that Δ_F, S_F, and D_F are, respectively, Green's functions for the free Klein–Gordon, Dirac, and Maxwell fields (see Exercises 3.2 and 3.9). Since these functions define propagation of virtual particles between two points of interaction in spacetime, the graphical representation of Fig. 3.1 is adopted for them.

The operators not paired by a contraction in a term will correspond to real rather than virtual particles. They are represented by the same sort of diagrams as the virtual particles of Fig. 3.1, but with only one end tied to a spacetime point. This is illustrated in Fig. 3.2. Depending on the direction

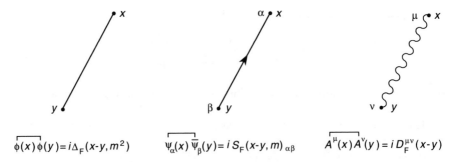

$$\overset{\overline{}}{\phi(x)\phi(y)}=i\Delta_F(x-y,m^2)\qquad \overset{\overline{\phantom{\psi_\alpha(x)\overline{\psi}_\beta(y)}}}{\psi_\alpha(x)\overline{\psi}_\beta(y)}=iS_F(x-y,m)_{\alpha\beta}\qquad \overset{\overline{}}{A^\mu(x)A^\nu(y)}=iD_F^{\mu\nu}(x-y)$$

Fig. 3.1   Graphical representation of contractions.

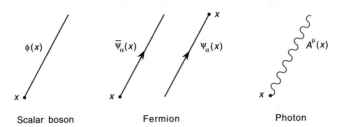

Scalar boson   Fermion   Photon

Fig. 3.2   Graphical representation of uncontracted fields.

of time, these diagrams correspond physically to either the absorption or the emission of particles or antiparticles at x.

Unless otherwise stated, time will be assumed to increase vertically in these diagrams. The endpoints labeled by spacetime coordinates correspond to the arguments of the field operators; they are called *vertices*. Since in the operator products to be considered more than one operator may share the same spacetime argument, the lines corresponding to these operators will meet at common vertices. These will represent points of interaction for local fields, and each vertex may be classified according to the number and kinds of lines tied to it. For example, Fig. 3.3 shows a vertex occurring in quantum electrodynamics (see §3.7). Notice that charged fields require directed lines to distinguish particle from antiparticle, while uncharged fields may be written without arrows on the lines.

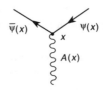

Fig. 3.3   A typical vertex in spinor electrodynamics.

3.4.6 Propagators in Momentum Space

Since the creation and annihilation operators are most conveniently defined in momentum space, and since we usually work with systems having energy and momenta as good quantum numbers, our equations are typically formulated in momentum space. The momentum representations for the propagators may be obtained by Fourier transformation (see Exercise 3.2):

$$\Delta_F(k) = \frac{1}{k^2 - m^2 + i\epsilon} \qquad \text{(Scalar)} \qquad (3.62)$$

$$S_F(k) = \frac{(\slashed{k} + m)}{k^2 - m^2 + i\epsilon} = \frac{1}{\slashed{k} - m + i\epsilon} \qquad \text{(Fermion)} \qquad (3.63)$$

$$D_F^{\mu\nu}(k) = \frac{-g^{\mu\nu}}{k^2 + i\epsilon} \qquad \text{(Massless vector)}, \qquad (3.64)$$

where k denotes the 4-momentum carried by the corresponding internal line. For tree diagrams this will be fixed by 4-momentum conservation at the vertices that terminate the internal line. For loop diagrams there may be additional free loop momenta over which integrations must be performed (§1.8). The propagator for *massive* spin-1 fields, corresponding to the Lagrangian density (2.49), will not enter our initial discussion; it will be given in eq. (6.49) and in Exercise 3.9a.

EXERCISE 3.2 (a) Solve the Dirac equation with gauge invariant electromagnetic coupling using the method of Green's functions. That is, show that a Dirac solution can be written

$$\psi(x) = q \int d^4x' \, G(x - x') \slashed{A} \psi(x')$$

if $G(x - x')$ is a solution of (3.60), and then show that $G(x - x')$ is given by the right side of (3.54). (b) Require that this Green's function propagate positive-energy solutions forward in time and negative-energy solutions backward in time. Show by contour integration of (3.54) in the complex k_0 plane that the prescription of adding the term $i\epsilon$ to the denominator of (3.54) satisfies these causality requirements. *Hint*: consider the following contours of integration

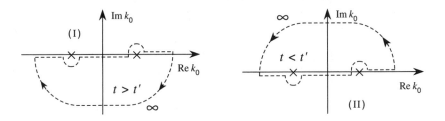

(c) Argue that results (b) and (a) show that the propagation of antiparticles is a direct consequence of wedding relativity to quantum mechanics. Argue that the proper helicity states are included in the positive-energy and negative-energy propagation.   (d) Demonstrate by contour integration of (3.56) that $i\Delta_F(x_1 - x_2)$ is equal to the right side of (3.47a) when $t_1 > t_2$, and the right side of (3.47b) when $t_1 < t_2$; thus establish (3.55). *Hint:* use the contours in part (b).

3.5 Example: Self-Coupled Scalar Field

Let us now illustrate these methods with some examples. We begin with a self-coupled, real spin-0 field, defined by a Hamiltonian density $\mathcal{H} = \mathcal{H}_0 + \mathcal{H}_{int}$ with [see eq. (2.60) and Lee (1981), §5.6]

$$\mathcal{H}_0 = \tfrac{1}{2}:\left(\Pi^2 + (\boldsymbol{\nabla}\phi)^2 + m^2\phi^2\right): \tag{3.65a}$$

$$\mathcal{H}_{int} = -\mathcal{L}_{int} = :\left(-\tfrac{1}{2}\delta m^2\phi^2 + \tfrac{1}{3!}g_0\phi^3 + \tfrac{1}{4!}f_0\phi^4\right): \quad . \tag{3.65b}$$

The parameters f_0 and g_0 are unrenormalized coupling constants, the physical mass m and the parameter m_0 are related by

$$\delta m^2 = m^2 - m_0^2 \tag{3.66}$$

(see the discussion in §6.4), and the factorials are chosen for later convenience. For weak coupling it is assumed that $\mathcal{O}\left(\delta m^2\right) \simeq \mathcal{O}\left(f_0^2\right) \simeq \mathcal{O}\left(g_0^2\right)$.

Such a self-coupled (pseudo)-scalar field might be a useful starting point for a field theory of pions.[‡] Normal ordering has been imposed in the Hamiltonian density to remove the infinite quantities associated with the vacuum state, so the unperturbed energy is measured relative to the vacuum (§2.4.4):

$$H_0 = \sum_{\mathbf{k}} \omega a_{\mathbf{k}}^\dagger a_{\mathbf{k}} \qquad \omega \equiv \omega_{\mathbf{k}} = \sqrt{\mathbf{k}^2 + m^2}. \tag{3.67}$$

3.5.1 $:\phi^4(x):$ Vertices

First we consider only the ϕ^4 term in the interaction (set $\delta m^2 = g_0 = 0$), and look just at the S-operator itself. In the following we will work in the interaction representation, and the meaning of the transformation (3.16) is

---

[‡]Pions are not fundamental—they are composed of quarks and antiquarks. Still, pions considered as discrete entities play an important role in nuclear physics, and the approximation of pions by a self-coupled pseudoscalar field is useful.

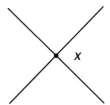

Fig. 3.4   Four-point vertex for $\phi^4(x)$ interaction.

that the field operators $\phi(0, \mathbf{x})$ appearing in the Schrödinger representation of $\mathcal{H}_{\text{int}}$ are to be replaced by *free Heisenberg fields* $\phi(x)$. Therefore, we set

$$V(t) = \int d^3x \, \mathcal{H}_{\text{int}}^{\text{I}} \simeq \int :\phi^4(x): d^3x.$$

The first-order term S_1 in the perturbation expansion (3.22) will lead to a time-ordered operator product $T(:\phi(x)\phi(x)\phi(x)\phi(x):)$. All fields operate at the same spacetime point in this term so the time ordering is immaterial. All contractions vanish because of the normal ordering (3.43), and by Wick's theorem (3.49) only the uncontracted term $\phi^4 = \phi(x)\phi(x)\phi(x)\phi(x)\phi(x)$ contributes in first order. This corresponds to a four-point vertex at the spacetime point x, as illustrated in Fig. 3.4.

Next we consider the second-order contributions S_2, which come from the time-ordered operators $T(:\phi^4(x)::\phi^4(y):)$. In applying Wick's theorem to reduce the time-ordered products only contractions of operators at *different* spacetime points need be considered; contractions between operators at the same points vanish because of the normal ordering and (3.43), as we have just seen for the first-order interaction. Such contractions yield a graphical loop starting and stopping at the same point, and the resulting "tadpole diagrams" (Fig. 3.5) may be omitted for normally ordered interactions. The physical reason is that such a diagram contributes to the structure of the vacuum, and

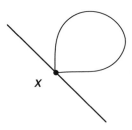

Fig. 3.5   Diagram that does not contribute to normally ordered interactions.

that is discarded by the normal-ordering prescription. Therefore, (3.49) yields

$$T\left(:\phi^4(x)::\phi^4(y):\right) =$$

$$:\phi^4(x)\phi^4(y):$$

$$+ 16\phi(x)\phi(y) :\phi^3(x)\phi^3(y):$$

$$+ 72\left(\phi(x)\phi(y)\right)^2:\phi^2(x)\phi^2(y):$$ (3.68)

$$+ 96\left(\phi(x)\phi(y)\right)^3:\phi(x)\phi(y):$$

$$+ 24\left(\phi(x)\phi(y)\right)^4$$

where the corresponding diagram has been sketched beside each term. Fig. 3.6 illustrates the single contractions that lead to the factor 16 in the second term of (3.68). Notice some important general features: the number of vertices is determined by the order of the interaction; the number of internal lines is determined by the number of contractions in a term; the number of external lines is determined by the number of uncontracted operators in a term; if the interaction Hamiltonian is normally ordered, contractions between field operators at the same spacetime point do not contribute.

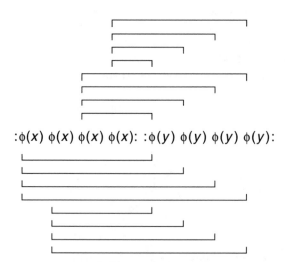

$$:\phi(x)\ \phi(x)\ \phi(x)\ \phi(x):\ :\phi(y)\ \phi(y)\ \phi(y)\ \phi(y):$$

Fig. 3.6   Single contractions contributing to the second term in eq. (3.68).

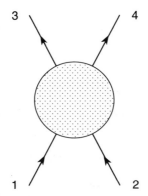

Fig. 3.7 A two-body scattering process $1 + 2 \to 3 + 4$.

3.5.2 Matrix Elements for a $:\phi^4:$ Interaction

Now we evaluate some S-matrix elements for a self-coupled scalar field. Consider the two-body scattering $1 + 2 \to 3 + 4$ shown in Fig. 3.7 (Lee, 1981, Ch. 5). First, we neglect all but the ϕ^4 term in (3.65b) and calculate to order f_0. The interaction is again given by $V \simeq \int :\phi^4(x):d^3x$, and the S-matrix element is, from (3.22) and (3.35)–(3.36),

$$\langle 34| \, S \, |12\rangle = \frac{-if_0}{4!} \int d^4x \, \langle 34| :\phi^4(x): |12\rangle . \tag{3.69}$$

All the fields are evaluated at the same spacetime point so they commute and the chronological operator has been omitted. Letting $a_i^\dagger \equiv a_{k_i}^\dagger$ we have

$$|12\rangle = a_1^\dagger a_2^\dagger |0\rangle \qquad \langle 34| = \langle 0| \, a_3 a_4,$$

and it is clear that only terms from $:\phi^4(x):$ proportional to $a_3^\dagger a_4^\dagger a_1 a_2$ will contribute in eq. (3.69). From the four fields in $:\phi^4(x):$ there are $4! = 24$ ways that we can select an equivalent such combination. Utilizing this and the momentum expansion of the field (3.28), we find in Exercise 3.3 that

$$\langle 34| \, S \, |12\rangle = \frac{-if_0}{4\sqrt{\omega_1\omega_2\omega_3\omega_4}} \frac{1}{\Omega^2} \int d^4x \, e^{i(k_1+k_2-k_3-k_4)\cdot x}, \tag{3.70}$$

where ω and k are the frequency and 4-momenta of the asymptotic particles. The integral is just $(2\pi)^4\delta^{(4)}(k_1 + k_2 - k_3 - k_4)$, so

$$\langle 34| \, S \, |12\rangle = \frac{(2\pi)^4}{\Omega^2}\delta^{(4)}(k_1 + k_2 - k_3 - k_4)\frac{1}{4}(\omega_1\omega_2\omega_3\omega_4)^{-1/2}\mathcal{M}, \tag{3.71}$$

Fig. 3.8 Feynman diagram corresponding to (3.71)–(3.72).

where an amplitude

$$\mathcal{M} \equiv -if_0 \tag{3.72}$$

has been separated from the less interesting factors, just as in (1.153). The corresponding Feynman diagram is shown in Figure 3.8. This is now a *momentum space graph*. The topology of the momentum space graph is the same as that of the corresponding coordinate space graph in Fig. 3.4, but the meaning is different. Figure 3.8 represents a complete matrix element, integrated over the spacetime coordinates.

Thus the Feynman rules giving the correspondence between the matrix element $\mathcal{M}$ and the diagrams for two-body scattering by the $\frac{1}{4!}f_0{:}\phi^4(x){:}$ interaction to first order in f_0 are

- A factor 1 for each external line
- A factor $-if_0$ for each 4-point vertex

EXERCISE 3.3 Evaluate eq. (3.69) using Wick's theorem and the field expansion (3.28). Show that (3.71) and Fig. 3.8 result.

Now we evaluate a second-order diagram for the reaction $1 + 2 \to 3 + 4$ with a ${:}\phi^4(x){:}$ interaction. The S-matrix element will be of the form

$$\langle 34| S_2 |12\rangle = -\frac{f_0^2}{2!4!4!} \int d^4x_1 d^4x_2 \, \langle 34| \, T\big({:}\phi^4(x_1){:}{:}\phi^4(x_2){:}\big) |12\rangle. \tag{3.73}$$

One of the many terms obtained by creating the initial and final states from the vacuum and applying Wick's theorem is shown in Fig. 3.9, along with the corresponding diagram. From (3.55) and (3.56)

$$\overline{\phi(x_1)\phi(x_2)} = \frac{i}{(2\pi)^4} \int \frac{e^{-ik\cdot(x_1-x_2)}}{k^2 - m^2 + i\epsilon} \, d^4k. \tag{3.74}$$

Therefore, this matrix element gives a contribution to (3.73) of

$$\frac{f_0^2}{2! \cdot 4! \cdot 4!} \int \frac{d^4q_1 d^4q_2}{(2\pi)^8} \int d^4x_1 d^4x_2 \frac{e^{-iq_1\cdot(x_1-x_2)}}{q_1^2 - m^2 + i\epsilon}$$

$$\cdot \frac{e^{-iq_2\cdot(x_1-x_2)}}{q_2^2 - m^2 + i\epsilon} \, \langle 0| \, a_3 a_4 {:}\phi(x_1)\phi(x_1){:}{:}\phi(x_2)\phi(x_2){:} a_1^\dagger a_2^\dagger |0\rangle.$$

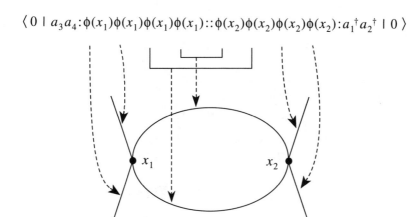

$$\langle 0 \mid a_3 a_4 {:} \phi(x_1)\phi(x_1)\phi(x_1)\phi(x_1){::}\phi(x_2)\phi(x_2)\phi(x_2)\phi(x_2){:} a_1{}^\dagger a_2{}^\dagger \mid 0 \rangle$$

Fig. 3.9   A term contributing to a second-order ϕ^4 interaction.

The remaining string of operators is an operator product to which we may also apply Wick's theorem, and since this is a vacuum expectation value only complete contractions survive. One possibility is

$$\langle 0| \, a_3 a_4 {:}\phi(x_1)\phi(x_1){::}\phi(x_2)\phi(x_2){:} a_1^\dagger a_2^\dagger \, |0\rangle$$
$$= \tfrac{1}{4}\Omega^{-2}(\omega_1\omega_2\omega_3\omega_4)^{-1/2} e^{i(k_3-k_1)\cdot x_1} e^{i(k_4-k_2)\cdot x_2},$$

where (3.48) and (3.28) have been used to carry out contractions such as

$$\phi(x)a_{\mathbf{k}}^\dagger = \langle 0| \, \phi(x)a_{\mathbf{k}}^\dagger \, |0\rangle = (2\omega\Omega)^{-1/2} e^{-ik\cdot x},$$

with $k = (\omega,\mathbf{k})$. This contributes to (3.73) a term

$$\mathfrak{J}_1 = \frac{f_0^2}{2!4!4!} \int d^4 q_1 d^4 q_2$$

$$\cdot \frac{1}{(2\pi)^8} \frac{1}{4\Omega^2} (\omega_1\omega_2\omega_3\omega_4)^{-1/2} \left(\frac{1}{q_1^2 - m^2 + i\epsilon} \right) \left(\frac{1}{q_2^2 - m^2 + i\epsilon} \right)$$

$$\int d^4 x_1 e^{i(k_3-k_1-q_1-q_2)\cdot x_1} \int d^4 x_2 e^{i(k_4-k_2+q_1+q_2)\cdot x_2}. \qquad (3.75)$$

But the integrals over x_1 and x_2 give [see (1.152)]

$$(2\pi)^8 \delta^{(4)}(k_3 - k_1 - q_1 - q_2)\delta^{(4)}(k_4 - k_2 + q_1 + q_2), \qquad (3.76)$$

and the second δ-function implies

$$q_2 = k_2 - k_4 - q_1, \tag{3.77}$$

in terms of which the first δ-function requires

$$k_4 + k_3 - k_2 - k_1 = 0, \tag{3.78}$$

which is overall 4-momentum conservation. Thus, using the condition (3.77) to do the integration over dq_2 and setting $k = q_1$, we finally obtain

$$\mathcal{J}_1(q) = \frac{f_0^2}{2!4!4!4\Omega^2}(\omega_1\omega_2\omega_3\omega_4)^{-1/2}\delta^{(4)}(k_4 + k_3 - k_2 - k_1)$$

$$\int d^4k \left(\frac{1}{k^2 - m^2 + i\epsilon}\right)\left(\frac{1}{(k-q)^2 - m^2 + i\epsilon}\right), \tag{3.79}$$

where $q = k_3 - k_1$. This is only one contraction; by considering the other possibilities we find that there are three independent second-order contributions $\mathcal{J}(q)$, as illustrated in Fig. 3.10 and given by (3.79) with $q = k_3 - k_1$, $q = k_3 - k_2$, and $q = k_3 + k_4$, respectively. These independent contributions are distinguished by different topologies in the *momentum space graphs* of Fig. 3.10, and correspond to the independent ways that the 4-momentum can be conserved at each vertex [the two δ-functions resulting from the integrals over spacetime (3.75)]. Since these graphs are in momentum space there is no time direction (we have integrated over spacetime). Therefore, the vertices do not have coordinate labels, but the lines are labeled by momenta. In Fig. 3.10 we have placed the entrance state momenta at the bottom and exit-state momenta at the top, with entrance-state momenta flowing into the vertices, exit state momenta out of the vertices, and internal momenta as indicated. Any arrangement of internal momenta consistent with the δ-functions in (3.75) would be equally satisfactory.

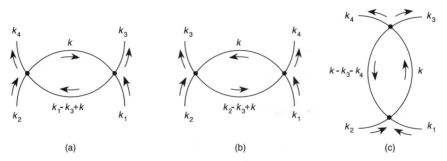

(a)   (b)   (c)

Fig. 3.10   Momentum space diagrams contributing to (3.80). The assumed direction of momentum flow is indicated by the arrows.

Other possible contractions give the same results as cases (a), (b), or (c). Therefore, the total contribution in second order becomes a combinatorial problem of determining how many different ways the three possibilities in Fig. 3.10 can be realized. The general rules are given in the solution to Exercise 3.4b; for the case at hand there are $(4!)^2$ contributions from each possibility, and the matrix element is given by (3.71) with

$$\mathcal{M} = f_0^2 \left[I(k_3 - k_1) + I(k_3 - k_2) + I(k_3 + k_4) \right] \tag{3.80}$$

$$I(q) \equiv \frac{1}{2} \int \frac{d^4k}{(2\pi)^4} \left(\frac{1}{k^2 - m^2 + i\epsilon} \right) \left(\frac{1}{(k-q)^2 - m^2 + i\epsilon} \right). \tag{3.81}$$

Comparison with the previous Feynman rules for first-order diagrams indicates that the diagrams in Fig. 3.10 correspond to eq. (3.80) provided our rules are expanded to include

- A factor $i/(q^2 - m^2 + i\epsilon)$ for an internal boson line with 4-momentum q
- A factor $\int d^4k /(2\pi)^4$ for each internal (boson loop) momentum k not fixed by total 4-momentum conservation

Before continuing we pause to note that (3.81) does not converge. It possesses an *ultraviolet divergence* (that is, a divergence at high energy or high k for the internal lines); as $k \to \infty$ the other terms in the denominator of the integral may be ignored and the integral grows logarithmically with the upper limit:

$$\mathcal{M} \to \int_0^\Lambda \frac{d^4k}{k^4} \simeq \log \Lambda.$$

The presence of such divergent integrals will be typical of diagrams containing loops in quantum field theory (see §1.8). We defer further consideration of the renormalization of these divergences until Ch. 6, but comment here that a self-interacting scalar field with terms no higher than fourth order belongs to that limited class of theories in which the infinities can be systematically eliminated from observable quantities.[‡] Such theories are called *renormalizable*.

3.5.3 Contributions from a Mass Counterterm

Now let us briefly illustrate the effect of the mass counterterm in (3.65b). For simplicity we set $g_0 = 0$ and consider an interaction

$$V = \int d^3x : -\tfrac{1}{2} \delta m^2 \phi^2(x) + \tfrac{1}{4!} f_0 \phi^4(x): \quad .$$

[‡]The renormalization of (3.80) by dimensional regularization (see §6.4.5) is discussed succinctly in Pokorski (1987), §4.3.

Suppose we examine a third-order interaction in (3.22); from the resulting terms there will be one first-order in the mass correction and second-order in f_0 of the form

$$T\big(:\phi^4(x_1)::\phi^4(x_2)::\phi^2(x_3):\big).$$

A possible contraction is displayed below, along with the corresponding diagram:

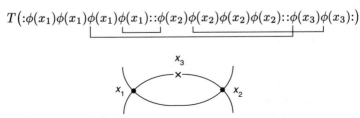

In this graph a cross is used to denote a mass counterterm vertex.

If the contributions to the S-matrix elements for terms such as these are evaluated along the same lines as the previous examples, the Feynman rule associated with the mass counterterm vertices is found to be

- A factor $i\delta m^2$ for each boson mass counterterm   —×—

With our assumptions such mass counterterms may be ignored to lowest order, but they are important in higher order calculations because of the renormalization between bare and physical masses (see §6.4.3).

3.5.4 Matrix Elements for a $:\phi^3(x):$ Interaction

We now consider a second-order contribution from the term $\frac{1}{3!}g_0:\phi^3(x):$ in the Hamiltonian density (3.65). This example is worked in Exercise 3.4; to order g_0^2, with $\delta m^2 = f_0 = 0$ in (3.65),

$$\langle 34|\,S_2\,|12\rangle = \frac{-g_0^2}{2!3!3!}\int d^4x_1\,d^4x_2\,\langle 34|\,T\big(:\phi^3(x_1)::\phi^3(x_2):\big)\,|12\rangle.$$

Using Wick's theorem as before, the invariant amplitude is given by

$$\mathcal{M} = -ig_0^2\left[\frac{1}{(k_1+k_2)^2 - m^2 + i\epsilon} + \frac{1}{(k_3-k_1)^2 - m^2 + i\epsilon}\right.$$
$$\left. + \frac{1}{(k_4-k_1)^2 - m^2 + i\epsilon}\right].\quad (3.82)$$

The corresponding momentum space diagrams are shown in Fig. 3.11, where arrows indicate the assumed direction of momentum flow. The internal momentum direction is a matter of convention since (3.82) is unaltered by re-

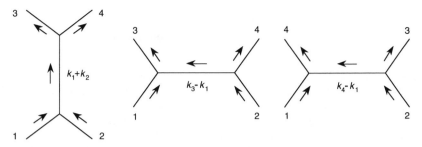

Fig. 3.11 Diagrams contributing in order g_0^2 for the reaction $1 + 2 \rightarrow 3 + 4$.

versing the sign of the transferred momentum. From this result we deduce a Feynman rule for three-point vertices:

- A factor $-ig_0$ for each three-point boson vertex

3.5.5 Feynman Rules for Scalar Fields

Collecting the preceding results, we may summarize the Feynman rules for a self-coupled, real spin-0 field defined by the Hamiltonian density (3.65): draw all topologically distinct momentum space diagrams and assign

1. A factor 1 for each external boson line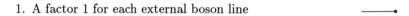

2. A factor $-ig_0$ for each 3-boson vertex

3. A factor $-if_0$ for each 4-boson vertex

4. A factor $i/(q^2 - m^2 + i\epsilon)$ for each internal boson line

5. A factor $i\delta m^2$ for each mass counterterm vertex

6. A factor $\int d^4k/(2\pi)^4$ for each internal (boson loop) momentum not fixed by 4-momentum conservation
   (Since these integrals typically diverge they have meaning only when a suitable renormalization scheme has been implemented.)

7. A factor $1/\nu!$, where ν is the number of equivalent internal loop lines once the vertices are fixed ($\nu = 2!$ for the diagrams in Fig. 3.10; $\nu = 3!$ for)

As illustrated in the preceding examples, this set of rules gives a one-to-one correspondence between the amplitude $\mathcal{M}$ and the diagrams. The connection of the resulting matrix elements with cross sections will be given in the following section. An efficient recipe for reading such Feynman rules directly from the Lagrangian density is illustrated in Exercises 3.9 and 3.10.

EXERCISE 3.4 (a) Derive the result (3.82) for a second-order ϕ^3 interaction. Show that the scalar field Feynman rules give the same result. (b) Find the general rules for the combinatorial weight of each graph in scalar field theory for $1 + 2 \rightarrow 3 + 4$ processes. Calculate the factor for a second-order ϕ^3, second-order ϕ^4, and first-order ϕ^4 interaction. (c) Derive the expression (3.83a) giving the laboratory-frame differential cross section for two-body boson scattering in terms of the invariant amplitude $\mathcal{M}$. *Hint:* Use a normalization of the plane wave states such that there are $2E$ particles in a volume Ω. Calculate a transition rate per unit volume, and define the cross section as

$$d\sigma = \left(\frac{\text{rate}}{\text{unit volume}} \right) \times \left(\frac{\text{number of final states}}{\text{initial particle flux} \times \text{target particle density}} \right).$$

3.6 Differential Cross Sections

In the preceding examples we have seen how to construct amplitudes for various processes. In this section we collect the formulas relating probability amplitudes to physically observable cross sections. For a two-body initial state and n-body final state of scalar particles $1 + 2 \rightarrow 1' + 2' + \ldots n'$, the differential cross section in the rest frame of particle 2 is [see Exercises 3.4c and 3.5a,b and Lee (1981), §5.7]

$$d\sigma_{\text{Lab}} = \frac{2\pi}{|\mathbf{v}_1|} \left(\frac{1}{8\pi^3} \right)^{n-1} (4\omega_1 \omega_2)^{-1} |\mathcal{M}|^2 \delta^{(4)} \left(\sum_{i=1}^{n} p_i' - p_1 - p_2 \right) \prod_{i=1}^{n} \frac{d^3 p_i'}{2\omega_i'}, \tag{3.83a}$$

which is to be integrated over undetected final momenta $p_1', p_2', \ldots$. All momenta and the velocity $\mathbf{v}_1$ of particle 1 are evaluated in the rest frame of particle 2. In the center-of-mass (CM) system the differential cross section takes the corresponding form

$$d\sigma_{\text{CM}} = \frac{2\pi}{|\mathbf{v}_1 - \mathbf{v}_2|} \left(\frac{1}{8\pi^3} \right)^{n-1} 2^{-n-2}$$

$$\cdot (\omega_1' \omega_2' \ldots \omega_n' \omega_1 \omega_2)^{-1} |\mathcal{M}|^2 \delta^{(4)} \left(\sum_{i=1}^{n} p_i' - p_1 - p_2 \right) \prod_{i=1}^{n} d^3 p_i'. \tag{3.83b}$$

In these equations the amplitude $\mathcal{M}$ is related to the S-matrix element by

$$\langle 1'2' \ldots n' | S | 12 \rangle = \frac{(2\pi)^4}{\Omega^{1+n/2}} \delta^{(4)} \left(\sum_{i=1}^{n} p_i' - p_1 - p_2 \right)$$

$$(\sqrt{2})^{-n-2} (\omega_1' \omega_2' \ldots \omega_n' \omega_1 \omega_2)^{-1/2} \mathcal{M}. \tag{3.84}$$

Similar formulas may be given to relate the amplitudes $\mathcal{M}$ to rates or lifetimes for decay processes. These equations are valid for scalar particles and photons. The corresponding expressions for fermions with the normalization (1.125c) are obtained by making the replacements $1/2E_i \to m_i/E_i$ for each fermion in (3.83a,b) and (3.84).

3.7 Example: Spinor Electrodynamics

From (2.202) and (2.203) we identify the fundamental interaction term for the quantum electrodynamics (QED) of fermions:

$$V = e\bar{\psi}A\!\!\!/\psi = e\bar{\psi}\gamma_\mu A^\mu \psi. \tag{3.85}$$

The corresponding vertex is shown in Fig. 3.3.

3.7.1 Some Diagrams in QED

The specific processes that such vertices describe may be found by expanding the interaction term in frequency components. Using (3.30)–(3.33) we have

$$V = e\!:\!(\bar{u}+\bar{v})\gamma^\mu(\Lambda_\mu^\dagger + \Lambda_\mu)(u+v)\!:\quad . \tag{3.86}$$

The vertices arising from this interaction to order n may be constructed by applying Wick's theorem to a product of n operators

$$T\big(V(x_1)V(x_2)V(x_3)\ldots V(x_n)\big).$$

In first order the uncontracted terms of (3.86) yield the diagrams in Fig. 3.12. In these diagrams a solid line directed upward corresponds to a positive-energy particle and a solid line directed downward corresponds to a positive-energy antiparticle; wiggly lines denote photons. The diagrams for fundamental processes in QED may be built up by combining the vertices in Fig. 3.12 in various ways.

3.7.2 Second-Order $e^- e^-$ Scattering

Let us calculate a two-electron scattering process (*Møller scattering*) in second order (in the interaction representation, as usual):

$$\langle 34| S_2 |12\rangle = \frac{i^2}{2}\int d^4x\,d^4y\,\langle 34|\,T\,(:V(x)::V(y):)\,|12\rangle\,. \tag{3.87}$$

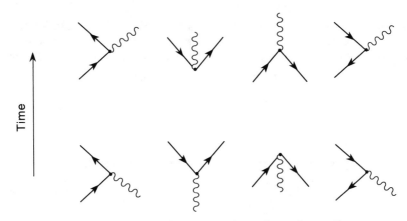

Fig. 3.12   First-order spinor electrodynamics vertices.

Let

$$|12\rangle = a^\dagger_{\mathbf{k}_1 s_1} a^\dagger_{\mathbf{k}_2 s_2} |0\rangle \equiv a^\dagger_1 a^\dagger_2 |0\rangle \qquad (3.88)$$

$$|34\rangle = a^\dagger_{\mathbf{k}_3 s_3} a^\dagger_{\mathbf{k}_4 s_4} |0\rangle \equiv a^\dagger_3 a^\dagger_4 |0\rangle \qquad (3.89)$$

$$V(x) = e{:}\overline{\psi}(x) A\!\!\!/(x)\psi(x){:}\;. \qquad (3.90)$$

Then

$$\langle 34| S_2 |12\rangle = \frac{i^2 e^2}{2} \int d^4x\, d^4y$$
$$\cdot \langle 0| a_4 a_3 T\big({:}\overline{\psi}(x) A\!\!\!/(x)\psi(x){:}{:}\overline{\psi}(y) A\!\!\!/(y)\psi(y){:}\big) a^\dagger_1 a^\dagger_2 |0\rangle. \qquad (3.91)$$

Feynman Diagrams. Proceeding as in eqns. (3.73) and the discussion that followed, only the complete contractions survive in the vacuum expectation value. Schematically, the independent nonvanishing possibilities are

$$\langle 34| S_2 |12\rangle = \frac{i^2 e^2}{2} \int d^4x\, d^4y$$
$$\cdot \Bigg[\langle 0| a_4 a_3 {:}\overline{\psi}(x) A\!\!\!/(x)\psi(x){:}{:}\overline{\psi}(y) A\!\!\!/(y)\psi(y){:} a^\dagger_1 a^\dagger_2 |0\rangle$$
$$+ \langle 0| a_4 a_3 {:}\overline{\psi}(x) A\!\!\!/(x)\psi(x){:}{:}\overline{\psi}(y) A\!\!\!/(y)\psi(y){:} a^\dagger_1 a^\dagger_2 |0\rangle \Bigg] \qquad (3.92)$$

which have graphs of the form shown in Fig. 3.13. Let us evaluate the first term. The operators A^μ commute with the fermion operators and with the γ-matrices since they represent independent degrees of freedom. However, care must be taken with the ordering of the γ-matrices and fermion bispinor factors. Carrying out the contractions and noting that an even number of permutations is required to bring the contracted fermion operators adjacent to each other [giving an overall phase $\delta_p = (-1)^p = 1$], we obtain

$$\langle 34| S_2 |12 \rangle_1 = \frac{i^2 e^2}{2\Omega^2} \frac{m^2}{\sqrt{E_1 E_2 E_3 E_4}} \int \frac{d^4 k}{(2\pi)^4} \frac{-ig^{\mu\nu}}{k^2 + i\epsilon}$$

$$\cdot (\overline{u}_{\mathbf{k}_4 s_4} \gamma_\mu u_{\mathbf{k}_2 s_2})(\overline{u}_{\mathbf{k}_3 s_3} \gamma_\nu u_{\mathbf{k}_1 s_1}) \int d^4 x\, d^4 y\, e^{i(k_4 - k_2 - k)\cdot x} e^{i(k_3 - k_1 + k)\cdot y}. \quad (3.93)$$

The integrals over x and y yield a factor $(2\pi)^8 \delta^{(4)}(k_4 - k_2 - k)\delta^{(4)}(k_3 - k_1 + k)$. Using the first δ-function to perform the momentum integration gives

$$\langle 34| S_2 |12 \rangle_1 = \frac{(2\pi)^4 e^2}{2\Omega^2} \frac{m^2}{\sqrt{E_1 E_2 E_3 E_4}} \frac{-i}{(k_4 - k_2)^2 + i\epsilon}$$

$$\cdot (\overline{u}_{\mathbf{k}_4 s_4} i\gamma_\mu u_{\mathbf{k}_2 s_2})(\overline{u}_{\mathbf{k}_3 s_3} i\gamma^\mu u_{\mathbf{k}_1 s_1}) \delta^{(4)}(k_4 + k_3 - k_2 - k_1), \quad (3.94)$$

where $\gamma^\mu = g^{\mu\nu}\gamma_\nu$ has been used, and the particular grouping of the i's with the γ-matrices is for consistency with the standard Feynman rules for spinor electrodynamics that will be listed shortly. Now we consider the exchange term (b) in Fig. 3.13. By analogy with the first term we must evaluate

$$\langle 34| S_2 |12 \rangle_2 = \frac{i^2 e^2}{2} \int d^4 x\, d^4 y$$

$$\cdot \langle 0| a_4 a_3 \overline{u}(x)\gamma_\mu u(x)\overline{u}(y)\gamma_\nu u(y) a_1^\dagger a_2^\dagger A^\mu(x) A^\nu(y) |0\rangle. \quad (3.95)$$

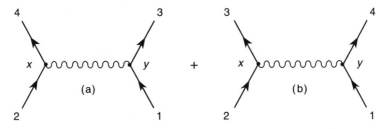

Fig. 3.13 Diagrams arising from the contractions in (3.92).

This will be the same as the first term if we switch the index 3 with the index 4 except for one important difference: in this case an *odd* number of permutations is required to bring the contracted fermion operators adjacent to each other. Thus the overall phase of this term will be $\delta_p = -1$, which is opposite that of the first graph. This difference in sign between the direct and exchange terms is a consequence of the required antisymmetry of the two-electron wavefunction. The permutation phase $\delta_p = (-1)^p$ for fermion operators appearing in connection with Wick's theorem stems from the anticommutators obeyed by the fermion operators; it is there to give the right quantum statistics. Compare this with eq. (1.157), where for first-order identical boson scattering the direct and exchange terms contribute with the same sign. Therefore, the total second-order S-matrix element is obtained by adding to (3.94) a term with the opposite sign and with $3 \leftrightarrow 4$ to account for the exchange graph (b) in Fig. 3.13, and multiplying by two to account for the contributions of the graphs obtained by permuting x and y in the diagrams:[‡]

$$\langle 34| S_2 |12 \rangle = \frac{(2\pi)^4}{\Omega^2} \frac{m^2}{\sqrt{E_1 E_2 E_3 E_4}} \delta^{(4)}(k_4 + k_3 - k_2 - k_1) \mathcal{M} \qquad (3.96)$$

$$\mathcal{M} = -ie^2 \left[\frac{\overline{u}_{k_4 s_4}(-i\gamma_\mu) u_{k_2 s_2} \overline{u}_{k_3 s_3}(-i\gamma^\mu) u_{k_1 s_1}}{(k_4 - k_2)^2 + i\epsilon} \right. $$
$$\left. - (\text{same term but } 3 \leftrightarrow 4) \right]. \qquad (3.97)$$

Now we sketch the path from the amplitude (3.97) to the differential cross section. From eqns. (3.83) the quantity

$$|\mathcal{M}|^2 = \mathcal{M}^\dagger \mathcal{M} \qquad (3.98)$$

is the essential ingredient. The calculation of (3.98) could become quite tedious because of the γ-matrices in (3.97), but there exists some mathematical machinery that can simplify this task (Feynman, 1961)

Trace Techniques and the Lepton Tensor. In many experiments polarization states of particles are not observed and expressions for the cross sections must be summed over final polarizations and averaged over initial

---

[‡]An nth-order graph has n vertices so there are $n!$ coordinate space graphs of a particular topology that differ only in the labeling of the vertices. Since the vertex coordinates will be integrated the labels on the vertices are dummy labels and we may evaluate one graph of each topology and multiply it by $n!$, which exactly cancels the $1/n!$ factor associated with the nth-order term in the expansion (3.22) (see also Exercise 3.4b).

polarizations:

$$|\mathcal{M}|^2 \to \overline{|\mathcal{M}|^2} \equiv \frac{1}{(2s_1 + 1)(2s_2 + 1)} \sum_{\text{spins}} |\mathcal{M}|^2. \tag{3.99}$$

Notice that this is an *incoherent sum*: the squares of amplitudes rather than the amplitudes have been added, and different polarization terms do not interfere. For unpolarized cross sections involving Dirac particles this will require evaluation of expressions such as

$$L_{\mu\nu} = \tfrac{1}{2} \sum_{ss'} \bar{u}_{\mathbf{k}'s'} \gamma_\mu u_{\mathbf{k}s} \bar{u}_{\mathbf{k}s} \gamma_\nu u_{\mathbf{k}'s'}. \tag{3.100}$$

This quantity, which is called the *lepton tensor*, is ubiquitous in fermion problems. In this expression recall that u is a 4-component column matrix, γ is a 4×4 matrix, and $\bar{u}$ is a 4-component row matrix. By attaching explicit matrix indices and using projection operators for the bispinors (Exercise 2.8c), the lepton tensor can be expressed as a matrix product summed over diagonal elements; that is, as a trace of a product of four matrices (Exercise 3.5c):

$$L_{\mu\nu} = \frac{1}{8m^2} \text{Tr} \left[(\rlap{/}{k}' + m)\gamma_\mu (\rlap{/}{k} + m)\gamma_\nu \right]. \tag{3.101}$$

Because the Dirac γ matrices obey the anticommutator (1.86) they have particularly simple trace properties, as have been summarized in Table 1.2. Using these properties the lepton tensor may be evaluated, with the result (Exercise 3.5c)

$$L_{\mu\nu} = \frac{1}{2m^2} \left(k'_\mu k_\nu + k'_\nu k_\mu + \tfrac{1}{2} q^2 g_{\mu\nu} \right), \tag{3.102}$$

where $q^2 = (k - k')^2$. Unpolarized fermion–fermion cross sections will typically involve the contraction of two tensors like (3.102) in their Lorentz indices:

$$\overline{|\mathcal{M}|^2} \simeq L^{\mu\nu} L'_{\mu\nu}.$$

Inserting (3.97) into (3.98), employing trace techniques of the sort just outlined to reduce the unpolarized matrix products, assuming the ultrarelativistic limit ($E \gg m$), and utilizing (3.83b) with appropriate care for fermion normalization, we obtain the second order center-of-mass cross section for unpolarized $e^- e^-$ scattering:

$$\left(\frac{d\sigma}{d\Omega} \right) = \frac{\alpha^2}{8E^2} \left(\frac{1 + \cos^4(\theta/2)}{\sin^4(\theta/2)} + \frac{2}{\sin^2(\theta/2)\cos^2(\theta/2)} + \frac{1 + \sin^4(\theta/2)}{\cos^4(\theta/2)} \right) \tag{3.103}$$

[see Bjorken and Drell (1964), §7.9, and Exercises 3.5–3.7]. In this expression E is the CM energy, θ is the CM scattering angle, and $\alpha = e^2/4\pi$ is the fine structure constant.

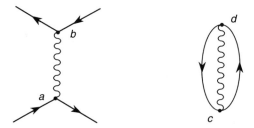

Fig. 3.14   A disconnected graph; these may generally be ignored.

3.8 Graphs That Are Excluded

In constructing amplitudes for the several examples investigated so far, tadpole graphs (Fig. 3.5) have been excluded by the normal ordering of the interaction. There is a second class of graphs that we may also omit from our considerations. A *disconnected graph* is one in which one or more isolated parts appear that are not connected by lines; an example is shown in Fig. 3.14. In this process a pair annihilates at *a* and the virtual photon propagates to *b* and creates another pair; meanwhile the vacuum undergoes the independent fluctuation shown in the right side of the figure. Since no lines connect the left and right diagrams these processes are uncorrelated, and the overall probability is the product of probabilities for the disjoint pieces. But the bubble in the right-hand portion of the graph contributes to the structure of the physical vacuum, so one finds that such disconnected graphs just provide a multiplicative factor for all the connected graph processes (see Ziman, 1969, §3.9). Hence, for the methods discussed in this chapter only *connected graphs* need be considered in evaluating amplitudes for transitions between particle or antiparticle states.

3.9 Feynman Rules for Electrodynamics

By considering many electrodynamic processes using the methods that have just been discussed, we conclude that there are rules of correspondence (Feynman rules) linking the diagrams to the invariant amplitudes, just as we found previously for the scalar field. Here we list the rules for electrodynamics (Bjorken and Drell, 1965, Appendix B). The reader may verify that they reproduce the matrix elements obtained from the application of Wick's theorem in quantum field theory.

1. Draw all momentum space diagrams consistent with the interaction that are topologically distinct. For momentum-space graphs the *lines* (not vertices) are labeled; equivalent graphs can be rotated or twisted into

each other while keeping the direction of external lines fixed. Exclude diagrams with disconnected vacuum bubbles and with lines beginning and ending on the same point. The ordering of factors in the matrix element is determined by following the arrows on the lines in the diagram.

2. Rules for external lines:

   A. A factor 1 for each external boson or antiboson

   B. A factor u_{ps} for each incoming (annihilated) fermion

   A factor v_{ps} for each outgoing (created) antifermion

   A factor $\bar{u}_{ps}$ for each outgoing (created) fermion

   A factor $\bar{v}_{ps}$ for each incoming (annihilated) antifermion

   C. A polarization factor ϵ_μ for each incoming photon

   A polarization factor ϵ_μ^* for each outgoing photon

3. Rules for internal lines (propagators for virtual particles):

   A. For each internal fermion with momentum p and mass m a factor

   $$= iS_F(p) = \frac{i(\not{p} + m)}{p^2 - m^2 + i\epsilon} = \frac{i}{\not{p} - m + i\epsilon}$$

   B. For each internal spin-0 particle of momentum k and mass m a factor

   $$= i\Delta_F(k) = \frac{i}{k^2 - m^2 + i\epsilon}$$

   C. For each internal photon with momentum q a factor

   $$= iD_F^{\mu\nu}(q) = \frac{i}{q^2 + i\epsilon}\left[-g^{\mu\nu} + (1 - \xi)\frac{q^\mu q^\nu}{q^2 + i\epsilon}\right]$$

   where ξ determines the gauge for the covariant propagator. Common terminology: $\xi = 1$ for "Feynman gauge", $\xi = 0$ for "Landau gauge". Most calculations are done in Feynman gauge where

   $$= iD_F^{\mu\nu}(q) = \frac{-ig^{\mu\nu}}{q^2 + i\epsilon} \qquad \text{(Feynman gauge)}.$$

   Physical results are independent of gauge.

4. Rules for loops and signs:

   A. For each internal loop momentum k not fixed by total momentum conservation a factor $\int d^4k/(2\pi)^4$

B. For each closed *fermion loop* a factor -1, and take the trace of γ-matrices

C. A factor -1 between graphs differing only by exchange of two identical *external fermion lines*; this includes exchange of initial particle and final antiparticle

D. A factor $\frac{1}{2}$ for each closed loop containing only 2 photon lines

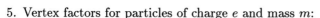

5. Vertex factors for particles of charge e and mass m:

   A. Spinor electrodynamics:

   i. $\qquad = -ie\gamma_\mu$ 	 ii. $\qquad = i\delta m$

   B. Spin-0 boson electrodynamics:

   i. $\qquad = -ie(p + p')_\mu$

   ii. $\qquad = i\delta m^2$ 	 iii. $\qquad = 2ie^2 g_{\mu\nu}$

In the application of these Feynman rules it is understood that 4-momentum is conserved at each vertex, and therefore is conserved in the entire diagram. The loop momentum integration over $\int d^4k$ of rule 4.A will generally be divergent. Therefore, for calculations beyond tree level these rules are meaningless until supplemented by a prescription for the removal of divergences in physical observables (see §1.8 and Ch. 6).

The origin of the trace in rule 4.B may be appreciated by writing out explicit spinor indices in a typical closed-loop matrix element (see also the solution of Exercise 3.5c); the physical meaning of the trace is a sum over all polarizations of the virtual fermions in the loop. The factors of -1 in rules 4.B and 4.C are required by Fermi–Dirac statistics. Finally, we note that a closed loop of fermion lines with an *odd* number of vertices vanishes identically because of the charge conjugation invariance of electrodynamics (*Furry's theorem*).

The Feynman rules for a self-interacting spin-0 field have already been given in §3.5.5. Feynman rules for other processes will be introduced as they are needed in the subsequent discussion. A recipe that allows Feynman rules to be read directly from a Lagrangian density is discussed in Exercises 3.9 and 3.10. Generally, these rules will provide a rapid and systematic way to make low-order calculations.

As an illustration of the Feynman rules for spinor electrodynamics, let us

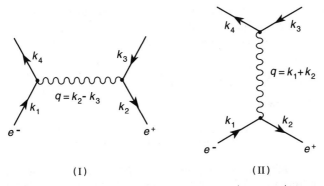

Fig. 3.15   Diagrams contributing to $e^+e^- \rightarrow e^+e^-$.

calculate the lowest order amplitude for electron–positron scattering (*Bhabha scattering*). From the momentum space vertices analogous to Fig. 3.12 we may construct two independent second-order contributions to the e^+e^- amplitudes, as displayed in Fig. 3.15. Applying the rules 2.B, 3.C, and 5.A, and the rule 4.C for the relative sign between graphs, we find

$$
\mathcal{M} = \bar{v}_{\mathbf{k}_2 s_2}(-ie\gamma_\mu)v_{\mathbf{k}_3 s_3}\frac{i}{(k_2 - k_3)^2}\bar{u}_{\mathbf{k}_4 s_4}(-ie\gamma^\mu)u_{\mathbf{k}_1 s_1}
$$
$$
- \bar{v}_{\mathbf{k}_2 s_2}(-ie\gamma_\mu)u_{\mathbf{k}_1 s_1}\frac{i}{(k_1 + k_2)^2}\bar{u}_{\mathbf{k}_4 s_4}(-ie\gamma^\mu)v_{\mathbf{k}_3 s_3}
$$

$$(3.104)$$

[eq. (3.104) can also be procured from (3.97) by crossing symmetry—see Exercise 3.8 and §1.7.3]. The interested reader may verify that this is the same result obtained from Wick's theorem (Exercise 3.6), and may calculate a differential cross section using the techniques outlined in §3.7.2 and Exercise 3.5. The final result for the center-of-mass system is (Exercise 3.8 and Bjorken and Drell, 1964, §7.9)

$$
\left(\frac{d\sigma}{d\Omega}\right) = \frac{\alpha^2}{8E^2}\left(\frac{1 + \cos^4(\theta/2)}{\sin^4(\theta/2)} - \frac{2\cos^4(\theta/2)}{\sin^2(\theta/2)} + \frac{1 + \cos^2\theta}{2}\right), \qquad (3.105)
$$

where the leptonic mass has been ignored.

EXERCISE 3.5   (a) Demonstrate that for two-body final states in the reaction $1 + 2 \rightarrow 3 + 4$, the cross section (3.83) can be written

$$
d\sigma = \frac{d\text{Lips}}{F}|\mathcal{M}|^2,
$$

where F is the incident flux factor defined in the solution to Exercise 3.4c, the "Lorentz invariant phase space" (Lips) is

$$
d\text{Lips} = \frac{1}{(4\pi)^2}\frac{p_3 dE_3}{E_4}\delta(-E_3 - E_4 + E_1 + E_2)d\Omega,
$$

and where $p_3 = |\mathbf{p_3}|$ and $d\Omega$ is the solid angle into which particle 3 scatters. *Hint*: use the $\delta^{(4)}$ function to eliminate one of the d^3p integrations in (3.83).
(b) For the center-of-mass frame, show that

$$d\text{Lips} = \frac{1}{(4\pi)^2} \frac{p}{E_1 + E_2} d\Omega \qquad F = 4p(E_1 + E_2)$$

where $p = |\mathbf{p_1}|$, so that

$$\left(\frac{d\sigma}{d\Omega}\right)_{\text{CM}} = \frac{|\mathcal{M}|^2}{[8\pi(E_1 + E_2)]^2}.$$

(c) Supply the missing steps between (3.100) and (3.102) in the evaluation of the lepton tensor. (d) Use the Feynman rules to evaluate the lowest order (unpolarized) $e^- \mu^- \to e^- \mu^-$ scattering:

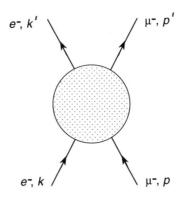

Show that in terms of the *Mandelstam variables*

$$s \equiv (k + p)^2 \qquad t \equiv (k - k')^2 \qquad u \equiv (k - p')^2,$$

the spin-averaged, squared invariant amplitude is

$$\overline{|\mathcal{M}(e^- \mu^- \to e^- \mu^-)|^2} = \frac{e^4}{4m_1^2 m_2^2} \left(\frac{s^2 + u^2}{t^2}\right),$$

where all masses have been neglected. (e) Use crossing symmetry to show that for the reaction $e^+ e^- \to \mu^+ \mu^-$ this implies

$$\overline{|\mathcal{M}(e^+ e^- \to \mu^+ \mu^-)|^2} = \frac{e^4}{4m_1^2 m_2^2} \left(\frac{t^2 + u^2}{s^2}\right).$$

(f) Use the results from parts (b) and (e) to show that for $e^+ e^- \to \mu^+ \mu^-$

$$\left(\frac{d\sigma}{d\Omega}\right)_{\text{CM}} = \frac{\alpha^2}{4s}(1 + \cos^2 \theta) \qquad \sigma = \frac{4\pi\alpha^2}{3s}.$$

Hint: show that if masses are neglected

$$s = 4k^2 \qquad t = -2k^2(1 - \cos\theta) \qquad u = -2k^2(1 + \cos\theta),$$

where θ is the CM scattering angle and $k = |\mathbf{k}_i|$ is the CM momentum of the incident electron. (g) Compare the experimental data on $e^+e^- \rightarrow \mu^+\mu^-$ (see Bartel et al., 1982) with the result derived in part (f).

EXERCISE 3.6 Verify that the Bhabha amplitude (3.104) results from the application of Wick's theorem in quantum field theory.

EXERCISE 3.7 Show that (3.103) follows from (3.97) in the ultrarelativistic limit. *Hint*: use the methods of Exercise 3.5.

EXERCISE 3.8 Use crossing symmetry and the results of Exercise 3.7 to show that in the ultrarelativistic limit (3.105) follows from (3.104).

EXERCISE 3.9 The propagators required in the Feynman rules for simple field theories may be read directly from the relevant Lagrangian density by the following recipe.

1. Eliminate all terms except those quadratic in the fields, and remove the numerical factors multiplying the quadratic terms. For example, this leaves for the scalar field (2.44) the terms $(\partial^\mu \phi)^\dagger (\partial_\mu \phi) - m^2 \phi^\dagger \phi$.

2. Replace the fields with plane waves $e^{ik \cdot x}$ and operate on them with the derivative operators to give factors of the form $ik_\mu \phi$; call the remaining expression with the plane waves eliminated K. For example, this prescription applied to the scalar field gives $K = k^2 - m^2$.

3. The momentum space propagator is the inverse, $P = K^{-1}$, of the matrix K. To find P, write it as the most general sum of Lorentz tensors with unknown coefficients and determine the coefficients by requiring that $PK = K^{-1}K = 1$. For a scalar field this gives

$$P = \Delta(k) = \frac{1}{k^2 - m^2},$$

which is (3.62), up to the $i\epsilon$ denominator term specifying the integration contour (see Exercise 3.2).

(a) Use this prescription to obtain the propagators (3.62), (3.63), and (6.49) starting from the Lagrangian densities (2.44), (2.46), and (2.49), respectively. (b) Show that if the Lagrangian density (2.48) is used this method yields no solution for the massless vector propagator because K has no inverse. (c) Add a "gauge-fixing term" to (2.48):

$$\mathcal{L} = -\tfrac{1}{4} F_{\mu\nu} F^{\mu\nu} - \tfrac{1}{2\xi} (\partial_\mu A^\mu)^2,$$

where ξ is a parameter. Show that the operator K now has an inverse that takes the the general form of the photon propagator given in Feynman rule 3.C for electrodynamics (see §3.9).

EXERCISE 3.10 The vertices required in the Feynman rules for simple field theories may be read directly from the appropriate interaction Lagrangian density by the following recipe.

1. Write $i\mathcal{L}_{\text{int}}$ and replace field operators with plane waves. If a field acts

more than once at a point, include terms with all permutations at the point.

2. Evaluate the resulting expression in momentum space and remove all factors associated with external lines to leave the vertex factor.

For example, applying this procedure to the first term of the interaction Lagrangian density derived in Exercise 2.4c for scalar electrodynamics gives

$$i\mathcal{L}_{\text{int}} = qA_\mu[\phi^*(\partial^\mu\phi) - (\partial^\mu\phi^*)\phi]$$
$$= q\,\underbrace{\epsilon^*_{k\mu}N_1N_2}\,(-ip_1^\mu - ip_2^\mu)\,\underbrace{e^{i(k-p_1+p_2)\cdot x}}$$

where we have used the plane waves

$$\phi = N_1 e^{-ip_1\cdot x} \qquad \phi^* = N_2 e^{ip_2\cdot x} \qquad A_\mu = \epsilon^*_{k\mu}e^{ik\cdot x},$$

and external line factors are indicated by the underbraces. Therefore, the corresponding Feynman rule is

$$\text{vertex} = -iq(p_1 + p_2)^\mu,$$

which is rule 5.B.i of §3.9 [see also eq. (1.144)].   (a) Use this recipe and the one of Exercise 3.9 to check the Feynman rules for quantum electrodynamics given in §3.9. *Hint*: see (2.201), (6.65), and Exercise 2.4c.   (b) Repeat part (a) for the scalar field rules given in §3.5.5. *Hint*: see (3.65b), and remember to include permutations of the fields.

3.10 Background and Further Reading

For general discussions of the material in this chapter see Bogoliubov and Shirkov (1983), Lee (1981), Bjorken and Drell (1965), Gasiorowicz (1966), Itzykson and Zuber (1980), Muirhead (1965), Mandl (1959), Roman (1969), and Jauch and Rohrlich (1955). The interaction representation Fourier expansions are based on the development in Ch. 5 of Lee (1981). Schweber (1961) and Bogoliubov and Shirkov (1983) have a good discussion of Wick's theorem. See Bogoliubov and Shirkov (1983), Ch. 18; and Bjorken and Drell (1965), Ch. 17, for relativistic propagators. The scalar field examples are treated in Lee (1981), Ch. 5; Bogoliubov and Shirkov (1983), Ch. 19; and Bjorken and Drell (1965), Ch. 17. Bogoliubov and Shirkov (1983), Ch. 19; Schweber (1961), Mandl (1959), and Ziman (1969) discuss graphical techniques in a clear fashion. The application of these methods in nonrelativistic field theory may be found in Mahan (1981) and Ziman (1969).

CHAPTER 4

Path Integral Quantization

In this chapter we discuss an alternative to canonical quantization that is based on early ideas of Dirac, but was developed primarily by Feynman. Quantization using the *path integral method* is equivalent to canonical quantization, but for many applications the path integral approach is preferable. In particular, quantization of non-Abelian gauge theories is accomplished most gracefully through the path integral. We first illustrate the method for one degree of freedom in nonrelativistic mechanics and then generalize to field theory.

4.1 Nonrelativistic Path Integral

The path integral method seeks to construct an expression for the quantum-mechanical transition amplitude solely in terms of a *classical Hamiltonian* $H(p, q)$, without explicit use of Hilbert space operators or states. In ordinary quantum mechanics the transition amplitude is given by

$$\langle q''t'' \mid q't' \rangle = \langle q''| \, e^{-i\widehat{H}(t''-t')} \, |q' \rangle \,, \tag{4.1}$$

where the Hamiltonian operator $\widehat{H}$ is assumed to be independent of time.[‡] The Hilbert space operators $\hat{p}$ and $\hat{q}$ and their corresponding eigenvalues obey the relations

$$
\begin{aligned}
[\hat{q}, \hat{q}] = [\hat{p}, \hat{p}] &= 0 & [\hat{q}, \hat{p}] &= i \\
\hat{q} \, |q\rangle &= q \, |q\rangle & \hat{p} \, |p\rangle &= p \, |p\rangle \\
\langle q'' \mid q' \rangle &= \delta(q'' - q') & \langle p'' \mid p' \rangle &= 2\pi\delta(p'' - p') \\
\langle p \mid q \rangle &= e^{-ipq} & \langle q| \, \hat{p} \, |p\rangle &= p \, \langle q \mid p \rangle \,.
\end{aligned}
\tag{4.2}
$$

[‡]Significant portions of this section and the next follow Huang (1982), Ch. VII. In this chapter we will sometimes deal with classical functions and the corresponding quantum-mechanical operators within the same expression. Where confusion might otherwise ensue the operators will be distinguished by placing a circumflex over the symbol.

135

We may divide the interval $t''-t'$ into N equal steps of length $\epsilon \equiv (t''-t')/N$. Then the transition amplitude may be written as

$$
\begin{aligned}
\langle q''t'' \mid q't' \rangle &= \langle q'' | e^{-i\epsilon N \widehat{H}} | q' \rangle \\
&= \langle q'' | \left(1 - i\epsilon \widehat{H} \right)^N | q' \rangle \\
&= \int dq_1 \, dq_2 \ldots dq_{N-1} \, \langle q'' | \left(1 - i\epsilon \widehat{H} \right) | q_{N-1} \rangle \cdot \\
&\quad \ldots \langle q_2 | \left(1 - i\epsilon \widehat{H} \right) | q_1 \rangle \langle q_1 | \left(1 - i\epsilon \widehat{H} \right) | q' \rangle ,
\end{aligned}
\tag{4.3}
$$

where completeness in the coordinate space,

$$
\int_{-\infty}^{+\infty} dq \, |q\rangle \langle q| = 1,
\tag{4.4a}
$$

has been used $N-1$ times. Completeness in momentum space takes the form

$$
\int_{-\infty}^{+\infty} \frac{dp}{2\pi} \, |p\rangle \langle p| = 1,
\tag{4.4b}
$$

and this may be used to express a representative integrand factor of (4.3) in the form

$$
\langle q_2 | \left(1 - i\epsilon \widehat{H} \right) | q_1 \rangle = \int \frac{dp_1}{2\pi} \langle q_2 \mid p_1 \rangle \langle p_1 | \left(1 - i\epsilon \widehat{H} \right) | q_1 \rangle .
\tag{4.5}
$$

Now the relation between the classical Hamiltonian $H(p,q)$ and the Hamiltonian operator $\widehat{H}$ is[‡]

$$
\langle p | \widehat{H} | q \rangle = \langle p \mid q \rangle H(p,q),
\tag{4.6}
$$

and this may be used to express the transition amplitude in terms of the classical Hamiltonian. From (4.5) and (4.6) we have

$$
\begin{aligned}
\langle q_2 | \left(1 - i\epsilon \widehat{H} \right) | q_1 \rangle &= \int \frac{dp_1}{2\pi} \langle q_2 \mid p_1 \rangle \langle p_1 \mid q_1 \rangle \left(1 - i\epsilon H(p_1, q_1) \right) \\
&= \int \frac{dp_1}{2\pi} e^{i p_1 (q_2 - q_1)} \left(1 - i\epsilon H(p_1, q_1) \right),
\end{aligned}
\tag{4.7}
$$

[‡]If $H(p,q)$ contains cross products of p and q an ordering prescription must be introduced since $\hat{p}$ and $\hat{q}$ do not commute (see Abers and Lee, 1973, §11). This case will not be considered.

where $\langle p \mid q \rangle = e^{-ipq}$ has been used. Inserting expressions such as (4.7) into the transition amplitude (4.3) yields

$$\langle q''t'' \mid q't' \rangle = \int \frac{dq_1\, dp_1}{2\pi} \int \frac{dq_2\, dp_2}{2\pi} \cdots \int \frac{dq_{N-1}\, dp_{N-1}}{2\pi}$$
$$\cdot \exp \left[i \sum_{n=0}^{N-1} p_n(q_{n+1} - q_n) \right] \prod_{n=1}^{N-1} \left[1 - i\epsilon H(p_n, q_n) \right], \quad (4.8)$$

where $q_0 \equiv q'$ and $q_N \equiv q''$. It can be shown by power series expansion in Z_n that

$$\lim_{N \to \infty} \prod_{n=1}^{N} \left(1 + \frac{Z_n}{N} \right) = e^x, \quad (4.9)$$

where

$$x \equiv \lim_{N \to \infty} \frac{1}{N} \sum Z_n.$$

Using this result we replace the factor $(1 - i\epsilon H)$ in (4.8) with $e^{-i\epsilon H}$,

$$\langle q''t'' \mid q't' \rangle = \int \frac{dq_1\, dp_1}{2\pi} \int \frac{dq_2\, dp_2}{2\pi} \cdots \int \frac{dq_{N-1}\, dp_{N-1}}{2\pi}$$
$$\cdot \exp \left[i\Delta t \sum_{n=1}^{N-1} \left(\frac{p_n(q_{n+1} - q_n)}{\Delta t} - H(p_n, q_n) \right) \right], \quad (4.10)$$

and adopt the notation

$$t_n \equiv t' + n\Delta t \qquad q_n \equiv q(t_n) \qquad p_n \equiv p(t_n), \quad (4.11)$$

so that in the limit $\Delta t \to 0$

$$\frac{q_{n+1} - q_n}{\Delta t} \xrightarrow[\Delta t \to 0]{} \dot{q}(t_n)$$

$$\sum_{n=1}^{N-1} f(t_n)\Delta t \xrightarrow[\Delta t \to 0]{} \int_{t'}^{t''} f(t)\, dt,$$

and the transition amplitude may then be written

$$\langle q''t'' \mid q't' \rangle = \int [Dq][Dp] \exp \left[i \int_{t'}^{t''} dt \left(p\dot{q} - H(p, q) \right) \right], \quad (4.12)$$

with $q(t') = q'$ and $q(t'') = q''$. The volume elements are

$$[Dq] = \prod_{n=1}^{N-1} dq(t_n) \qquad [Dp] = \prod_{n=1}^{N-1} \frac{dp(t_n)}{2\pi}, \quad (4.13)$$

and the integral is a *path integral* (or *functional integral*) over all paths in phase space between the fixed endpoints (q', t') and (q'', t''):

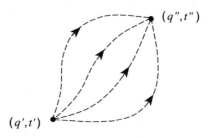

Generalization of the formula for the transition amplitude to more than one degree of freedom is simple; just redefine q and p as vectors:

$$\langle q_1'' q_2'' \ldots q_n'', t'' \mid q_1' q_2' \ldots q_n', t' \rangle$$
$$= \int \prod_i [Dq_i][Dp_i] \exp \left[i \int_{t'}^{t''} \left(\sum_i p_i \dot{q}_i - H(p, q) \right) dt \right]. \quad (4.14)$$

For the particular case of a classical Hamiltonian with the form

$$H(p, q) = \frac{p^2}{2m} + V(q), \quad (4.15)$$

the momentum integrations can be done explicitly and

$$\int [Dp] \exp \left(i \int_{t'}^{t''} dt \, (p\dot{q} - H) \right) = \left(\frac{m}{2\pi} \right)^{N/2} \exp \left(i \int_{t'}^{t''} L(q, \dot{q}) \, dt \right), \quad (4.16)$$

where the classical Lagrangian is given by

$$L(q, \dot{q}) = \tfrac{1}{2} m \dot{q}^2 - V(q). \quad (4.17)$$

On defining the action by

$$S = \int_{t'}^{t''} L(q, \dot{q}) \, dt, \quad (4.18)$$

we obtain the Feynman path integral formula for the transition amplitude

$$\langle q'' t'' \mid q' t' \rangle = N_0 \int [Dq] e^{iS}. \quad (4.19)$$

The constant N_0 is usually infinite, but irrelevant for physical results because it cancels in normalized matrix elements.

The Feynman amplitude for the matrix element of a time-ordered set of operators can be shown by the same techniques to be

$$\langle q''t'' |\, T\big(\hat{q}(t_1)\hat{q}(t_2)\dots\hat{q}(t_n)\big)\, |q't'\rangle = N_0 \int [Dq]\, q(t_1)q(t_2)\dots q(t_n) e^{iS}. \quad (4.20)$$

Notice that in (4.20) the left side contains time-ordered Hilbert space operators, but the right side involves only c-numbers. The price we have paid for this is that the integral over purely classical quantities on the right side of (4.20) is infinite-dimensional.

4.2 Path Integral for Field Theory

As in previous discussions, this formalism may be extended to a field theory by allowing the number of variables to approach infinity. For simplicity we illustrate for a single boson field. Extension to many boson fields is straightforward, but the path integral for fermions is complicated by integrations over Grassmann variables (anticommuting c-numbers) and will be dealt with later (see §4.5–§4.7).

For a single boson field the role of a coordinate is played by the field operator $\hat{\phi}(\mathbf{x})$, which has eigenstates $|\phi\rangle$

$$\hat{\phi}(\mathbf{x})\,|\phi\rangle = \phi(\mathbf{x})\,|\phi\rangle\,, \quad (4.21)$$

where $\phi(\mathbf{x})$ is a *c-number function* of the spatial coordinates. The transition amplitude from one field eigenstate at time t_1 to another eigenstate at t_2 is

$$\langle \phi_2 t_2 \,|\, \phi_1 t_1 \rangle = \langle \phi_2 |\, e^{-i\hat{H}(t_2 - t_1)}\, |\phi_1\rangle\,, \quad (4.22)$$

where the operator $\hat{H}$ is assumed to be independent of time. By analogy with the previous discussion for discrete quantum mechanics, the Feynman path integral formula for the amplitude is

$$\langle \phi_2 t_2 \,|\, \phi_1 t_1 \rangle = N_0 \int_{\phi_1}^{\phi_2} [D\phi]\exp\left(i \int_1^2 \mathcal{L}\, d^4 x \right), \quad (4.23)$$

where the classical Lagrangian density is $\mathcal{L} = \mathcal{L}\big(\phi(x), \partial^\mu \phi(x)\big)$, and where we employ a notation

$$\int_1^2 d^4 x \equiv \int_{t_1}^{t_2} dx_0 \int_{\text{space}} d^3 x \quad (4.24)$$

$$\phi_2(\mathbf{x}) \equiv \phi(\mathbf{x}, t_2) \qquad \phi_1(\mathbf{x}) \equiv \phi(\mathbf{x}, t_1).$$

The integration $\int [D\phi]$ is a *functional integration*. Integrations over functional spaces are in some sense a product of an infinite number of ordinary integrals [see (4.13)], and we shall have to define approximations that give practical meaning to (4.23) before it can be used.

For a time-ordered set of field operators $\hat{\phi}(x_n)$ the Feynman amplitude is analogous to (4.20):

$$\langle \phi_2 t_2 | T\big(\hat{\phi}(x_1)\hat{\phi}(x_2)\ldots\hat{\phi}(x_n)\big) | \phi_1 t_1 \rangle$$
$$= N_0 \int_{\phi_1}^{\phi_2} [D\phi]\, \phi(x_1)\phi(x_2)\ldots\phi(x_n) \exp\left(i \int_1^2 \mathcal{L}\, d^4 x\right), \quad (4.25)$$

where the $\phi(x_n)$ are c-number functions. A notational distinction between c-numbers and the corresponding operators has been maintained in expressions such as this, but we note that it is also common practice to use the same symbols on the two sides of equations like (4.25), with the context distinguishing operators and c-numbers.

These formulas are not applicable in the present form for field theories where there are constraints imposed by conjugate momenta that vanish identically. Since this is generally true of gauge fields (recall the canonical quantization of the Maxwell field in §2.7), there is cause for some concern. However, one finds that the present formalism may be applied in that case provided the Lagrangian density is augmented by a *gauge-fixing term*, as will be discussed in Ch. 7 (see also Exercise 3.9b,c).

4.2.1 Evaluation of Integrals in Euclidian Space

In evaluating path integrals it will often be useful to rotate the time to the imaginary axis; this corresponds to a transformation from Minkowski to Euclidian space (the metric changes from $g_{\mu\nu}$ to $\delta_{\mu\nu}$). The operator $\exp(-i\hat{H}t)$ propagates a quantum state through a real time interval t. The transformation from Minkowski space to Euclidian space involves the replacement $t \to -i\tau$, with $\tau = x_4$ a real number [see (4.27)]. Therefore, the operator that propagates a state through an imaginary time $t = -i\tau$ is

$$e^{-i\hat{H}(-i\tau)} = e^{-\hat{H}\tau}.$$

This is not a physical propagation, but the operator $\exp(-\hat{H}\tau)$ is well defined mathematically and can be used to calculate the eigenvalues $\exp(-E\tau)$ of $\exp(-\hat{H}\tau)$ (Wilson, 1977). Once this is done the eigenvalues E of $\hat{H}$ may be determined from

$$E = -\frac{1}{\tau} \log\left(e^{-E\tau}\right). \quad (4.26)$$

Thus finding eigenvalues and eigenvectors of $\exp(-\widehat{H}\tau)$ is equivalent to finding eigenvalues and eigenvectors of $\widehat{H}$ itself—no information is gained or lost by transforming the problem to Euclidian space (assuming no poles are encountered in rotating the time to the imaginary axis; see below). Once the eigenvalues and eigenvectors of $\exp(-\widehat{H}\tau)$ are known we may use them to calculate the corresponding quantities for the operator $\exp(-i\widehat{H}t)$ in the original Minkowski space.

In field theory this procedure is called the *Wick rotation*. The relations among the Euclidian and Minkowski coordinates and momenta are (Coleman, 1975; Huang, 1982)

$$
\begin{aligned}
x_{\rm E} &= (\mathbf{x}, x_4) & d^4x &= -i\, d^4x_{\rm E} \\
k_{\rm E} &= (\mathbf{k}, k_4) & d^4k &= i\, d^4k_{\rm E} \\
x_{\rm E}^2 &= x_1^2 + x_2^2 + x_3^2 + x_4^2 = -x^2 \\
k_{\rm E}^2 &= k_1^2 + k_2^2 + k_3^2 + k_4^2 = -k^2
\end{aligned}
\tag{4.27}
$$

$$
x_4 \equiv ix_0 \qquad k_4 \equiv -ik_0 \qquad (x_4, k_4 \in \text{real numbers}),
$$

where Euclidian quantities are written with a subscript E, while Minkowski quantities have no subscript. After a Wick rotation the Lorentz invariance in Minkowski space becomes a corresponding $O(4)$ invariance under orthogonal transformations in four-dimensional Euclidian space. We may also note that since the metric in Euclidian space is $\delta_{\mu\nu}$ (rather than $g_{\mu\nu}$), there is no distinction between upper and lower Euclidian indices. For the transformation of fields we find (Huang, 1982):

1. Real Lorentz scalar Minkowski fields continue to real $O(4)$ Euclidian scalar fields

$$
\phi(x) \to \phi(x_{\rm E}). \tag{4.28a}
$$

2. The components of *massive* vector fields continue according to

$$
\begin{aligned}
A^k(x) &\to A^k(x_{\rm E}) & (k = 1, 2, 3) \\
A^0(x) &\to iA_4(x_{\rm E}).
\end{aligned}
\tag{4.28b}
$$

3. For gauge fields the continuation is complicated by unphysical degrees of freedom introduced by the gauge invariance (Ch. 7), but in covariant gauges (4.28b) is valid.

The form of the Euclidian propagators may be inferred directly from these prescriptions. For example, the Feynman propagator for a free scalar field in Minkowski space is [eq. (3.56)]

$$\Delta(x) = \int \frac{d^4k}{(2\pi)^4} \frac{e^{-ik\cdot x}}{k^2 - m^2 + i\epsilon} \qquad (\epsilon \to 0^+). \qquad (4.29a)$$

In Euclidian space the corresponding propagator is

$$\Delta_E(x_E) = -i \int \frac{d^4k_E}{(2\pi)^4} \frac{e^{-ik_E\cdot x_E}}{k_E^2 + m^2}. \qquad (4.29b)$$

The Euclidian propagator no longer requires the $i\epsilon$ term because the integrand of (4.29b) has no poles on the real axis: the rotation between Minkowski and Euclidian space may be viewed as providing the correct $i\epsilon$ prescription for the propagators (see Exercise 3.2). We may also note that the Euclidian d'Alembertian operator is

$$\Box_E = \nabla^2 + \partial_4^2 = -\Box, \qquad (4.30)$$

with no distinction between upper and lower indices, and the propagator (4.29b) is a solution of the Euclidian wave equation

$$(\Box_E - m^2)\Delta_E(x) = i\delta^{(4)}(x). \qquad (4.31)$$

4.2.2 Coupling to External Sources

In quantum field theories it can be demonstrated that *all dynamical informa-tion about a system may be extracted by studying the response of the vacuum to an arbitrary external source* (see, e.g., Schwinger, 1969). This response is specified by the vacuum to vacuum transition amplitude

$$\langle 0^+ \,|\, 0^- \rangle_J \equiv \langle \text{vacuum}\,(t = +\infty) \,|\, \text{vacuum}\,(t = -\infty) \rangle_J \qquad (4.32)$$

which measures the amplitude for the system to be in the ground state at $t = +\infty$ when it was known to be in the ground state at $t = -\infty$, while in the presence of an external source J that is turned off in the remote past and future. Normally the source J is taken to be a local c-number function of the coordinates.

Generating Functionals. To introduce the effect of an external source $J(x)$ into the path integral it is only necessary to add a term $\phi(x)J(x)$ to the Lagrangian density,

$$\langle \phi_2 t_2 \,|\, \phi_1 t_1 \rangle_J \equiv N_0 \int_{\phi_1}^{\phi_2} [D\phi] \exp\left[i \int_1^2 d^4x \left(\mathcal{L} + \phi(x)J(x) \right) \right], \qquad (4.33)$$

where we assume that the Feynman form (4.23) of the path integral is valid. Because of unitarity the vacuum–vacuum amplitude can only be a phase factor and we may introduce two *quantum generating functionals* $Z[J]$ and $W[J]$ by the definitions:

$$Z[J] \equiv e^{iW[J]} \equiv \langle 0^+ \mid 0^- \rangle_J . \qquad (4.34)$$

If Z is expanded in a functional Taylor series

$$Z = \sum_{n=0}^{\infty} \frac{i^n}{n!} \int d^4x_1 d^4x_2 \ldots d^4x_n \, G^{(n)}(x_1 x_2 \ldots x_n) J(x_1) J(x_2) \ldots J(x_n), \qquad (4.35a)$$

it is a standard result (see below) that the coefficients $G^{(n)}(x_1 x_2 \ldots x_n)$ are the n-point Green's functions (the sum of Feynman diagrams with n external lines) in coordinate space. Likewise, we may write

$$iW = \sum_{n=0}^{\infty} \frac{i^n}{n!} \int d^4x_1 d^4x_2 \ldots d^4x_n \, G_c^{(n)}(x_1 x_2 \ldots x_n) J(x_1) J(x_2) \ldots J(x_n), \qquad (4.35b)$$

where the coefficients $G_c^{(n)}(x_1 x_2 \ldots x_n)$ are the *connected n-point Green's functions*. That is, $G_c^{(n)}$ is the sum of all connected Feynman diagrams having n external lines.

Thus, the vacuum–vacuum amplitude

$$\langle 0^+ \mid 0^- \rangle_J = Z[J]$$

is the generating functional for the Green's functions of the system, while

$$W[J] = -i \log Z[J]$$

is the generating functional for the connected Green's functions (see Exercise 4.1).[‡] Since the Green's functions define the evolution of the system, eqns. (4.35) indicate explicitly that the dynamics of a field theory may be established by studying the vacuum–vacuum amplitude in the presence of external sources. It should be emphasized that this is a general property of quantum field theory and is not peculiar to path integral methods. However, we are about to see that a path integral is particularly well equipped to exploit this idea.

EXERCISE 4.1 Show that $W[J]$ generates only connected graphs for two- and four-point functions in scalar ϕ^4 theory. *Hint:* do Exercise 4.3 first; then consult Ryder (1985), §6.6.

The key to obtaining Feynman diagrams from functional integrals is that

---

[‡]The reader is warned that many authors reverse this notation and use $Z[J]$ to represent the connected generating functional and $W[J]$ the total generating functional.

there is a simple connection between the quantum generating functional for the Green's functions and the classical action integral:

$$Z[J] = \langle 0^+ \mid 0^- \rangle_J = e^{iW[J]} = N \int [D\phi] e^{iS[\phi, J]}$$

$$S[\phi, J] = \int d^4x \left[\mathcal{L} + \phi(x)J(x) \right], \tag{4.36}$$

where the normalization N is chosen so that W vanishes when the source J vanishes. The functional integral (4.36) is to be evaluated in Euclidian space; this yields the Euclidian generating functional, and the physical Minkowski results may then be procured by analytical continuation.[‡]   From (4.27) the Euclidian functional integral for the vacuum–vacuum amplitude is (Huang, 1982)

$$Z[J] = N \int [D\phi] \exp \left[-\frac{1}{\hbar} \left(S_{\mathrm{E}}[\phi] - \int d^4x_{\mathrm{E}} \phi(x_{\mathrm{E}}) J(x_{\mathrm{E}}) \right) \right], \tag{4.37}$$

where the subscripts E denote Euclidian quantities, we have made explicit only for this expression the normally suppressed $\hbar$, and the Euclidian action is (see Exercise 13.7c)

$$S_{\mathrm{E}}[\phi] \equiv -iS[\phi], \tag{4.38}$$

where $S[\phi]$ is the Minkowski action with $x_0 \to -ix_4$. In this form we see that the Euclidian functional integral looks like the *partition function* of a four-dimensional classical statistical mechanics at temperature $\hbar$. Thus, in the Minkowski space each path contributing to the path integral has a phase that oscillates rapidly except near the path(s) of stationary phase (classical paths). In the Euclidian-space formulation each path instead has a Boltzmann factor attached to it, and the classical paths make the dominant contribution because they are the ones with minimal Euclidian action.

Functional Calculus. Because of (4.36) and (4.35) the Green's functions may be obtained from the path integral by functional differentiation. Before proceeding we digress briefly to review functional derivatives.

A functional integral associates with each function a number (the integral over the function space), just as an ordinary function associates a number

---

[‡]At our level of discussion it is common to gloss over deep mathematical questions such as whether this continuation always exists. The assumption that the Minkowski space Green's functions are uniquely related by analytical continuation to a corresponding set of well-behaved Euclidian space Green's functions is sometimes called the *Euclidicity postulate* (see Abers and Lee, 1973, §12).

$f(x)$ with a number x. The functional (or variational) derivative represents a generalization of the partial derivative. In the sum

$$df = \sum_i C_i dz_i = \sum_i \left(\frac{\partial f}{\partial z_i}\right) dz_i \tag{4.39}$$

the coefficients C_i of dz_i are the partial derivatives. Likewise, if we consider a variation $\delta F(u)$ of a functional $F(u)$, this variation can be represented by an integral:

$$\delta F(u) = \int C(x, u)\delta u \, dx \tag{4.40}$$

[see eqns. (2.4) and (2.5)]. We then define the functional derivative as

$$\frac{\delta F(u)}{\delta u(x)} \equiv C(x, u), \tag{4.41}$$

where $C(x, u)$ is a functional of u that depends on the continuous integration variable x, just as the partial derivative C_i depends on the discrete summation index i. Because of this correspondence, functional derivatives have properties that are analogous to those of normal derivatives. For example, a limiting procedure similar to that for ordinary derivatives may be used to construct functional derivatives:

$$\frac{\delta F[f]}{\delta f(y)} \equiv \lim_{\epsilon \to 0} \left\{ \frac{F[f(x) + \epsilon\delta^{(4)}(x-y)] - F[f(x)]}{\epsilon} \right\}. \tag{4.42}$$

Some examples are given in Exercise 4.2.

EXERCISE 4.2 Use the definition (4.42) to show that

$$\frac{\delta \int f(x)dx}{\delta f(y)} = 1 \qquad \frac{\delta J(y)}{\delta J(x)} = \delta^{(4)}(x-y)$$

$$\frac{\delta \int G(x,y)f(y)dy}{\delta f(z)} = G(x,z)$$

$$\frac{\delta Z[J]}{\delta J(y)} = i\int [D\phi]\phi(y)\exp\left[i\int d^4x \left(\mathcal{L} + \phi(x)J(x)\right)\right]$$

$$\frac{\delta^{(2)} Z[J]}{\delta J(y)\delta J(z)} = (i)^2 \int [D\phi]\phi(y)\phi(z)\exp\left[i\int d^4x \left(\mathcal{L} + \phi(x)J(x)\right)\right],$$

where $f(x)$, $\phi(x)$, $J(x)$, and $G(x,y)$ are arbitrary functions, and $Z[J]$ is defined in (4.36)–(4.37).

Green's Functions. From (4.35) and (4.36) we conclude that the matrix element of a time-ordered set of operators (Green's function) may be obtained from the path integral by functional differentiation with respect to the source, followed by setting $J(x)$ equal to zero. In particular [see (4.35), (3.55), and (3.59)]

$$i^{-n}\left[\frac{\delta^{(n)}Z[J]}{\delta J(x_1)\delta J(x_2)\ldots\delta J(x_n)}\right]_{J=0} \propto \langle 0|\, T\big(\hat{\phi}(x_1)\hat{\phi}(x_2)\ldots\hat{\phi}(x_n)\big)\,|0\rangle$$

$$= G^{(n)}(x_1 x_2 \ldots x_n), \qquad (4.43)$$

where the constant of proportionality is not physically significant, and

$$i^{1-n}\left[\frac{\delta^{(n)}W[J]}{\delta J(x_1)\delta J(x_2)\ldots\delta J(x_n)}\right]_{J=0} = \langle 0|\, T\big(\hat{\phi}(x_1)\hat{\phi}(x_2)\ldots\hat{\phi}(x_n)\big)\,|0\rangle_{\mathrm{c}}$$

$$= G^{(n)}_{\mathrm{c}}(x_1 x_2 \ldots x_n), \qquad (4.44)$$

where the subscript c means that only connected Feynman diagrams are to be admitted in the perturbation series. Comparing with (4.36), we see that path integrals may be used to construct Green's functions; in general,

$$G^{(n)}(x_1 x_2 \ldots x_n) \simeq \int [D\phi]\,\phi(x_1)\phi(x_2)\ldots\phi(x_n)\exp\left(i\int \mathcal{L}\,d^4x\right), \quad (4.45)$$

where is is understood implicitly that any ambiguities in evaluating the integrals are to be resolved by performing the integrations in Euclidian space. In §4.4 we will consider the evaluation of Green's functions by perturbation theory for interacting fields.

The Feynman propagator picture [left side of (4.45)] views physical processes as being dominated by classical propagation (represented by tree diagrams), with the quantum effects associated with virtual excitation in internal loops. In the functional integral picture [right side of (4.45)] the amplitudes are dominated by classical paths (those that extremize the action) and quantum effects come through excursions from the classical paths. As we have noted, these departures are penalized by small Boltzmann factors in Euclidian space, or by rapid integrand oscillations that suppress the path integral in Minkowski space.

4.3 Evaluation of Path Integrals

The path integral formulation of quantum mechanics is intuitive and elegant, but not very practical unless we can evaluate the effectively infinite-dimensional integrals. In the remainder of this chapter we consider some ways to accomplish this.

If the path integrals involve quadratic exponentials they may be calculated by a generalization of the well-known formula for a one-dimensional Gaussian integral (see Exercise 4.5). For a vector $x = (x_1, x_2, x_3, \ldots, x_n)$ we may define the general quadratic form

$$Q(x) = \tfrac{1}{2}(x, Ax) + (b, x) + c, \tag{4.46}$$

where the notation (u, v) specifies the scalar product of vectors

$$(u, v) \equiv \sum_{i=1}^{n} u_i v_i, \tag{4.47}$$

b is a constant vector, c is a number, and A is a complex, symmetric, $n \times n$ matrix with positive definite real part. Then (Coleman, 1975)

$$\int (dx) e^{-Q(x)} = e^{-Q(\bar{x})} (\operatorname{Det} A)^{-1/2}$$
$$= \exp\left[\tfrac{1}{2}(b, A^{-1}b) - c\right] (\operatorname{Det} A)^{-1/2}, \tag{4.48a}$$

where

$$(dx) \equiv d^n x \, (2\pi)^{-n/2}, \tag{4.48b}$$

and where $\bar{x}$ is the minimum of $Q(x)$:

$$\bar{x} = -A^{-1}b. \tag{4.48c}$$

The factor $\operatorname{Det} A$ is equal to the product of eigenvalues a_i of the matrix A,

$$\operatorname{Det} A = \prod_{i=1}^{n} a_i. \tag{4.49}$$

As an example we may consider a free scalar field (Huang, 1982, §7.6). The Lagrangian density is

$$\mathcal{L} = \tfrac{1}{2}\left[\partial_\mu \phi(x)\right]\left[\partial^\mu \phi(x)\right] - \tfrac{1}{2}m^2 \phi^2(x), \tag{4.50}$$

leading to a classical action functional

$$S[\phi] = -\frac{1}{2} \int d^4x \, \phi(x) \left(\Box + m^2\right) \phi(x). \tag{4.51}$$

Going to Euclidian space [see eqns. (4.27) and (4.30)],

$$S_{\mathrm{E}}[\phi] = \frac{i}{2} \int d^4x_{\mathrm{E}} \, \phi(x_{\mathrm{E}}) \left(-\Box_{\mathrm{E}} + m^2\right) \phi(x_{\mathrm{E}}), \tag{4.52}$$

and from (4.36) the Euclidian generating functional is

$$Z[J] = e^{iW[J]} = N \int [D\phi] e^{-Q[\phi,J]}, \qquad (4.53)$$

where

$$Q[\phi, J] = \tfrac{1}{2}(\phi, A\phi) - (b, \phi)$$
$$A = (-\Box_{\mathrm{E}} + m^2) \qquad b = J(x_{\mathrm{E}}), \qquad (4.54)$$

and where the generalization of (4.47) to continuous field variables is

$$(\phi_1, \phi_2) = \int d^4x \, \phi_1^*(x)\phi_2(x).$$

Using (4.48a) to evaluate the integral and taking the logarithm of both sides of (4.53) we obtain the generating functional for connected Green's functions

$$iW[J] = \tfrac{1}{2}(b, A^{-1}b) + \log\left[N(\operatorname{Det} A)^{-1/2}\right]. \qquad (4.55)$$

Now, (4.49) in momentum space gives

$$\operatorname{Det} A = \prod_{k_{\mathrm{E}}} \left(k_{\mathrm{E}}^2 + m^2\right). \qquad (4.56)$$

This is divergent but independent of J, and we may choose $N = (\operatorname{Det} A)^{1/2}$ to cancel it from $W[J]$. Then from this result, (4.55), (4.54), and (4.31),

$$W[J] = \tfrac{1}{2} \int d^4x_{\mathrm{E}} \, d^4y_{\mathrm{E}} \, J(x_{\mathrm{E}})\Delta_{\mathrm{E}}(x_{\mathrm{E}} - y_{\mathrm{E}})J(y_{\mathrm{E}}), \qquad (4.57)$$

or on returning to Minkowski space

$$W[J] = -\tfrac{1}{2} \int d^4x \, d^4y \, J(x)\Delta(x - y)J(y). \qquad (4.58)$$

Comparing with the functional expansion (4.35), only the $n = 2$ term contributes in this case and we identify the Green's function with the free Feynman propagator:

$$G^{(2)}(x, y) = i\Delta(x - y). \qquad (4.59)$$

Thus, for a free scalar field the path integral formalism yields the same result as the canonical method of Chs. 2–3. The explicit form of the Euclidean propagator is given by (4.29b). Upon rotating back to Minkowski space this takes the familiar form (4.29a), with the $i\epsilon$ prescription fixed by the requirement that the Wick rotation not encounter any poles in the k plane [see Exercise 3.2 and Ramond (1981), §IV.8].

4.4 Feynman Diagrams

Suppose that the Lagrangian density can be separated into two parts plus a source term

$$\mathcal{L} = \mathcal{L}_0 + \mathcal{L}' + \phi(x)J(x), \tag{4.60}$$

where $\mathcal{L}'$ is an interaction that may be considered a perturbation relative to $\mathcal{L}_0$. The corresponding action may be split into two parts

$$S = S_0^J + S' = \int d^4x \left(\mathcal{L}_0 + \phi J \right) + \int d^4x \, \mathcal{L}', \tag{4.61}$$

and the generating functional may be written

$$Z[J] = e^{iW[J]} = \exp\left[i \int d^4x \, \mathcal{L}' \left(-i \frac{\delta}{\delta J} \right) \right] \exp\left(iW_0[J] \right), \tag{4.62}$$

where we indicate explicitly that the classical fields are defined by functional differentiation $\delta/\delta J(x)$ with respect to the sources and

$$e^{iW_0} \equiv \int [d\phi] e^{iS_0^J}. \tag{4.63}$$

A perturbation series with the usual Feynman diagrams is obtained if these exponentials are expanded. For example, if

$$\mathcal{L}_0 = \tfrac{1}{2}(\partial_\mu \phi)(\partial^\mu \phi) - \tfrac{1}{2}m^2 \phi^2 \qquad \mathcal{L}' = \frac{-f_0}{4!}\phi^4, \tag{4.64}$$

the generating functional is

$$e^{iW[J]} = \exp\left[\frac{-if_0}{4!} \int d^4z \left(\frac{1}{i} \frac{\delta}{\delta J(z)} \right)^4 \right]$$

$$\times \exp\left[-\frac{i}{2} \int d^4x \, d^4y \, J(x)\Delta(x-y)J(y) \right], \tag{4.65}$$

where the results of the last section have been utilized for $W_0[J]$. If this is expanded as a power series in the coupling constant f_0, we obtain scalar ϕ^4 perturbation theory (see Exercise 4.3).

EXERCISE 4.3 In §3.5 we used canonical quantization to derive Feynman rules for a scalar ϕ^4 field. Show that an equivalent perturbation theory results from path integral methods by expanding (4.65) in powers of the coupling f_0.

4.5 Grassmann Variables

There is a fundamental difference between boson and fermion fields if we attempt to construct a classical limit. Recall that boson and fermion operators are defined by algebras of the form

$$[b, b^\dagger] = bb^\dagger - b^\dagger b = \hbar \qquad \text{(boson)}$$
$$\{a, a^\dagger\} = aa^\dagger + a^\dagger a = \hbar \qquad \text{(fermion)}. \tag{4.66}$$

Formally, the classical limit results as $\hbar \to 0$ (more precisely, in the classical limit $\hbar$ is much smaller than the relevant classical action). Therefore, the classical limit for boson fields gives *commuting c-number fields*

$$bb^\dagger = b^\dagger b \qquad (\hbar \to 0), \tag{4.67a}$$

and a macroscopic collection of bosons becomes an excellent approximation to a classical field. The classical electromagnetic and gravitational fields attest to the validity of this limit.

On the other hand, there is no simple limit as $\hbar \to 0$ for Fermi fields since the classical limit of the fermion anticommutator yields *anticommuting c-numbers*

$$aa^\dagger = -a^\dagger a \qquad (\hbar \to 0). \tag{4.67b}$$

Even in the "classical limit" fermion fields cannot be described by the classical mechanics of commuting c-numbers. The anticommuting c-numbers defined by the $\hbar \to 0$ limit of the fermion commutator are called *Grassmann variables* by mathematicians, and the peculiar algebra that results is called a *Grassmann algebra* (see Exercises 4.4 and 4.5).

EXERCISE 4.4 Grassmann numbers have some unusual properties that are examined in this exercise (see Cheng and Li, 1984, §1.3). For n independent Grassmann generators θ_i we have $\{\theta_i, \theta_j\} = 0$. Consider first the simple case of one Grassmann variable θ: (a) Show that functions can be expanded in a *finite* series, $p(\theta) = p_0 + \theta p_1$. (b) Show that differentiation acting from the left and integration give the same result:

$$\int d\theta\, p(\theta) = \frac{\overrightarrow{d}}{d\theta} p(\theta) = -p(\theta) \frac{\overleftarrow{d}}{d\theta} = p_1 \qquad \text{(Grassmann number)}.$$

Hint: Define formal "differentiation" by requiring that

$$\frac{\overrightarrow{d}}{d\theta} \theta = \theta \frac{\overleftarrow{d}}{d\theta} = 1;$$

define formal "integration" by requiring that

$$\int d\theta\, p(\theta) = \int d\theta\, p(\theta + \lambda),$$

where λ is a Grassmann number independent of θ. Notice that since Grassmann numbers anticommute we must distinguish whether differentiation acts from the left

$$\frac{\vec{d}}{d\theta} p(\theta),$$

or from the right

$$p(\theta) \frac{\overleftarrow{d}}{d\theta}.$$

(c) Show that if the integration rule in part (b) is to be preserved the "Jacobian" associated with a change of Grassmann integration variables must be the inverse of what we would expect for ordinary numbers:

$$\int d\theta'\, p(\theta') = \int d\theta \left(\frac{d\theta'}{d\theta} \right)^{-1} p\left[\theta'(\theta)\right].$$

Now consider the general case of n Grassmann generators θ_i. Convince yourself that **(d)** Derivatives are given by the rule: the derivative

$$\frac{\vec{d}}{d\theta_i} \left(\frac{\overleftarrow{d}}{d\theta_i} \right)$$

of $\theta_1 \theta_2 \ldots \theta_k$ is determined by commuting θ_i all the way to the left (right) in the product, and then dropping that θ_i. **(e)** Integration is defined by

$$\{d\theta_i, d\theta_j\} = 0 \qquad \int d\theta_i = 0 \qquad \int d\theta_i \theta_j = \delta_{ij}.$$

(f) A change of integration variable is defined by

$$\int d\theta'_n \ldots d\theta'_1\, p(\theta') = \int d\theta_n \ldots d\theta_1 \left(\mathrm{Det}\, \frac{d\theta'}{d\theta} \right)^{-1} p\left[\theta'(\theta)\right].$$

That is, for Grassmann variables integration is formally equivalent to differentiation, and the Jacobian of a transformation of variables is the inverse of that expected for ordinary variables.

4.6 Fermions in the Path Integral

By analogy with the discussion of path integrals for Bose fields, we might expect that generating functionals involving Fermi fields will utilize functional integrals over classical limits of those fields. However, as $\hbar \to 0$ we have seen that Fermi fields become the elements of a Grassmann algebra. These are "notoriously objects that make strong men quail" (Coleman, 1975), but it is possible to make a kind of end run around the problem of defining a functional integral on Grassmann variables.

If a Fermi field is denoted by η and its conjugate field by η^*, the part of the action involving fermions will be at most quadratic in the fields for the theories that interest us:

$$S_\mathrm{f} = (\eta^*, A\eta). \tag{4.68}$$

By analogy with the corresponding method for bosons, it would be nice if we could write for a functional integral defined over Fermi fields [see (4.36)]

$$\langle 0^+ \mid 0^- \rangle \equiv e^{iW} = N \int [d\eta^*][d\eta] e^{iS_\mathrm{f}}. \tag{4.69}$$

With Coleman (1975) we adopt the philosophy that a field theory is defined by its Feynman diagrams. Now for *complex Bose fields*, if the Hermitian scalar product is defined by (z^*, y), then

$$\int [dz^*][dz] e^{-i(z^*, Az)} = [\mathrm{Det}\,(iA)]^{-1} \tag{4.70}$$

(see Exercise 4.5), where the change in the power of $\mathrm{Det}\,(iA)$ relative to (4.55) for the real field comes because each eigenvalue of A contributes twice, once for $\int [dz]$ and once for $\int [dz^*]$. For actions of the form (4.68) the connected Feynman diagrams contributing to W contain single closed loops, and the only change introduced if bosons are replaced by fermions in the perturbative expansion is a factor -1 for each closed fermion loop (see Feynman rule 4.B in §3.9). Since every graph contributing to W has one loop, W is replaced by $-W$, and this is equivalent to the replacement

$$[\mathrm{Det}\,(iA)]^{-1} \to \mathrm{Det}\,(iA)$$

(see also Exercise 4.5b,c). Thus, by these cavalier arguments we conclude that the correct Feynman rules result from the Fermi field integration if we *define* (up to a constant that can be absorbed in the normalization)

$$\int [d\eta^*][d\eta] e^{i(\eta^*, A\eta)} \equiv \mathrm{Det}\,(iA). \tag{4.71}$$

A more formal justification of these results is given in Exercises 4.4 and 4.5.

EXERCISE 4.5 Convince yourself that for a Gaussian path integral with a nonsingular matrix A the following properties hold (Cheng and Li, 1984):

(a) For commuting real variables $x = (x_1, x_2, \ldots, x_n)$

$$\int \frac{dx_1}{\sqrt{2\pi}} \cdots \frac{dx_n}{\sqrt{2\pi}} \exp\left[-\tfrac{1}{2}(x, Ax)\right] = \frac{1}{\sqrt{\text{Det } A}}.$$

(b) For commuting complex variables z

$$\int \frac{dz_1}{\sqrt{\pi}} \cdots \frac{dz_n}{\sqrt{\pi}} \frac{dz_1^*}{\sqrt{\pi}} \cdots \frac{dz_n^*}{\sqrt{\pi}} \exp\left[-(z^*, Az)\right] = \frac{1}{\text{Det } A}.$$

(c) For Grassmann variables $\theta = (\theta_1, \theta_2, \ldots, \theta_n)$ and $\bar{\theta} = (\bar{\theta}_1, \bar{\theta}_2, \ldots, \bar{\theta}_n)$, where θ_i and $\bar{\theta}_i$ are independent generators of the algebra,

$$\int d\theta_1 d\bar{\theta}_1 \ldots d\theta_n d\bar{\theta}_n \, e^{(\bar{\theta}, A\theta)} = \text{Det } A.$$

Hint: Generalize the evaluation of the Gaussian integral

$$\int dx \, e^{-\frac{1}{2}ax^2} = \sqrt{\frac{2\pi}{a}}.$$

4.7 Ghost Fields

The only path integral that is easy to evaluate is a Gaussian path integral. If the main part of the action can be coerced into a form that approximates a Gaussian path integral, that integral can be computed and the remainder of the action treated as a perturbation. However, in many cases of interest the path integral cannot be cast in Gaussian form because there is a pre-exponential factor in the path integrand that cannot be ignored or absorbed in a normalization. In some instances the results just obtained for fermion integrals may be used to advantage by elevating a pre-exponential factor to an exponent at a cost of introducing fictitious *ghost fields*. As an example of this trick let us consider a theory that contains derivative interactions, with a Lagrangian that is no more than quadratic in the time derivatives. The corresponding functional integral is of the form (Coleman, 1975)

$$e^{iW} = N \int [dq] (\text{Det } K)^{1/2} e^{iS}, \tag{4.72}$$

where K is a linear operator. This expression is not easily handled as it stands because of $\text{Det } K$. However, we may remedy that by the following trick: eq. (4.71) may be viewed as a (novel!) definition of a determinant. Introducing a set of complex Fermi variables η_a and ignoring a multiplicative constant

$$(\text{Det } K)^{1/2} = \int [d\eta^*][d\eta] \exp\left[i(\eta_a^*, K_{ab}^{1/2} \eta_b)\right], \tag{4.73}$$

where $K^{1/2}$ denotes the matrix square root of K. Thus

$$e^{iW} = N \int [dq][d\eta^*][d\eta] \exp \left[iS + i\big(\eta_a^*, K_{ab}^{1/2}\eta_b\big) \right], \qquad (4.74)$$

which is a form (path integral of an exponential) better suited than (4.72) to manipulations such as the construction of a perturbation series. In this example the variables η are not true dynamical quantities (they do not appear on any external lines of the resulting diagrams); they are just a subterfuge for elevating the effect of a determinant to an exponent for mathematical convenience. In field theory such fictitious variables are called *ghost fields*. Similar fields (the Faddeev–Popov ghosts) play important roles in the quantization of non-Abelian gauge theories (§7.4).

Physically, the role of fermions is to impose constraints in path integrals; for actual fermions one example of such a constraint is the Pauli principle (see the discussion in Cvitanovic, 1983, Ch. 4). As another example, the fictitious Faddeev–Popov ghost fields (spinless fermion fields) to be introduced in §7.4 play the role of "negative degrees of freedom," with the minus signs coming from fermion loops. These negative degrees of freedom serve as a constraint to cancel the spurious boson degrees of freedom associated with unphysical longitudinal polarizations that propagate on internal lines of gauge field diagrams. Neither the particles associated with the longitudinal polarizations, nor the ghosts, are physically significant in themselves since they cancel each other in contributions to physical observables. However, in gauges haunted by ghosts the ghosts will be essential to the unitarity of the S-matrix.

4.8 Background and Further Reading

The discussion in this chapter has followed Abers and Lee (1973); Huang (1982), Ch. 7; Ryder (1985); Taylor (1976), Ch. 10; and Coleman (1975) (which is reprinted as Ch. 5 of Coleman, 1985). The book by Ramond (1981) provides a systematic exposition of field theory from a path integral perspective, with little recourse to the canonical method. Additional useful material is presented in Cvitanovic (1983), Lee (1981), Faddeev (1976), Itzykson and Zuber (1980), Rivers (1987), and Negele and Orland (1988). The conceptual basis of the method of path integrals may be found in Feynman and Hibbs (1965) and Dirac (1958).

Part II

GAUGE FIELD THEORIES

CHAPTER 5

Symmetry and Group Theory

Symmetry has always played an important role in physics. However, within the last half century principles of symmetry have moved from providing a useful pedagogical framework to the forefront of physics. As we shall see, this has largely been a consequence of the discovery that there are considerably more profound symmetries than those associated with spacetime (symmetries operating on internal dynamical degrees of freedom), and that those internal symmetries can be "broken" in more subtle ways than the obvious one of adding an explicit symmetry-breaking term to the Hamiltonian (*hidden symmetry*, or *spontaneous symmetry breaking*).

5.1 Introduction to the Theory of Groups

The root of a symmetry principle is the assumption that certain quantities are unobservable; this in turn implies an invariance under a related mathematical transformation, and the invariance under this transformation (if it is unitary, as is usually the case in quantum mechanics) implies a conservation law or selection rule. For example, the assumption that an absolute direction in space is not observable (isotropy of space) implies an invariance under transformations that rotate the system, and this in turn yields a conservation law for angular momentum. We discussed such conservation laws in terms of conserved Noether tensors in §2.8. Table 5.1 lists such relationships for some of the most important symmetries we will be encountering. In all cases, notice the relations among nonobservables, symmetry transformations, and conservation laws.

Because symmetries play such an important role we need quantitative methods for dealing with them. The natural mathematical technique for analyzing symmetries is the *theory of groups*. Because of their importance in the material we are considering, particular attention will be devoted in the subsequent discussion to *Lie groups* and their associated *Lie algebras* (Georgi, 1982; Wybourne, 1974).

Table 5.1
Some Symmetries and Conservation Laws

Nonobservable	Symmetry Transformation	Conservation Law or Selection Rule
Difference between identical particles	Permutation	Bose–Einstein or Fermi–Dirac statistics
Absolute position	Space translation $\mathbf{x} \rightarrow \mathbf{x} + \delta(\mathbf{x})$	Momentum
Absolute time	Time translation $t \rightarrow t + \delta t$	Energy
Absolute direction	Rotation $\Theta \rightarrow \Theta'$	Angular momentum
Absolute velocity	Lorentz transformation	Generators of the Lorentz group
Absolute right or left	$\mathbf{x} \rightarrow -\mathbf{x}$	Parity
Absolute sign of electrical charge	$e \rightarrow -e$	Charge conjugation
Relative phase between states of different charge Q	$\psi \rightarrow e^{iQ\theta}\psi$	Charge
Relative phase between states of different baryon number N	$\psi \rightarrow e^{iN\theta}\psi$	Baryon number
Relative phase between states of different lepton number L	$\psi \rightarrow e^{iL\theta}\psi$	Lepton number
Difference between coherent mixtures of p and n states	$\begin{pmatrix} p \\ n \end{pmatrix} \rightarrow U \begin{pmatrix} p \\ n \end{pmatrix}$	Isospin

Table adapted from Lee (1981).

5.1.1 Basic Principles

A group is a set G for which a multiplication operation ($\cdot$) is defined, and that has the following properties:

1. If x and y are in G, then $x \cdot y$ is also an element of G.
2. An identity element, e, exists such that $e \cdot x = x \cdot e = x$ for any x in G.
3. Every x in G has an inverse x^{-1}, also an element of G, such that $x \cdot x^{-1} = x^{-1} \cdot x = e$.
4. The multiplication is associative; that is, $(x \cdot y) \cdot z = x \cdot (y \cdot z)$ for every x, y, and z in G.

Note that the multiplication operation need not be ordinary multiplication; it is a law of combination that is called "multiplication" for convenience. Note also that associativity is required by the group definition but *commutativity is not*. If the members of a group mutually commute the group is termed *Abelian*; if not the group is *non-Abelian*. We shall have much to say about non-Abelian groups in our discussion. As a trivial example of a group, the set of all integers $G = \{\cdots -3, -2, -1, 0, 1, 2, 3, \ldots\}$ forms a group under the operation of ordinary addition (*Note:* the group "multiplication" operation is addition!), because

- The sum of any two integers is an integer.
- The identity element 0 exists.
- An inverse exists for each element because $-I + I = 0$.
- Ordinary addition of integers is obviously associative.

Because addition of integers is commutative this group (called the *additive group of the integers*) is Abelian.

EXERCISE 5.1 If the operator c_4 rotates a system by $\pi/2$ about a specified axis, demonstrate that the operator set $\{1, c_4, c_4^2, c_4^3\}$ constitutes an Abelian group under products of rotations.

The number of elements in a group is called the *order* of the group; the order may be finite or infinite. A particular kind of infinite group that will be of interest is the *continuous group*, for which the number of elements is not denumerable and the notation $G(a, b, c, \ldots)$ inappropriate. In that case, each group element is labeled by a continuously varying parameter α, which may be complex. Often the specification of a continuous group element requires a set of continuous parameters, $\boldsymbol{\alpha} \equiv (\alpha_1, \alpha_2, \ldots, \alpha_n)$; we then say that the continuous group is an *n-parameter group*. Notice the distinction between the order of a continuous group (infinite) and the number of continuous parameters required to define it (usually finite). Generally, a continuous group will be specified by the notation $G(\alpha_1, \alpha_2, \ldots, \alpha_n)$, where the n parameters α_i vary continuously.

As another simple (but ubiquitous) example, let us consider the *permutation group* S_n for a finite number of objects n. As illustration, we investigate the permutations for three objects labeled a, b, c. There are $n! = 6$ independent ways the objects can be rearranged. These may be specified by an obvious notation (Georgi, 1982):

1. $(\ ) \equiv \mathbf{1}$ do nothing, so that $(abc) \rightarrow (abc)$
2. (12) swap the objects in positions 1 and 2, so that $(abc) \rightarrow (bac)$
3. (23) so that $(abc) \rightarrow (acb)$
4. (13) so that $(abc) \rightarrow (cba)$

5. (321) cyclic permutation $(abc) \rightarrow (cab)$
6. (123) anticyclic permutation $(abc) \rightarrow (bca)$

Therefore, we have a set of six transformations with a natural multiplication law: $A \cdot B \equiv$ the transformation obtained by *first* making the transformation B, *then* the transformation A (the order matters!). For example,

$$(12) \cdot (13) = (123),$$

because (13) generates $(abc) \rightarrow (cba)$, and subsequent application of (12) gives $(cba) \rightarrow (bca)$, which is the same result as the direct application of (123) to (abc).

By considering all combinations we find that the product of any two permutations is equivalent to a permutation (Exercise 5.3). The multiplication law is clearly associative, because at any step in the product of transformations $A \cdot B \cdot C$ the system has a well-defined state on which subsequent permutations can act. The identity transformation is ( ) $\equiv \mathbf{1}$ (doing nothing); undoing a transformation is the inverse, $(123) \cdot (321) = \mathbf{1}$, and so on. Thus the set of six permutations on three identical objects constitutes the group S_3, which is finite and of order $n! = 6$. The permutation operations do not commute [for example, $(12) \cdot (13) \neq (13) \cdot (12)$], so S_3 is *non-Abelian*.

The permutation groups, or the group discussed in Exercise 5.1, are examples of groups of *transformations*. These are the most important kinds we will consider because of the central role of transformations in quantum mechanics.

5.1.2 Subgroups

Within a set of objects constituting a group, one or more subsets of objects may satisfy the requirements to form a group under the same multiplication operation as defined for the whole group. We then call the subset a *subgroup* of the larger group. As a rather trivial example, consider the set of real numbers, which obviously forms a group under addition with zero as the identity. The set of all integers, including zero, is a subset of this set and is also a group. Therefore, under addition the real numbers form a group and the integers form a subgroup of this group. As another simple example, the permutation group on three objects S_3 has as a subgroup the permutation group on two objects S_2. It also has a subgroup $A_3 \equiv \{1, (123), (321)\}$, which is the group of even permutations on three objects (see Exercise 5.3).

EXERCISE 5.2 Prove that for groups of order two the multiplication table is unique. Hence, all groups of order two are isomorphic to S_2 (isomorphism is defined in the following section).

EXERCISE 5.3 Construct the multiplication table for all products of elements in the permutation group S_3. Use this to demonstrate explicitly that S_3 is a group and that it is non-Abelian. Show that S_3 has two nontrivial Abelian subgroups: the group S_2 of permutations on two objects and a group (called the *alternating group*) $A_3 \equiv \{1, (123), (321)\}$. Define an operator c_3 that rotates a system by $120°$ about a given axis. Show that the set $C_3 \equiv \{1, c_3, c_3^2\}$ constitutes an Abelian group that is isomorphic to A_3.

5.1.3 Isomorphism and Representations

If two groups $G(\alpha_1, \alpha_2, \alpha_3, \ldots, \alpha_N)$ and $G'(\alpha_1', \alpha_2', \alpha_3', \ldots, \alpha_N')$ have elements that can be put into correspondence with each other when their multiplication tables are compared, we say that the two groups are *homomorphic*; for the special case that the correspondence is one-to-one the two groups are said to be *isomorphic*.

Among the groups that may be isomorphic to a given group there are particularly useful ones called *matrix groups*. Consider the permutation group S_3 again. If we denote the three objects by a vector (a, b, c), the reader may easily verify that the following set of matrices constitutes a group in one-to-one correspondence with the group S_3

$$D(1) = \begin{pmatrix} 1 & 0 & 0 \\ 0 & 1 & 0 \\ 0 & 0 & 1 \end{pmatrix} \quad D(12) = \begin{pmatrix} 0 & 1 & 0 \\ 1 & 0 & 0 \\ 0 & 0 & 1 \end{pmatrix} \quad D(13) = \begin{pmatrix} 0 & 0 & 1 \\ 0 & 1 & 0 \\ 1 & 0 & 0 \end{pmatrix}$$

$$D(23) = \begin{pmatrix} 1 & 0 & 0 \\ 0 & 0 & 1 \\ 0 & 1 & 0 \end{pmatrix} \quad D(321) = \begin{pmatrix} 0 & 0 & 1 \\ 1 & 0 & 0 \\ 0 & 1 & 0 \end{pmatrix} \quad D(123) = \begin{pmatrix} 0 & 1 & 0 \\ 0 & 0 & 1 \\ 1 & 0 & 0 \end{pmatrix}$$

$$(5.1)$$

For example,

$$D(12) \begin{pmatrix} a \\ b \\ c \end{pmatrix} = \begin{pmatrix} 0 & 1 & 0 \\ 1 & 0 & 0 \\ 0 & 0 & 1 \end{pmatrix} \begin{pmatrix} a \\ b \\ c \end{pmatrix} = \begin{pmatrix} b \\ a \\ c \end{pmatrix},$$

and $D(12) \leftrightarrow (12)$, where the symbol $\leftrightarrow$ denotes a one-to-one mapping. Further, by explicit matrix multiplication we find that the matrices D have the same multiplication table as the group S_3. For example, the multiplication $(12) \cdot (23) = (321)$ is mapped onto the matrix multiplication

$$D(12) \cdot D(23) = \begin{pmatrix} 0 & 1 & 0 \\ 1 & 0 & 0 \\ 0 & 0 & 1 \end{pmatrix} \begin{pmatrix} 1 & 0 & 0 \\ 0 & 0 & 1 \\ 0 & 1 & 0 \end{pmatrix} = \begin{pmatrix} 0 & 0 & 1 \\ 1 & 0 & 0 \\ 0 & 1 & 0 \end{pmatrix} = D(321). \quad (5.2)$$

Therefore, the group of permutations on three objects and the matrices (5.1) are isomorphic: we say that the set of six matrices D is a (matrix) *representation*[‡] of the permutation group S_3.

It is here that the abstract theory of groups makes contact with the more concrete world of physics: a group is in essence a multiplication table satisfying the postulates of group theory; a representation is a specific realization of that multiplication. For the cases of primary interest the representations will be in terms of finite- or infinite-dimensional orthogonal or unitary matrices. From now on we will work with representations of groups in terms of matrices. The foregoing considerations show that this is adequate, since the matrix representation has the same multiplication table as the group. The power of group theory lies in the fact that it is possible to determine many properties of any representation from the abstract properties of the group; thus, the group mathematics can be worked out once and for all, independent of the particular physical application.

The *dimension* (or *degree*) of a matrix representation for a group is equal to the dimension of each matrix representative. A representation in terms of $N \times N$ matrices is termed an N-dimensional representation of the group. For example, the representation (5.1) is a 3-dimensional representation of the group S_3. The *fundamental matrix representation* of a group is the lowest-dimension faithful representation.

Axiomatic quantum mechanics is formulated in terms of operators acting in linear vector spaces. Therefore, it is also convenient to view representations of quantum mechanical groups in terms of abstract linear operators whose concrete manifestation in a given basis is a matrix operator. If $|j\rangle$ is a complete orthonormal basis in the space in which $D(g)$ is a linear operator, then

$$D(g)\,|i\rangle = \sum_j |j\rangle \,\langle j|\, D(g)\,|i\rangle, \qquad (5.3)$$

which provides the connection between linear operators and their matrix realizations.

For a representation we require that for every x in G there is a unitary operator $D(x)$, and that the mapping preserve the group multiplication law:

$$D(x)D(y) = D(x \cdot y). \qquad (5.4a)$$

For matrices the representation property (5.4a) is (see Exercise 5.4)

$$\sum_j D_{ij}(x)D_{jk}(y) = D(xy)_{ik}. \qquad (5.4b)$$

In the previous example of a matrix representation for the permutation group S_3 this is seen to hold [eq. (5.2)].

[‡]More generally a representation of a group G is a homomorphism of G onto a group of linear operators. For the special case that the homomorphism is one-to-one (isomorphism) the representation is said to be *faithful*. Thus (5.1) is a faithful matrix representation of S_3.

5.1.4 Reducible and Irreducible Representations

The matrix realization of a quantum-mechanical operator depends on the basis employed, and it is clear that a matrix representation of a group is generally not unique. Two representations D and D' of a group are said to be *equivalent* if they are related by a *similarity transformation*:

$$D'(x) = SD(x)S^{-1} \quad \text{(operators)}$$
$$\phi' = S\phi \quad \text{(wavefunctions)},$$
(5.5)

with a fixed operator S for all elements x of the group G. A representation D is *reducible* if each of its matrices can be similarity transformed by the same operator S to an equivalent matrix representation D' that is in *block-diagonal form*:

$$D'(x) = SD(x)S^{-1} = \begin{pmatrix} D'_1(x) & 0 & \cdots & 0 \\ 0 & D_2'(x) & \cdots & 0 \\ \vdots & \vdots & \ddots & \vdots \\ 0 & 0 & \cdots & D'_k(x) \end{pmatrix}$$
(5.6)

If a representation is not reducible it is an *irreducible representation* (an *irrep*, for short): it cannot be brought to block diagonal form by any similarity transformation. If the matrix representation $D(x)$ is reducible the vector space on which D' acts breaks up into k orthogonal spaces, each mapped onto itself by the operators $D'_i(x)$. The representation $D'(x)$ is then termed the *direct sum* of $D'_1, D'_2, \ldots, D'_k$ (denoted by $\oplus$),

$$D' = D'_1 \oplus D'_2 \oplus \ldots D'_k.$$

The resulting spectrum will exhibit a degenerate multiplet structure. A pedagogical discussion of symmetry in physics, with particular emphasis on the multiplets of physical states corresponding to group representations for semisimple Lie groups, may be found in McVoy (1965).

5.2 Lie Groups and Lie Algebras

Of particular interest will be the class of groups known as *Lie groups*.[‡] For our purposes a Lie group is a group with elements labeled by a finite number

---

[‡]We will primarily be interested in *compact* Lie groups. Texts dealing with group theory may be consulted for a precise definition of the term compact (see Lichtenberg, 1978, §2.3). Essentially, a compact group has a parameter space with a finite volume. The group $SO(2)$ of rotations $0-2\pi$ about a single axis is an example of a compact group. The Lorentz group is an example of a noncompact group. All of our discussion of internal symmetries will involve compact Lie groups.

of continuous parameters and a multiplication law depending smoothly on the parameters; one sometimes refers to these groups as *continuous analytical groups*.

From the preceding discussion a continuous one-parameter group $G(\alpha)$ satisfies

$$G(a) \cdot G(b) = G(c), \tag{5.7}$$

where a, b, and c are particular values of the continuous parameter α. This continuous group is also a one-parameter Lie group if it obeys the further restriction that c be an analytic function of a and b. In a similar way N-parameter Lie groups may be defined.

Compact Lie groups have some remarkable properties that are closely associated with the analyticity requirement on the continuous parameters of the group (Georgi, 1982):

- Any representation of a compact Lie group is equivalent to a representation by unitary operators; therefore, only these need be considered.
- Most of the local properties of Lie groups can be determined by considering elements differing infinitesimally from the identity. These properties are defined by a concise set of commutation relations called a *Lie algebra* (global properties of Lie groups that are not specified by the group algebra are taken up in §5.4 and §13.2).

5.2.1 Commutator Algebra

Since compact Lie groups may always be represented by unitary operators we adopt the notation $U(\alpha_1, \alpha_2, \ldots, \alpha_N)$ for an element of an N-parameter Lie group. Then any group element continuously related to the identity may be written

$$U(\alpha_1, \alpha_2, \ldots, \alpha_N) = e^{i\alpha_1 X_1 + i\alpha_2 X_2 + \ldots + i\alpha_N X_N} \equiv e^{i\alpha_a X_a}, \tag{5.8}$$

where a *sum over the repeated index a is implied* and the X_a are linearly independent Hermitian operators.[‡] The set of all combinations $\alpha_a X_a$ spans a linear vector space and the X_a form a basis in that space. The X_a are called

---

[‡]Any unitary matrix may be written in the form $U = \exp(ih)$ where h is a Hermitian matrix; this is proved below. The identity element of a Lie group is associated with the set of null parameters $U(\alpha_1, \alpha_2, \ldots) = U(0, 0, \ldots) = 1$. In our discussions of Lie algebras in this chapter we adopt a convention of summing over any repeated indices, whether subscript or superscript. This convention will be maintained in the remainder of the book for internal symmetries. We will continue to distinguish between upper and lower Lorentz indices.

the *generators* of the group, and there are as many independent generators as there are parameters for the Lie group.

The N-dimensional linear vector space of the generators X_a should not be confused with the Hilbert space in which each generator acts as a quantum mechanical operator. Because we deal with compact groups, these Hilbert spaces can always be taken as finite-dimensional. Therefore, each of the N operators (group generators) X_a may be represented by a *finite matrix* operating in a particular Hilbert space.

An extremely important property of an N-parameter Lie group is that the N group generators satisfy the commutator algebra

$$[X_a, X_b] = i f_{abc} X_c. \tag{5.9}$$

This requirement that the commutator of any two group generators be a linear combination of group generators (we say that the generators are closed under commutation) is called a *Lie algebra*, and the complex numbers f_{abc} are called *structure constants* of the group.

As will become apparent in subsequent discussion, the structure constants are generally not unique for a group since we are free to construct linear combinations of the generators that are also generators. These new generators will satisfy (5.9), but with a different set of f_{abc} [compare (5.32) and (5.40)–(5.42) below]. However, once we have agreed on a basis for the generators the corresponding set of structure constants specifies the local group properties. For most of the examples to be discussed the structure constants f_{abc} carry a great deal of information about the group, because specification of the structure constants of a Lie group is analogous to giving the multiplication table for a finite group.

We may, in fact, define Lie groups in terms of Lie algebras: a Lie group is any continuous group with elements given by (5.8) and generators X_a satisfying the commutator (5.9). It will prove convenient to introduce Lie groups in this way, since in many physical applications we may be more interested in the Lie algebras than in the Lie groups themselves. For example, working in terms of the Lie algebras will allow us to cast equations in a form independent of Lie group representation.

5.2.2 Simple and Semisimple Lie Groups

It may happen that a subset of group generators is closed under commutation with the whole set of generators. That is, commutation of members of the subset with members of the entire algebra yields a linear combination of subset members only, or zero. Such an algebra is call an *invariant subalgebra* of the total algebra. Every algebra has two trivial subalgebras, the whole algebra and the set with only the identity element. A non-Abelian Lie algebra that has no nontrivial continuous invariant subalgebra is called a *simple algebra*, and the corresponding group is a *simple group*.

A subalgebra with generators that commute with every generator is called an *Abelian invariant subalgebra*;[‡] for unitary groups each such generator is associated with a "$U(1)$ factor" of the group, where $U(1)$ is the group of phase transformations (see §5.3.1). Algebras that do not contain continuous Abelian invariant subalgebras are called *semisimple*, and these give rise to *semisimple groups*. Simple algebras are necessarily also semisimple, but the converse may not be true.

The structure constants carry minimal information for Abelian invariant subalgebras, since if X_a is an Abelian subalgebra generator $f_{abc} = 0$ for any b and c. In contrast, each of the group generators for semisimple algebras has non-zero commutators with some other generator and the structure constants are rich in information.

5.2.3 Direct Product Groups

Suppose a group G has subgroups H_1 and H_2 with the following properties:

1. Every element of H_1 commutes with every element of H_2.
2. Every element g of G can be written uniquely as $g = h_1 h_2$, where $h_1 \in H_1$ and $h_2 \in H_2$.

Then G is said to be the *direct product* of H_1 and H_2, which is expressed as $G = H_1 \times H_2$. Simple groups cannot be expressed as a direct product; semisimple groups can be written as a direct product of simple groups without Abelian $[U(1)]$ factors. The generator algebra of a semisimple Lie group can be completely reduced to a sum of invariant simple Lie subalgebras, with the generators in each subalgebra commuting with those in any other subalgebra. Some important examples of direct product groups are the unified electroweak $SU(2) \times U(1)$ gauge group (Ch. 9) and the chiral $SU(3) \times SU(3)$ group (§12.4) [see also the footnote preceding eq. (5.58)].

5.2.4 Structure Constants and Adjoint Representations

Since commutators satisfy

$$[A, B] = -[B, A], \tag{5.10}$$

[‡]An invariant subalgebra is also called an *ideal*. An Abelian invariant subalgebra is sometimes termed a *maximal ideal* or *center of the algebra*. Thus, the center of an algebra commutes with every element of the algebra.

and the *Jacobi identity*,

$$[[A, B], C] + [[B, C], A] + [[C, A], B] = 0, \tag{5.11}$$

we have the useful relations for the structure constants

$$f_{bac} = -f_{abc} \tag{5.12}$$

$$f_{bcd}f_{ade} + f_{abd}f_{cde} + f_{cad}f_{bde} = 0. \tag{5.13}$$

In addition, for *compact, semisimple* groups the f_{abc} may be chosen to be antisymmetric under the exchange of any two indices (see Exercise 5.10b). This will be the case for all the structure constants that we will employ.

Suppose we define a set of matrices T_a with matrix elements

$$(T_a)_{bc} \equiv -if_{abc}; \tag{5.14}$$

then, from (5.13)

$$[T_a, T_b] = if_{abc}T_c. \tag{5.15}$$

Since this is a Lie algebra [eq. (5.9)] and the generators T_a are given by (5.14), we see that *the structure constants themselves generate a representation of the algebra*. This representation plays a pivotal role; it is called the *adjoint representation* or the *regular representation*. The dimension of a representation is equal to the dimension of the vector space on which it acts, so *the adjoint representation has a dimension equal to the number of group generators*.

EXERCISE 5.4 Show that the matrices associated with classical rotations in two dimensions form a one-parameter Lie group.

EXERCISE 5.5 (a) Given eqns. (5.9)–(5.11) and (5.14), prove the relations (5.12), (5.13), and (5.15). (b) Derive the Lie algebra (5.9) by expanding the group elements written in the canonical exponential form (5.8) about the origin. *Hint*: define the commutator of two group elements $U(\alpha)$ and $U(\beta)$ lying near the origin $U(0)$ by

$$U(\alpha)^{-1}U(\beta)^{-1}U(\alpha)U(\beta) \equiv U(\gamma),$$

where $U(\gamma)$ must, by the definition of a Lie group, also be an element of the group lying near the origin.

5.3 Unitary Symmetries

The most interesting Lie groups for our purposes are the $SU(N)$ groups. These correspond to groups with a fundamental representation in terms of

$N^2 - 1$ unitary $N \times N$ matrices [the $U(N)$ portion of the designation] of unit determinant (the S portion of the designation). Such groups are simply connected, compact, and depend on $N^2 - 1$ real parameters.[‡] The elements of such groups are canonically written in the form (5.8), and the specification of the group is often given in terms of the traceless, Hermitian group generators X_a and their Lie algebras, rather than in terms of the unitary matrices $U(\alpha_1, \alpha_2, \ldots, \alpha_N)$ themselves.

5.3.1 $U(1)$ Symmetries

The simplest unitary symmetry is rather trivial. If we choose in (5.8) a single constant generator J (independent of spacetime coordinates), the Lie group is the one-parameter group of global phase transformations

$$U(\theta) = e^{iJ\theta}, \tag{5.16}$$

where θ is any real number. This is obviously an Abelian group. Symmetry under $U(1)$ implies an invariance under global phase transformations, which is generally linked to a global conservation law (§2.8). For example, the QED Hamiltonian is invariant under the unitary transformations

$$e^{iL\theta} H e^{-iL\theta} = H, \tag{5.17}$$

where L is the lepton number operator

$$L \equiv \int :\psi^\dagger(x)\psi(x): d^3x \,. \tag{5.18}$$

This invariance is associated with lepton number conservation in QED (see Exercises 5.6a and 2.14). Some other important $U(1)$ symmetries follow from

---

[‡]A complex $N \times N$ matrix has $2N^2$ real parameters; $U^\dagger U = 1$ imposes N^2 conditions; Det $U = 1$ imposes one condition. Thus, $SU(N)$ matrices are specified by $N^2 - 1$ real parameters. As noted earlier, compactness means (loosely) that the volume of the parameter space is finite. A group is connected if for any group element the identity may be reached by continuous variation of the parameters. For example, the group $O(n)$ of orthogonal $n \times n$ matrices is not connected because it has two disjoint pieces with Det $= +1$ and -1 respectively, and we cannot pass continuously between them by variation of the parameters. On the other hand, the group $SO(n)$ of Det $= 1$ (special), orthogonal $n \times n$ matrices is connected; the part of $O(n)$ connected to the identity is $SO(n)$. A connected group is *simply connected* if there is topologically only one kind of closed path in the group manifold, *multiply connected* if there is more than one kind (see §13.2).

baryon number (B) and charge (Q) conservation, which entail invariance of the Hamiltonian under the unitary transformations

$$e^{iB\alpha} H e^{-iB\alpha} = H \qquad e^{iQ\beta} H e^{-iQ\beta} = H, \tag{5.19}$$

respectively, with α and β arbitrary real numbers.

5.3.2 $SU(2)$ Symmetries

The simplest non-Abelian special unitary group is $SU(2)$, with a fundamental representation corresponding to the group of transformations

$$\psi' = U\psi, \tag{5.20}$$

where U is a 2×2 unitary matrix of unit determinant acting on a doublet

$$\psi = \begin{pmatrix} u \\ d \end{pmatrix}. \tag{5.21}$$

For the moment we need not attach physical significance to the components of the doublet, but they might represent spin up or spin down for an electron in an angular momentum context, or neutron and proton states in an isospin context (see below). The matrices U are composed of the elements of a linear operator U acting in a Hilbert space spanned by the basis vectors $|u\rangle$ and $|d\rangle$. This two-dimensional space is the (Pauli) spinor space in the first example cited above, or the isospinor space in the second example.

Now for a matrix A [see eq. (13.43)]

$$\text{Det}\,(e^A) = e^{\,\text{Tr}\,A}, \tag{5.22}$$

where Tr denotes the trace (the sum over diagonal elements). For $SU(2)$ the transformation matrices U are traditionally parameterized as

$$U \equiv e^{i\sigma_j \theta_j/2}, \tag{5.23}$$

where θ_j are continuous parameters and the three matrices $\sigma_j/2$ are the group generators. Therefore, since $SU(2)$ matrices are restricted to those with unit determinant,

$$\text{Det}\,U = e^{\,\text{Tr}\,(i\sigma_j \theta_j/2)} = 1, \tag{5.24}$$

and the generators are necessarily traceless,

$$\text{Tr}\,(\sigma_j) = 0. \tag{5.25}$$

Further, the inverse and Hermitian conjugates of U are

$$U^{-1} = e^{-i\sigma_j \theta_j/2} \tag{5.26}$$

$$U^\dagger = e^{-i\sigma_j^\dagger \theta_j/2}. \tag{5.27}$$

But the $SU(2)$ matrices are unitary, $U^\dagger = U^{-1}$, and this requires that

$$\sigma_j^\dagger = \sigma_j. \tag{5.28}$$

Therefore, we have established that the matrices σ_j for the fundamental representation are $N \times N = 2 \times 2$, traceless, and Hermitian. There are only three independent possibilities for these matrices. They are usually chosen to be the Pauli matrices [eq. (1.78)]

$$\sigma_1 = \begin{pmatrix} 0 & 1 \\ 1 & 0 \end{pmatrix} \quad \sigma_2 = \begin{pmatrix} 0 & -i \\ i & 0 \end{pmatrix} \quad \sigma_3 = \begin{pmatrix} 1 & 0 \\ 0 & -1 \end{pmatrix}, \tag{5.29}$$

which satisfy the commutator

$$\left[\frac{\sigma_i}{2}, \frac{\sigma_j}{2} \right] = i\epsilon_{ijk} \frac{\sigma_k}{2}. \tag{5.30}$$

The commutators (5.30) generally do not vanish and $SU(2)$ is a non-Abelian group.

Comparing with (5.9) we see that this commutator relation defines the Lie algebra of the group $SU(2)$, with the structure constants in this basis given by the completely antisymmetric third-rank tensor,

$$f_{ijk}^{SU(2)} = \epsilon_{ijk}. \tag{5.31}$$

Thus the Pauli matrices may be chosen as generators satisfying (5.30), and they operate on the fundamental (two-dimensional) representation (5.21) of $SU(2)$. Generalizing, N-dimensional representations of $SU(2)$ may be constructed using three $N \times N$ matrices J_i satisfying the algebra

$$[J_i, J_j] = \epsilon_{ijk} J_k. \tag{5.32}$$

As in the fundamental representation these matrices are Hermitian and traceless, and the N-member multiplets infinitesimally transformed by the J_i constitute N-dimensional representations of $SU(2)$.

EXERCISE 5.6 (a) The QED Hamiltonian density is given by (2.87) and (2.202). Show that it is invariant under the $U(1)$ transformation (5.17), and that this leads to conservation of lepton number (5.18). *Hint*: First show that $[L, \psi(x)] = -\psi(x)$, from which

$$e^{iL\theta}\psi(x)e^{-iL\theta} = e^{-i\theta}\psi(x).$$

(b) Show that the most general form for $SU(2)$ generators in the fundamental representation is

$$\sigma = \begin{pmatrix} \alpha & \beta \\ \beta^* & -\alpha \end{pmatrix}$$

with α real [compare (5.29)]. Prove that the most general form for 2×2 unitary matrices of unit determinant is

$$U = \begin{pmatrix} a & b \\ -b^* & a^* \end{pmatrix} \qquad |a|^2 + |b|^2 = 1,$$

where a and b are complex numbers; thus the group manifold of $SU(2)$ may be parameterized by a 3-dimensional sphere.

Group Theory of Angular Momentum. The group theory of $SU(2)$ is independent of the particular physical application, but we will illustrate it with concrete examples. The first is angular momentum—the commutator (5.32) is the familiar angular momentum algebra, and angular momentum in quantum mechanics can be discussed in terms of $SU(2)$ representations.[‡] The strategy to be used is termed the method of *roots and weights*, or the *Cartan–Dynkin method*. Although we will find that most concepts are already familiar from the elementary quantum mechanics of angular momentum, the same techniques form the basis of a powerful and systematic procedure to analyze the structure of more complicated groups.

The essence of the roots and weights method is to divide the generators of a group into two sets by taking judicious linear combinations. The first set consists of Hermitian operators that can be diagonalized to give quantum numbers to label members of irreducible multiplets. The second set consists of stepping operators (sometimes called ladder operators) that allow us to move through an irreducible multiplet in such a manner that from any member of the multiplet it is possible to reach any other member of the multiplet by the application of successive stepping operators, and no combination of stepping operators takes us out of the multiplet. The procedure may be outlined as follows:

1. For non-Abelian groups the generators generally do not commute, but it is possible to find nonlinear functions of generators that do commute with

---

[‡]See Brink and Satchler (1968) for a concise summary of angular momentum in quantum mechanics.

each generator of the algebra. These nonlinear functions are called *Casimir operators*. In general, a semisimple Lie algebra has l independent Casimir operators, where l is termed the *rank of the algebra*. For the compact, semisimple algebras the generators may be chosen such that the lowest order Casimir operator is equal to the sum of the squares of the group generators. The rank can usually be expressed in terms of simple formulas for non-Abelian groups. For example, the $SU(N)$ groups are of rank $l = N - 1$. The significance of the Casimir operators is expressed by Schur's Lemma: *a matrix that commutes with every matrix of an irreducible representation is a constant multiple of the unit matrix*. This means that the Casimir operators give *eigenvalue equations*, with quantum numbers that can be used to label the irreps of a group.

2. The generators X_a are the basis vectors of a linear vector space, and we are permitted to construct linear combinations such that as many operators as possible are (a) Hermitian, (b) mutually commuting, and (c) diagonal. *The maximum number of such operators is equal to l, the rank of the algebra*, and this maximal set of commuting operators

$$H_i(i = 1, 2, \ldots, l) \tag{5.33}$$

is termed the *Cartan subalgebra*. The diagonal generators of the Cartan subalgebra provide l eigenvalue equations

$$H_i \psi_{m_i}^{(j)} = m_i \psi_{m_i}^{(j)} \qquad (i = 1, 2, \ldots, l), \tag{5.34}$$

where j denotes all labels necessary to specify the representation and m_i is termed a *weight* of the eigenstate $\psi_{m_i}^{(j)}$. These eigenvalue equations give "additive" quantum numbers, corresponding to components of vectors, while the Casimir operators give quantum numbers that are not simply additive. For example, we shall see below that the Casimir operator of the angular momentum algebra has an eigenvalue proportional to $J(J + 1)$, while the single quantum number coming from the Cartan subalgebra is the magnetic quantum number m.

It is convenient to assemble the weights into a *weight vector*,

$$\mathbf{m} = (m_1, m_2, \ldots, m_l). \tag{5.35}$$

The components of $\mathbf{m}$ span a space that is termed the *weight space*, and the l-dimensional plot of all the weights for a representation is called the *weight diagram*. The particular case of the weight diagram for the adjoint representation plays an important role in the analysis of Lie groups, and the weights in the adjoint representation are called the *roots*. The *multiplicity* or the *degeneracy* of a weight is the number of different eigenvectors in a representation having that weight.

The special unitary groups $SU(N)$ have $N^2 - 1$ generators, and $N - 1$ of these may be diagonalized simultaneously. Therefore, the rank of $SU(N)$ is $l = N - 1$ and it has $N - 1$ Casimir operators. The number of independent stepping operators that may be constructed to move through the l-dimensional weight space of $SU(N)$ is

$$N^2 - 1 - (N - 1) = N^2 - N.$$

For example, $SU(2)$ has $N^2 - 1 = 3$ generators [the familiar Pauli matrices (1.78) in the fundamental representation] and $N - 1$ of them may be diagonalized simultaneously, corresponding to the familiar observation that only one projection of the angular momentum may be chosen as a good quantum number (conventionally taken to be the "magnetic quantum number", $m = J_z$). Therefore, $SU(2)$ is a rank-one Lie group. Specifically, the angular momentum algebra is (5.32) and the Cartan subalgebra of $SU(2)$ may be chosen to be the single element

$$H_1 = J_3, \tag{5.36}$$

corresponding to an eigenvalue equation

$$J_3 \psi_m^{(J)} = m \psi_m^{(J)}, \tag{5.37}$$

where the weight is $m = J_3$ and the Casimir operator is

$$C = J^2 = J_1^2 + J_2^2 + J_3^2, \tag{5.38}$$

which yields an eigenvalue equation ($\hbar = 1$)

$$C\psi^{(J)} = J(J+1)\psi^{(J)}. \tag{5.39}$$

Since $SU(2)$ is rank-1, this is the only Casimir operator and the quantum number J, which we know secretly to be the total angular momentum quantum number, is seen to label different $SU(2)$ irreducible representations. There are $(2J + 1)$ weights $m = J_3$ corresponding to an angular momentum J, so the irreps of $SU(2)$ are $(2J + 1)$-dimensional. The $N^2 - N = 2$ stepping operators that may be constructed from the generators not in the Cartan subalgebra are familiar from elementary quantum mechanics:

$$J_+ \equiv J_1 + iJ_2 \tag{5.40}$$
$$J_- \equiv J_1 - iJ_2, \tag{5.41}$$

as are the algebraic properties,

$$[J_3, J_\pm] = \pm J_\pm \qquad J^2 = \tfrac{1}{2}(J_+J_- + J_-J_+) + J_3^2$$
$$[J_+, J_-] = 2J_3 \qquad [J^2, J_i] = 0 \tag{5.42}$$

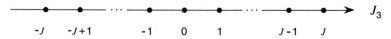

Fig. 5.1   Weight diagram for $SU(2)$ representation of integer spin J.

$$J_\pm \left| J, J_3 \right\rangle = \sqrt{(J \pm J_3 + 1)(J \mp J_3)} \left| J, J_3 \pm 1 \right\rangle, \qquad (5.43)$$

except that

$$J_+ \left| J_3 = J \right\rangle = 0 \qquad J_- \left| J_3 = -J \right\rangle = 0. \qquad (5.44)$$

Therefore the successive application of stepping operators takes us between all members of a $(2J + 1)$-dimensional multiplet corresponding to an irrep of $SU(2)$, but not out of the irreducible representation, and all states accessible to the stepping operators are labeled by the Casimir eigenvalue of the state from which we start.

The multiplicity of each weight in an $SU(2)$ irreducible representation is unity, since each eigenvector has a separate value of J_3 within an angular momentum multiplet. Thus the weight diagram corresponding to an irrep of $SU(2)$ is one-dimensional and may be represented by a line with sites that are singly occupied at integer steps from $-J$ to $+J$. Figure 5.1 illustrates for an integer angular momentum.

The construction of an $SU(2)$ irreducible multiplet is most easily implemented using the *highest-weight algorithm*, which rests on the following assertions, valid for any compact, semisimple group:

1. For each irreducible representation there is a weight that can in some sense be designated the "highest weight" of the representation; it is unique, corresponding to a singly occupied site in the weight diagram.
2. Two equivalent irreducible representations have the same highest weight; two irreducible representations with the same highest weight are equivalent.

Let us now use this algorithm to construct the weight space for the direct product of some $SU(2)$ representations.

Direct Product of Representations. The direct product of matrices may be defined as follows. If A is an $m \times m$ matrix and B is an $n \times n$ matrix, their direct product is

$$A \otimes B = C, \qquad (5.45)$$

where C is an $mn \times mn$ matrix with elements

$$c_{ik,j\ell} = a_{ij} b_{k\ell}. \qquad (5.46)$$

For example, if A and B are 2×2 matrices we may write

$$A \otimes B = \begin{pmatrix} a_{11}B & a_{12}B \\ a_{21}B & a_{22}B \end{pmatrix}$$

$$= \begin{pmatrix} a_{11}b_{11} & a_{11}b_{12} & a_{12}b_{11} & a_{12}b_{12} \\ a_{11}b_{21} & a_{11}b_{22} & a_{12}b_{21} & a_{12}b_{22} \\ a_{21}b_{11} & a_{21}b_{12} & a_{22}b_{11} & a_{22}b_{12} \\ a_{21}b_{21} & a_{21}b_{22} & a_{22}b_{21} & a_{22}b_{22} \end{pmatrix}. \tag{5.47}$$

The space in which $A \otimes B$ operates is called a *direct-product space*; it occurs in quantum mechanics when two uncorrelated degrees of freedom supply independent state labels. A specific example is afforded by the coupling of independent angular momenta. Since the weights are additive quantum numbers, the direct product representation weights are the sums of the weights for the direct product factors.

A representation $D^{(1)} \otimes D^{(2)}$ is generally reducible. For finite or simple and compact groups the direct product representation may be decomposed into a direct sum of irreducible representations

$$D^{(1)} \otimes D^{(2)} = \sum_i \Gamma_i D^{(i)} = \Gamma_1 D^{(1)} \oplus \Gamma_2 D^{(2)} \oplus \ldots, \tag{5.48}$$

where Γ_i is the number of times the irreducible representation $D^{(i)}$ appears in the sum. This is called a *Clebsch–Gordan series*; for $SU(2)$ it leads to the familiar expression for the product of two spherical harmonics or Wigner D-functions, since the D-functions D^J_{MK} are $(2J+1)$-dimensional representations of the angular momentum algebra (Brink and Satchler, 1968).

If we consider the direct product of two irreducible angular momentum representations the sites on the weight diagram may be multiply occupied, since the weight J_3 for the direct product is the sum of weights for the product vectors. For example, if the weight multiplicity is denoted by placing the symbol $\times$ above the weight for each occurrence of the weight in the representation, the diagram for the direct product of a $J = \frac{3}{2}$ and $J = 1$ representation is given in Fig. 5.2. These multiplicities may be obtained by considering the number of ways the weights $J_3 = \pm\frac{3}{2}, \pm\frac{1}{2}$ for $J = \frac{3}{2}$, and $J_3 = \pm 1, 0$ for $J = 1$ can be added to give a total J_3. For example, $J_3 = \frac{1}{2}$ has a multiplicity of three because

$$\tfrac{1}{2} = \tfrac{3}{2} - 1 = -\tfrac{1}{2} + 1 = \tfrac{1}{2} + 0.$$

We may find the irreducible representations contained in this direct product by using the highest weight algorithm. Take the highest value of J_3; this must be the highest weight state in a $J = J_3$ multiplet of dimension $2J + 1$.

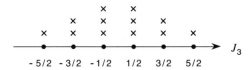

Fig. 5.2   Weight diagram and multiplicity for the direct product of $SU(2)$ representations $J = \frac{3}{2}$ and $J = 1$.

Generate the other $2J$ states of the multiplet by successive applications of J_- and remove them from the diagram. [Recall that the sequence will terminate after $2J + 1$ states because of (5.44)]. Take the largest remaining value of J_3; this must be the highest weight state of another multiplet whose $2J + 1$ members can be generated by successive application of J_- and removed from the diagram. This procedure is repeated until no states remain. For the example in Fig. 5.2 we conclude that the direct product of a $J = 1$ and a $J = \frac{3}{2}$ irreducible representation gives one irreducible $SU(2)$ representation of dimension $2(\frac{5}{2})+1 = 6$, one of dimension $2(\frac{3}{2})+1 = 4$, and one of dimension $2(\frac{1}{2}) + 1 = 2$. This is just what would be expected from the vector coupling model, since for $J_1 = \frac{3}{2}$ and $J_2 = 1$ we obtain resultant values of $J = \frac{5}{2}, \frac{3}{2}$, and $\frac{1}{2}$. In a common notation,

$$4 \otimes 3 = 6 \oplus 4 \oplus 2, \tag{5.49}$$

where $\otimes$ and $\oplus$ denote direct product and direct sum, respectively, and the boldface numbers are dimensions $(2J+1)$ of irreducible $SU(2)$ representations.

We have belabored a number of points concerning angular momentum that should be familiar from elementary quantum mechanics. Our purpose was twofold: first, some fancy terminology has been attached to some familiar ideas; second, and more important, most of the procedures illustrated for angular momentum $SU(2)$ will carry over with appropriate generalization to $SU(N)$ symmetries.

Before turning to those higher unitary symmetries we examine two more aspects of $SU(2)$: the adjoint representation, and $SU(2)$ as an internal symmetry of the strong interactions (isospin symmetry). These will have important counterparts in the higher $SU(N)$ symmetries as well.

Adjoint Representation. From the foregoing considerations we see that for angular momentum 1 there will be a $2J + 1 = 3$-dimensional representation of $SU(2)$, and the corresponding members of the multiplet may be labeled by the eigenvalues of one of the generators. Choosing J_3 to be diagonal, it is readily verified that the following matrices obey the $SU(2)$ algebra (5.32)

$$J_1 = \frac{1}{\sqrt{2}} \begin{pmatrix} 0 & -1 & 0 \\ -1 & 0 & 1 \\ 0 & 1 & 0 \end{pmatrix} \quad J_2 = \frac{1}{\sqrt{2}} \begin{pmatrix} 0 & i & 0 \\ -i & 0 & -i \\ 0 & i & 0 \end{pmatrix} \quad J_3 = \begin{pmatrix} 1 & 0 & 0 \\ 0 & 0 & 0 \\ 0 & 0 & -1 \end{pmatrix}.$$

$$(5.50)$$

These act on a multiplet $|J, J_3\rangle$ that may be labeled by the eigenvalues of J_3,

$$|1,1\rangle = \begin{pmatrix} 1 \\ 0 \\ 0 \end{pmatrix} \qquad |1,0\rangle = \begin{pmatrix} 0 \\ 1 \\ 0 \end{pmatrix} \qquad |1,-1\rangle = \begin{pmatrix} 0 \\ 0 \\ 1 \end{pmatrix}. \qquad (5.51)$$

Since $SU(2)$ is a three-parameter group and hence has three generators, this representation is the adjoint representation (the representation with dimension equal to the number of group generators). As previously noted (5.14), the generators of the adjoint representation are just the structure constants multiplied by $-i$. This fact could have been used to construct the three-dimensional $SU(2)$ representation (Exercise 5.7). Alternatively, direct products of lower dimensional representations may be used to build the adjoint and other higher dimensional irreducible representations (Exercise 5.12). For example, the direct product of the fundamental doublet with itself yields the decomposition

$$\mathbf{2} \otimes \mathbf{2} = \mathbf{1} \oplus \mathbf{3}, \qquad (5.52)$$

corresponding to the well-known result that two spin-$\frac{1}{2}$ objects may be vector coupled to give a spin-0 singlet and spin-1 triplet. This is a specific example of *tensor methods* in group theory; we will see below how to use such methods to determine the irreducible representations contained in the direct product of arbitrary representations of unitary groups.

EXERCISE 5.7 Use the observation that the generators of the adjoint representation are the structure constants of the group multiplied by $-i$ [see (5.14)] to construct a three-dimensional matrix representation of $SU(2)$. Verify explicitly that the resulting matrices satisfy the $SU(2)$ algebra. Transform this set of matrices to a new set J_1, J_2, J_3, where J_3 is diagonal.

Isotopic Spin. Consider the creation and annihilation operators for neutrons and protons, where to simplify notation we use the particle's symbol to stand for the corresponding operator. For example,

$$P^\dagger \equiv a^\dagger_{\text{proton}} \qquad N \equiv a_{\text{neutron}} \qquad (5.53)$$

create a proton and annihilate a neutron, respectively. Since these are Fermi operators they obey anticommutation relations [eq. (2.96)]. If we limit attention to combinations that do not change the number of particles in the system there are four independent bilinear products of neutron and proton operators:

$$P_\alpha^\dagger N_\alpha \quad N_\alpha^\dagger P_\alpha \quad P_\alpha^\dagger P_\alpha \quad N_\alpha^\dagger N_\alpha, \tag{5.54}$$

where the index α (implied sum) stands for all required labels. From these bilinear products four independent operators may be constructed (Lipkin, 1966, Ch. 2)

$$B = P_\alpha^\dagger P_\alpha + N_\alpha^\dagger N_\alpha \tag{5.55}$$

$$T_+ = P_\alpha^\dagger N_\alpha, \qquad T_- = N_\alpha^\dagger P_\alpha \tag{5.56a}$$

$$T_3 = \tfrac{1}{2}(P_\alpha^\dagger P_\alpha - N_\alpha^\dagger N_\alpha). \tag{5.56b}$$

If we restrict attention to a neutron–proton system these operators have a simple interpretation. The operator B counts the total number of nucleons, so it is the baryon number operator for our simple system; B commutes with each of the other operators so the baryon number is conserved. The operator T_+ destroys a neutron and creates a proton, while T_- destroys a proton and creates a neutron; T_3 counts half the excess of protons over neutrons so it may also be written

$$T_3 = Q - \frac{B}{2}, \tag{5.57}$$

where $Q = P_\alpha^\dagger P_\alpha$ is the charge operator. It is easy to verify that the T operators satisfy the commutators[‡]

$$[T_3, T_\pm] = \pm T_\pm \qquad [T_+, T_-] = 2T_3. \tag{5.58}$$

Comparing with eq. (5.42) we see that this is just the Lie algebra of angular

---

[‡]The complete algebra of (5.55)–(5.56) corresponds to a $U(2)$ symmetry. The operators $T_\pm$, T_3 alone are closed under commutation and form the $SU(2)$ subgroup of $U(2)$. The baryon number operator corresponds to a $U(1)$ subgroup of phase transformations (see Lichtenberg, 1978, §6.2). Notice from the commutators (5.55)–(5.56) that the group $U(2)$ may be written as the direct product $SU(2) \times U(1)$, so $U(2)$ is neither simple nor semisimple (see §5.2). The group $SU(2)$ is an example of a simple group, for it has no continuous invariant subgroup. It does have a *discrete* invariant subgroup consisting of the 2×2 matrices ± 1 that commute with all other $SU(2)$ matrices (see also §13.2). However, our definition of a simple Lie group requires only the absence of *continuous* invariant subgroups.

momentum. By analogy with angular momentum $SU(2)$ there exists one Casimir operator commuting with all group generators

$$\widehat{T}^2 = T_1^2 + T_2^2 + T_3^2 = \tfrac{1}{2}(T_+T_- + T_-T_+) + T_3^2, \tag{5.59}$$

where

$$T_\pm = T_1 \pm iT_2. \tag{5.60}$$

Continuing the analogy with angular momentum $SU(2)$, we expect multiplets of dimension $2T + 1$, where the members ψ of the multiplet satisfy

$$\widehat{T}^2\psi = T(T + 1)\psi, \tag{5.61a}$$

and multiplet members may be labeled by the quantum number T_3

$$\widehat{T}_3\psi = T_3\psi. \tag{5.61b}$$

If the Hamiltonian is invariant under this $SU(2)$ symmetry, which we call $SU(2)$ isospin, the stepping operators $T_\pm$ transform between members of a degenerate multiplet [irreducible representation of $SU(2)$ in the isospace]. Since $T_\pm$ transforms neutrons to protons and *vice versa*, $SU(2)$ isospin symmetry means that the energy of the system is unchanged when neutrons and protons are interconverted, subject to antisymmetry requirements. This is the group-theoretical implementation of the *charge independence hypothesis* of nuclear physics. Since the neutron and proton are opposite "spin" projections in some abstract space, we may formulate $SU(2)$ isospin symmetry as a rotational invariance in that space: isospin invariance implies the isotropy of the isospace, just as rotational invariance implies the isotropy of ordinary space (see Table 5.1).

The group theory of angular momentum is the same as that of isospin, and the technical machinery developed for the former may be carried over intact to the latter. For example, we may vector couple isospins just as for angular momentum, and the $SU(2)$ vector coupling coefficients are just the familiar Clebsch–Gordan coefficients. This will lead to selection rules for isospin transitions analogous to those for angular momentum transitions, and so on. The reader not already familiar with these ideas is urged to consult standard texts in nuclear and particle physics where they are discussed (for example, see Perkins, 1987, Ch. 4).

Isotopic spin is found to be a very good symmetry of the strong interactions, but it is broken by the weak and electromagnetic interactions. We will see later that isospin is no longer considered to be a primordial symmetry. The strong interactions are explained in terms of a more fundamental symmetry $[SU(3)_{\text{color}}]$ in the theory of quantum chromodynamics (Ch. 10). Isospin is a good symmetry by what could be an accident of nature—the near equivalence of the up and down quark masses, and the smallness of their masses relative to the characteristic scale for strong interactions. Nevertheless, isospin remains an excellent phenomenological symmetry of the strong interactions, and it occupies an important historical niche: it was the first symmetry to focus attention in particle physics on symmetry principles for internal (rather than spacetime) degrees of freedom. Considering the subsequent development of theories of fundamental interactions, this was an achievement of considerable import.

Just as for the angular momentum algebra, we may expect physical realization of higher-dimensional representations of $SU(2)_{\text{isospin}}$. For example, the adjoint (or vector) representation is generated by the 3×3 matrices (5.50) satisfying the algebra (5.32). Just as the 2-dimensional representation of $SU(2)$ could accommodate a $T = \frac{1}{2}$ isodoublet corresponding to a neutron and proton of nearly degenerate mass, we expect the adjoint representation to accommodate degenerate three-member $T = 1$ (isovector) multiplets. The members of these multiplets differ only in T_3, so for fixed baryon number they may be distinguished by their charge [eq. (5.57)]. In the hadronic spectrum we observe a nearly degenerate triplet $(\pi^{\pm}, \pi^0)$ near 140 MeV. These may be viewed as the components of a 3-dimensional irreducible representation of isospin $SU(2)$; that is, as an isovector multiplet.[‡]

Conjugate Representation. Suppose antinucleons are used to construct a fundamental representation of isospin $SU(2)$. Then there are two possible two-dimensional representations of $SU(2)$, which may be labeled

$$\mathbf{2} \equiv \begin{pmatrix} p \\ n \end{pmatrix} \qquad \bar{\mathbf{2}} \equiv \begin{pmatrix} -\bar{n} \\ \bar{p} \end{pmatrix},$$

where the **2** denotes a dimensionality and bars signify antinucleons. (It is also common to denote the second representation by an asterisk: $\bar{\mathbf{2}} = \mathbf{2}^*$.) The ordering and phasing of the $\bar{\mathbf{2}}$ representation components is a conventional choice so that the **2** and $\bar{\mathbf{2}}$ transform in the same way under $SU(2)$ (see Halzen and Martin, 1984, §2.7). The $\bar{\mathbf{2}}$ is called the *conjugate representation.* It is not difficult to prove that $\bar{\mathbf{2}}$ and **2** are equivalent, so there is only one fundamental representation for $SU(2)$. This is not generally true; $SU(N)$ has representations of dimension $\mathbf{N}$ and $\overline{\mathbf{N}}$, and for $N > 2$ these representations are not equivalent. For example, in $SU(3)$ the fundamental representations are **3** and $\bar{\mathbf{3}}$, and they are distinguishable (Exercise 5.9). We will be loose in our terminology and sometimes refer to both the **3** and $\bar{\mathbf{3}}$ as fundamental

---

[‡]Do not confuse the isospace with the normal space. Pions are Lorentz pseudoscalars ($J^P = 0^-$), but vectors ($T = 1$) in the isospin degrees of freedom. In the jargon they are pseudoscalar, isovector particles.

representations for $SU(3)$; when more precise terminology is required we will call them the fundamental and conjugate representations, respectively.

A representation is said to be *real* if it is equivalent to its conjugate representation and *complex* if it is not. Thus, groups like $SU(3)$ admit complex representations, but $SU(2)$ does not. This property plays an important role in the discussion of grand unified theories in Ch. 11.

EXERCISE 5.8 (a) Verify the isospin commutators (5.58). (b) Show that the Lagrangian density

$$\mathcal{L} = (\partial_\mu \phi)^\dagger (\partial^\mu \phi) - \mu^2 (\phi^\dagger \phi) - \lambda (\phi^\dagger \phi)^2 \qquad \phi \equiv \begin{pmatrix} \phi_1 \\ \phi_2 \end{pmatrix}$$

is invariant under $SU(2)$ rotations of the two-component fields ϕ. Derive the conserved (isospin) current and charge.

5.3.3 $SU(3)$ Symmetries

Generators. The next unitary symmetry that we consider is $SU(3)$, which has $N^2 - 1 = 8$ independent generators. In the fundamental representation these will be traceless 3×3 Hermitian matrices, and since this is a rank-2 Lie algebra two of the eight generators may be simultaneously diagonalized. The canonical representation is

$$U = e^{i\lambda_k \alpha_k / 2} \qquad (k = 1, 2, 3, \dots, 8), \tag{5.62}$$

where the group generators are $\lambda_k/2$, with the standard choice for the λ_k (Gell-Mann matrices)

$$\lambda_1 = \begin{pmatrix} 0 & 1 & 0 \\ 1 & 0 & 0 \\ 0 & 0 & 0 \end{pmatrix} \qquad \lambda_2 = \begin{pmatrix} 0 & -i & 0 \\ i & 0 & 0 \\ 0 & 0 & 0 \end{pmatrix} \qquad \lambda_3 = \begin{pmatrix} 1 & 0 & 0 \\ 0 & -1 & 0 \\ 0 & 0 & 0 \end{pmatrix}$$

$$\lambda_4 = \begin{pmatrix} 0 & 0 & 1 \\ 0 & 0 & 0 \\ 1 & 0 & 0 \end{pmatrix} \qquad \lambda_5 = \begin{pmatrix} 0 & 0 & -i \\ 0 & 0 & 0 \\ i & 0 & 0 \end{pmatrix} \qquad \lambda_6 = \begin{pmatrix} 0 & 0 & 0 \\ 0 & 0 & 1 \\ 0 & 1 & 0 \end{pmatrix}$$

$$\lambda_7 = \begin{pmatrix} 0 & 0 & 0 \\ 0 & 0 & -i \\ 0 & i & 0 \end{pmatrix} \qquad \lambda_8 = \frac{1}{\sqrt{3}} \begin{pmatrix} 1 & 0 & 0 \\ 0 & 1 & 0 \\ 0 & 0 & -2 \end{pmatrix}$$

$$\tag{5.63}$$

with λ_3 and λ_8 chosen as the two diagonal generators. The λ_k satisfy the Lie algebra

$$\left[\frac{\lambda_i}{2}, \frac{\lambda_j}{2}\right] = i f_{ijk}\left(\frac{\lambda_k}{2}\right),$$ (5.64a)

and the non-vanishing structure constants in this representation are

$$f_{123} = 1$$
$$f_{147} = f_{246} = f_{257} = f_{345} = \tfrac{1}{2}$$
$$f_{156} = f_{367} = -\tfrac{1}{2}$$ (5.64b)
$$f_{458} = f_{678} = \tfrac{\sqrt{3}}{2},$$

with the f_{ijk} completely antisymmetric under exchange of indices (see Exercise 5.10b). The generators $\lambda_k/2$ are traceless and normalized such that

$$\text{Tr}\left(\frac{\lambda_i}{2} \cdot \frac{\lambda_j}{2}\right) = \frac{1}{2}\delta_{ij}.$$ (5.65)

This is a standard normalization that we will employ uniformly for the $SU(N)$ groups [notice from (1.78) that the $SU(2)$ generators $\sigma_i/2$ also have this normalization]. Generally a compact, semisimple Lie group with generators τ_a may be normalized

$$\text{Tr}\left(\tau_a\tau_b\right) = \text{constant} \times \delta_{ab},$$

with the constant chosen for convenience. The generators (5.63) also satisfy the anticommutation relation

$$\left\{\frac{\lambda_i}{2}, \frac{\lambda_j}{2}\right\} = \frac{1}{3}\delta_{ij} + d_{ijk}\left(\frac{\lambda_k}{2}\right).$$ (5.66a)

The coefficients d_{ijk} are symmetric under exchange of indices; those with non-zero values are

$$d_{118} = d_{228} = d_{338} = -d_{888} = \tfrac{1}{\sqrt{3}}$$
$$d_{146} = d_{157} = d_{256} = d_{344} = d_{355} = \tfrac{1}{2}$$ (5.66b)
$$d_{247} = d_{366} = d_{377} = -\tfrac{1}{2}$$
$$d_{448} = d_{558} = d_{668} = d_{778} = -\tfrac{1}{2\sqrt{3}},$$

and permutations.

It is convenient to form eight new operators from the generators (5.63) in the following way. To facilitate notation we first introduce

$$F_k \equiv \frac{\lambda_k}{2}, \tag{5.67}$$

from which we may construct two diagonal operators

$$T_3 = F_3 \qquad Y = \frac{2}{\sqrt{3}} F_8, \tag{5.68a}$$

and six additional operators

$$T_\pm = F_1 \pm i F_2 \qquad U_\pm = F_6 \pm i F_7 \qquad V_\pm = F_4 \pm i F_5. \tag{5.68b}$$

The commutation relations obeyed by these operators are derived in Exercise 5.10a and summarized in Table 5.2. The $SU(3)$ Casimir operator corresponding to J^2 in the $SU(2)$ case is

$$F^2 = F_i F_i, \tag{5.69}$$

which is an invariant because it commutes with all group generators. For a rank-l algebra there are l independent Casimir operators, so there is a second $SU(3)$ Casimir operator (which is cubic in the generators). As we do not require it, we do not write it down; it may be found in Lichtenberg (1978).

Table 5.2
Some $SU(3)$ Commutators

$[T_3, T_\pm] = \pm T_\pm$	$[Y, T_\pm]$	$= 0$
$[T_3, U_\pm] = \mp \frac{1}{2} U_\pm$	$[Y, U_\pm] = \pm U_\pm$	
$[T_3, V_\pm] = \pm \frac{1}{2} V_\pm$	$[Y, V_\pm] = \pm V_\pm$	

$$[T_+, T_-] = 2T_3$$
$$[U_+, U_-] = \tfrac{3}{2} Y - T_3 \equiv 2U_3$$
$$[V_+, V_-] = \tfrac{3}{2} Y + T_3 \equiv 2V_3$$
$$[T_+, V_+] = [T_+, U_-] = [U_+, V_+] = 0$$

$[T_+, V_-] = -U_-$	$[T_+, U_+] = V_+$	
$[U_+, V_-] = T_-$	$[T_3, Y] = 0$	

The unlisted commutation relations may be obtained from
$$T_+ = (T_-)^\dagger \qquad U_+ = (U_-)^\dagger \qquad V_+ = (V_-)^\dagger$$

Since the two Casimir operators of $SU(3)$ commute with each group generator, their eigenvalues could be used to label $SU(3)$ representations. In particle physics we often choose some alternative ways of labeling representations (see below) and disregard the $SU(3)$ Casimir operators. They are very useful, however, if we attempt to construct dynamical operators for systems with $SU(3)$ symmetries.

EXERCISE 5.9 Show that the **2** and $\bar{\mathbf{2}}$ representations are equivalent for $SU(2)$, but $\mathbf{3} \neq \bar{\mathbf{3}}$ for $SU(3)$. *Hint*: examine the weight space.

EXERCISE 5.10 (a) Verify the commutators in Table 5.2.   (b) Given (5.64a) and (5.65), prove that the corresponding $SU(3)$ structure constants (5.64b) are antisymmetric in all indices. *Hint*: first prove that

$$\text{Tr}\left(\lambda_k[\lambda_i, \lambda_j]\right) = 4if_{ijk},$$

and then use the cyclic property of the trace.

Labeling of Representations. Although the group theory of $SU(3)$ may be developed quite independent of physics, it is easiest for our purposes to discuss it in terms of a definite application. Let us assume that the fundamental representation **3** of $SU(3)$ is defined by a flavor triplet of quarks

$$q = \begin{pmatrix} u \\ d \\ s \end{pmatrix} = \begin{pmatrix} \text{"up"} \\ \text{"down"} \\ \text{"strange"} \end{pmatrix}$$

that have the quantum numbers summarized in Table 5.3, and the conjugate representation $\bar{\mathbf{3}}$ by a corresponding triplet of antiquarks, (we ignore the c, b, and t quarks for the present discussion).

The empirical relationship between charge Q, baryon number B, strangeness S, hypercharge Y, and the third component of isospin T_3 is

$$Q = T_3 + \frac{B+S}{2} = T_3 + \frac{Y}{2} \tag{5.70}$$

$$Y \equiv B + S. \tag{5.71}$$

The corresponding $SU(3)$ charge operator is

$$Q = \frac{\lambda_3}{2} + \frac{\lambda_8}{2\sqrt{3}}, \tag{5.72}$$

so the $SU(3)$ multiplet members may be labeled by two additive quantum numbers (the third component of isospin T_3, and the hypercharge Y), and the weight space is 2-dimensional (Y vs. T_3). Upon examining the commutation relations of $U_\pm$, $V_\pm$, and $T_\pm$, we find that these six operators constitute a set of

Table 5.3
Quantum Numbers for Quarks

	Up	Down	Strange	Charm	Bottom	(Top)
Symbol	u	d	s	c	b	t
Baryon number (B)	$\frac{1}{3}$	$\frac{1}{3}$	$\frac{1}{3}$	$\frac{1}{3}$	$\frac{1}{3}$	$\frac{1}{3}$
Spin	$\frac{1}{2}$	$\frac{1}{2}$	$\frac{1}{2}$	$\frac{1}{2}$	$\frac{1}{2}$	$\frac{1}{2}$
Charge (Q)	$\frac{2}{3}$	$-\frac{1}{3}$	$-\frac{1}{3}$	$\frac{2}{3}$	$-\frac{1}{3}$	$\frac{2}{3}$
Isospin (T)	$\frac{1}{2}$	$\frac{1}{2}$	0	0	0	0
T_3	$\frac{1}{2}$	$-\frac{1}{2}$	0	0	0	0
Strangeness (S)	0	0	-1	0	0	0
Charm (c)	0	0	0	1	0	0
Beauty (b)	0	0	0	0	-1	0
Truth (t)	0	0	0	0	0	1

The additive quantum numbers Q, T_3, S, c, B, b, and t of the corresponding antiquarks are the negative of those for the quarks. The sixth quark ($t = top$ or $truth$) is expected, but has not yet been found below about 90 GeV. The charge is $Q = T_3 + \frac{1}{2}(B + S + c + b + t)$; this reduces to (5.70) for hadrons made from u, d, and s quarks.

stepping operators in the Y and T_3 plane (see Exercise 5.11b). In particular, any member of the multiplet can be reached from any other member through some combination of these operators, and application of one of these operators to an $SU(3)$ multiplet member gives either another member of the multiplet, or zero. In Fig. 5.3 we indicate sites for possible multiplet members and the general action of the stepping operators in the weight space. For example, U_+ increases Y by one unit and decreases T_3 by $\frac{1}{2}$ unit when applied to a member of the multiplet, unless this action would lead to a site outside the multiplet; in this case the application of U_+ gives zero.

EXERCISE 5.11 (a) Demonstrate that $J_\pm |jm\rangle \simeq |j, m \pm 1\rangle$ for the group $SU(2)$. (b) Use the algebra of Table 5.2 to show that the operators $U_\pm$, $V_\pm$, and $T_\pm$ have the properties depicted in Fig. 5.3. *Hint*: generalize the proof of part (a) to the $SU(3)$ group.

Notice that T_+, U_-, and V_+ all increase the value of T_3, so that for $SU(3)$ representations there must exist a *maximally stretched state* ϕ_{max} satisfying

$$T_+\phi_{max} = V_+\phi_{max} = U_-\phi_{max} = 0. \qquad (5.73)$$

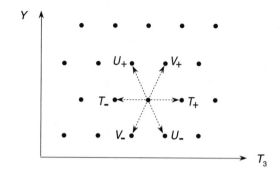

Fig. 5.3 Weight space for $SU(3)$ and the action of the stepping operators defined in eq. (5.68b).

As we did for $SU(2)$ representations, the other members of an $SU(3)$ representation may be generated by judicious application of the stepping operators to the highest weight state ϕ_{max}.

The weight diagram for $SU(3)$ is 2-dimensional, so members of the multiplet correspond to points in a plane; by connecting these points with lines we obtain a geometric figure that uniquely specifies the representation. It can be shown using the $SU(3)$ commutation relations that this figure has a convex boundary in T_3–Y space (see Gasiorowicz, 1966, Ch. 17), which suggests a simple way to construct and label the representations of $SU(3)$. Starting with the state ϕ_{max} we apply the operator V_- repeatedly to generate new states until the application of V_- gives zero; this defines a corner of the weight diagram. Hence, if p applications of V_- are required to reach the corner

$$(V_-)^{p+1}\phi_{max} = 0. \tag{5.74}$$

Next we operate with T_- repeatedly (q times) until another corner is reached:

$$(T_-)^{q+1}(V_-)^p\phi_{max} = 0. \tag{5.75}$$

The $SU(3)$ representation and corresponding weight diagram are now specified by the integers p and q because of the symmetries implied by the $SU(2)$ subgroups associated with U, V, and T (see Fig. 5.3). For example, there is a 27-dimensional representation of $SU(3)$ that has the weight diagram show in Fig. 5.4. Starting at the state with maximal T_3 we see that two applications of V_- bring us to a corner ($p = 2$), and two subsequent applications of T_- bring us to another corner ($q = 2$), so this representation may be specified by the ordered integers $(p,q) = (2,2)$. In the diagram each point corresponds to a degeneracy of one, and each circle about a point indicates an additional degeneracy of one. The general rule is that the sites on the outer boundary have unit degeneracy, and the degeneracy increases by one for each layer inward until a layer with a triangular shape is encountered; then the degener-

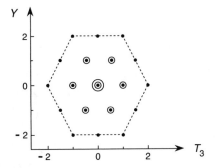

Fig. 5.4   Weight diagram for (2,2) or **27** in $SU(3)$. The circles around a point signify degeneracy for states of the corresponding weight.

acy remains constant for subsequent layers. For the $(2,2)$ representation this implies a triply occupied central site. Counting the total degeneracy in the diagram we obtain **27**, as expected.

The first Casimir operator (5.69) can be expressed in terms of p and q,

$$F^2 = \tfrac{1}{3}(p^2 + pq + q^2) + p + q, \tag{5.76}$$

and can also be used to specify the representation. It is more common to label the representation by the ordered pair (p, q), or to give the dimension of the representation (see below), than to give the eigenvalues of F^2 and the other Casimir operator [probably because the Casimir operators have no obvious physical interpretation in flavor $SU(3)$, unlike in angular momentum $SU(2)$].

In Table 5.4 we list some low-dimensional representations of $SU(3)$, labeled by (p, q) and dimension; also shown are the expectation value of the quadratic Casimir operator and the corresponding Young diagram (these will be discussed below). The general relation between the ordered numbers (p, q) and the dimension D of an $SU(3)$ representation is (Gasiorowicz, 1966)

$$D = \tfrac{1}{2}(p+1)(p+q+2)(q+1). \tag{5.77}$$

The state $\phi_{\max}$ of maximal T_3 has

$$T_3(\phi_{\max}) = \frac{p+q}{2} \qquad Y(\phi_{\max}) = \frac{(p-q)}{3}, \tag{5.78}$$

so we may construct the weight diagrams for the representations listed in Table 5.4 starting from this state. These are shown in Fig. 5.5 for a few cases.

As an example, let us sketch the diagram for the 10-dimensional $(p, q) = (3, 0)$ representation. The state of maximal T_3 has

$$T_3 = \tfrac{1}{2}(p+q) = \tfrac{3}{2} \qquad Y = \tfrac{1}{3}(p-q) = 1.$$

Table 5.4
Some $SU(3)$ Representations

Young Diagram	Dimension	(p, q)	$\langle F^2 \rangle$
	1	(0, 0)	0
	3	(1, 0)	$\frac{4}{3}$
	$\overline{3}$	(0, 1)	$\frac{4}{3}$
	8	(1, 1)	3
	6	(2, 0)	$\frac{10}{3}$
	$\overline{6}$	(0, 2)	$\frac{10}{3}$
	10	(3, 0)	6
	$\overline{10}$	(0, 3)	6
	27	(2, 2)	8

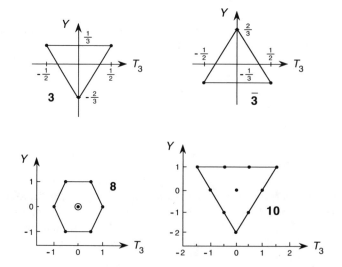

Fig. 5.5   Some weight diagrams for $SU(3)$ representations.

Starting from this state, $p = 3$ applications of V_- take us to a corner. Then $q = 0$ applications of T_- leave us at the same point, which has $T_3 = 0$ and $Y = -2$, since each application of V_- decreases Y by one and T_3 by $\frac{1}{2}$. The $SU(3)$ diagrams are symmetric about the vertical axis at $T_3 = 0$ (because horizontal lines are isospin multiplets); therefore, the outer boundary of the $(3, 0)$ must be a triangle. In the horizontal direction sites must be separated by one unit in T_3, which requires that there be four sites in the top row of the triangle and that the only occupied interior site is the one at $T_3 = Y = 0$. Because the outer boundary is triangular all sites are singly degenerate and this is a 10-dimensional representation (decuplet), as may be seen by counting the occupied sites.

Graphical Construction of Direct Products. Graphical means may be used to construct and decompose direct products (see Gasiorowicz, 1966, Ch. 17). In Fig. 5.6 we illustrate the graphical construction of $\mathbf{3} \otimes \bar{\mathbf{3}} = \mathbf{8} \oplus \mathbf{1}$ in $SU(3)$. First place a $\mathbf{3}$ (Υ) at the origin, and to the tip of each $\mathbf{3}$ arrow attach a $\bar{\mathbf{3}}$ ($\curlywedge$); the tips of the $\bar{\mathbf{3}}$ arrows now denote occupied sites in the product. Similar methods can be used for other representations (Exercise 5.12c). The validity of this graphical procedure is based on the additivity of the weights.

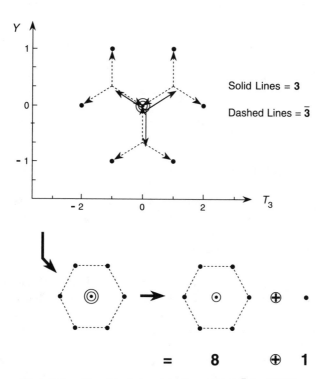

Fig. 5.6 Graphical construction of $\mathbf{3} \otimes \bar{\mathbf{3}}$ in $SU(3)$.

5.3.4   Representation Dimensionality: Young Diagrams

Higher dimensional representations may be constructed by taking the direct products of lower dimensional ones, and we require efficient methods to calculate the dimensions of the resulting representations. Except for simple cases the graphical method just presented is cumbersome. Fortunately, there exist recipes based on tensor analysis techniques that make the determination of representation dimensionalities rather easy. Such a recipe is given here, utilizing the method of *Young diagrams*. The reader is urged to consult one of the standard group theory texts to understand *why* the method works [for example, see Lichtenberg (1978)]—we concern ourselves only with *how* it works, using as an outline the presentation in Close (1979) and Georgi (1982).

Allowed Young Diagrams. For $SU(N)$ the fundamental representation $\mathbf{N}$ is denoted by a box,

$$\mathbf{N} = \square$$

and the conjugate representation $\overline{\mathbf{N}}$ is denoted by a column of $N-1$ boxes.

$$\overline{\mathbf{N}} = \left. \begin{array}{c} \square \\ \square \\ \vdots \\ \square \end{array} \right\} \quad (N-1 \text{ boxes})$$

Examples:

$$SU(2): \quad \square = \mathbf{2} = \overline{\mathbf{2}} \qquad SU(3): \quad \square = \mathbf{3} \quad \begin{array}{c}\square\\\square\end{array} = \overline{\mathbf{3}},$$

displaying explicitly that the conjugate and fundamental representations are equivalent in $SU(2)$, but not in $SU(3)$. Young diagrams with only one row are associated with *totally symmetric representations;* those with one column correspond to *totally antisymmetric representations;* all others are termed representations of *mixed symmetry.*

A wavefunction of mixed symmetry is symmetrized under interchange of some particles and then antisymmetrized under the interchange of some of these with others. The total wavefunction of a physical system of identical particles must either be completely symmetric (bosons) or completely antisymmetric (fermions). However, if the $SU(N)$ symmetry is an internal symmetry describing only part of the available degrees of freedom, an $SU(N)$ representation of mixed symmetry may be part of a total wavefunction that is symmetric or antisymmetric under exchange of *all* coordinates of identical particles (see Exercise 5.14b).

The integers (p, q) introduced in the graphical construction of $SU(3)$ representations (Table 5.4) can be directly related to the pattern of boxes in the corresponding Young diagram. $SU(3)$ diagrams require no more than three rows; p is the difference between the number of boxes in the first and the second rows; q is the corresponding difference between rows 2 and 3. For example,

$$\text{(diagram)} = (0,0) = \mathbf{1} \qquad \text{(diagram)} = (1,1) = \mathbf{8} \qquad \text{(diagram)} = (3,0) = \mathbf{10}.$$

The direct products of $SU(N)$ representations may be found by the multiplication of Young diagrams according to the following set of rules.

1. Draw the two diagrams corresponding to the representations to be multiplied and label the second diagram with numbers in the boxes giving the row of the diagram in which the box occurs. For example

$$\square\square \otimes \begin{smallmatrix}1\\2\end{smallmatrix}.$$

   Attach the individual boxes of the second diagram to those of the first in all possible ways such that
   a. All diagrams are *proper diagrams* (rows do not increase in length as we go down the diagram).
   b. No column has more than N boxes for $SU(N)$.
   c. On a path through each row from the right, and moving from top to bottom through the entire diagram, at any point on the path the number of boxes encountered containing the number i must be less than or equal to the number of boxes encountered containing the number $i - 1$.
   d. The numbers in the boxes do not decrease from left to right in a row.
   e. The numbers in a column increase from top to bottom.
   For example,

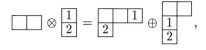

but the diagrams

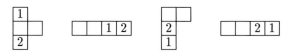

each violate a rule and so do not contribute to this product.

2. Each diagram so constructed corresponds to an irreducible representation of $SU(N)$ occurring in the direct product.

As illustration of the use of these rules we find

$$\square \otimes \square = \square\square \oplus {\begin{array}{c}\square\\\square\end{array}} \qquad (5.79)$$

$$\square\square \otimes \square = \square\square\square \oplus {\begin{array}{c}\square\square\\\square\end{array}} \qquad (5.80)$$

$$\begin{array}{c}\square\\\square\end{array} \otimes \square = \begin{array}{c}\square\square\\\square\end{array} \oplus \begin{array}{c}\square\\\square\\\square\end{array} \qquad (5.81)$$

These examples are simple and require only a portion of the rules. More complicated ones are considered in Exercises 5.12d and 12.2.

Dimensionality of Diagrams. The dimension of the representation specified by a particular Young diagram may be determined by the following procedure (Close, 1979):

1. For an $SU(N)$ diagram, insert an N in each diagonal box starting at the upper left corner. Insert $N+1$ and $N-1$ along the diagonals immediately above and below the main diagonal, respectively, $N+2$ and $N-2$ along the next diagonals above and below, and so on. Continue until each box has a number.

Example:

N	$N+1$	$N+2$
$N-1$	N	$N+1$
$N-2$	$N-1$	N
$N-3$		

We now define a numerator n as the product of all these numbers,

$$n = \prod_{\text{boxes}} (\text{numbers in boxes}). \qquad (5.82)$$

Example:

For $SU(3)$:
3	4	5
2	3	

$n = 3 \times 4 \times 5 \times 2 \times 3 = 360$

2. For each box in the diagram we define a "hook" and an associated number in the following way. Draw a line from the right, entering the row in which the box is found and proceeding to the box in question. On entering the box the line turns downward and passes down the column until it exits the diagram. The value of the hook for this box is the total number of boxes the line passes through. A denominator d is defined by the product of all the hooks, one for each box:

$$d \equiv \prod_{\text{boxes}} (\text{hooks}). \tag{5.83}$$

Example:

The value of this hook is 3; the product of hooks for the diagram is $d = 1 \times 3 \times 4 \times 1 \times 2 = 24$.

3. The dimension of the representation is given by the ratio of the numerator n and denominator d,

$$\text{Dim} = \frac{n}{d}. \tag{5.84}$$

Let us use these rules to work out a few simple examples. Consider the direct product of the fundamental representation of $SU(N)$ with itself [eq. (5.79)]. What are the dimensions of the two representations that result? In a notation inspired by the preceding rules,

$$\text{Dim} \left(\square\square \right) = \frac{\boxed{N \ \vert \ N+1}}{\boxed{2 \vert 1}} = \frac{N(N+1)}{1 \times 2} = \frac{N(N+1)}{2}$$

$$\text{Dim} \left(\begin{array}{c} \square \\ \square \end{array} \right) = \frac{\begin{array}{c} \boxed{N} \\ \boxed{N-1} \end{array}}{\begin{array}{c} \boxed{2} \\ \boxed{1} \end{array}} = \frac{N(N-1)}{2 \times 1} = \frac{N(N-1)}{2},$$

where the hook numbers are displayed in the denominator boxes. For $SU(2)$ this corresponds to the familiar result

$$\mathbf{2} \otimes \mathbf{2} = \mathbf{3} \oplus \mathbf{1}, \tag{5.85}$$

while for $SU(3)$ (see also Exercise 5.12c)

$$\mathbf{3} \otimes \mathbf{3} = \mathbf{6} \oplus \overline{\mathbf{3}}, \tag{5.86}$$

and in general, for $SU(N)$,

$$\square \otimes \square = \frac{N(N+1)}{2} \oplus \frac{N(N-1)}{2}. \tag{5.87}$$

We may form the direct product of three objects in $SU(N)$ by first multiplying two of them, and then multiplying by the third. Let us take the direct product of three fundamental representations of $SU(N)$. We have already found in (5.79) the result of the direct product of the fundamental representation with itself. Multiplying by the fundamental again, subject to the rules for allowed diagrams, gives

$$\square \otimes \square \otimes \square = \left(\square\square \oplus \begin{array}{c}\square\\\square\end{array} \right) \otimes \square$$

$$= \square\square\square \oplus \begin{array}{c}\square\square\\\square\end{array} \oplus \begin{array}{c}\square\\\square\\\square\end{array} \oplus \begin{array}{c}\square\square\\\square\end{array} \tag{5.88}$$

For a particular $SU(N)$ symmetry we may use the rules just discussed to find the dimensions of the irreducible representations contained in this product. In general (Exercise 5.12a)

$$\square \otimes \square \otimes \square = \tfrac{1}{3}N(N+1)(N-1)$$
$$\oplus \tfrac{1}{3}N(N+1)(N-1)$$
$$\oplus \tfrac{1}{6}N(N+1)(N+2)$$
$$\oplus \tfrac{1}{6}N(N-1)(N-2). \tag{5.89}$$

For $SU(2)$ this gives[‡]

$$\mathbf{2} \otimes \mathbf{2} \otimes \mathbf{2} = \mathbf{4} \oplus \mathbf{2} \oplus \mathbf{2}, \tag{5.90}$$

corresponding to the familiar result that three spin-$\frac{1}{2}$ objects may couple to a resultant of $S = \frac{3}{2}$ (one way) and $S = \frac{1}{2}$ (two ways), with degeneracy $2S+1$ in each case. The corresponding result for $SU(3)$ is

$$\mathbf{3} \otimes \mathbf{3} \otimes \mathbf{3} = \mathbf{10} \oplus \mathbf{8} \oplus \mathbf{8} \oplus \mathbf{1}. \tag{5.91}$$

Notice the appearance of the dimension **1** representation (singlet) in $SU(3)$,

---

[‡]From (5.84) we may verify that any $SU(N)$ diagram with more than N boxes in a column has dimension zero, and that omission of columns containing N boxes does not affect the dimension of an $SU(N)$ diagram. Thus for $SU(2)$ the third diagram of (5.88) does not contribute, while diagrams two and four have the same dimension as a single box (**2**). These observations follow because of the antisymmetrization implied by the columns in the diagram.

corresponding to the representation $(p, q) = (0, 0)$ in Table 5.4. It will play an important role in the discussion of quantum chromodynamics.

As another example, the direct product of the conjugate and fundamental representations of $SU(3)$ is given by (5.81). Applying the rules (Exercise 5.12b) we find that for $SU(3)$

$$\bar{\mathbf{3}} \otimes \mathbf{3} = \mathbf{8} \oplus \mathbf{1}, \tag{5.92}$$

which is equivalent to the result obtained by graphical means in Fig. 5.6.

EXERCISE 5.12 (a) Use Young diagrams to find the dimension of each representation that results from the product of three fundamental representations in $SU(N)$. (b) Use Young diagrams to find the dimension of each representation resulting from the direct product of the **3** and $\bar{\mathbf{3}}$ in $SU(3)$. (c) Use the graphical methods illustrated in Fig. 5.6 to find the irreducible $SU(3)$ representations contained in $\mathbf{3} \otimes \mathbf{3}$. In particular, show that a 3-dimensional representation is obtained that is $\bar{\mathbf{3}}$ rather than **3** [cf. eq. (5.86)]. (d) Use Young diagrams to construct $\mathbf{8} \otimes \mathbf{8}$ in $SU(3)$.

5.3.5 $SU(3)$ Flavor Multiplets and the Quark Model

Some properties of selected baryons and mesons are tabulated in Appendix B. The hypercharge Y and the third component of isospin T_3 are know for these hadrons (see Aguilar-Benitez et al., 1988) and we may plot them as points in the Y–T_3 plane, grouping them according to baryon number and spin. Some diagrams that result are shown in Fig. 5.7. These diagrams are familiar,

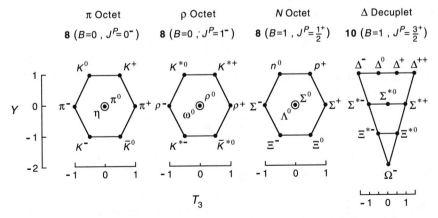

Fig. 5.7 Some $SU(3)$ multiplets labeled by spin J, parity P, and baryon number B. The baryon Σ^* is now called $\Sigma(1385)$ and Ξ^* is now $\Xi(1530)$.

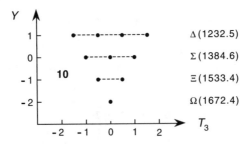

Fig. 5.8   Average masses for isospin multiplets in the Δ decuplet.

of course. They correspond to irreducible representations of $SU(3)$, which suggests that $SU(3)$ might be a good symmetry for classifying these states provided the mass splittings within a multiplet are not too large [if $SU(3)$ were a good symmetry, all members of a multiplet would be degenerate].

In fact, the masses within these multiplets are rather similar. As an example, we show the $J = \frac{3}{2}$ baryon decuplet in Fig. 5.8 (see also Fig. 12.9). The states lying on a horizontal line correspond to members of isospin multiplets; these typically have masses within a few percent of each other. Shown to the right are the average masses for each isospin multiplet; these are seen to differ by no more than about 20% from the average mass of the $SU(3)$ multiplet. Thus we conclude that flavor $SU(3)$ may be a useful approximate symmetry of the strong interactions, representing a generalization of isospin to include the hypercharge degree of freedom. We are further led by this identification and the preceding discussion to consider the quarks of §5.3.3 as the fundamental building blocks of the hadrons, because all $SU(3)$ representations can be obtained from products of fundamental representations.

Recall that the fundamental $SU(3)$ triplet **3** consists of u, d, and s quarks, the $\bar{\mathbf{3}}$ consists of the corresponding antiquarks, and

$$\mathbf{3} \otimes \mathbf{3} \otimes \mathbf{3} = \mathbf{10} \oplus \mathbf{8} \oplus \mathbf{8} \oplus \mathbf{1} \qquad \bar{\mathbf{3}} \otimes \mathbf{3} = \mathbf{8} \oplus \mathbf{1}.$$

Since **8**'s and **10**'s are required for baryon representations and **8**'s and **1**'s for mesons, this suggests for the quark structure of the hadrons

$$\text{baryons} = qqq \qquad \text{mesons} = q\bar{q}, \tag{5.93}$$

where q stands for quarks of various flavors. The specific quark content of the baryon and meson **8** and the baryon **10** are shown in Fig. 5.9 (see Exercise 5.13).

The $SU(3)$ flavor model is empirical, and we shall see in Ch. 10 that this symmetry may in a sense be accidental: the strong interactions originate in

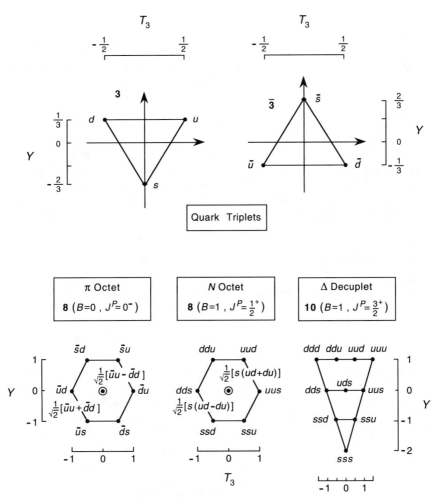

Fig. 5.9 Quark content of $SU(3)$ flavor **8** and **10** and the fundamental quark flavor triplets. The ρ octet of Fig. 5.7 has the same quark content as the π octet. The notation is shorthand for properly symmetrized expressions; for example, udd in the baryon **10** is actually the symmetric flavor combination $(udd + ddu + dud)/\sqrt{3}$ (see Exercises 5.13 and 12.14).

a more fundamental symmetry operating on non-flavor degrees of freedom [color $SU(3)$], and flavor $SU(3)$ symmetry is a consequence of the u, d, and s quarks being relatively light and not too different in mass. The QCD color fields are tasteless: for massless quarks they couple indiscriminately to all flavor of quark; only the masses for the various quarks (which originate in the electroweak interactions—see Ch. 9) carry flavor information in the QCD

Lagrangian. In general, the flavor indices will be more relevant for weak interactions than for strong ones (Ch. 9).

EXERCISE 5.13 Derive the quark content of the $B = 1$ decuplet shown in Fig. 5.7. Construct flavor wavefunctions with the appropriate permutation symmetry for these states. *Hint*: see Table 5.3 and the solution of Exercise 12.14.

EXERCISE 5.14 (a) Show that the ground states violate the Pauli principle if we assume space, spin, and flavor to be the relevant degrees of freedom for the baryon **10** (Figs. 5.7 and 5.9). Show that if we conjecture an additional $SU(3)$ degree of freedom (color), with the baryons all $SU(3)$ color singlets, the Pauli principle is satisfied for the **10**. (b) Construct a flavor–spin wavefunction for the proton consistent with the Pauli principle. *Hint*: see Exercise 12.14.

5.3.6 $SU(4)$ and $SU(6)$ Quark Models

Isospin $SU(2)$ is an excellent phenomenological symmetry, good to a few percent for particle masses. The generalization to flavor $SU(3)$ is also a usable phenomenological symmetry, even though it is broken at the 20% level. With the discovery of a third additive quantum number, charm, it was tempting to extend flavor symmetry to incorporate a rank-3 algebra. The natural candidate would be $SU(4)$, and such extensions are discussed in various places (see Close, 1979). We will refrain from doing so: this symmetry is broken severely because the charmed quark is considerably heavier than the u, d, s quarks; therefore, it is not particularly useful. For the same reason we dismiss any flavor symmetries that attempt to incorporate top and bottom quarks.

Another quark model that has been quite successful attempts to give the u, d, s quarks a spin consistent with their fermion nature. We may motivate this by noting that the spin-$\frac{1}{2}$ baryon octet and the spin-$\frac{3}{2}$ baryon decuplet differ in energy by an amount that is comparable with the splitting between members of each $SU(3)$ multiplet (see Appendix B and Figs. 5.8 and 12.9). This suggests that the two multiplets might profitably be combined into a single multiplet transforming as an irreducible representation of a group larger than $SU(3)$. If we assume that the approximate flavor $SU(3)$ symmetry and the $SU(2)$ symmetry associated with the spin of the quarks are independent degrees of freedom, the group structure of such objects involves the direct product $SU(3)_{\text{flavor}} \times SU(2)_{\text{spin}}$. This is a subgroup of the group $SU(6)$, and the $SU(6)$ *quark model* employs a fundamental sextet consisting of u, d, and s quarks, each with spin up and spin down.

From the Young diagrams (5.89) we find that in $SU(6)$

$$\mathbf{6} \otimes \mathbf{6} \otimes \mathbf{6} = \mathbf{56} \oplus \mathbf{70} \oplus \mathbf{70} \oplus \mathbf{20}, \tag{5.94}$$

so that a symmetric **56** appears naturally in an $SU(6)$ model based on three quarks with two possible values of spin. This is just what is needed to accommodate the spin-$\frac{1}{2}$ octet and spin-$\frac{3}{2}$ decuplet baryon representations mentioned above, since each octet member has $2S+1 = 2$ possible spins, and each decuplet member has $2(\frac{3}{2}) + 1 = 4$ possible spins, giving $2 \times 8 + 4 \times 10 = 56$ states. Symbolically, this decomposition of the **56** of $SU(6)$ with respect to its $SU(3)$ and $SU(2)$ subgroups may be written

$$\mathbf{56} \rightarrow (\mathbf{8}, \mathbf{2}) \oplus (\mathbf{10}, \mathbf{4}).$$

In this notation the number on the left side is the dimension of the $SU(6)$ representation, while the right side is expressed in the form $\big($Dim $SU(3)$, Dim $SU(2)\big)$.

By a similar token we might expect meson states in the $SU(6)$ model to correspond to

$$\mathbf{6} \otimes \bar{\mathbf{6}} = \mathbf{35} \oplus \mathbf{1}. \tag{5.95}$$

This is precisely what is needed to accommodate the pseudoscalar meson octet $[8 \times (2S+1) = 8$ states] and vector meson $\mathbf{8} \oplus \mathbf{1}$ (nonet) $[9 \times (2S+1) = 27$ states] in the **35** representation of $SU(6)$. (The mixing of the center state of the octets with the singlets is called nonet structure; see Perkins, 1987, §5.5 and §5.6.)

The $SU(6)$ model has had some notable successes, such as the prediction of magnetic moments for baryons. It is not completely understood why this should be the case because the model is formally incorrect at a relativistic level. For example, the separation of orbital and intrinsic spin angular momenta is untenable in a relativistic theory (see Exercise 12.10). An extensive discussion of the $SU(6)$ model can be found in Close (1979). We will return to such nonrelativistic descriptions of hadrons when the quark potential model is introduced in Ch. 12.

5.4 Topological Properties

Our discussion of Lie groups has touched briefly on topological characteristics of group manifolds; that is, on the nature of the space defined by variation of the group parameters. For example, the concept of connectedness is topological. The Lie algebra (5.9) defines the *local* properties of a Lie group, but global properties such as connectedness are not uniquely specified by the commutator algebra. The topology of group manifolds plays an important role in gauge field theories, in particular for those theories exhibiting soliton solutions. Such theories are introduced in Ch. 13, and we postpone further discussion of topological concepts until then.

5.5 Background and Further Reading

The content of this chapter has been influenced strongly by Georgi (1982), Close (1979), Gasiorowicz (1966), and Lichtenberg (1978). Readable general introductions to symmetry and group theory may be found in Georgi (1982), Hamermesh (1962), Lichtenberg (1978), Wybourne (1974), McVoy (1965), Schensted (1976), Chen (1989), and Elliott and Dawber (1979). Discussions concentrating on applications of Lie groups (mostly unitary groups) in particle physics can be found in Georgi (1982); Close (1979), Chs. 2–3; O'Raifeartaigh (1986); Lee (1981), Chs. 9–16; Lipkin (1966); Itzykson and Nauenberg (1966); and Gasiorowicz (1966), Chs. 17–18. Greiner and Müller (1989) give a variety of worked exercises relevant to our discussion of symmetries. Gilmore (1974) is a more rigorous text than many listed above, but less readable for the novice. Treatises completely unintelligible to the average physicist may be found in any library.

CHAPTER 6

Weak and Electromagnetic Interactions

Besides gauge bosons such as the photon, the particles studied by high-energy physics fall into two broad categories: the *leptons* and the *hadrons*. The charged leptons interact by gravitational, weak, and electromagnetic interactions, while uncharged leptons undergo only the first two. All are spin-$\frac{1}{2}$ objects, and are pointlike at the current level of experimental sophistication. There appear to be three *generations* (or *families*) of leptons: the electron and its neutrino, the muon and its neutrino, and the tau with its neutrino. Each of these particles has a corresponding antiparticle. The masses and lifetimes of these six particles are summarized in Table 6.1.

Lepton family conservation laws appear to hold, in addition to total lepton conservation (see Exercise 6.1). For example, within the electron family the sum of e^- plus ν_e minus the number of e^+ minus the number of $\overline{\nu}_e$ is conserved; similar conservation laws are thought to hold for the muon and tau generations.

Recent precise measurements of the width for the neutral intermediate vector boson suggest that there are only three generations of leptons, which is a conclusion in accord with limits placed by astrophysical and cosmological arguments. The existence of three generations of leptons, the mass pattern, and the conservation of lepton number separately within each generation have no convincing explanation.

In contrast to leptons, the hadrons experience all four fundamental interactions. Thus, they are distinguished by their *strong interactions*. There are many hadrons, with a broad range of spins and masses; they appear to be composite objects, rather than pointlike as is the case for leptons. The hadrons of half-integer spins are called *baryons*, and baryon number appears to be conserved: the number of baryons minus the number of antibaryons is invariant in all known reactions (however, see Ch. 11). Just as for lepton conservation laws, no fundamental reason is known that requires baryon conservation; both lepton and baryon conservation may be only approximate, although the experimental limit on the amount of violation is extremely small. The hadrons with baryon number zero have integer spin and are called *mesons*; the number

Table 6.1
Lepton Masses and Lifetimes

Lepton	Mass (MeV/c²)	Mean Life (seconds)
e^-	0.511	Stable
ν_e	$< 1.8 \times 10^{-5}$	Stable
μ^-	105.66	2.2×10^{-6}
ν_μ	< 0.25	Stable
τ^-	1784	3.4×10^{-13}
ν_τ	< 35	?

of mesons is not generally conserved in the collisions of strongly interacting particles.

In these notes we are only peripherally concerned with the gravitational interaction of elementary particles, because it is irrelevant to our immediate concerns, and because it is not well understood even if it were relevant. Quantum electrodynamics has already been introduced as the prototype of a quantum field theory. In Ch. 10 the strong interactions of the hadrons will be considered. This chapter and Ch. 9 will address weak interactions of both leptons and hadrons, and the relation between the weak and electromagnetic interactions. We begin by briefly refreshing our memory concerning quantum electrodynamics (QED).

6.1 QED: Prototype Quantum Field Theory

In much of what follows we will be discussing quantum field theories of the weak and strong interactions. The development of these theories has not been an easy task, and perhaps would not have been accomplished at all without guidance from the best field theory that we have, quantum electrodynamics. Before proceeding, it will be useful to collect the essential ingredients of QED. Much of this we have already discussed, or is likely to be familiar from previous experience, but time is taken to present it here in a form that will be relevant for subsequent discussion.

If we consider a theory of charged leptons the equation of motion for the free field is the Dirac equation, which follows by Hamilton's principle from the Lagrangian density (2.85):

$$\mathcal{L}_0 = \overline{\psi}(x)(i\not{\partial} - m)\psi(x). \tag{6.1}$$

This Lagrangian density is invariant under a *global* $U(1)$ phase transformation (§5.3.1)

$$\psi(x) \rightarrow e^{i\alpha}\psi(x) \qquad \overline{\psi}(x) \rightarrow \overline{\psi}(x)e^{-i\alpha}, \tag{6.2}$$

where α is independent of the spacetime coordinates x. Proceeding as in §2.8 (see Exercise 2.14), we determine by Noether's theorem that this continuous symmetry of the Lagrangian density is associated with a conservation law; in particular, the Dirac current

$$j^\mu(x) = \overline{\psi}(x)\gamma^\mu\psi(x) \tag{6.3}$$

is conserved:

$$\partial_\mu j^\mu(x) = 0, \tag{6.4}$$

and the electrical charge (2.194)

$$Q = \int d^3x\, j^0(x) = \int d^3x\, \psi^\dagger(x)\psi(x) \tag{6.5}$$

is a constant of motion. Thus a global symmetry yields global (integral over all space) charge conservation.

The free Lagrangian density is not invariant under a *local U(1)* transformation of the form

$$\psi(x) \rightarrow e^{i\alpha(x)}\psi(x), \tag{6.6}$$

because of the derivative terms in (6.1). There is a way to generate a local $U(1)$ phase invariance, but the price to be paid is that an additional field must be introduced with transformation properties that compensate for the term $\partial_\mu\alpha(x)$ appearing in the Lagrangian density when the field is transformed as in (6.6). In §2.7 we saw how this could be accomplished: the derivatives appearing in the Lagrangian density must be replaced by covariant derivatives (the *minimal substitution*)

$$\partial^\mu \rightarrow \partial^\mu + iqA^\mu \equiv D^\mu, \tag{6.7}$$

where q is the particle charge and A^μ is the vector potential. The Lagrangian density becomes

$$\begin{aligned}\mathcal{L}(x) &= i\overline{\psi}(x)\gamma^\mu[\partial_\mu + iqA_\mu(x)]\psi(x) - m\overline{\psi}(x)\psi(x) \\ &= \overline{\psi}(x)(i\slashed{D} - m)\psi(x),\end{aligned} \tag{6.8}$$

which is now invariant under the *local gauge transformation*

$$\psi(x) \rightarrow e^{i\alpha(x)}\psi(x) \qquad \overline{\psi}(x) \rightarrow \overline{\psi}(x)e^{-i\alpha(x)}$$

$$A_\mu(x) \rightarrow A_\mu(x) - \frac{1}{q}\partial_\mu\alpha(x), \tag{6.9}$$

where (6.9) and (2.147) are related through $\alpha(x) \equiv q\chi(x)$. Comparing (6.1) and (6.8), the term

$$\mathcal{L}_{\text{int}} = -q\overline{\psi}A\psi \equiv -q\overline{\psi}(x)\gamma^{\mu}A_{\mu}(x)\psi(x) \qquad (6.10)$$

[see (2.203)] may be identified as the interaction of the lepton and photon fields, and upon quantization A_{μ} will act as a photon creation and destruction operator (§2.7.5). Thus two important principles emerge from these simple considerations:

1. Imposing a *local* phase invariance on the fermion field demands the introduction of a new *massless vector field*, in this case the electromagnetic field entering through the vector potential A_{μ}.
2. The local phase invariance has a *dynamical content:* it specifies the nature of the interaction term between the fermion field and the vector boson (gauge) field A_{μ}.

The new field must be a vector field because of the terms in the Lagrangian arising from the derivative of the local phase that are to be canceled; it must be massless to preserve gauge invariance. (However, see the discussion of spontaneous symmetry breaking in Chs. 8–9). The extension of a global symmetry to a local gauge invariance is termed *gauging the symmetry*, and will soon be seen to be one of the most important instruments in the modern field theorist's tool kit. The new fields that are introduced are called gauge boson fields, and the quanta of these fields are called gauge particles. Thus the photon is the gauge boson of the electromagnetic field.

A natural question arises: "when is it useful to gauge a symmetry?" Two general comments are in order. The first is that experimental observation is the arbiter: gauging the $U(1)$ charge symmetry of QED forces the introduction of a gauge boson field with exactly the properties of the experimentally observed photon field. On the other hand, if we gauge the global $U(1)$ symmetry associated with baryon number [make (5.19a) a local phase transformation] a gauge field must be introduced that is coupled to baryon number. This field has the same group structure as for the photon, so it is not difficult to determine its properties. For example, a "Coulomb's law" should result for a force acting between two objects that is proportional to the product of their baryonic charges. Since baryonic charges are generally not in the same ratio as the masses for two objects, this would have observable consequences [see Okun (1982) for a discussion where it is concluded that if "baryonic photons" exist their coupling to baryonic matter is $< 10^{-45}$ the strength of photon coupling to electrical charges]. There is little substantial evidence for such effects, so we conclude that the gauging of baryon number is not useful. Similar considerations argue against gauging lepton number, isotopic spin, or $SU(3)$ flavor symmetries.

The second comment is that gauging a symmetry is useful only if the ungauged symmetry is exact, because the renormalizability of the corresponding theory requires such a symmetry. We will discuss this further, but for now let us note that the charge symmetry gauged in QED is thought to be exact.

When QED is quantized using the methods of Chs. 2–4, it is found that physical quantities may be calculated as perturbation series in powers of the fine structure constant α. In higher order there appear diagrams for which integrations over free loop momenta lead to ultraviolet divergences. It is a remarkable fact that there is a consistent way to tame these divergences in QED by absorbing their effects into renormalizations (more precisely: redefinitions) of the wavefunctions and parameters of the theory. This renormalization process is outlined later in this chapter, and we will see that QED is renormalizable precisely because it possesses a local gauge invariance. This is also one of the fundamental discoveries of modern field theory: *theories that are locally gauge invariant are renormalizable theories* (provided they do not contain anomalies—see §9.3).

With this renormalization procedure the perturbation series in α usually converges rapidly, and extremely precise calculations are possible for QED. When these are compared with the corresponding measurements the agreement is stunning: QED is the best microscopic theory that physicists have yet managed to conjure. Table 6.2 gives some examples of the remarkable experimental tests of QED.

Now that the beacon of QED is shining brightly before us we are ready to tackle the weak interactions. The trail leading to a consistent microscopic theory of the weak interactions is every bit as entertaining as a good mystery.

Table 6.2

Comparison between Theory and Experiment in Quantum Electrodynamics

Physical Quantity	Theory	Experiment
Fine structure constant	input	$[137.035\,987(29)]^{-1}$
Photon mass	0	$< 1.2 \times 10^{-21} m_e$
Deviation from gyromagnetic ratio: $a = (g-2)/2$		
$a_e \times 10^{12}$	$1\,159\,652\,570(150)$	$1\,159\,652\,200(40)$
$a_\mu \times 10^9$	$1\,165\,921(8.3)$	$1\,165\,924(8.5)$
Muonium hyperfine structure in kHz	$4\,463\,297.9(3.0)$	$4\,463\,302.35(.52)$
Positronium fine structure in MHz	$203\,400.3$	$203\,384.9(1.2)$ $203\,387.0(1.6)$

Source: Gasiorowicz and Rosner (1981).

To avoid losing our way in the twists and turns of this story we will often need illumination from our experience with QED.

6.2 Phenomenology of the Weak Interactions

The weak interactions have been studied extensively since Fermi's pioneering theoretical work in the 1930s. They have been a continual source of surprises, two notable examples being the discovery of parity violation and of CP violation. More recently, they have served as a laboratory for the construction and testing of non-Abelian gauge theories, and (with electrodynamics) as an important waypoint along the path to a unified description of the fundamental forces. We now turn our attention to gauge theories of weak interactions and the partial unification of weak and electromagnetic forces. To appreciate the desirability of such a theory, let us first examine our understanding of the weak interactions before the advent of non-Abelian gauge theories; in so doing we follow closely Ch. 1 of Leader and Predazzi (1982).

6.2.1 Fermi Current–Current Lagrangian

In the mid-1950s it was concluded that all weak interaction phenomenology could be described by the Fermi Lagrangian density

$$\mathcal{L}_{\mathbf{F}} = -\frac{G}{\sqrt{2}} J^{\alpha}(x) J^{\dagger}_{\alpha}(x), \tag{6.11}$$

where

$$G = 1.166 \times 10^{-5} \text{ GeV}^{-2} \tag{6.12}$$

is the universal weak coupling constant (*Fermi constant*). Weak interaction theory based on this Lagrangian is often called the *current–current theory*, because of the two local currents appearing in (6.11). The weak current J^{α} is the analog of the electromagnetic current (6.3) and consists of a leptonic part l^{α} and a hadronic part h^{α},

$$J^{\alpha}(x) = l^{\alpha}(x) + h^{\alpha}(x). \tag{6.13}$$

Thus the Fermi Lagrangian density contains three kinds of interactions, each of which is pointlike since each current has the same spacetime coordinate in (6.11):

1. Leptonic, mediated by $l^{\alpha} l^{\dagger}_{\alpha}$ and corresponding to processes such as

$$\mu^{-} \rightarrow e^{-} + \bar{\nu}_{e} + \nu_{\mu},$$

2. Hadronic (or nonleptonic), mediated by $h^\alpha h_\alpha^\dagger$ and corresponding to processes such as

$$\Lambda \to p + \pi^-,$$

3. Semileptonic, mediated by $l^\alpha h_\alpha^\dagger + h^\alpha l_\alpha^\dagger$ and corresponding to processes such as

$$n \to p + e^- + \bar{\nu}_e.$$

Let us concentrate first on the leptonic and semileptonic interactions. To accommodate maximal parity and charge conjugation violation in weak decays the leptonic current must be of the vector minus axial vector $(V - A)$ form (Feynman and Gell-Mann, 1958; Sudarshan and Marshak, 1958)

$$l^\alpha(x) = \bar{e}(x)\gamma^\alpha(1 - \gamma_5)\nu_e(x) + \bar{\mu}(x)\gamma^\alpha(1 - \gamma_5)\nu_\mu(x)$$
$$+ \bar{\tau}(x)\gamma^\alpha(1 - \gamma_5)\nu_\tau(x), \tag{6.14}$$

(see Table 1.3) where e, ν_e, μ, ν_μ, τ, and ν_τ are used to denote, respectively, the field operators for the electron and its neutrino, the muon and its neutrino, and the tau and its neutrino. The leptonic current separately conserves electron, muon, and tau number since the terms of eq. (6.14) generate no transitions across family lines (see Exercise 6.1). Taking the matrix element of the leptonic current between neutrino and electron states gives a representative matrix element[‡]

$$\langle e|\, l^\alpha\, |\nu_e\rangle \simeq \bar{u}_e\gamma^\alpha(1 - \gamma_5)u_\nu, \tag{6.15}$$

where $\bar{u}_e$ and u_ν are Dirac spinors for the electron and neutrino, and we have assumed essentially free field operators, consistent with the empirical weakness of the interaction. Although relativistic invariance alone would allow scalar, pseudoscalar, vector, axial vector, and tensor couplings in (6.14), it is an empirical observation (next section) that only left-handed leptons participate in the charged-current weak interactions; this restricts the possibilities to the combination $V - A$ and is responsible for the $(1 - \gamma_5)$ factors in (6.14).

Unlike the electromagnetic interactions, weak interactions induced by leptonic and hadronic currents may cause transitions in which the charge Q is altered by one unit in the hadronic and leptonic sections (the total charge is conserved, of course). For example, in the electromagnetic scattering of electrons from protons there is no change in the hadronic or leptonic charge, but in the weak decay $n \to p + e^- + \bar{\nu}_e$ the hadronic charge increases by one unit and the leptonic charge decreases by one unit. In addition, weak interactions may cause changes in flavor quantum numbers such as strangeness (S). It is

---

[‡]The Feynman prescription for dealing with particles and antiparticles (§1.6.4) permits decay processes to be written in more symmetrical form. For example, $n \to p + e^- + \bar{\nu}_e$ is equivalent to $n + \nu_e \to p + e^-$.

usual to classify semileptonic weak reactions according to what happens in the hadronic part of the transition. For example,

$$n \rightarrow p + e^- + \bar{\nu}_e \qquad (\Delta Q = 1, \; \Delta S = 0)$$
$$\Lambda \rightarrow p + e^- + \bar{\nu}_e \qquad (\Delta Q = 1, \; \Delta S = 1).$$

Currents that cause transfer of charge, such as those observed in many weak interactions, are called *charged currents;* currents that do not alter the charge, such as in QED, are called *neutral currents.* Both charged and neutral currents will eventually be required to describe the weak interactions, but the neutral weak currents were discovered only in 1973 and do not yet enter our discussion of pre-gauge weak interactions.

EXERCISE 6.1 (a) Explain why the weak decay $\mu^+ \rightarrow e^+ + \nu_e + \bar{\nu}_\mu$ occurs, but $\mu^+ \rightarrow e^+ + \gamma$ has never been observed. *Hint:* what quantities must be conserved? (b) Neutrinoless double β-decay of nuclei,

$$(Z, A) \rightarrow (Z + 2, A) + 2e^-,$$

is not observed, but it would be kinematically favored over the observed

$$(Z, A) \rightarrow (Z + 2, A) + 2e^- + 2\bar{\nu}_e.$$

Propose a simple mechanism for the neutrinoless decay; show that it is forbidden by a leptonic conservation law.

6.2.2 Left-Handed Neutrinos

For a Dirac particle with energy much larger than its rest mass the bispinors become approximate eigenstates of the *chirality operator* γ_5. For *massless particles* they are exact eigenstates (Exercise 6.2):

$$\begin{aligned} \gamma_5 u_{\rm R} &= u_{\rm R} & \gamma_5 u_{\rm L} &= -u_{\rm L} \\ \gamma_5 v_{\rm R} &= -v_{\rm R} & \gamma_5 v_{\rm L} &= v_{\rm L}, \end{aligned} \tag{6.16}$$

where right-handed particles (R) correspond to helicity $+\frac{1}{2}$, and left-handed particles (L) correspond to helicity $-\frac{1}{2}$. Therefore,

$$\begin{aligned} \tfrac{1}{2}(1 - \gamma_5)\, u &= u_{\rm L} & \tfrac{1}{2}(1 + \gamma_5)\, u &= u_{\rm R} \\ \tfrac{1}{2}(1 - \gamma_5)\, v &= v_{\rm R} & \tfrac{1}{2}(1 + \gamma_5)\, v &= v_{\rm L} \\ \tfrac{1}{2}\bar{u}(1 + \gamma_5) &= \bar{u}_{\rm L} & \tfrac{1}{2}\bar{u}(1 - \gamma_5) &= \bar{u}_{\rm R} \\ \tfrac{1}{2}\bar{v}(1 + \gamma_5) &= \bar{v}_{\rm R} & \tfrac{1}{2}\bar{v}(1 - \gamma_5) &= \bar{v}_{\rm L}, \end{aligned} \tag{6.17}$$

where $u = u_{\rm L} + u_{\rm R}$ and the adjoint relations follow from steps such as

$$\bar{u}_{\rm R}\gamma_5 = u_{\rm R}^\dagger \gamma_0 \gamma_5 = (\gamma_5 u_{\rm R})^\dagger \gamma_0 \gamma_5$$
$$= u_{\rm R}^\dagger \gamma_5^\dagger \gamma_0 \gamma_5 = u_{\rm R}^\dagger \gamma_5 \gamma_0 \gamma_5$$
$$= -u_{\rm R}^\dagger \gamma_0 \gamma_5^2 = -\bar{u}_{\rm R},$$

in which properties of the γ-matrices summarized in Table 1.2 have been used.

Therefore, the $(1 - \gamma_5)$ factor responsible for the $V - A$ structure of the charged-current weak interactions is proportional to a projection operator for left-handed neutrinos, or right-handed antineutrinos. Only left-handed neutrinos (spin projection antiparallel to the direction of motion) and right-handed antineutrinos participate in the charged-current weak interactions [eq. (6.14)], assuming the neutrinos to be massless. In fact, only left-handed components of all fermions play a role in the charged-current weak interactions if masses are negligible (Exercise 6.2d). Unlike the charged currents, the neutral currents of modern weak interactions to be introduced in Ch. 9 generally will not have a pure $V - A$ structure.

Searches for finite neutrino mass are reviewed in Robertson and Knapp (1988) and Boehm and Vogel (1984). The present upper limits for the masses of the electron and muon neutrinos are 18 eV and 0.25 MeV, respectively. Theoretical expectations for neutrino mass are discussed in Kuo and Pantaleone (1989).

A *massive* particle that is distinguishable from its antiparticle must have both left- and right-handed helicity components, because the helicity of a massive particle is not a relativistic invariant: imagine a particle with a particular momentum, spin, and helicity in a given reference frame. On boosting to a frame with a velocity larger than that of the particle the spin remains the same, but the sign of the helicity changes because of the change in the sign of the relative momentum. A *massless* particle need not have both helicity components since it travels at light velocity: there is no Lorentz boost to a frame moving at higher velocity than the particle, and helicity is an invariant concept for massless particles.

Notice that we can always make the wavefunction decomposition

$$u = u_{\rm L} + u_{\rm R} = \left(\frac{1 - \gamma_5}{2}\right)u + \left(\frac{1 + \gamma_5}{2}\right)u.$$

However, only when $E \gg m$ can the components $u_{\rm L}$ and $u_{\rm R}$ be identified with negative and positive helicities, respectively (see Exercise 6.2a,b).

Finally, we conclude this digression by noting that there are three possibilities for the structure of free spin-$\frac{1}{2}$ particles (Ramond, 1981, §V.2). *Weyl spinors* are two-component spinors related by a discrete CP transformation that describe left-handed massless particles and right-handed massless antiparticles. *Majorana spinors* are two-component spinors corresponding to massive spin-$\frac{1}{2}$ particles that are their own antiparticles;

they have two spin degrees of freedom (Kayser, 1985). *Dirac spinors* describe massive particles with two degrees of freedom and a distinct antiparticle; they conserve P and CP, and are four-component spinors with twice as many degrees of freedom as Weyl or Majorana particles.

Since the currents entering the weak interaction Lagrangian density (6.11) are of the $V - A$ form, the structure of $\mathcal{L}_F$ is

$$\mathcal{L}_F \simeq VV^\dagger + AA^\dagger - VA^\dagger - AV^\dagger. \tag{6.18}$$

The vector (V) and axial vector (A) currents transform oppositely under P, and we see explicitly the source of parity violation in the weak Lagrangian: the cross terms $VA^\dagger$ and $AV^\dagger$ change sign under the P operation and the weak Lagrangian fails to conserve parity. The vector and axial vector terms occur with equal weight in (6.14), so the parity violation is *maximal*.

The charge conjugation operation C carried out on a left-handed neutrino yields a left-handed antineutrino. But in the $V - A$ theory only *right-handed* antineutrinos participate, so the interaction (6.11) also violates C symmetry maximally. However time reversal (T) and CP, and therefore CPT, symmetries are respected. Time reversal inverts both spin and momentum and hence has no effect on helicity. The combined operation CP changes a left-handed neutrino into a right-handed antineutrino, so CP is conserved. Thus the observation of parity violation in the weak interactions (Wu et al., 1957; Lee and Yang, 1956) is accommodated by (6.11), but not the small violations of T or CP symmetry that are documented in the neutral kaon system (Christenson et al., 1964).

EXERCISE 6.2 The massless Dirac equation is appropriate for massless or nearly massless particles such as neutrinos, or for very high energy fermions where $E \gg m$. (a) Show that the helicity can be chosen as a constant of motion for a free Dirac particle, but chirality (eigenvalue of γ_5) is a good quantum number only if the particle is massless. Prove that for massless fermions the chirality equals twice the helicity for positive-energy solutions, and minus twice the helicity for negative-energy solutions. *Hint*: multiply (1.100) from the left by $\gamma^5\gamma^0$ and use γ-matrix algebra and the Pauli–Dirac representation to show that [see (1.127)]

$$\boldsymbol{\Sigma} \cdot \hat{\mathbf{p}}\psi = \pm\gamma^5\psi \qquad \boldsymbol{\Sigma} \equiv \begin{pmatrix} \boldsymbol{\sigma} & 0 \\ 0 & \boldsymbol{\sigma} \end{pmatrix}.$$

(b) Write the Dirac equation in the chiral representation [see the footnote following eq. (1.88)]. Express ψ in terms of a pair of two-component spinors ϕ and χ, and show that the Dirac equation becomes two coupled equations for ϕ and χ. Show that if the mass vanishes the equations decouple (Weyl equations), with ϕ and χ eigenfunctions of helicity. (c) Show that the matrices $(1 \pm \gamma_5)/2$ are projection operators. (d) Show that only left-handed particles contribute to (6.14). (e) For a wavefunction decomposed into right- and left-handed chiral components, $\psi = u_R + u_L \equiv R + L$, prove that

$$\overline{\psi}\psi = \overline{L}R + \overline{R}L \qquad \overline{\psi}\gamma^\mu\psi = \overline{L}\gamma^\mu L + \overline{R}\gamma^\mu R$$

$$\overline{\psi}\gamma^\mu\gamma^5\psi = \overline{L}\gamma^\mu\gamma^5 L + \overline{R}\gamma^\mu\gamma^5 R.$$

Thus scalar interactions and mass terms (each proportional to $\overline{\psi}\psi$) do not respect chiral symmetry, but vector and axial vector interactions do.

6.2.3 Hadronic Current

What about the structure of the hadronic current in the Lagrangian density (6.11)? This seems a difficult question because the hadrons will be affected by strong interactions and would not appear likely to behave as structureless particles. Nevertheless, classical β-decay is described well by matrix elements *analogous to* (6.15):

$$\langle p| \, h^\alpha \, |n\rangle \simeq \overline{u}_p\gamma^\alpha(g_V - g_A\gamma_5)u_n, \tag{6.19}$$

with $g_V \simeq 0.98$ and $g_A \simeq 1.24$. This is surprising! The vector coefficient g_V in (6.19) is almost the same as in (6.15), whereas we might expect strong interactions to alter it significantly. In addition, the axial vector coefficient g_A is not much larger than would be expected for a point-like particle.

Conserved Vector Current Hypothesis. The similarity of the vector coupling constants for leptonic and hadronic weak decays was first explained by Feynman and Gell-Mann (1958) using the *conserved vector current hypothesis* (CVC). They noted that in QED (also a vector theory) a related situation occurs: the electric charge is rigorously conserved, even for strongly interacting particles. No matter what the detailed structure of the charge cloud surrounding the bare particle, the total charge is always the same; it is not renormalized by the presence of interactions among the hadronic constituents. In the final analysis this is because the electromagnetic 4-current is a conserved current, and suggests that some conserved current may be responsible for the protection of hadronic vector matrix elements; a logical candidate is a conserved isospin current.

As we saw in §2.8, symmetry under global isospin rotations implies the existence of three conserved isospin currents. If attention is restricted to nucleons for a moment, in terms of the isospin doublet

$$N = \begin{pmatrix} p \\ n \end{pmatrix} \tag{6.20}$$

the three currents are

$$V_j^\alpha(x) = \overline{N}(x)\gamma^\alpha \frac{\tau_j}{2} N(x), \tag{6.21}$$

where j labels isospin components, α is a Lorentz index, and τ_j denotes the three Pauli matrices. But $(1 + \tau_3)/2$ is a projection operator for protons and the nucleon electromagnetic current is (Leader and Predazzi, 1982)

$$J_{\rm EM}^\alpha(x) = \overline{N}(x)\gamma^\alpha \left(\frac{1 + \tau_3}{2} \right) N(x)$$

$$= J_{\rm EM}^\alpha(\text{isoscalar}) + J_{\rm EM}^\alpha(\text{isovector}), \tag{6.22}$$

where

$$
\begin{aligned}
J_{\rm EM}^\alpha(\text{isoscalar}) &= \tfrac{1}{2}\overline{N}\gamma^\alpha N = \tfrac{1}{2}(\overline{p}\gamma^\alpha p + \overline{n}\gamma^\alpha n) \\
J_{\rm EM}^\alpha(\text{isovector}) &= \tfrac{1}{2}(\overline{N}\gamma^\alpha \tau_3 N) = \tfrac{1}{2}(\overline{p}\gamma^\alpha p - \overline{n}\gamma^\alpha n).
\end{aligned}
\tag{6.23}
$$

Thus, for the nucleons

$$V_3^\alpha(x) = J_{\rm EM}^\alpha(\text{isovector}). \tag{6.24}$$

Suppose that the vector part of the hadronic weak current h^α is of the form

$$V_\pm^\alpha(x) = \overline{N}(x)\gamma^\alpha \left(\frac{\tau_1 \pm i\tau_2}{2} \right) N(x). \tag{6.25}$$

The CVC hypothesis then assumes that the currents (6.24) and (6.25) are components of a conserved common isovector current,

$$\mathbf{J}^\alpha(\text{isovector}) = \overline{N} \left(\tfrac{1}{2}\gamma^\alpha \boldsymbol{\tau} \right) N. \tag{6.26}$$

It is the conservation of this current that prevents renormalization of the hadronic weak vertex by the strong interactions under the CVC hypothesis, leading to the first term of the result (6.19).

The CVC hypothesis has a variety of experimental implications that are supported by the data. A discussion may be found in Cheng and O'Neill (1979), Chs. 7–8.

Partially Conserved Axial Current. In a similar way as for the above discussion of CVC, we may consider the deviation of the axial vector coefficient $g_{\rm A}$ in (6.19) by only about 25% from the leptonic value to arise from a *partially conserved axial-vector current* (PCAC). This will be discussed in §12.2, but we note here that such an approach links the failure of axial current conservation to the finite mass of the pion; in particular, the divergence of the axial current is proportional to a pseudoscalar operator, and the constant of proportionality vanishes as the mass of the pion tends to zero [see Exercise 8.3 and eq. (12.18)]. The divergence of the axial current will play an important role in later discussions of the bag model for hadrons, and of the general significance of chiral symmetry for the strong interactions (Chs. 12 and 13).

Cabibbo Modification. The CVC hypothesis has considerable experimental verification, but g_V in eq. (6.19) differs by about 2% from unity, which means that this cannot be the whole story. In addition, we have so far neglected strangeness-changing weak interactions such as $\Lambda \to p + e^- + \bar{\nu}_e$. When such $\Delta S = 1$ hadronic weak interactions are compared with $\Delta S = 0$ interactions they are found to be strongly suppressed, typically by factors of about 20. Cabibbo (1963) noted that if we write

$$\langle p | h^\alpha | \Lambda \rangle = \bar{u}_p \gamma^\alpha \left(g_V^{\Delta S=1} - g_A^{\Delta S=1} \gamma_5 \right) u_\Lambda \tag{6.27}$$

and so on for other $\Delta S = 1$ reactions, then the data indicate that

$$\left(g_V^{\Delta S=0} \right)^2 + \left(g_V^{\Delta S=1} \right)^2 \simeq 1, \tag{6.28}$$

suggesting that the vector part of h^α be parameterized as

$$V^\alpha = \cos\theta_c V^\alpha(\Delta S = 0) + \sin\theta_c V^\alpha(\Delta S = 1), \tag{6.29}$$

where $V^\alpha(\Delta S = 0)$ and $V^\alpha(\Delta S = 1)$ are, respectively, strangeness conserving and strangeness nonconserving currents. Then

$$\langle p | V^\alpha | n \rangle = \cos\theta_c \bar{u}_p \gamma^\alpha u_n \tag{6.30}$$

$$\langle p | V^\alpha | \Lambda \rangle = \sin\theta_c \bar{u}_p \gamma^\alpha u_\Lambda. \tag{6.31}$$

The *Cabibbo angle* θ_c is readily determined by comparing $\Delta S = 0$ and $\Delta S = 1$ weak interactions. One finds

$$\sin\theta_c \simeq 0.230 \tag{6.32}$$

or $\cos\theta_c \simeq 0.973$. Because $\cos^2\theta_c \gg \sin^2\theta_c$, matrix elements proportional to $\sin\theta_c$ are sometimes termed *Cabibbo forbidden*, and those proportional to $\cos\theta_c$, *Cabibbo allowed*. Thus this bit of phenomenology accounts at once for the small deviation of g_V from unity, *and* for the suppression of the strangeness-changing vector current, by the simple hypothesis that the hadronic weak interaction strength is shared between $\Delta S = 0$ and $\Delta S = 1$ transitions.

In the quark picture the weak interactions of the hadrons originate in the corresponding interactions of the quarks. For example, neutron β-decay results from the weak decay of a down quark, as illustrated in Fig. 6.1. At the quark level it is assumed that there is an octet of conserved $SU(3)$ vector currents

$$V_j^\alpha(x) = \bar{q}(x)\gamma^\alpha \left(\frac{\lambda_j}{2} \right) q(x), \tag{6.33}$$

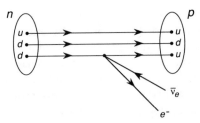

Fig. 6.1   Neutron β-decay at the quark level in the Fermi theory.

and an octet of partially conserved $SU(3)$ axial vector currents

$$A_j^\alpha(x) = \bar{q}(x)\gamma^\alpha\gamma_5\left(\frac{\lambda_j}{2}\right)q(x), \tag{6.34}$$

where $q = (u, d, s)$ represents a flavor triplet of quarks and the $\lambda_j/2$ are $SU(3)$ generators [eq. (5.63)]. Writing the matrices (5.63) out explicitly in (6.33) and (6.34), one sees that the currents $V_4^\alpha \pm iV_5^\alpha$ and $A_4^\alpha \pm iA_5^\alpha$ correspond to $|\Delta S| = 1$ transitions. For example, $V_4^\alpha + iV_5^\alpha = \bar{u}\gamma^\alpha s$. Thus the second term of (6.29) (and an analogous one for the axial current) may be identified with these currents. Likewise, $V_1^\alpha \pm iV_2^\alpha$ and $A_1^\alpha \pm iA_2^\alpha$ correspond to $\Delta S = 0$ transitions, and the first term of (6.29) and its axial analog are associated with them. Then, if charm is neglected the charged hadronic weak currents may be expressed as

$$\begin{aligned} h^\alpha &= \bar{u}\gamma^\alpha(1 - \gamma_5)d_c \\ &= \bar{u}\gamma^\alpha(1 - \gamma_5)(d\cos\theta_c + s\sin\theta_c), \end{aligned} \tag{6.35}$$

where the *Cabibbo rotated quark* d_c is defined through the transformation

$$\begin{pmatrix} d_c \\ s_c \end{pmatrix} = \begin{pmatrix} \cos\theta_c & \sin\theta_c \\ -\sin\theta_c & \cos\theta_c \end{pmatrix} \begin{pmatrix} d \\ s \end{pmatrix} = \begin{pmatrix} d\cos\theta_c + s\sin\theta_c \\ -d\sin\theta_c + s\cos\theta_c \end{pmatrix}, \tag{6.36}$$

and where d and s are operators for the mass eigenstate down and strange quarks (the orthogonal combination s_c and the inclusion of charmed quarks will be discussed in §9.3).[‡]   Then there appears to be a formal similarity

---

[‡]If neutrinos are massless a mixing angle in the leptonic sector would be superfluous because the corresponding mixings have no observable consequence. The choice of the d and s quark sector for the mixing (6.36) rather than the u and c sector (or a combination of the two) is conventional: because absolute phases are not observable, any Cabibbo-like mixing for four quark flavors is equivalent to (6.36). Extension of mixing to more than four quark flavors is addressed briefly in §9.3. For six flavors the mixing matrix (*Kobayashi–Maskawa matrix*) requires three real angles and a phase. If neutrinos have finite mass a corresponding mixing matrix could be required in the leptonic sector as well.

between multiplets such as

$$
\begin{pmatrix} \nu_e \\ e^- \end{pmatrix} \quad \begin{pmatrix} \nu_\mu \\ \mu^- \end{pmatrix} \quad \begin{pmatrix} u \\ d_c \end{pmatrix} \quad \begin{pmatrix} c \\ s_c \end{pmatrix} \tag{6.37}
$$

with respect to the weak interaction. These are called *weak isospin doublets*, but they have nothing to do with normal isospin. We will have much more to say about the importance of weak isospin in Ch. 9. There we will see that the doublets (6.37) are representations of a spontaneously broken gauge symmetry.

Thus the basic idea of the Cabibbo theory is that the particle states carrying the "weak charge" are a linear combination of the mass-eigenstate strange and nonstrange hadronic states. There presently is no fundamental explanation of this mixing, but by introducing the empirical Cabibbo angle we may retain the concept of *universality* in the weak interactions at the quark level. That is, matrix elements of the charged weak currents entering (6.11) may be written in the general form

$$
\langle e | J_\alpha | \nu \rangle \propto G \bar{u}_e \gamma_\alpha (1 - \gamma_5) u_\nu
$$
$$
\langle u | J_\alpha | d \rangle \propto G \cos \theta_c \bar{u}_u \gamma_\alpha (1 - \gamma_5) u_d \tag{6.38}
$$
$$
\langle u | J_\alpha | s \rangle \propto G \sin \theta_c \bar{u}_u \gamma_\alpha (1 - \gamma_5) u_s.
$$

In these expressions all differences between the leptonic and hadronic weak interaction strengths lie hidden in the Cabibbo angle, and all matrix elements are governed by a single interaction strength G. This is phenomenology, but not garden-variety phenomenology. A single empirically-determined parameter θ_c is sufficient to correlate all weak interaction data for those hadrons containing only up, down, and strange quarks. There clearly is physics here, and a strong indication of weak interaction universality. This is critical for later discussions of unified theories; before we can unify the weak interactions with anything else we must first be certain that the weak interactions themselves constitute a whole fabric, unified through a single weak interaction coupling constant.

6.3 Problems with the Fermi Phenomenology

6.3.1 Unitarity Violation

The phenomenology of the Fermi current–current Lagrangian correlates a large amount of data in an extremely economical way. However, the Fermi theory has some hidden diseases that seem relatively innocuous at low energies, but that become lethal at higher energies. Consider the reaction $\nu_e + e \rightarrow e + \nu_e$

in the CM system; if the electron mass is neglected,

$$\frac{d\sigma}{d\Omega} = \frac{G^2 k^2}{\pi^2} \tag{6.39a}$$

$$\sigma = \frac{4G^2 k^2}{\pi} \simeq G^2 s, \tag{6.39b}$$

where s is the square of the center-of-mass energy: $s = E_{\text{CM}}^2 \simeq 2m E_{\text{Lab}}.$[‡]
In the current–current theory the interaction is point-like and must proceed
by the S-wave in a partial wave decomposition of the cross section. Neglecting intrinsic spin as an unimportant complication, the cross section may be
expanded in partial waves

$$\sigma = \frac{\pi}{k^2} \sum_\ell (2\ell + 1)|f_\ell|^2, \tag{6.40}$$

where k is the magnitude of the center-of-mass momentum and where unitarity
requires $|f_\ell| \leqslant 1$. Thus the S-wave ($\ell = 0$) cross section is bounded by

$$\sigma(\ell = 0) \lesssim \frac{\pi}{k^2} \simeq \frac{1}{s}. \tag{6.41}$$

Comparing (6.39) and (6.41), the Fermi theory will violate partial-wave unitarity when

$$E_{\text{CM}} \simeq G^{-1/2} \simeq 300 \text{ GeV}. \tag{6.42}$$

This energy is high, but the current–current theory must be modified if we
wish to find an internally consistent model of the weak interactions.

6.3.2 Intermediate Vector Bosons

The initial attempt to cure this problem proceeds by direct analogy with QED.
At least part of the difficulty is the point nature of the Fermi Lagrangian.
If the analogs of photons are introduced as exchange particles to mediate

---

[‡]This result can be obtained from field theory using the methods of Ch. 3 (see
Renton, 1990, §5.1.3), but we can also construct it on dimensional and invariance
grounds. The cross section has dimension $[L]^2 = [M]^{-2}$ (see appendix A). It must
contain the weak coupling constant squared since it is the square of a matrix element, but $[G]^2 = [M]^{-4}$ [eq. (6.12)]. At high energies lepton masses are negligible,
and the only relativistically invariant quantity of dimension $[M]^2$ available to yield
cross section units when multiplied by G^2 is the square of the center-of-mass energy; thus, $\sigma \simeq G^2 s$. Obtaining the constant of proportionality in (6.39) requires
considerably harder work (see Lee, 1981, Ch. 8, for an introduction to such quick
order-of-magnitude estimates).

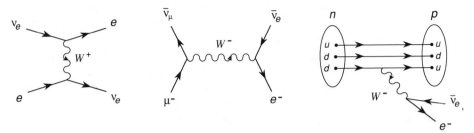

Fig. 6.2   Diagrams including charged intermediate vector bosons $W^{\pm}$.

the interactions, the currents can interact over a finite range. The requisite particles are called *intermediate vector bosons* (IVBs), and the corresponding weak Lagrangian that replaces (6.11) is [compare eq. (2.203)]

$$\mathcal{L} = g_{\mathrm{w}}\left[J^{\alpha}(x)W_{\alpha}^{+}(x) + J^{\alpha\dagger}(x)W_{\alpha}^{-}(x)\right], \tag{6.43}$$

where $W_{\alpha}(x)$ denotes the field of the vector boson. Some typical diagrams are shown in Fig. 6.2.

The Fermi weak interaction theory as amended by the inclusion of intermediate vector bosons is inspired by analogy with QED, but there are some obvious differences between the two:

- Photons are neutral, but some of the intermediate vector bosons must be charged since they participate in charge-changing interactions. In fact, two of them are charged ($W^{\pm}$) and one is neutral (Z^{0}).
- Because the weak interaction is known to be almost point-like, the intermediate vector bosons must be very massive (Yukawa–Wick relation between mass of exchanged particle and range of force). The IVBs are indeed massive, of order 100 GeV; conversely, the photon appears to be massless.
- Electromagnetic interactions respect C and P symmetries to high precision; weak interactions violate them with abandon, and even break CP or T symmetries to a small extent. The IVB interactions must incorporate these features.
- Because the intermediate vector bosons are massive, gauge invariance cannot be imposed on this field as in QED (see §2.7.3). The consequences of this simple fact for the high-energy behavior of the Fermi theory will turn out to be catastrophic.

The differences would appear to be crucial, and to say that QED and the weak interactions are similar except for these differences seems at first to be a ridiculous proposition. For instance, if photons were charged the linear superposition principle for the electromagnetic field (and all that it implies) would no longer hold, and if photons mediated only short-range forces because of a large mass there would likely be no atoms (and no physicists!) in the universe. However, it may be fair to say that the weak and electromagnetic interactions are similar except that they differ in almost all their obvious characteristics, but it would be far too hasty to conclude that they are unrelated. Indeed, we will see that they may in a certain sense be the *same* interaction. The justification of the preceding two statements will clearly require extracting more than one rabbit from more than one hat, and will constitute the subject matter of Chs. 7–9.

6.3.3 Renormalizability

Now let us consider the process $\nu_e + e \rightarrow e + \nu_e$, as mediated by intermediate vector boson exchange. A typical diagram is illustrated in Fig. 6.2; to lowest order the CM differential cross section is (Leader and Predazzi, 1982)

$$\frac{d\sigma}{d\Omega} = \frac{2g_w^4 k^2}{\pi^2 m_W^4} \left(\frac{1}{1 - 2q^2/m_W^2 + q^4/m_W^4} \right) \tag{6.44a}$$

$$q^2 \simeq -2k^2(1 - \cos\theta), \tag{6.44b}$$

where we neglect electron masses, g_w is the coupling constant appearing in eq. (6.43), and m_W is the mass of the W^+. This expression differs from the current–current cross section by the appearance of a different coupling constant, and by the denominator factor that comes from the propagator for a massive vector boson [see (6.49)]. It reduces to the corresponding Fermi result (6.39) if we take the limit $q^2/m_W^2 \rightarrow 0$ and identify

$$\frac{g_w^2}{m_W^2} = \frac{G}{\sqrt{2}}. \tag{6.45}$$

Comparing (6.12) and (6.45) we see that the constant g_w coupling the weak current to the vector boson field is *dimensionless*. In contrast, the Fermi constant G has dimension $[M]^{-2}$. This difference will play an important role in the subsequent discussion of renormalization in weak interaction theories.

The IVB-mediated interaction is no longer pointlike (it is of range $\simeq 1/m_W$), and now higher partial waves contribute [hence the angular dependence in (6.44)]. The high-energy behavior is improved since the W propagator will suppress the growth of the cross section at high energies:

$$\sigma = \frac{4G^2 k^2}{\pi} \bigg/ \left(1 + \frac{4k^2}{m_W^2} \right), \tag{6.46}$$

and at large k,

$$\sigma \xrightarrow[k\to\infty]{} \frac{G^2 m_W^2}{\pi} = \text{constant}. \tag{6.47}$$

This looks more promising, but it is still not satisfactory. First, this expression too will violate unitarity, although at extremely high energies compared with the current–current theory. We might be tempted to overlook this, but there are more serious difficulties associated with the requirement that a massive vector field such as that corresponding to $W^{\pm}$ possess a longitudinal state of polarization in addition to the two transverse states permitted the massless photon.

To illustrate the problem we analyze the reaction $\nu_e + \bar{\nu}_e \to W^+ + W^-$, which is impractical but should give sensible results if the theory is to be trusted. The sum over the three polarization states of a vector particle of mass M gives a factor

$$\sum_\lambda \epsilon_{k\lambda}^{\mu*} \epsilon_{k\lambda}^{\nu} = -g^{\mu\nu} + \frac{k^\mu k^\nu}{M^2}, \tag{6.48}$$

leading to a massive vector propagator (Exercise 3.9)

$$W^{\mu\nu}(q) = \frac{-g^{\mu\nu} + q^\mu q^\nu / M^2}{q^2 - M^2 + i\varepsilon}. \tag{6.49}$$

The second term in (6.48) is associated with the *longitudinal state of polarization* for the massive particle. Heuristically, this may be inferred by comparing (6.49) for the massive vector field with (3.64) for the photon field, and recalling that the longitudinal polarization state makes no physical contribution for the free photon. A term of the form $q^\mu q^\nu / q^2$ may be added to the photon propagator, but it is possible to eliminate it by a gauge transformation and recover (3.64): see the covariant photon propagator in §3.9 and Exercise 3.9. Alternatively, a term $q^\mu q^\nu / q^2$ makes no contribution when evaluated between conserved currents in a matrix element because the momentum-space form of (1.82b) implies $q^\nu j_\nu = 0$. The production of W's in transverse (T) and longitudinal (L) states gives the high-energy behavior (Quigg, 1983, §6.2; Leader and Predazzi, 1982, §2.1)

$$\sigma(\nu\bar{\nu} \to W_T \overline{W}_T) \underset{s\to\infty}{\simeq} \text{constant} \qquad \text{(transverse)} \tag{6.50}$$

$$\sigma(\nu\bar{\nu} \to W_L \overline{W}_L) \underset{s\to\infty}{\simeq} G^2 s \qquad \text{(longitudinal)}, \tag{6.51}$$

and the uncontrolled growth of W production in longitudinal polarization states will lead to unitarity violation at high energies.

A related high-energy disease is also suggested by the form of the propagator

(6.49). At large momentum this propagator is dominated by the second term of the numerator and tends to a constant. In high-order weak interaction diagrams involving loops, integrals over internal momenta will appear with vector boson propagators in the integrand. Since the propagator tends to a finite constant, such integrals will possess *ultraviolet divergences*. Further, these divergences are quite ferocious; they cannot be suppressed without the introduction of an infinite number of arbitrary constants. Here at last is the nub of the matter: *the Fermi theory, even with intermediate vector bosons, is not renormalizable*, for we will shortly define a renormalizable theory to be one in which all ultraviolet divergences may be removed by the introduction of a finite number of parameters taken from experiments.

This is also the real reason that unitarity violation in our previous examples is a serious problem: in principle, unitarity violation at a particular order in perturbation theory need not signal a fundamental inconsistency if unitarity can somehow be restored by higher order terms in the perturbative expansion. However, the higher order terms in the Fermi theory are *infinite*, so they cannot be used for this purpose and the theory must violate probability conservation at sufficiently high energy. These two problems are intimately related, as the preceding discussion suggests: poor high-energy behavior and the associated unitarity violation in low order diagrams manifests itself as divergences in higher order diagrams, because they contain integrals over lower order diagrams.

Before pursuing this matter it is instructive to see more clearly why the related problems of unitarity violation and non-renormalizability plague the weak interaction phenomenology. We do so by comparing with QED, which has managed to avoid such embarrassments. The first point is that the current–current theory immediately lands itself in trouble because *the Fermi coupling constant G is not dimensionless* [eq. (6.12)]. Consequently, dimensionality arguments require that the cross section grow with s in (6.39), and this must inevitably violate unitarity. In contrast, the QED coupling constant α is dimensionless and QED cross sections fall rapidly at large energy; for example (Exercise 3.5f),

$$\sigma(e^+ e^- \to \mu^+ \mu^-) \simeq \frac{4\pi\alpha^2}{3s}. \tag{6.52}$$

Therefore, one may surmise that perturbative renormalizability and unitarity properties depend critically on the dimensionality of the relevant coupling constant. This is indeed so, although it is not the complete story. There is a simply stated criterion for determining whether a quantum field theory is renormalizable:

- If the coupling constant has negative mass dimension the theory is not renormalizable (see Exercise 6.4).

- If the coupling constant has positive mass dimension the theory is less divergent than QED and is renormalizable.
- If the coupling constant is dimensionless the theory may, or may not, be renormalizable; more information, such as the high-energy behavior of the propagators, will be required to decide which situation is realized.

The introduction of intermediate vector bosons with mass m_W stops the growth of the weak interactions at energies comparable to m_W, and the coupling constant g_w of the W boson to the weak current is dimensionless, in contrast to the Fermi coupling G, but similar to the coupling of the electrical charge to the electromagnetic current. Therefore, one might hope that the weak interactions with massive W bosons included would be renormalizable. Unfortunately they are not, because of the crucial differences between W particles and photons. In essence, the longitudinal polarization states accessible to the massive vector propagators conspire at high energy to reintroduce an effective coupling constant with dimension $[M]^{-2}$ [see eq. (6.51)]. Thus, the hope that the dimensionless constant appearing in (6.43) will produce a renormalizable theory is illusory; the effective high-energy coupling in *both* (6.43) and (6.11) is of negative mass dimension, and the theory cannot be renormalized as it stands.

This difficulty is avoided in QED because the freedom of a gauge transformation renders harmless any contribution from longitudinally polarized photon states. This yields cross sections that fall obediently at high energies and leads to divergences sufficiently meek to be eliminated by importing a few parameters from the experiments. In contrast, the massive gauge quanta $W^{\pm}$ and Z^0 preclude gauge invariance if they obtain their masses through conventional mass terms in the Lagrangian. Thus, the crucial question is how to achieve a renormalizable theory of weak interactions. With that in mind, we now take a closer look at how renormalization is carried out in quantum electrodynamics.

6.4 Renormalization in QED

In relativistic quantum field theories ultraviolet divergences are the norm for any calculation that goes beyond the tree level. These divergences must be eliminated if perturbative calculations are to have a sensible interpretation. This statement applies even if we restrict attention to first-order calculations: we have no justification for trusting low order terms—even those in impressive agreement with observations—unless higher order terms are under control. The systematic process by which the infinities are removed is called renormalization, and a theory that admits a consistent algorithm for removing divergences is said to be renormalizable. To expedite the discussion of renormalization in field theories a scheme for classifying the various types of divergences that may appear is useful; to this end we introduce the idea of a *degree of divergence* for a Feynman graph.

6.4.1 Classification of Divergences

A *primitively divergent graph* is a divergent graph that becomes convergent when any momentum integration variable (internal loop momentum) is held constant. Graphically, this corresponds to the cutting of an internal line. Tree graphs are generally convergent since they involve no internal momentum integrations, so it is clear that any graph can be made finite by severing enough internal lines.

The *superficial degree of divergence* $D(g)$ for a graph g is defined as the number of internal momentum powers in the integrand numerator of the corresponding matrix element (counting 4 for each d^4k), minus the number in the denominator. For QED in four spacetime dimensions

$$D = 4k - 2b - f, \tag{6.53}$$

where

$$k = \text{number of loop momentum integrations}$$
$$b = \text{number of internal photon lines}$$
$$f = \text{number of internal electron lines.}$$

This may be rewritten (Exercise 6.5a; Bjorken and Drell, 1965, §19.10)

$$D = 4 - \tfrac{3}{2}F - B, \tag{6.54}$$

where

$$B = \text{number of external photon or boson lines}$$
$$F = \text{number of external fermion lines.}$$

Thus QED possesses the noteworthy property that the degree of divergence for its graphs is independent of the detailed structure of the internal lines, and depends only on the number and kind of external lines; in particular, the superficial degree of divergence is independent of the number of vertices (perturbative order). This may be traced to the observation that the coupling constant is dimensionless, and is ultimately what makes QED renormalizable.

Determination of the superficial degree of divergence for a primitively divergent graph is called "naive power counting". The index $D(g)$ is termed superficial because the actual degree of divergence may differ from that im-

Table 6.3
Primitively Divergent Graphs in QED, the Superficial Degree of Divergence
(6.54), and the Actual Degree of Divergence

Graph	B	F	D	Actual Divergence
electron self-energy	0	2	1	Logarithmic
photon self-energy	2	0	2	Logarithmic
vertex	1	2	0	Logarithmic

plied by $D(g)$. For example, we will see below that photon self-energy graphs
are superficially quadratically divergent $(D = 2)$, but gauge invariance reduces the actual divergence to a logarithmic one. On the other hand, a graph
for which the actual degree of divergence is (considerably) greater than the
superficial degree of divergence may be found in Exercise 6.3c.[‡]

A viable renormalization procedure requires that a theory may have only a
finite number of primitively divergent graphs: *the systematic renormalization
of all such graphs leads to an order-by-order cancellation of divergences, and to
finite predictions from the perturbation series.*[‡‡] QED is such a theory, because
it has only three primitively divergent diagrams: (1) the proper electron self-
energy, (2) the proper photon self-energy, and (3) the proper vertex. These
are illustrated with their superficial and actual degrees of divergence in Table
6.3. The adjective "proper" in these expressions means *one-particle irreducible*
(1PI), and denotes a connected graph that cannot be made disconnected by
cutting a single internal line. For example,

[‡]Despite these remarks, in a systematic order-by-order renormalization scheme it is
the *superficial* rather than the actual degree of divergence that is the most important
quantity. See Coleman (1985), Ch. 4, and Collins (1984), §§3.2, 5.1–5.3, and 2.3.
[‡‡]Order-by-order cancellation ensures the stability of the renormalization against
small changes in the coupling parameters.

are 1PI diagrams, but

is not, because it falls into two separate diagrams if cut in the middle (it is termed a reducible diagram). Diagrams that are 1PI, and 1PI Green's functions (Green's functions receiving contributions only from 1PI diagrams), play a central role in perturbative renormalizability because one-particle reducible diagrams can be decomposed into 1PI diagrams without additional loop integrations. Hence the ability to remove UV divergences in the 1PI diagrams ensures that they can be removed in all diagrams. (Coleman, 1985, Ch. 4; Collins, 1984, §3.2 and §§5.1–5.3; Cheng and Li, 1984, Ch. 2).

EXERCISE 6.3 (a) Show that for scalar field theory with a ϕ^r interaction the superficial degree of divergence for a typical graph is

$$D = d - \left(\frac{d}{2} - 1\right) E - \left[d - \frac{r}{2}(d-2)\right] n,$$

where d is the dimension of spacetime, E is the number of external lines, and n is the number of vertices. (b) What does this imply for ϕ^3, ϕ^4, and ϕ^6 theories in four dimensions? What about ϕ^r in two dimensions? (c) For ϕ^4 theory in four spacetime dimensions, what is D for the following graph:

Is this a convergent diagram?

EXERCISE 6.4 For scalar field theory in d dimensions with an interaction $g\phi^r$, show that

$$[\phi] = [M]^{d/2-1} \qquad [g] = [M]^\delta \qquad \delta = d + r - \frac{rd}{2}.$$

Hint: consider the dimension of terms in the Lagrangian density. Use this result and Exercise 6.3 to show that the superficial degree of divergence for graphs in scalar ϕ^r theory can be written

$$D = d - \left(\frac{d}{2} - 1\right) E - \delta n,$$

where E is the number of external lines and n the number of vertices. Hence a necessary condition for renormalizability is that the mass dimension δ of the coupling constant g satisfy $\delta \geqslant 0$.

EXERCISE 6.5 (a) Derive eq. (6.54) for the superficial degree of divergence in a QED diagram. (b) Show that for the Fermi weak interaction theory described by (6.11) the degree of divergence for graphs in four dimensions is

$$D = 2n + 4 - \tfrac{3}{2}F,$$

where n is the number of vertices and F is the number of external fermion lines. Thus the degree of divergence increases with perturbative order and the theory cannot be renormalized.

6.4.2 Example: Charge and Mass Renormalization

A renormalization algorithm may be divided into two general parts: (1) a *regularization procedure* that makes integrals well-defined and allows formal manipulations, and (2) a *subtraction procedure* to cancel divergences in the physical matrix elements. To illustrate the philosophy, we consider the renormalization of charge in QED for the process $e^-\mu^- \to e^-\mu^-$ (Aitchison and Hey, 1982, §7.3). The lowest order diagram is shown in Fig. 6.3. Employing the Feynman rules of §3.9, this corresponds to an amplitude of the form

$$\mathcal{A}^{(1)} \simeq e^2 \bar{u}(k_2)\gamma_\mu u(k_1) \frac{-g^{\mu\nu}}{q^2} \bar{u}(p_2)\gamma_\nu u(p_1), \qquad (6.55)$$

where spinor labels are suppressed. In the next order a contributing diagram is shown in Fig. 6.4. This diagram will involve a divergent loop momentum integral

$$i\Pi^{\mu\nu}(q) \equiv - \int \frac{d^4k}{(2\pi)^4}\, \mathrm{Tr}\left(ie\gamma^\mu \frac{i}{\not{k} - m} ie\gamma^\nu \frac{i}{\not{k} - \not{q} - m} \right), \qquad (6.56a)$$

where m is the mass of the electron. This may be viewed as giving a modified photon propagator. In Feynman gauge, to second order in the unrenormalized

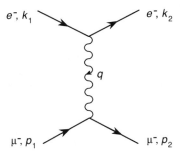

Fig. 6.3 Lowest order diagram for $e^-\mu^- \to e^-\mu^-$.

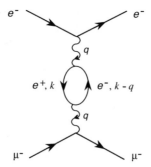

Fig. 6.4   Vacuum polarization diagram contributing to $e^-\mu^- \to e^-\mu^-$.

charge,

$$iD_{\mu\nu}(q) = \frac{-ig_{\mu\nu}}{q^2} + \frac{-ig_{\mu\alpha}}{q^2} i\Pi^{\alpha\beta}(q)\frac{-ig_{\beta\nu}}{q^2}$$

$$= \text{\small(diagram)} + \text{\small(diagram)} \qquad (6.56b)$$

By power counting the degree of divergence in (6.56a) is two, since at large momentum there are four powers of k in the numerator, but only two in the denominator. However, a more careful examination shows that the integral actually diverges only logarithmically (see Bjorken and Drell, 1964, §8.2). We will regulate the amplitude by the *cutoff method*: the divergent integral is truncated at some finite value M of k, yielding results that are functions of M (*the regularization*). We then attempt to modify the result in such a way that the observables are finite and independent of the cutoff (*the subtraction*). This strategy will succeed only if the divergences can be buried in quantities that are not observables.

The sum of the diagrams in Figs. 6.3 and 6.4 is of the form (Bjorken and Drell, 1964, Ch. 8; Aitchison and Hey, 1982, §7.3)

$$\mathcal{A} \simeq e^2 \left[1 - \frac{\alpha}{3\pi}\log\left(\frac{M^2}{m^2}\right) + \chi(q^2)\right] \bar{u}(k_2)\gamma_\mu u(k_1)\frac{-g^{\mu\nu}}{q^2}\bar{u}(p_2)\gamma_\nu u(p_1), \quad (6.57)$$

where $\chi(q^2)$ is finite and tends to zero as $q \to 0$. The essential point for the renormalization scheme is the realization that (6.55) and (6.57) are of exactly the same form provided we identify an effective charge

$$e_{\mathrm{R}} = e\left[1 - \frac{\alpha}{3\pi}\log\left(\frac{M^2}{m^2}\right) + \chi(q^2)\right]^{1/2}$$

$$\underset{q^2 \to 0}{=} e\left[1 - \frac{\alpha}{3\pi}\log\left(\frac{M^2}{m^2}\right)\right]^{1/2}. \qquad (6.58)$$

To obtain this expression we have considered a particular correction to e that was generated by one vacuum polarization loop. There will be an infinite series of such corrections and generally

$$e_R = e \left[1 + \sum_{n=1}^{\infty} \alpha^n C_n(M^2) \right]. \tag{6.59}$$

We now surmise that the "bare charge" e is not a measurable quantity, since we cannot remove the electrons from the polarizable medium (the vacuum); in fact, the measured electron charge is e_R, and this is a number that can be taken from experiments.

Let us now come to the point and demonstrate the removal of divergences in the photon propagator to order α. An explicit expression for $\chi(q^2)$ valid for low q is (see Halzen and Martin, 1984, §7.2)

$$\chi(q^2) \simeq -\frac{\alpha}{15\pi} \frac{q^2}{m^2}. \tag{6.60}$$

Since the electrical charge is a measurable quantity with a classical meaning, it should be possible to define it in the long-wavelength ($q \to 0$) limit. To this order in the $q \to 0$ limit the physically measurable electronic charge is

$$e_R^2 = e^2 \left[1 - \frac{\alpha}{3\pi} \log \left(\frac{M^2}{m^2} \right) + \mathcal{O}\left(\alpha^2\right) \right]. \tag{6.61}$$

Inverting this expression gives

$$e^2 = e_R^2 \left[1 + \frac{\alpha_R}{3\pi} \log \left(\frac{M^2}{m^2} \right) + \mathcal{O}\left(\alpha_R^2\right) \right], \tag{6.62}$$

and on inserting this in (6.57),

$$\mathcal{A} \simeq e_R^2 \left[1 - \frac{\alpha_R}{15\pi} \frac{q^2}{m^2} + \mathcal{O}\left(\alpha_R^2\right) \right] \bar{u}(k_2)\gamma_\mu u(k_1) \frac{-g^{\mu\nu}}{q^2} \bar{u}(p_2)\gamma_\nu u(p_1), \tag{6.63}$$

where $\alpha_R \equiv e_R^2/4\pi$. This expression for the amplitude to order α is *finite* and *independent of the cutoff*. By an analogous inversion procedure explicit forms for the other coefficients $C_n(M^2)$ in eq. (6.59) may be found and infinities in the physical amplitudes removed systematically, order-by-order. Thus a large number of QED infinities are exorcised by the introduction of a single physically measurable quantity, the electronic charge.

In a similar way, the bare and physical electron masses may be related by a power series

$$m_R = m \left[1 + \sum_{n=1}^{\infty} \alpha^n D_n(M^2) \right], \tag{6.64}$$

with the corrections to the bare mass associated with electron self-energy graphs (Table 6.3). The coefficients D_n can be determined by considering all graphs contributing to a given order in perturbation theory. By an inversion procedure similar to that for charge renormalization the physical mass may be introduced in amplitudes to make them finite and independent of momentum cutoff order-by-order; this removes another set of QED divergences. Thus we see in some very simple examples how infinities may be systematically eliminated by relating "bare" quantities to physical quantities. Renormalization is more properly "reparameterization" of the equations in terms of such finite quantities, and we can now give a precise operational definition of (perturbative) renormalizability: *a theory is renormalizable if all divergences can be eliminated by the introduction of a finite number of empirical quantities.* QED is such a theory: by introducing a few measured parameters like electron mass and charge, all QED amplitudes become finite and in excellent agreement with the experimental data (Table 6.2).

> Notice that by this definition a theory that is not renormalizable might still be made finite, but only by employing new parameters at each order of perturbation theory; the point is that such a theory has no predictive power. Notice also that our discussion is framed in the context of perturbation theory. We do not investigate whether a theory that is not perturbatively renormalizable might yet make sense non-perturbatively, but the present mood is that the absence of perturbative renormalizability is a portent of fundamental pathologies in the full theory—not just a failure of the perturbation technique. At any rate, in our subsequent discussion we consider only renormalizable theories as legitimate candidates for theories of fundamental interactions.

6.4.3 Counterterms

The preceding ideas are usually implemented in a slightly different way in actual calculations. Replacing divergent integrals with renormalized integrals is equivalent to adding some terms to the original Lagrangian; the requisite terms are called *counterterms*. For example, a perturbation theory could be based on the Dirac equation

$$(i\not\partial - m_{\rm R} + e\not A + \delta m)\psi = 0, \tag{6.65}$$

where the bare mass m and physical mass $m_{\rm R}$ are related through the mass counterterm δm,

$$\delta m = m_{\rm R} - m. \tag{6.66}$$

The term δm may now be viewed as an additional interaction giving rise to graphs that contribute to the amplitudes for various processes (see §3.5.3 where mass counterterms in a scalar theory are discussed, and Feynman rule

5.A.ii in §3.9). In the resulting perturbation theory the same infinities appear as before, but now there are additional pieces involving the counterterms. Just as for the bare masses or charges the counterterms are unknown to begin with; a theory is renormalizable if a *finite number* of counterterms can be selected to cancel all divergences order-by-order. In practice, this means that a theory is renormalizable if the counterterms required to cancel divergences at each order of perturbation theory have the same form as the terms in the original Lagrangian density (Coleman, 1975, §3.2).

6.4.4 Running Coupling Constant

It will also prove useful to discuss the renormalization of the charge in QED from a slightly different perspective. Notice that there are two consistent interpretations of charge renormalization: (1) the bare charge is "dressed" to the physical charge in expressions such as (6.61); (2) the coupling constant α is momentum dependent. A momentum dependent coupling constant is frequently called a *running coupling constant*. For QED one finds for momentum transfers such that $Q^2 \equiv -q^2 \gg m^2$ in the *leading log approximation* (Quigg, 1983)[‡]

$$\alpha(Q^2) = \alpha(m^2) \left\{ 1 + \frac{\alpha(m^2)}{3\pi} \log\left(\frac{Q^2}{m^2}\right) + \left[\frac{\alpha(m^2)}{3\pi} \log\left(\frac{Q^2}{m^2}\right) \right]^2 + \dots \right\},$$

$$(6.67)$$

or, recognizing the series $\sum\limits_{n=0}^{\infty} x^n = \dfrac{1}{1-x}$,

$$\alpha(Q^2) = \frac{\alpha(m^2)}{1 - \dfrac{\alpha(m^2)}{3\pi} \log\left(\dfrac{Q^2}{m^2}\right)},$$

$$(6.68)$$

[‡]The leading log approximation sums over the largest terms in each order of perturbation theory [an economical discussion may be found in Close (1982b) and a simple example is presented in Kogut (1983), §II.C]. These terms generally involve powers of coupling constants multiplied by logarithmic factors, $\alpha^n \log^n$, whence the name. For QED the leading log corrections to the photon propagator are of the form:

The leading log series solves the one-loop approximation to the renormalization group equations (see Close, 1982b), so (6.68) represents a summation of part of the perturbation series to all orders.

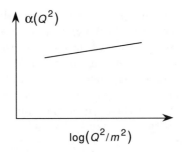

$\log(Q^2/m^2)$

Fig. 6.5   Evolution of the QED running coupling constant.

where questions of convergence have been ignored. The slow increase of the QED running coupling constant with momentum transfer is illustrated in Fig. 6.5. This has a simple physical interpretation, as we have already suggested in §1.6.2: the bare charge polarizes the vacuum so low momentum probes see only the dressed particle and the vacuum screens the charge (Fig. 1.4). The effective coupling increases with momentum transfer because the scattering particles come closer together and the screening is less efficient. One says that the QED vacuum is a *screening vacuum*. In contrast, we will see in Chs. 10 and 12 that the QCD vacuum is *antiscreening* because of nonlinear interactions among the gluons. This means that the QCD running coupling constant *decreases* at large momentum transfer.

The QED coupling $\alpha \simeq \frac{1}{137}$ is often thought of as "constant" because it has a very slow momentum dependence. For example, at momentum transfers comparable with the masses of the intermediate vector bosons the running coupling calculated including all known lepton and quark loops increases to

$$\alpha\left(Q^2 = (90 \text{ GeV})^2\right) \simeq \frac{1}{130} \tag{6.69}$$

[the preceding example considered only an e^+e^- loop; we must also consider contributions from $\mu^+\mu^-$ loops, $q\bar{q}$ loops, ...; see eq. (11.2c)]. Thus, for presently accessible momentum transfers the effect of a momentum dependent α in QED appears only in small effects such as the Lamb shift (see Halzen and Martin, 1984, §7.3 and §7.4). The momentum dependence of the running coupling in QCD is more spectacular, as we will see in Chs. 10–12.

6.4.5   Dimensional Regularization

In the heuristic example just discussed we have chosen to regulate divergent integrals through a simple cutoff of the momentum scale. This is sufficient to illustrate basic ideas, but a regularization scheme chosen carefully with respect to the natural symmetries of a problem will be essential for more complicated situations. The most powerful method of demonstrating renormalizability for gauge fields relies on *dimensional regularization*, which employs analytical continuation in the number of spacetime dimensions for a field theory ('t Hooft and Veltman, 1972). The impetus for this approach is the observation that the factor of four appearing in the first term of (6.53) originates in a loop momentum integration measure $d^d k$, where d is the dimension of spacetime: the degree of divergence for a graph depends on the number of spacetime dimensions (see Exercise 6.3).

In the dimensional regularization procedure the internal momenta over which we must integrate in a Feynman diagram are taken to have d components. Results are interpreted as holding for arbitrary complex d, and for small enough d the momentum integrals converge; indeed, for the logarithmically divergent integrals typical of gauge theories the integrals become convergent in $4 - \varepsilon$ dimensions with ε infinitesimal. Analytically continuing back to $d = 4$, the original divergences show up as poles; these can be systematically eliminated by using infinite counterterms to cancel the divergences associated with the poles in $d = 4$ dimensions. See Pokorski (1987), §4.3, for a simple introduction, or Collins (1984) and Leibbrandt (1975) for a more detailed exposition.

Dimensional regularization is particularly attractive for gauge theories because it respects those algebraic relations among Green's functions that do not depend on the dimensionality of spacetime. This is critical for gauge theories with their high degree of symmetry because relations among Green's functions following from the gauge symmetry— *Ward identities* for QED and generalized Ward identities (*Taylor–Slavnov identities*) for non-Abelian gauge theories—play a crucial role in the renormalization of these theories by restricting the number of independent UV divergences.

> To give one example, the alert reader may have begun to be uneasy about the preceding discussion of the renormalization of electrical charge. It is known with high experimental precision that the physical charges of the e^- and μ^- are equivalent. If the physical charges correspond to renormalized bare quantities, how can we be certain that the renormalizations of the e^- and μ^- charges lead to *exactly* the same physical charges? An important consequence of Ward identities is that the renormalized charge of the electron and muon *are* equivalent, as required by experiment. This comes about because of an exact cancellation conspiracy (to all orders) among those loop diagrams that depend on the nature of the external lines (electron or muon), leaving only vacuum polarization graphs, which do not depend on the nature of external lines, to renormalize the charge (see Sakurai, 1967, §4.7).

The technical rigors of a dimensional regularization renormalization scheme would take us far from the main theme of this book. We are content with the present heuristic discussion, since our aim is not to become expert in the technique of renormalization, but rather to gain a broad conceptual background in elementary particle physics. However, we will appropriate the most important result from the use of these techniques: *if a theory possesses a local gauge symmetry it will necessarily be a renormalizable theory.*[‡] The proof will not be reproduced here, but the literature reference ('t Hooft, 1971) represents a landmark achievement in quantum field theory.

6.5   Renormalization and Weak Interactions

QED is a renormalizable theory. This is not true of the weak interaction phenomenology, even with the addition of the intermediate vector bosons. The crucial difference is local gauge invariance: QED has it, the Fermi theory does not. In Exercise 6.5b the divergences of the current–current theory based on (6.11) are analyzed as for QED, and one finds that the degree of divergence is proportional to the order of the diagram for high orders. In each order of perturbation theory there are new divergences, and no finite number of counterterms can remove them: as we had already surmised, the Lagrangian (6.43) is an improvement over the original Fermi theory, but still is not renormalizable. We must search elsewhere for a renormalizable theory of the weak interactions. A tempting avenue is suggested by the QED paradigm, for if a local gauge invariance could be imposed on the weak interaction phenomenology we might expect the resulting theory to be renormalizable. But this appears to be impossible: the short range of the weak interactions requires *massive* exchange particles, and it is precisely the finite mass of the IVBs that spoils the gauge invariance of the theory.

However, quantum field theory is a subtle discipline in which the obvious should always be viewed with some suspicion. We will now demonstrate that it is possible to construct a theory of weak interactions having both massive vector bosons *and* gauge invariance bearing the promise of renormalizability. The reader is warned, however, that as in all entertaining adventures things will appear to get considerably worse in this story before they get better!

As a first step we consider a theoretical construction that initially seemed to be only an interesting mathematical exercise with little practical significance: the imposition on a field theory of a local non-Abelian gauge invariance. The original idea was introduced by Yang and Mills (1954), and non-Abelian gauge fields are now commonly called *Yang–Mills fields*. Because of subsequent

---

[‡]However, a theory may be renormalizable even though it does not possess a local gauge invariance. An example is a scalar field with a ϕ^4 self-interaction. Also, as will be discussed in §9.3, the renormalizability of even a local gauge theory may be forfeit if the theory is beset by anomalies (the breakdown of a classical symmetry at the quantum level). In particular, if the theory under consideration contains axial and vector currents, the Ward identities may not be secure because of axial anomalies.

developments these theories have moved from the periphery to center stage in modern theories of fundamental interactions.

6.6 Background and Further Reading

The discussion of the Fermi theory, the Cabibbo modification, the need for intermediate vector bosons, and the shortcomings of the Fermi theory even with the vector bosons follows Leader and Predazzi (1982), Chs. 1–2; and Aitchison and Hey (1982), Chs. 5–7. Okun (1982) discusses weak interactions in considerable detail and has an extensive bibliography. Kayser (1985) gives a pedagogical discussion of Majorana and Dirac neutrinos. The discussion of renormalization is qualitative and borrows from Coleman (1985), Ch. 4; Huang (1982), Ch. IX; Aitchison and Hey (1982), Ch. 7; Bjorken and Drell (1965), Ch. 19; Chaichian and Nelipa (1984), Ch. 9; and Ryder (1985), Ch. 9; see also Halzen and Martin (1984), Ch. 7. Close (1982b) gives a compact heuristic discussion of running coupling constants and the renormalization group; see also Kogut (1983) in this connection. More extensive treatments of renormalization may be found in Itzykson and Zuber (1980), Collins (1984), and Bjorken and Drell (1965).

CHAPTER 7

Yang–Mills Fields

Yang and Mills (1954) suggested that the concept of global phase invariance was inconsistent with the principles underlying local field theories. For example, in the isotopic spin formalism there are two "spin states" for the nucleon, conveniently denoted by up and down. In the normal implementation of isospin one is free to choose which corresponds to a neutron and which to a proton, but the choice is global. If we choose spin-up to be a proton at one spacetime point, then that phase convention applies to all protons at all other spacetime points. But should we not have the freedom to alter the phase in the internal isospin space at different spacetime points as long as the bookkeeping is done properly? That is, can we extend the concept of a global isospin invariance to a local isospin invariance?

This extension from global to local symmetry is possible in QED, and our experience there suggests a procedure to convert a global non-Abelian symmetry into a local one. In QED the use of the covariant derivative (minimal substitution) makes the Lagrangian invariant under local $U(1)$ phase rotations, but only at the expense of introducing a vector boson field (the photon). In this chapter we investigate the extension of this procedure to non-Abelian fields.

7.1 Local Non-Abelian Gauge Invariance

We assume a set of group generators T_i that obey a Lie algebra

$$[T_j, T_k] = f_{jkl}T_l, \tag{7.1}$$

and introduce a set of fields

$$\Psi = \begin{pmatrix} \Psi_1 \\ \Psi_2 \\ \vdots \\ \Psi_n \end{pmatrix}, \tag{7.2}$$

transforming as

$$\Psi(x) \rightarrow \Psi'(x) = e^{-i\tau \cdot \theta(x)} \Psi(x) \equiv U(\boldsymbol{\theta})\Psi(x), \tag{7.3}$$

where $\tau_j (j = 1, 2, \ldots, N)$ are $n \times n$ matrix representations of the group generators, and $\theta_j(x)$ with $j = 1, 2 \ldots, N$ are functions of spacetime that play the role of generalized angles. The original formulation of Yang and Mills was for $SU(2)$ isospin, but the procedure applies generally to compact, semisimple groups.

We now introduce as many vector fields $A_\mu^j(x)$ as necessary to construct a Lagrangian density that is invariant under the local gauge transformation specified by the angles $\theta_j(x)$. Thus, the $A_\mu^j(x)$ will give rise to the analogs of photons when the fields are quantized, but because of the more complicated structure of the non-Abelian groups we will find that more than one gauge boson is required (the index j), and that the properties of the "non-Abelian photons" may be markedly different from those of ordinary photons. From (7.3) the gradients of the fields transform as

$$\partial_\mu \Psi(x) \to U(\boldsymbol{\theta}) \partial_\mu \Psi(x) + \big(\partial_\mu U(\boldsymbol{\theta})\big) \Psi(x). \qquad (7.4)$$

Just as for QED, it is the transformation of the derivatives that will cause difficulties. By analogy with that case, we require a covariant derivative D_μ that transforms as

$$D_\mu \Psi(x) \to D'_\mu \Psi'(x) = U(\boldsymbol{\theta}) D_\mu \Psi(x), \qquad (7.5)$$

where D_μ is to be understood as an $n \times n$ matrix carrying a Lorentz index μ and operating on the n-component "internal" field $\Psi(x)$. Then, if the Lagrangian contains gradients only through the particular combination D_μ, the portion of the Lagrangian depending on the fields $\Psi(x)$ and their gradients will be invariant under the local transformation (7.3).

If there are N group generators we introduce one vector field $A_\mu^j(x)$ for each, and define [see (2.138)]

$$D_\mu \Psi(x) \equiv \big(\partial_\mu + ig A_\mu(x)\big) \Psi(x), \qquad (7.6)$$

where

$$A_\mu(x) \equiv \boldsymbol{\tau} \cdot \mathbf{A}_\mu(x) = \tau_i A_\mu^i(x)$$
$$\mathbf{A}_\mu(x) \equiv \big(A_\mu^1(x), A_\mu^2(x), \ldots, A_\mu^N(x)\big) \qquad (7.7)$$
$$\boldsymbol{\tau} \equiv (\tau_1, \tau_2, \ldots, \tau_N),$$

and g will be seen later to play the role of a coupling constant. From (7.3), (7.5), and (7.6), and short-circuiting some algebra (see Exercise 7.1), the required transformation law for the matrix potential $A_\mu(x)$ is found to be

$$A'_\mu = U A_\mu U^{-1} + \frac{i}{g} (\partial_\mu U) U^{-1}$$
$$= U A_\mu U^{-1} - \frac{i}{g} U \partial_\mu U^{-1} \qquad (7.8)$$

(see the solution of Exercise 13.7a for the last step), which reduces to (2.147) for a $U(1)$ field. For an infinitesimal transformation we have $U(\boldsymbol{\theta}) \simeq 1 - \boldsymbol{\tau}{\cdot}\boldsymbol{\theta}$, and some more algebra yields

$$A_\mu'^j(x) = A_\mu^j(x) + \frac{1}{g}\partial_\mu\theta^j(x) + f_{jkl}\theta^k(x)A_\mu^l(x). \tag{7.9}$$

Comparing with the corresponding expression for photons [eq. (2.147)], we see that the third term is new; its dependence on the structure constants f_{jkl} shows explicitly that it is a consequence of the non-Abelian symmetry. From (7.9) we see that the transformation properties depend on the structure constants of the group, but not on the representation.

EXERCISE 7.1 Supply the missing steps between eqns. (7.6) and (7.9) in the proof of the transformation properties for non-Abelian vector potentials.

Now let us consider kinetic energy and mass terms in the Lagrangian. We relegate the algebra to Exercise 7.2 and state that

- A generalized field tensor $F_{\mu\nu}$ defined by

$$F_{\mu\nu} \equiv F_{\mu\nu}^j\tau_j = \partial_\mu A_\nu - \partial_\nu A_\mu + ig[A_\mu, A_\nu] \tag{7.10a}$$

$$F_{\mu\nu}^j = \partial_\mu A_\nu^j - \partial_\nu A_\mu^j - gf_{jkl}A_\mu^k A_\nu^l \tag{7.10b}$$

  plays the role of $F_{\mu\nu}$ in the electromagnetic theory (Exercise 7.2c,d). Although $F_{\mu\nu}$ is not itself gauge invariant, $\mathbf{F}_{\mu\nu}{\cdot}\mathbf{F}^{\mu\nu} = F_{\mu\nu}^j F_j^{\mu\nu}$ is gauge invariant.

- Mass terms for the fields A_μ^j are not permitted since they must be proportional to $\mathbf{A}_\mu{\cdot}\mathbf{A}^\mu$, which is not gauge invariant (see §2.7.3).

Thus we finally conclude that a Lagrangian density is invariant under a local non-Abelian gauge transformation provided it is of the form

$$\mathcal{L} = -\tfrac{1}{4}\mathbf{F}_{\mu\nu}{\cdot}\mathbf{F}^{\mu\nu} + \mathcal{L}_{\text{matter}} + \mathcal{L}_{\text{int}}(\Psi^j, D_\mu\Psi^j), \tag{7.11}$$

where the factor $\tfrac{1}{4}$ is conventional and

$$\mathbf{F}_{\mu\nu} \equiv (F_{\mu\nu}^1, F_{\mu\nu}^2, \ldots, F_{\mu\nu}^N), \tag{7.12a}$$

$$\mathbf{F}_{\mu\nu}{\cdot}\mathbf{F}^{\mu\nu} = F_{\mu\nu}^j F_j^{\mu\nu} = 2\operatorname{Tr} F_{\mu\nu}F^{\mu\nu}. \tag{7.12b}$$

[The last form of (7.12b) follows from the normalization (5.65), which may be adopted for any $SU(N)$ group.] The term $\mathcal{L}_{\text{matter}}$ is the gauge invariant Lagrangian density of the free matter fields Ψ, and the coupling $\mathcal{L}_{\text{int}}$ between the gauge and matter fields is a function of the fields and their *covariant derivatives*.

EXERCISE 7.2 (a) Prove that the transformation law (7.8) is consistent with eq. (2.147) for quantum electrodynamics. (b) Prove that (7.10b) follows from (7.10a). (c) Show that the $U(1)$ field strength tensor (2.127) can be expressed as

$$F_{\mu\nu} = \frac{1}{iq}[D_\mu, D_\nu],$$

where D_μ is the covariant derivative and q is the charge, and that this suggests the form (7.10a) for the non-Abelian field tensor. (d) Use (7.5) and the result of Exercise 7.2c to demonstrate that $F_{\mu\nu}$ in QED is gauge invariant, but in a non-Abelian theory $F_{\mu\nu}^i$ is not gauge invariant—instead, it transforms like the adjoint representation.

7.2 Properties of Yang–Mills Fields

Some important general properties of Yang–Mills fields follow immediately from the preceding equations:

1. As for electromagnetic $U(1)$ gauge invariance, the local symmetry prescribes the form of the interaction between the gauge fields $A_\mu(x)$ and the matter fields $\Psi(x)$.
2. Even for pure Yang–Mills fields (no matter fields), the Lagrangian density contains interactions because of the self-couplings of the gauge fields entering through $\mathbf{F}_{\mu\nu}\cdot\mathbf{F}^{\mu\nu}$.

By inserting (7.10) into (7.11) with $\mathcal{L}_{\text{int}} = \mathcal{L}_{\text{matter}} = 0$, we see that there are trilinear and quadrilinear couplings of pure non-Abelian gauge fields giving rise to diagrams such as those shown in Fig. 7.1, where coils are used to signify non-Abelian gauge propagators.

Thus the Yang–Mills photons are "self-radiating" and the theory is expected to be highly nonlinear: *Yang–Mills quanta carry the charges of the Yang–Mills*

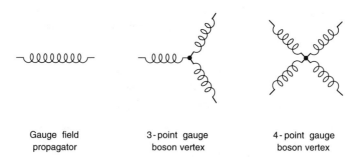

| Gauge field | 3-point gauge | 4-point gauge |
| propagator | boson vertex | boson vertex |

Fig. 7.1   Propagator and self-interactions for a pure non-Abelian gauge theory.

fields. This is in sharp contrast to the (uncharged) photon field that acquires nonlinearities only by interaction with matter fields.

3. Non-Abelian gauge fields transform according to the *adjoint representation* of the group in question because the number of gauge fields required is equal to the number of group generators (see §5.2.4).
4. Only *one* gauge coupling constant g appears in the Yang–Mills formulation if the gauge symmetry group G cannot be factored into a direct product of simple groups.[‡] This depends crucially on the non-Abelian nature of the fields: the formulation does not work if we arbitrarily change the relative scale of operators with finite commutators. Thus the Yang–Mills fields transforming under some group G *couple universally to matter fields and to themselves*, provided G cannot be factored into direct products. That is quite unlike $U(1)$ electromagnetic theory, where each matter field can couple to A_μ with its own charge.
5. If the group G can be factored into k direct products

$$G = G_1 \times G_2 \times \ldots G_k,$$

then k independent coupling constants g_i specify the interaction of the Yang–Mills fields with the matter fields and with themselves. For example, the Glashow–Salam–Weinberg electroweak theory to be discussed in Ch. 9 contains gauge bosons transforming according to the product group $SU(2) \times U(1)$ and there are two coupling constants, with the relation between them specified by the empirical Weinberg angle. In that sense the electroweak theory is only a partial unification of the electromagnetic and weak interactions, which is one motivation to embed a group of the type $SU(2) \times U(1)$ in a larger group [such as $SU(5)$] that does not factor in this way. These theories are called *grand unified theories (GUTs)*, and will be dealt with in Ch. 11; in such a theory a single coupling constant would describe all gauge interactions.
6. Because a Yang–Mills theory possesses a local gauge invariance it is renormalizable, except for possible difficulties with anomalies that will be discussed in §9.3.
7. Yang–Mills fields must be massless vector fields since mass terms would spoil the gauge invariance if explicitly included in the Lagrangian.

It is this last property that appears to reduce Yang–Mills theories to mere curiosities. By analogy with the photon the vector fields A_μ^j should be associated with *massless gauge bosons*, but nature has been extremely parsimonious

---

[‡]The group direct product $G_1 \times G_2$ was defined in §5.2.3. An example was previously encountered in eqns. (5.55)–(5.56): $U(1)_{\text{baryon}} \times SU(2)_{\text{isospin}}$. However, in that example the symmetries were global, not local.

with massless gauge bosons: the photon is the only one directly observed. (Gluons are massless, but the evidence for them is indirect; see Ch. 10.) The simplest non-Abelian symmetries involve at least three generators, so Yang–Mills theories suffer an embarrassment of riches. No one ordered all those massless particles! The original formulation of Yang and Mills attempted to construct a theory of strong interactions by gauging (converting to a local symmetry) global isotopic spin; it foundered on a surfeit of massless vector bosons. We now understand the reasons for the initial failure: (1) The wrong symmetry was gauged—strong interaction isotopic spin appears to be a global rather than local symmetry. (2) The intricacies and implications of spontaneous symmetry breaking in the Goldstone and Higgs modes were not yet appreciated (see Ch. 8).

Once these points are understood we will see that the path is cleared for a Yang–Mills theory of the electroweak and the strong interactions. Accordingly, in the next chapter we turn our attention to the pivotal concept of spontaneous symmetry breaking in quantum field theories. However, we first use the rest of this chapter to give a brief introduction to path-dependent representations and to path integral quantization for Yang–Mills fields. These concepts will play an important role in our subsequent discussions.

7.3 Path-Dependent Representations

Consider a real field $\psi(x)$ with no internal symmetries; invariance under the displacement $x \to x + dx$ requires that

$$dx_\mu \frac{\partial \psi(x)}{\partial x_\mu} \equiv dx_\mu \partial^\mu \psi(x) = 0. \tag{7.13}$$

For a complex $\psi(x)$ coupled to a $U(1)$ gauge field the corresponding relation is obtained by replacing ∂^μ with a covariant derivative (Huang, 1982; Creutz, 1985),

$$dx_\mu D^\mu \psi(x) \equiv dx_\mu [\partial^\mu + ig A^\mu(x)] \psi(x) = 0. \tag{7.14}$$

This defines the idea of a *parallel displacement* (see the discussion in §13.1.2), which may be extended to a Yang–Mills field by introducing the matrix potential $A^\mu(x)$ [see (7.7) and Exercise 7.3]:

$$A^\mu(x) = A_i^\mu(x) \tau^i \tag{7.15}$$

$$A_i^\mu(x) = 2 \operatorname{Tr}(\tau_i A^\mu), \tag{7.16}$$

where the $A_i^\mu(x)$ are vector potentials with a Lorentz index μ and internal index i, and the τ^i are matrix generators for the gauge group.

Thus we see that the theory of gauge fields is a kind of "general relativity" applied to an internal space (the gauge or charge space): the gauge group plays the role of a group of general coordinate transformations, and there is no physical procedure to distinguish fields A_μ and A'_μ if they are related by a gauge transformation. Stated somewhat differently, if ψ is a matter field transforming under a representation of the gauge group there is no way to define a preferred frame for measuring the internal-space components of ψ. Ryder (1985), §3.6 gives a concise discussion of the parallels between general relativity and the geometry of gauge fields.

Gauge field theories may also be defined in terms of differential geometry and the theory of *fiber bundles* (intuitively, fiber bundles are topological spaces that are locally the product of two spaces). In this context gauge fields are called *connections*. Such a description is elegant and powerful, but highly mathematical, and we will not use it. An introduction may be found in Nash and Sen (1983).

The solution of the differential equation (7.14) for a Yang–Mills field (7.15) may be written

$$\psi(x_1) = P e^{ig \int_\gamma A_\mu dx^\mu} \psi(x_0), \qquad (7.17)$$

where γ labels a path between x_0 and x_1,

and P is a *path-ordering operator* enjoining us to order the matrices $A_\mu(x)$ in the sequence encountered on the path for every term in the power series expansion of the exponential. This is analogous to the chronological operator introduced in (3.20), which is necessary because $H(t)$ and $H(t')$ may not commute. Here the matrices $A_\mu(x)$ and $A_\mu(x')$ do not necessarily commute, so a path ordering is required. Thus every path γ with endpoints x_0 and x_1 is associated with a matrix

$$U_\gamma(x_0, x_1) = P e^{ig \int_\gamma A_\mu^j \tau_i dx^\mu}, \qquad (7.18)$$

which is a *path-dependent representation of a gauge group element*. The matrix U is an element of the gauge group because it is a product of group elements associated with infinitesimal segments of the path. It is significant because it defines a gauge transformation that transports the system from one spacetime point to another along a particular path γ ("parallel displacement")

$$\psi(x_1) = U_\gamma(x_0, x_1)\psi(x_0). \qquad (7.19)$$

Under a local gauge transformation U_γ is sensitive to the transforming gauge function only at the endpoints of the path (see Exercise 13.1),

$$U_\gamma(x_0, x_1) \to U'_\gamma(x_0, x_1) = G(x_1)U_\gamma(x_0, x_1)G^{-1}(x_0), \qquad (7.20)$$

from which it may be demonstrated that for a closed path (Exercise 7.3)

$$\operatorname{Tr} U_\gamma(x, x) = \text{gauge invariant.} \qquad (7.21)$$

We may use (7.18) to reduce any spacetime component of $A_i^\mu(x)$ to zero by a continuous local gauge transformation (Huang, 1982, §4.4). If a spatial component of the Yang–Mills vector potential is set to zero the resulting gauge is called an *axial gauge*, while if the time component is set to zero the gauge is termed a *temporal gauge*[‡]

$$A_i^0 = 0 \qquad \text{(temporal gauge)} \qquad (7.22a)$$

$$A_i^3 = 0 \qquad \text{(axial gauge).} \qquad (7.22b)$$

EXERCISE 7.3 (a) Prove the relations (7.20) and (7.21). (b) Invert eq. (7.15) to obtain eq. (7.16).

EXERCISE 7.4 Write the matrix vector potentials (7.15) explicitly for the gauge groups $SU(2)$ and $SU(3)$ in the standard representations (1.78) and (5.63).

7.4 Path Integral Quantization

The standard method of quantizing Yang–Mills fields is through the path integral (Ch. 4). There are several reasons for this, but the primary one is that quantization of a non-Abelian gauge theory requires a quantum version of a dynamical system subject to gauge-fixing constraints (see the discussion in Itzykson and Zuber, 1980, §9–3). This is an arduous task, and the path integral method is better suited to it than the traditional canonical procedure because the path integral is connected more transparently to classical mechanics. It is not our intention to give a comprehensive discussion of the path integral method for gauge fields (for that, consult Itzykson and Zuber, 1980; Ramond, 1981; or Abers and Lee, 1973). However, in this section some concepts associated with quantization of Yang–Mills fields that we will need are introduced. For simplicity we illustrate with $SU(2)$ gauge fields, but similar procedures apply for more complicated gauge groups.

[‡]In Euclidian space, where the convenient mathematics lives, these gauge conditions may be considered equivalent since there is no distinction between space and time [invariance under $O(4)$ rotations]. The distinction comes in the physical Minkowski space with its associated invariance under Lorentz transformations.

7.4.1 Gauge-Fixing Requirements

Consider an $SU(2)$ Yang–Mills field described by a Lagrangian density $\mathcal{L} = -\frac{1}{4}F^i_{\mu\nu}F^{\mu\nu}_i$ with ($i = 1, 2, 3$) and $F^i_{\mu\nu}$ given by (7.10b) with $f_{ijk} = \epsilon_{ijk}$. Comparing with the action functionals for the scalar field (4.53), we conclude that the free field generating functional is of the form (Cheng and Li, 1984)

$$Z[J] = \int [d\mathbf{A}_\mu] \exp\left(i\int d^4x\, \mathcal{L}_0 + i\int d^4x\, \mathbf{J}_\mu(x) \cdot \mathbf{A}^\mu(x)\right), \qquad (7.23)$$

where the functional integral is over the gauge fields, the free Lagrangian is obtained by neglecting the self-coupling,

$$\mathcal{L}_0 = -\tfrac{1}{4}(\partial_\mu A^i_\nu - \partial_\nu A^i_\mu)(\partial^\mu A^\nu_i - \partial^\nu A^\mu_i)$$

$$= \tfrac{1}{2}A^i_\mu(g^{\mu\nu}\Box - \partial^\mu\partial^\nu)A^i_\nu,$$

and the source fields $\mathbf{J}_\mu(x)$ have components J^i_μ with μ the spacetime index and $i = 1, 2, 3$ the $SU(2)$ index. Defining

$$K_{\mu\nu} \equiv g_{\mu\nu}\Box - \partial_\mu\partial_\nu$$

and proceeding by analogy with (4.50)–(4.58), we would like to write the generating functional in the form

$$Z[J] \simeq \frac{1}{\sqrt{\text{Det}\,K}}\exp(J, K^{-1}J),$$

and then construct the Green's functions as in §4.3 and §4.4. However, this is not possible for the gauge symmetry under consideration because *the operator K does not have an inverse* (see Exercise 3.9b,c). The immediate reason is that it is a projection operator, satisfying $K_{\mu\nu}K^\nu_\lambda \simeq K_{\mu\lambda}$. It projects the transverse (physical) degrees of freedom for the gauge field, and nontrivial projection operators do not have inverses ($\text{Det}\,K$ means the product of the eigenvalues of K, some of which are zero, so $1/\sqrt{\text{Det}\,K}$ is undefined).

The ultimate source of this difficulty lies in the integral we have attempted to perform over the gauge fields in (7.23). The Yang–Mills action S is invariant under local gauge transformations (7.8) on the vector potentials,

$$S(\mathbf{A}_\mu) = S(\mathbf{A}'_\mu). \qquad (7.24)$$

For a given spacetime point x, the set of all fields $\mathbf{A}'_\mu(x)$ related to some $\mathbf{A}_\mu(x)$ by variation of the parameters in a gauge transformation constitute an *orbit* within the group manifold. But because of the gauge invariance each point on a given orbit corresponds to the same physical situation, and a naive integration over gauge fields is destined to overcount physical configurations; as Ramond (1981) quips, this makes the path integral even more infinite than

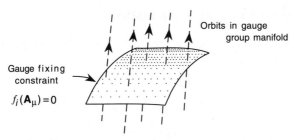

Fig. 7.2   Gauge orbits and a gauge-fixing condition.

usual! To quantize a gauge theory we must restrict this gauge freedom so that the unphysical contributions to the integration measure in (7.23) can be isolated and canceled by an appropriate normalization; this restriction is termed *gauge fixing*. The situation is illustrated schematically in Fig. 7.2, with an $SU(2)$ gauge-fixing condition of the form

$$f_i(\mathbf{A}_\mu) = 0 \qquad (i = 1, 2, 3). \tag{7.25}$$

[Equations (2.121) illustrate two such constraints for the Abelian gauge field.] An integration procedure must be devised that corresponds to counting each orbit only once, where it crosses a surface defined by the gauge-fixing condition (7.25). In mathematics this is known as the determination of the *Haar measure* (or *Hurwicz measure*), and roughly corresponds to dividing out the redundant integrations. The most common recipe for suitably restricting the path integral measure of gauge fields is called the *Faddeev–Popov method*, which we define in the next section.

The difficulty of quantizing a gauge field is expressed here in path integral language, but the same problems appear in a somewhat different guise if one attempts to impose canonical quantization on a gauge theory. In the functional integral approach a naive sum over paths counts each path an infinite number of times because paths related by a gauge transformation are physically equivalent. In the canonical quantization method a corresponding difficulty appears because we must find a complete set of coordinates and momenta such that specification of their initial values determines their values for all time. In this case a consistent quantum theory results because imposing canonical commutation relations at $t = 0$ guarantees that those commutation relations hold at all subsequent times. This cannot be done in an arbitrary gauge theory because it is always possible to make gauge transformations that vanish at $t = 0$, but not at other times. Only by eliminating the freedom of gauge transformations (gauge fixing) can such theories be quantized (see Coleman, 1975, and Exercises 3.9b,c).

7.4.2 Faddeev–Popov Procedure

The Faddeev–Popov (1967) *ansatz* corresponds to the prescription (Cheng and Li, 1984; Coleman, 1975; Faddeev, 1976)

$$e^{iW[J]} = \int [d\mathbf{A}_\mu] \operatorname{Det} \left(\frac{\delta f}{\delta \theta} \right) \delta[f_i(\mathbf{A}_\mu)]$$
$$\times \exp \left(i \int d^4x \, [\mathcal{L} + \mathbf{J}_\mu \cdot \mathbf{A}^\mu] \right) \tag{7.26}$$

for the generating functional of the gauge field $\mathbf{A}_\mu$. That is, the redundant integrations are removed by insertion in the path integral of a factor

$$\operatorname{Det} \left(\frac{\delta f}{\delta \theta} \right) \delta[f_i(\mathbf{A}_\mu)],$$

where $\delta[f_i(\mathbf{A}_\mu)]$ implements the gauge-fixing condition (7.25),[‡] and the initial factor is the functional Jacobian determinant with

$$\left(\frac{\delta f}{\delta \theta} \right)_{ij} \equiv \frac{\delta f_i}{\delta \theta_j}, \tag{7.27}$$

with θ_i the group parameters. Thus $\delta f / \delta \theta$ is a matrix giving the response of the gauge-fixing functionals to an infinitesimal gauge transformation.

EXERCISE 7.5 Show that for the special cases of Abelian gauge fields, and non-Abelian fields in axial gauge, the Faddeev–Popov determinant is independent of the gauge fields. Hence it plays no role in the dynamics of the field theory.

For Abelian gauge theories, or for non-Abelian theories in axial gauge, the Faddeev–Popov determinant can be brought outside the path integral (Exercise 7.5). However, for non-Abelian theories in arbitrary gauges the determinant will generally depend on the gauge fields. We saw in §4.7 that a convenient way to handle a determinant appearing in the numerator of a path integrand is to introduce Fermi–Dirac ghost fields η and η^* through the definition

$$\int [d\eta][d\eta^*] e^{i(\eta^*, K\eta)} \equiv \operatorname{Det}(iK). \tag{7.28}$$

The mathematical effect of this trick applied to (7.26) is to replace the determinant by local interactions of fictitious fields. For non-Abelian theories in axial gauges and for Abelian theories these Faddeev–Popov ghosts are in-

---

[‡]The "δ-functional" satisfies $\int [d\phi] G[\phi] \delta[\phi] = G[0]$ for any functional G. It may be interpreted as a product of Dirac delta functions, one for each spacetime point.

nocuous, but for Yang–Mills theories in covariant gauges they couple to the other fields in internal loops of the Feynman diagrams. The ghosts are not physical particles, as confirmed by their absence in axial gauges for gauge-invariant Yang–Mills theory. They do not propagate on external lines of the diagrams, and they have the peculiar attribute of being Lorentz scalars that exhibit Fermi–Dirac statistics, because the Faddeev–Popov determinant appears in the numerator rather than the denominator—see eqns. (4.70)–(4.71) and Exercise 4.5. Their purpose is to cancel the contributions from unphysical polarizations of gauge fields and enforce unitarity of the S-matrix in nonaxial gauges.

Therefore, a complete set of Feynman rules for a Yang–Mills field in an arbitrary gauge must include rules for the ghost propagators and vertices (for example, see Fig. 10.16). Generally there will be a ghost loop for each loop corresponding to gauge fields, and the ghost lines will have arrows to distinguish ghosts and antighosts, as for normal Fermi fields. Since closed fermion loops insert factors of (-1) into the matrix elements relative to the corresponding boson loop (§4.6), it is clear that the Fermi–Dirac nature of the ghosts is essential in their task of canceling contributions from unphysical polarizations.

The axial gauges are ghost free (Exercise 7.5) and general diagrammatic arguments are often carried out in such gauges. However, these gauges are not very practical for higher-order calculations because of the lack of manifest Lorentz invariance and the appearance of complicated expressions for the propagators [see eq. (12–96) in Itzykson and Zuber, 1980]. Therefore, serious calculations are often performed in gauges haunted by ghosts. It is then a delicate undertaking to demonstrate explicitly that the ghosts and unphysical gauge polarizations decouple from the physical spectrum.

7.5 Background and Further Reading

A discussion of Yang–Mills fields may be found in any recent book on quantum field theory. Good general introductions appear in Cheng and Li (1984), Huang (1982), and Ryder (1985). More detailed expositions with particular emphasis on path integral methods may be found in Abers and Lee (1973), Ramond (1981), and Itzykson and Zuber (1980).

CHAPTER 8

Spontaneously Broken Symmetry and the Higgs Mechanism

One of the most important developments in quantum field theory during the past three decades has been the realization that there is more than one way that symmetries and broken symmetries can manifest themselves in physical systems. In this chapter we discuss three important modes of symmetry realization: (1) the *Wigner–Weyl mode (Wigner mode)*, (2) the *Nambu–Goldstone mode (Goldstone mode)*, and (3) the *Higgs mode*. In the next chapter we will elaborate on the Higgs mode and introduce yet another method of symmetry realization: (4) *anomalous symmetries*, which are symmetries realized at the classical level but broken at the quantum level.

8.1 Classical Symmetry Modes

The Wigner mode is the usual form of symmetry and symmetry breaking that is encountered in elementary quantum mechanics. Its characteristic signature is degenerate multiplet structure for the spectrum [see the comments following eq. (5.6)], and the violation of this kind of symmetry involves explicit symmetry-breaking terms in the Hamiltonian that lift the multiplet degeneracies. It is typical of the Wigner implementation of symmetry that the original multiplet structure is easily recognized for small symmetry breaking perturbations. A good example is provided by a spherical quantum-mechanical system such as an atom. In the absence of external fields the states form degenerate $SU(2)$ multiplets as a consequence of the conservation of angular momentum. If we now place a magnetic field along the z-axis the rotational symmetry is lost since a preferred direction has been selected in space; the corresponding nondegenerate multiplet structure is well known from the Zeeman effect, as Fig. 8.1 illustrates. In group-theoretical language the original $SU(2)$ symme-

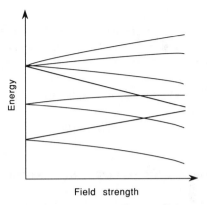

Fig. 8.1   Loss of symmetry by addition of an explicit symmetry-breaking field to the Hamiltonian.

try has been broken down to $U(1)$, since the system is still invariant under rotations about a single axis (chosen to be the z-axis in this example).

Another familiar illustration is the $SU(2)$ multiplet structure for isospin or, more generally, the $SU(N)$ flavor symmetry discussed in Ch. 5. For example, $SU(3)$ flavor multiplet symmetry is broken to $SU(2)$ isospin by terms in the Hamiltonian that depend on hypercharge. The isospin symmetry is further broken to $U(1)$ charge symmetry by terms such as Coulomb interactions that select a preferred direction in isospace, but the $U(1)$ symmetry remains intact since all known interactions conserve charge. By such an analysis a *hierarchy of symmetry breakings* may be established for symmetries implemented in the Wigner mode.

Of more interest in the present discussion are the other two symmetry modes. The manifestation of a symmetry in the Goldstone or Higgs mode is commonly termed *spontaneous symmetry breaking*. This is a picturesque but somewhat misleading expression. A more descriptive name is *hidden symmetry:* in spontaneous symmetry breaking the original symmetry is still present, but nature manages to camouflage the symmetry in such a way that its presence can be glimpsed only indirectly through relations among coupling constants, or by the unexpected appearance of massless bosons. The difference between the Goldstone and Higgs modes is simply that the spontaneous symmetry breaking occurs in the presence of a *local* gauge symmetry for the Higgs mode; as we shall see, this simple proviso has enormous consequences for the particle spectrum of such theories.

The crucial distinction between symmetry implementation in the Wigner, Goldstone, and Higgs modes lies in the structure of the vacuum (the lowest energy state). The Lagrangian of a system may be invariant under transformation by some unitary representation U of a symmetry group. However, for a perturbative quantum field theory we build states from the vacuum and the symmetry properties of such a theory require specification of the symmetry for the vacuum state, as well as that of the Lagrangian. If the Lagrangian is invariant under a set of transformations the symmetry is implemented in the

1. *Wigner mode* if the vacuum $|0\rangle$ is also invariant:

$$U|0\rangle = |0\rangle,$$

2. The *Goldstone mode* if the vacuum is not invariant and the Lagrangian symmetry is global:

$$U|0\rangle \neq |0\rangle \qquad \text{(global symmetry)},$$

3. The *Higgs mode* if the vacuum is not invariant and the Lagrangian symmetry is a *local* gauge symmetry:

$$U|0\rangle \neq |0\rangle \qquad \text{(local symmetry)}.$$

We now present some illustrations.

8.2   A Simple Example

Consider a self-interacting real scalar field ϕ with a Lagrangian density

$$\mathcal{L} = \tfrac{1}{2}(\partial_\mu \phi)(\partial^\mu \phi) - \tfrac{1}{2}\mu^2 \phi^2 - \tfrac{1}{4}\lambda\phi^4 \qquad (\lambda > 0). \qquad (8.1)$$

(Omitting powers of ϕ higher than four ensures a renormalizable perturbation theory—see Exercise 6.3b; the coefficient λ is required to be positive so that the energy is bounded from below.) This Lagrangian is invariant under the discrete transformation

$$\phi \to -\phi. \qquad (8.2)$$

Two qualitatively different cases may be distinguished, depending on the sign of the coefficient μ^2. The potential for $\mu^2 > 0$ is shown in Fig. 8.2a, and that for $\mu^2 < 0$ is shown in Fig. 8.2b. The case (a) with $\mu^2 > 0$ corresponds to the usual situation (Wigner mode). From Fig. 8.2a we have for the vacuum expectation value of the field

$$\langle \phi \rangle_0 \equiv \langle 0|\phi|0\rangle = 0 \qquad (\mu^2 > 0).$$

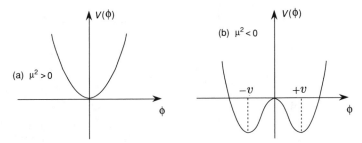

Fig. 8.2 Effective potentials $V(\phi) = \frac{1}{2}\mu^2\phi^2 + \frac{1}{4}\lambda\phi^4$ for the Lagrangian density (8.1) with differing signs for the coefficient μ^2.

Expanding about $\langle \phi \rangle_0$ to second order

$$\mathcal{L} \simeq \tfrac{1}{2}(\partial_\mu\phi)(\partial^\mu\phi) - \tfrac{1}{2}\mu^2\phi^2, \tag{8.3}$$

which is the Lagrangian density of a free scalar field of mass μ [see eq. (2.41)]. Thus we may interpret small quantized oscillations of the field about the origin as particles, and for the symmetric case shown in Fig. 8.2a the parameter μ plays the role of a mass.

Now consider the case $\mu^2 < 0$. This situation is depicted in Fig. 8.2b, and the potential has minima at

$$\langle \phi \rangle_0 = \pm\sqrt{\frac{-\mu^2}{\lambda}} \equiv \pm v. \tag{8.4}$$

Now there are *two degenerate vacuum states.* The difference between the situations in Fig. 8.2a and 8.2b is characteristic of a *phase transition*, with μ^2 playing the part of an order parameter. The minima at $\langle \phi \rangle_0 = \pm v$ are equivalent, and either may be chosen as the classical ground (vacuum) state of the system.[‡] The original Lagrangian is invariant under (8.2), so the physical results must be independent of this choice; however, once the vacuum is chosen as either $+v$ or $-v$ it is no longer invariant under the transformation (8.2). This is a typical case of spontaneous symmetry breaking or, more descriptively,

---

[‡]For ordinary particle mechanics the degeneracy of the vacua in Fig. 8.2b would be lifted by tunneling between the two minima (see §8.5.1 and §13.1.7). However, in the present case the abscissa is the field ϕ, not a spatial coordinate. The tunneling probability is $P \simeq \exp(iS)$ where S is the action (see Exercise 13.7c). For a tunneling process S is imaginary; further, in this example it is *infinite*, since it is found by integrating over all space. Thus the tunneling probability between the field vacua in Fig. 8.2b is $P \simeq \exp(-\infty) = 0$.

hidden symmetry: the Lagrangian is invariant under a symmetry operation, the vacuum is not.

Let us choose

$$\langle \phi \rangle_0 = +v \tag{8.5}$$

as the vacuum state on which to construct our quantum theory. In the previous example ($\mu^2 > 0$) we examined the particle spectrum by expanding the field about the minimum at $\langle \phi \rangle_0 = 0$. In the present case of a spontaneously broken symmetry this is no longer suitable as an expansion point since it is a maximum of the potential energy (it is also obvious that the interpretation of μ as a mass is untenable since $\mu^2 < 0$ when the symmetry is spontaneously broken); an infinitesimal fluctuation is sufficient to drive the system into either of the minima at $\pm v$, and it is clear that the corresponding particle spectrum should be examined by expansion about the minima at $\pm v$. To facilitate this let us define a *shifted field*

$$\xi(x) \equiv \phi(x) - \langle \phi \rangle_0 = \phi(x) - v. \tag{8.6}$$

In terms of this new variable the vacuum state is $\langle \xi \rangle_0 = 0$, and the Lagrangian density is (neglecting constant terms)

$$\mathcal{L} = \tfrac{1}{2}(\partial_\mu \xi)(\partial^\mu \xi) - \lambda v^2 \xi^2 - \lambda v \xi^3 - \tfrac{1}{4}\lambda \xi^4, \tag{8.7}$$

which has no apparent reflection symmetry. In fact, the symmetry is there because the original Lagrangian possessed such a symmetry, but it has been hidden. For small oscillations about the classical vacuum

$$\mathcal{L} \simeq \tfrac{1}{2}(\partial_\mu \xi)(\partial^\mu \xi) - \lambda v^2 \xi^2, \tag{8.8}$$

which is the Lagrangian density of a free scalar field of mass $m_\xi = \sqrt{-2\mu^2}$ [see (2.41) and (8.4)]. The mass is real and positive since $\mu^2 < 0$.

This is an exceedingly simple example, but it contains most of the features that characterize spontaneous symmetry breaking:

1. There is a nonzero expectation value of some field in the vacuum state.
2. The resulting classical theory has a degenerate vacuum, with the choice among the equivalent vacua completely arbitrary.
3. The transition from a symmetric vacuum to a degenerate vacuum typically occurs as a phase transition as some order parameter (μ^2 in the above example) is varied.
4. The chosen vacuum state does not possess the same symmetry as the Lagrangian.
5. On expansion around the chosen vacuum the original symmetry of the Lagrangian is no longer apparent. The degenerate vacua are related to

each other by symmetry operations [eq. (8.2)], which tells us that the symmetry is still there, but it is not manifest; it is hidden.

6. The masses of the particles appearing in the theory with and without the spontaneous symmetry breaking may differ substantially. We say that the masses have been acquired spontaneously in the latter case.

7. Once the theory develops degenerate vacua the origin becomes an unstable point. Thus the symmetry may be "broken spontaneously" in the absence of external intervention.

However, there are two important aspects of spontaneous symmetry breaking that do not appear in this simple model. They will occur only when the symmetry that is spontaneously broken is a *continuous* one. Briefly stated,

8. If the spontaneously broken symmetry is a *continuous global symmetry*, one massless scalar field (*Goldstone boson*) must appear in the theory for each group generator that has been broken.

9. If a *continuous local gauge symmetry* is spontaneously broken no Goldstone bosons are produced, and the gauge bosons may acquire a mass without spoiling gauge invariance (*Higgs mechanism*).

To appreciate the importance of the new features (8) and (9), we now examine the spontaneous breaking of continuous symmetries.

8.3 Goldstone Bosons

Consider a Lagrangian density involving a complex scalar field

$$\mathcal{L} = (\partial_\mu \phi)^\dagger (\partial^\mu \phi) - \mu^2 \phi^\dagger \phi - \lambda (\phi^\dagger \phi)^2, \tag{8.9}$$

where $\lambda > 0$. This Lagrangian density is invariant under the group $U(1)$ of global phase transformations,

$$\phi(x) \rightarrow \phi'(x) = e^{i\theta} \phi(x), \tag{8.10}$$

where θ is independent of x (see Exercise 2.15). Defining

$$\rho = \phi^\dagger \phi, \tag{8.11}$$

we may identify a potential

$$V(\rho) = \mu^2 \rho + \lambda \rho^2. \tag{8.12}$$

As before, we may distinguish two cases: (a) $\mu^2 > 0$, the minimum is at $\rho = \phi = 0$, and the classical ground state is symmetric, as illustrated in

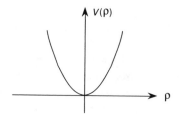

Fig. 8.3   Manifest symmetry for the Lagrangian density (8.9).

Fig. 8.3. (b) $\mu^2 < 0$, and the minima occur in the complex ϕ plane on a circle of radius

$$|\phi| = \sqrt{\frac{-\mu^2}{2\lambda}} \equiv \frac{v}{\sqrt{2}}, \tag{8.13}$$

as illustrated in Fig. 8.4. Case (b) is the one of primary interest, and we recognize immediately the characteristic features of a spontaneously broken symmetry. There is now an *infinity of degenerate ground states*, corresponding to different positions on the ring of minima in the complex ϕ plane, and the symmetry operation (8.10) relates one to another. Proceeding as before we choose as the vacuum the point in the minimum on the real ϕ axis, $\text{Re}(\phi) = v/\sqrt{2}$, and expand around it to investigate the spectrum. We may write [see (8.26)]

$$\phi(x) = \frac{1}{\sqrt{2}}\left[v + \xi(x) + i\chi(x)\right]. \tag{8.14}$$

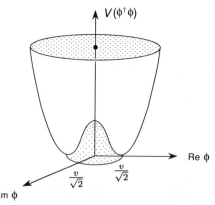

Fig. 8.4   Spontaneously broken symmetry for the Lagrangian density (8.9).

Substituting in the Lagrangian density (8.9) yields

$$\mathcal{L} = \tfrac{1}{2}(\partial_\mu \xi)^2 + \tfrac{1}{2}(\partial_\mu \chi)^2 - \lambda v^2 \xi^2 - \lambda v \xi(\xi^2 + \chi^2)$$
$$- \tfrac{1}{4}\lambda(\xi^2 + \chi^2)^2 + \text{constants.} \tag{8.15}$$

Now this resembles a Lagrangian density for a quantum field theory with two basic fields, ξ and χ. If interpreted in this way the χ field is massless, but the ξ field is massive by virtue of the term $-\lambda v^2 \xi^2$:

$$m_\xi = \sqrt{2\lambda v^2} = \sqrt{-2\mu^2} \qquad (\mu^2 < 0). \tag{8.16}$$

The physical interpretation of these degrees of freedom is illustrated in Fig. 8.5. The massive mode ξ corresponds to "radial oscillations" against a restoring potential; we may say that the field ξ has acquired its mass spontaneously. The massless mode χ corresponds to angular motion about the bottom of the circular valley, for which there is no restoring force.

The appearance of the massless scalar field χ is a specific example of a general phenomenon in the spontaneous breaking of global symmetries that is important enough to have achieved the status of a theorem:

> **Goldstone Theorem**: *If a continuous global symmetry is broken spontaneously, for each broken group generator there must appear in the theory a massless particle.*

(Goldstone, 1961; Nambu, 1960; Nambu and Jona-Lasinio, 1961; Goldstone, Salam, and Weinberg, 1962; Bludman and Klein, 1962).

The massless particles for the theories that interest us here are quanta of

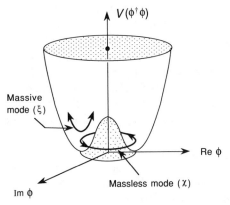

Fig. 8.5 Modes of a complex scalar field with spontaneously broken symmetry.

scalar or pseudoscalar fields that are termed *Goldstone bosons.*[‡]   In the example just discussed $U(1)$ has a single generator that is broken. That is, once the symmetry is hidden by choosing a particular vacuum state from the infinite number of equivalent possibilities the manifest symmetry under phase rotations by the $U(1)$ generator is obscured. As a consequence, there appears in the theory a massless field associated with the same motion as that induced by the generator that was broken (circular motion in the valley of the potential). We say that the corresponding massless particle (the Goldstone boson) carries the quantum numbers of the broken generator.

In this simple example only a single generator is broken, but the symmetry breaking generally may involve more than one group generator, and a Goldstone particle associated with each. For example, let us consider a Lagrangian density containing n real scalar fields ϕ_i in the form (Abers and Lee, 1973)

$$\mathcal{L} = \tfrac{1}{2}(\partial_\mu \phi_i)(\partial^\mu \phi_i) - \tfrac{1}{2}\mu^2 \phi_i \phi_i - \tfrac{1}{4}\lambda(\phi_i \phi_i)^2. \tag{8.17}$$

This Lagrangian density is invariant under the group $O(n)$ of orthogonal transformations in n dimensions, which has $\tfrac{1}{2}n(n-1)$ generators (we deal with orthogonal transformations rather than unitary ones because the fields are real—see Exercise 2.15). If the symmetry is spontaneously broken by choosing $\mu^2 < 0$, a ring of minima appears satisfying $\phi_i \phi_i = -\mu^2/\lambda \equiv v$. The fields may be viewed as the components of a vector ϕ, in which case the equation for the minimum defines the magnitude but not the direction of ϕ. The vacuum state can be chosen as

$$\langle \phi \rangle_0 \equiv \begin{pmatrix} \phi_1 \\ \phi_2 \\ \vdots \\ \phi_n \end{pmatrix}_{\text{vac}} = \begin{pmatrix} 0 \\ 0 \\ \vdots \\ -\mu^2/\lambda \end{pmatrix} \tag{8.18}$$

and all other vacuum states are related to this one by $O(n)$ rotations. In contrast to our earlier example, this vacuum state is *invariant under a subgroup of the original group*: the group $O(n-1)$, which does not mix the last field with the others. The group $O(n)$ has $\tfrac{1}{2}n(n-1)$ generators, so $O(n-1)$ has $\tfrac{1}{2}(n-1)(n-2)$ generators and the difference between the number of generators for the original group $O(n)$ and the residual group $O(n-1)$ is $n-1$; thus, there are $n-1$ broken generators. An analysis similar to that given for the previous example then shows that only one field acquires a mass, and $n-1$ scalar fields appear in the theory with no mass (see Exercise 8.2). These are

---

[‡]Goldstone particles are not always bosons: in spontaneously broken supersymmetries there are spin-$\tfrac{1}{2}$ Goldstone fields. However, we will only consider theories for which the Goldstone quanta are bosons.

the Goldstone bosons, corresponding to the $n-1$ broken symmetry generators of the original group.

The largest subgroup of G that leaves the vacuum invariant is termed the *little group* or the *stability subgroup* (O'Raifeartaigh, 1986, Chs. 8 and 11). If that subgroup is denoted by H, we say that G has been spontaneously broken down to H. For example, in Exercise 8.2 we consider a Lagrangian invariant under $G = SO(3)$, with a ground state that is only invariant under the subgroup $H = SO(2)$. The resulting theory has $n - 1 = 2$ Goldstone bosons.

EXERCISE 8.1 (a) Show that when there is spontaneous symmetry breaking at least one of the generators for the symmetry group fails to annihilate the vacuum. (b) Derive the minima of Fig. 8.2 from eq. (8.1). Obtain the result (8.7) for $\mu^2 < 0$ and show that

$$m_\xi = \sqrt{2\lambda v^2} = \sqrt{-2\mu^2} \qquad (\mu^2 < 0)$$

is the mass of the ξ field after spontaneous symmetry breaking.

EXERCISE 8.2 Investigate the Lagrangian density (8.17) for an isovector Lorentz scalar field $\phi_i (i = 1, 2, 3)$, invariant under the global internal group $SO(3)$. Show that for $\mu^2 < 0$ the symmetry is spontaneously broken from $SO(3)$ to $SO(2)$, and that the particle spectrum corresponds to two massless Goldstone fields and one massive scalar.

EXERCISE 8.3 A concept that will be important when we consider chiral symmetry and the nature of the pion in subsequent chapters is that of an *approximate Goldstone boson*. Consider a complex scalar field with a potential

$$V(\phi) = \mu^2 \phi^\dagger \phi + \lambda(\phi^\dagger \phi)^2 - \epsilon\phi_1 \equiv V_0(\phi) - \epsilon\phi_1,$$

where $\lambda > 0$, the parameter ϵ is small and positive, and $\phi \equiv (\phi_1 + i\phi_2)/\sqrt{2}$. First, set $\epsilon = 0$, choose the μ and λ parameters to break the symmetry spontaneously, sketch the potential $V(\phi)$, identify a classical vacuum state, and investigate the free-particle spectrum for small oscillations of the fields about this vacuum. Now, demonstrate that for $\epsilon \neq 0$ the field that would be a Goldstone boson for $\epsilon = 0$ acquires a small mass proportional to $\epsilon^{1/2}$ and proportional to the divergence $\partial_\mu J^\mu$ of a current

$$J_\mu = \phi_1 \partial_\mu \phi_2 - \phi_2 \partial_\mu \phi_1$$

that is conserved if $\epsilon = 0$ (*Hint:* see §12.3). These ideas will be employed in Ch. 12 to understand the partially conserved axial current (PCAC). The almost massless pion will be interpreted as an approximate Goldstone boson, with its finite mass related to the divergence of the axial current (Exercise 12.4).

8.4 The Higgs Mechanism

The Goldstone theorem and the paucity of massless scalar or pseudoscalar particles in nature would seem to preclude the use of spontaneous symmetry breaking in realistic quantum field theories. However, there is a loophole in this argument: the Goldstone theorem applies to any field theory obeying the "normal postulates" such as locality, Lorentz invariance, and positive definite norm on the Hilbert space. But gauge field theories do not fit into that category: there is no single gauge in which such theories simultaneously fulfill each of these conditions! For example, in Ch. 2 it was shown that the quantization of the electromagnetic field (the archetypical gauge field) is nontrivial. We chose to quantize in a manner sacrificing manifest Lorentz covariance; conversely, if the Maxwell field is quantized in a manifestly covariant fashion the notion of a positive definite metric must be sacrificed [see the Gupta–Bleuler mechanism below and the discussion in Ryder (1985), §4.4].

Thus we are led to investigate whether the Goldstone theorem is operative in a theory possessing a local gauge invariance. We will find a remarkable result: there is an unexpected collusion between the massless gauge fields and the Goldstone bosons produced by the spontaneous symmetry breaking that can be arranged so as to eliminate the massless Goldstone bosons and give a mass to the gauge quanta without spoiling the gauge invariance or renormalizability of the theory. This (seemingly) miraculous state of affairs is called the *Higgs mechanism* (Higgs, 1964; Guralnik, Hagen, and Kibble, 1964; Englert and Brout, 1964; Anderson, 1963).

The simplest example of the Higgs mechanism is the extension of the global $U(1)$ symmetry just discussed to a local $U(1)$ symmetry. This gauge-invariant $U(1)$ theory is often called the *Abelian Higgs Model*. In the absence of spontaneous symmetry breaking it would describe the ordinary electrodynamics of charged scalars; when the symmetry is broken spontaneously it will describe something quite different.

The Lagrangian density is (Quigg, 1983; Leader and Predazzi, 1982)

$$\mathcal{L} = (D_\mu \phi)^\dagger (D^\mu \phi) - \mu^2 \phi^\dagger \phi - \lambda (\phi^\dagger \phi)^2 - \tfrac{1}{4} F_{\mu\nu} F^{\mu\nu}, \tag{8.19}$$

where λ is positive and

$$\phi = \tfrac{1}{\sqrt{2}}(\phi_1 + i\phi_2) \tag{8.20}$$

$$D^\mu = \partial^\mu + iq A^\mu \tag{8.21}$$

$$F_{\mu\nu} = \partial_\mu A_\nu - \partial_\nu A_\mu. \tag{8.22}$$

The Lagrangian is invariant under global $U(1)$ rotations and under the local gauge transformations

$$\phi(x) \rightarrow e^{iq\alpha(x)} \phi(x) \tag{8.23a}$$

$$A_\mu(x) \rightarrow A_\mu(x) - \partial_\mu \alpha(x). \tag{8.23b}$$

As in the example with global $U(1)$ symmetry, two possibilities may be distinguished: (a) $\mu^2 > 0$, and the potential has a minimum at $\phi = \phi^\dagger = 0$ that is unique. The symmetry of the Lagrangian is also the symmetry of the ground state, and the spectrum consists of a massless photon A^μ and a pair of scalar fields ϕ and $\phi^\dagger$ with a common mass μ. (b) $\mu^2 < 0$ corresponds to a spontaneously broken local symmetry. Because spontaneous local symmetry breaking is subtle, we must analyze this situation carefully.

The absolute minima (degenerate vacua) occur at

$$|\phi|^2 = -\frac{\mu^2}{2\lambda} \equiv \frac{v^2}{2}. \tag{8.24}$$

Choosing the vacuum as

$$\langle \phi \rangle_0 = \frac{v}{\sqrt{2}} \tag{8.25}$$

with v real and positive, and expanding in polar coordinates,

$$\phi(x) = \tfrac{1}{\sqrt{2}}\left[v + \eta(x)\right]e^{i\xi(x)/v}$$

$$= \tfrac{1}{\sqrt{2}}\left[v + \eta(x) + i\xi(x) + \dots\right]. \tag{8.26}$$

Substituting in (8.19) and retaining low order terms gives

$$\mathcal{L} \simeq \tfrac{1}{2}(\partial_\mu \eta)(\partial^\mu \eta) + \mu^2 \eta^2 + \tfrac{1}{2}(\partial^\mu \xi)(\partial_\mu \xi)$$

$$+ qvA_\mu(\partial^\mu \xi) + \tfrac{1}{2}q^2 v^2 A_\mu A^\mu - \tfrac{1}{4}F_{\mu\nu}F^{\mu\nu} + \dots \tag{8.27}$$

Now this looks like the Lagrangian density of a quantum field theory with three fields: η, ξ, and A^μ. By inspection, the η field has a mass [eq. (2.41)]

$$m_\eta = \sqrt{-2\mu^2}, \tag{8.28}$$

as implied by the term $\mu^2 \eta^2$. Surprisingly, the photon appears to have gained a mass

$$m_A = qv, \tag{8.29}$$

as implied by the term $\tfrac{1}{2}q^2 v^2 A_\mu A^\mu$ [see eq. (2.49)], and the ξ field seems to be massless. However, we should count degrees of freedom. Originally we had

2	(complex scalar field)
+ 2	(transverse field for massless photon)
4	

After the spontaneous symmetry breaking we have

$$
\begin{array}{ll}
1 & (\eta \text{ field}) \\
1 & (\xi \text{ field}) \\
+\ 3 & (massive \text{ vector field } A_\mu) \\
\hline
5 &
\end{array}
$$

So all is not as it appears—a degree of freedom seems to have been gained in the spontaneous symmetry breaking. This is an illusion, as can be made obvious by an appropriate change of gauge. Implementing the local transformation

$$
\phi(x) \to e^{-i\xi(x)/v}\phi(x) = \frac{v + \eta(x)}{\sqrt{2}}
$$
$$
A_\mu(x) \to A_\mu(x) + \frac{1}{qv}\partial_\mu\xi(x) \equiv A'_\mu(x),
$$

(8.30)

and dropping the primes on A_μ and $F_{\mu\nu}$, the Lagrangian density takes the form

$$
\mathcal{L} = \tfrac{1}{2}(\partial_\mu\eta)(\partial^\mu\eta) + \mu^2\eta^2 + \tfrac{1}{2}q^2v^2 A_\mu A^\mu - \tfrac{1}{4}F_{\mu\nu}F^{\mu\nu}.
$$

(8.31)

Now the particle spectrum is clear: a scalar particle η with mass $\sqrt{-2\mu^2}$ and a massive vector field A_μ with mass qv. The ξ field has disappeared—we say that it has been gauged away—and the number of degrees of freedom has been reduced to the required four: one for η and three for A_μ.

Thus no massless fields appear, and *the Goldstone theorem does not apply to a local gauge theory.* What has happened to the massless field ξ that was gauged away? It contrives to reappear effectively as a *longitudinal polarization degree of freedom for the vector field,* giving it a mass. The massive scalar field η is called the physical *Higgs field,* and the special gauge in which the particle spectrum is manifest for spontaneous breaking of a local gauge symmetry is called the *Unitary Gauge* or *U-gauge.*

It is common to say that the gauge field absorbs the Goldstone boson and becomes massive, or that the Goldstone field becomes the third state of polarization for the massive vector boson. One way to visualize this is the following. If one attempts to quantize a vector field A^μ in a manifestly covariant fashion (retaining all four components), the indefinite Minkowski metric $g_{\mu\nu}$ leads to a *negative norm* for the timelike component A^0. For massless fields the A^0 contribution exactly cancels with the longitudinal spacelike component $\mathbf{p}\cdot\mathbf{A}$, leaving two physical transverse components $\mathbf{p}\times\mathbf{A}$. This is called the *Gupta–Bleuler mechanism.* The Higgs mechanism may be viewed as a kind of generalized Gupta–Bleuler mechanism in which the *Goldstone scalar* cancels the timelike component of the gauge field, leaving the three spacelike components of A_μ intact, so A_μ behaves like a massive vector boson (O'Raifeartaigh, 1986, §8.5).

Although unitary gauge exhibits the particle spectrum clearly, it is not a good gauge for tasks such as demonstrating renormalizability. For that purpose a set of gauges called R-gauges is normally used (see, e.g., Cheng and Li, 1984, Ch. 9). In U-gauges the spurious degrees of freedom are transformed away, but the propagators are ill-behaved at high energy; in R-gauges the particle spectrum is confused by spurious degrees of freedom, but the propagators have a more benign high-energy behavior.

This type of symmetry realization is called the *Higgs mode*. The Wigner mode was characterized by degenerate multiplet structure and the Goldstone mode by the appearance of one massless Goldstone boson for each spontaneously broken generator. The distinctive signature of the Higgs mode is the acquisition of mass by gauge bosons at the expense of would-be Goldstone bosons, which vanish from the theory.

This analysis may be adapted to other Abelian or non-Abelian gauge theories. For the gauge fields to acquire masses we must break the vacuum symmetry with scalar fields. Some pieces of the scalar fields disappear, only to reappear as the longitudinal polarization states for the gauge bosons that acquire an effective mass; the remaining pieces become physical scalar fields, the Higgs bosons.[‡] In a theory with multiple gauge bosons it is possible to arrange for some to acquire a mass and for others to remain massless by this procedure.

A necessary consequence of symmetry realization in this mode will be the appearance of the physical Higgs bosons, but the Higgs particles have a distinct advantage over massless gauge quanta or Goldstone bosons in our theory: their appearance may violate our sense of economy, but they enter with adjustable masses at the present level of understanding and the failure to observe them can be attributed to their masses being too large.

As a preview of subsequent developments, we note that there appear to be two ways that nature has contrived to reduce the number of massless bosons that might have been expected from our initial discussions of Yang–Mills fields and the Goldstone theorem. The first we have just met: the *Higgs mechanism* does double duty in converting the Goldstone bosons into effective longitudinal polarization states for the gauge bosons, which become massive in the process. The second we shall encounter in Ch. 10: the nonlinear interactions of the Yang–Mills gauge quanta, which themselves carry the charges of the gauge field, may in some cases cause an absolute *confinement* of the gauge fields and the matter fields coupled to them. This appears to be the case in QCD, where the $SU(3)$ color symmetry is unbroken so that the eight associated gauge bosons (gluons) are massless, but are confined to the interior of hadrons along with the quark fields.

[‡] The Higgs field is required to transform as a scalar to prevent spontaneous breaking of the Lorentz invariance, which would not be in accord with observation.

8.5 Some General Remarks

The concept of spontaneous symmetry breaking is an important but subtle one in modern theoretical physics. In this section we make some comments that may help clarify the nature of the phenomenon.

First we emphasize that spontaneous symmetry breaking is not confined to relativistic fields. It was known in many guises in condensed matter and nuclear physics before its introduction into relativistic field theory. If a nonrelativistic, many-body Lagrangian is invariant under some set of transformations G and the ground state is nondegenerate, then the ground state transforms as a singlet under the group G; this is the Wigner mode. However, if the ground state is degenerate it may transform as some finite-dimensional multiplet of the group G. If *one* member of this multiplet is arbitrarily selected as the ground state the symmetry is spontaneously broken; this is the Goldstone mode.

8.5.1 Heisenberg Ferromagnet

A well-known example is the infinite ferromagnet, often termed the Heisenberg ferromagnet. The model corresponds to an infinite crystalline array of spin-$\frac{1}{2}$ dipoles with spin–spin interactions between nearest neighbors. Above the critical temperature there is short-range order because of nearest-neighbor interactions, but long-range order is suppressed by thermal fluctuations; below the critical temperature there is also long-range order with macroscopic alignment of spins. The Hamiltonian for the interaction of neighboring spins is rotationally invariant, but below the critical temperature the ground state has a net spin alignment in a particular direction that breaks rotational invariance.

The infinite degeneracy of the possible ground states, corresponding to an infinite number of possible directions for the aligned ground state spin, depends crucially on the assumption of an unlimited spatial extent for the system. Mathematically, the infinite size of the system means that there is no unitary operator that can connect the different ground states and they lie in different Hilbert spaces. This is because the ferromagnet has an infinite moment of inertia, implying that no finite amount of energy can rotate one ground state into another.

If the ferromagnet is of limited spatial extent the different ground states are separated by finite energy barriers and may tunnel into each other; the situation then resembles band structure in the solid state. The degeneracy of the ground state is lifted since one of the linear combinations resulting from the degenerate vacuum will lie lower in energy than all others and will be the (nondegenerate) physical ground state. For a relativistic field theory exhibiting spontaneous symmetry breaking we may expect a related situation.

The vacua are degenerate only if the universe is of infinite spatial extent, for only then will the vacua be orthogonal (§13.1.7). However, from the radius of the known universe one estimates that the frequency of rotation from one vacuum to another is negligible (Taylor, 1976, §5.3).

8.5.2 Superconductivity and the Meissner Effect

The Higgs mechanism described in §8.4 is sometimes expressed in an alternative way: Goldstone bosons can be made to disappear in the presence of long-range forces (Anderson, 1963; Guralnik, Hagen, and Kibble, 1968). The connection between the two descriptions comes through the Yukawa–Wick interpretation that long-range forces like the Coulomb interaction are mediated by massless exchange particles (gauge fields are massless). In this picture the long-range force is shielded and becomes short-ranged, which is equivalent to the generation of an effective mass for the gauge boson.

One nonrelativistic example of the Higgs mechanism for which this shielding description is illuminating is the Meissner effect in superconductivity. Aitchison and Hey (1982, Ch. 9) give a nice heuristic discussion of how the condensate of electron pairs in the ground state of a superconductor plays the role of a Higgs field, and how this gives photons an effective mass inside the superconductor. As a direct consequence a magnetic field can penetrate only exponentially into the superconductor, with a range proportional to the inverse of the effective photon mass. Thus a superconductor expels a magnetic field from its interior (the Meissner effect—see Fig. 12.3), except for a thin layer at the surface (the London penetration depth) over which the field decreases exponentially.

The microscopic origin of the Higgs phenomenon in this case lies in resistanceless screening currents that are produced in the superconductor to compensate the external field. The finite range of the magnetic field that results is just what would be expected if the photon had acquired a mass inside the superconductor. Furthermore, it can be shown that the superconductor with a "massive" photon is gauge invariant, despite our remarks in connection with eq. (2.149). That is because the photon mass has come by the Higgs mechanism—not by an explicit mass term in the Lagrangian. This is our first example of a real physical system in which gauge bosons acquire an effective mass without breaking gauge invariance. Analogies with this example will have important implications in subsequent chapters, particularly in connection with the gauge theory of the weak interactions.

8.5.3 Multiplets and Coupling Constants

The various schemes of spontaneous symmetry breaking in the Higgs mode lead to renormalizable theories only if the original symmetry before gauging is exact. Then the gauge bosons acquire a mass if the exact symmetry is hidden (spontaneously broken). A necessary consequence is that the degenerate multiplet structure characteristic of the Wigner realization of a symmetry is lost. Thus, precisely because we *see* approximately degenerate multiplet structure in the isospin and flavor degrees of freedom we must doubt the possibility of hidden symmetry there. This, and the lack of evidence for massless gauge bosons coupled to flavor degrees of freedom, appear to rule out the gauging of such degrees of freedom; they are thought to be associated with approximate *global* symmetries.

A fundamental consequence of underlying local gauge invariance in a theory with hidden symmetry is the appearance of definite relations between the parameters of the theory. Such theories are typically characterized by a pronounced economy of adjustable constants relative to the predictivity of the theory, because of the stringent requirements imposed by a local gauge invariance. Although the symmetry is hidden, experimental verification of relations among the parameters provides a method to identify the hidden symmetry. It has also been demonstrated in many realistic examples that these relations are *precisely* what is required to cause cancelations among graphs with bad high-energy behavior and render such theories renormalizable (see Chs. 6 and 9).

8.5.4 Improved Perturbation Theory

When a symmetry is broken spontaneously the true vacuum cannot be reached by perturbative expansion from the normal one. Spontaneous symmetry breaking is a *phase transition* and is manifestly a *non-perturbative effect.*[‡] As such, it is not easily handled in relativistic quantum field theories and heavy reliance must still be placed on models.

We can, however, develop a consistent perturbation theory about one of the degenerate vacua if a potential with such degenerate vacua is introduced by hand, as in the Landau–Ginsberg theory of phase transitions. If done at the macroscopic level this is equivalent to using phenomenology to approximate the non-perturbative effects that generate the new vacuum, and perturbation theory within the new vacuum. Obviously this is not a completely satisfactory state of affairs from a microscopic point of view, but this approach allows calculations that would otherwise be impossible.

[‡]For example, the acquisition of mass by the gauge bosons through the agency of a third state of polarization must be a phase transition—we cannot change the number of degrees of freedom for a particle in infinitesimal steps. An analogous situation occurs in superconductivity: the superconducting solution cannot be obtained by perturbative expansion about free-particle states because a phase transition is involved. For an excellent discussion of this see Mattuck (1976), §5.4 and Ch. 17.

Similar approximations are also employed in nonrelativistic many-body physics. For example, in the Nilsson model of nuclear physics (Bohr and Mottelson, 1975; Ring and Schuck, 1980) we may assume the nucleons to move in an axially-symmetric deformed potential

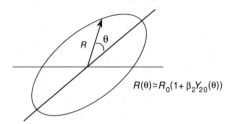

$$R(\theta) \simeq R_0(1 + \beta_2 Y_{20}(\theta))$$

with the potential energy of the nucleus a minimum at a finite value of the *deformation parameter* β_2 (for illustration we consider only $\beta_2 \geqslant 0$):

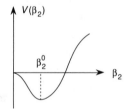

Then β_2 plays the role of an order parameter (we use the language of phase transitions loosely; these are finite systems), and the symmetry is spontaneously broken when $\langle \beta_2 \rangle = \beta_2^0 \neq 0$. For the nucleus this whole situation can sometimes be calculated reasonably well. Starting from a spherical shell model with appropriate residual interactions among the particles, we may find approximate solutions (by numerical methods, or by group theory) that have $\langle \beta_2 \rangle = 0$, but above critical values of the residual interaction strengths we may find solutions with $\langle \beta_2 \rangle \neq 0$ that are associated with a phase transition to a deformed state:

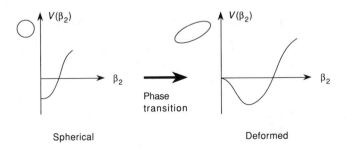

However, in the Nilsson model one *assumes* the symmetry to be broken spontaneously, determines $\langle \beta_2 \rangle \equiv \beta_2^0$ from phenomenology, neglects all fluctuations about $\langle \beta_2 \rangle$, and solves a deformed shell model at that average deformation (higher order corrections involve fluctuations of β_2 about $\langle \beta_2 \rangle$ and represent collective vibrations of the deformed potential). This is highly phenomenological but quite successful, and clearly is easier than a microscopic calculation of all the non-perturbative dynamics giving the spontaneously broken symmetry. In this illustration the vacuum (ground) state has a finite value of an order parameter β_2; in the many-body context such states are called *intrinsic states*. An intrinsic state typically violates fundamental symmetries—in this example it does not conserve angular momentum. However, symmetries can be restored by employing projection integrals over the intrinsic states (see Ring and Schuck, 1980).

8.5.5 Dynamical Symmetry Breaking

In nonrelativistic theories more progress has been made in understanding the microscopic origin of spontaneous symmetry breaking; for example, we have seen that in superconductivity one has an understanding of the Higgs field as a condensate of Cooper pairs. For the relativistic field theories used in particle physics we do not know whether the Higgs fields correspond to fundamental scalar fields, or whether (as in the case of superconductivity) the Higgs fields are composite and arise from the interactions of more elementary fields. The latter possibility is termed *dynamical symmetry breaking*, in that the nonzero expectation value of the scalar field in the vacuum state originates from the dynamics of fields appearing in a more fundamental Lagrangian.

The idea of the Higgs field as a composite object generated by "pairing" of more fundamental fields is attractive, but the specific models that have been advanced are not yet very convincing. One such model is that of *technicolor*, which attempts to construct composite scalars from QCD-like gauge interactions operating among "technifermions" on a mass scale $M \simeq 1000$ GeV (1 TeV). Chanowitz (1988) may be consulted for a simple introduction and references.

8.5.6 Renormalizability and Hidden Symmetry

Finally, a question of some urgency is whether field theories with spontaneously broken symmetries are renormalizable. As Coleman (1975) makes plausible, secret (hidden) symmetry begets secret renormalizability: the divergence structure of renormalizable theories is unaffected by spontaneous symmetry breaking, and the renormalization counterterms needed in the spontaneously broken theory remain those of the symmetric theory. Therefore, gauge fields that are renormalizable before spontaneous symmetry breaking retain their renormalizability in the presence of the Higgs phenomenon.

8.6 Background and Further Reading

A chapter on spontaneous symmetry breaking is now obligatory in any discussion of elementary particle physics. Aitchison and Hey (1982), Ch. 9; Quigg (1983), Ch. 5; Coleman (1975); Abers and Lee (1973); Leader and Predazzi (1982), Ch. 3; and Aitchison (1982), Chs. 2 and 6 are particularly recommended. Ring and Schuck (1980) and Aitchison and Hey (1982) give useful discussions of nonrelativistic spontaneous symmetry breaking.

CHAPTER 9

Standard Electroweak Model

We now have at our disposal the tools required to construct a microscopic theory of the weak interactions. The essential ingredients are Yang–Mills fields and the Higgs mechanism to spontaneously break a local gauge symmetry without compromising renormalizability. A general recipe can be given for constructing such a spontaneously broken gauge theory of the weak interactions:

- Choose the gauge group.
- Assign fermions to suitable representations of the group in a manner consistent with low-energy phenomenology and with renormalizability requirements.
- Introduce scalar fields to give masses eventually to all gauge bosons except the photon, and to all massive fermion fields.
- Choose parameters to break symmetries spontaneously in a manner consistent with low-energy phenomenology.

The simplest and most successful implementation of this prescription is the Glashow–Salam–Weinberg (GSW) theory, which we now consider. This theory will be found to reproduce the successes of the Fermi theory, be renormalizable, make new predictions that have subsequently been verified, and provide a partial unification of electromagnetic and weak interactions.

9.1 The Electroweak Gauge Group

First, recall that the low-energy phenomenology of the weak interactions discussed in Ch. 6 suggests that the weak field quanta are *massive vector particles*. Charged weak currents require that two of the intermediate vector bosons be charged ($W^{\pm}$); the existence of neutral weak currents will require the uncharged Z^0 as well. Further, let us remember that there is a strong suggestion of *universality* in the weak interactions. These observations argue for a theory of the weak interactions based on a non-Abelian gauge symmetry (Yang–Mills fields), with spontaneous symmetry breaking implemented in the Higgs mode. This is precisely the basis of the GSW theory, also often termed the *Standard Electroweak Model* (Glashow, 1961; Weinberg, 1967; Salam, 1968).

Having settled on a strategy, the initial question to be addressed is that of the appropriate symmetry to be gauged—what is the *weak gauge group*? First, we must anticipate that the relevant symmetry is hidden by the spontaneous symmetry breaking. It is not likely to leap out at us as in the Wigner implementation of symmetry with its degenerate multiplet structure: the masses of multiplet members could be drastically altered by the symmetry breaking.

A suggestive clue is provided by the observation that the weak charged currents induce transitions within the same families, but not across family lines (see Exercise 6.1). Since each lepton family consists of a charged lepton and its associated neutrino, a reasonable guess is that the left-handed matter fields should transform as *doublets* under some non-Abelian group. The simplest possibility is $SU(2)$, which is usually termed *weak isospin* (not to be confused with hadronic isospin). Using t and t_3 to denote the weak isospin quantum number and its third component, respectively, we make the lepton assignments

$$t = \tfrac{1}{2} \text{ doublets} \quad \begin{cases} t_3 = +1/2 \\ t_3 = -1/2 \end{cases} \quad \begin{pmatrix} \nu_e \\ e^- \end{pmatrix}_{\mathrm{L}} \quad \begin{pmatrix} \nu_\mu \\ \mu^- \end{pmatrix}_{\mathrm{L}} \cdots \qquad (9.1a)$$

and the quark assignments

$$t = \tfrac{1}{2} \text{ doublets} \quad \begin{cases} t_3 = +1/2 \\ t_3 = -1/2 \end{cases} \quad \begin{pmatrix} u \\ d_c \end{pmatrix}_{\mathrm{L}} \quad \begin{pmatrix} c \\ s_c \end{pmatrix}_{\mathrm{L}} \cdots \qquad (9.1b)$$

[u = up, d = down, s = strange, c = charmed; subscript c means Cabibbo rotated—see eq. (6.36)]. In these expressions the subscript L denotes explicitly that only the left-handed fermions participate in the charged-current weak interactions. Because right-handed electrons and quarks do not enter the charged-current interactions, it is assumed that they transform as $SU(2)$ singlets:

$$t = 0 \text{ singlets} \quad \begin{cases} (e^-)_{\mathrm{R}} \quad (\mu^-)_{\mathrm{R}} \\ (u)_{\mathrm{R}} \quad (d_c)_{\mathrm{R}} \quad (c)_{\mathrm{R}} \quad (s_c)_{\mathrm{R}}. \end{cases} \qquad (9.1c)$$

Notice that these multiplets are definitely not mass degenerate, as anticipated by the previous comments on spontaneous symmetry breaking.

Now we gauge this non-Abelian symmetry. The group $SU(2)$ has three generators so three gauge fields are necessary, transforming under the adjoint representation ($t = 1$). The members of the weak isodoublets differ by one unit in charge so two of the gauge bosons ($W^\pm$) must be charged [the gauge bosons transform as $SU(2)$ generators; since linear combinations of generators allow us to step through a multiplet (Exercise 5.11 and §5.3.2), one gauge boson must be able to raise, and one must be able to lower, the charge of a multiplet member]. The third $SU(2)$ boson will be electrically neutral, as required by the existence of weak neutral currents.

The experimental observation that neutral weak currents do not have pure $V - A$ structure, and the prospect of unifying the weak and the electromagnetic interactions, argues for enlarging the group to include another neutral gauge boson (we proceed pedagogically; historically, the inclusion of the second neutral gauge field preceded the experimental discovery of weak neutral currents). Then, by taking appropriate orthogonal linear combinations of the two neutral gauge fields we may hope to obtain a photon and a neutral IVB (Z^0), with the currents of the Z^0 and the photon having a structure in accord with electroweak phenomenology.

Glashow suggested enlarging the gauge group to $SU(2)_w \times U(1)_y$, where the Abelian $U(1)_y$ group is associated with a new quantum number y called the *weak hypercharge* (as with weak isospin, weak hypercharge shares only a name and analogous mathematics with the corresponding strong-interaction symmetry). Suppose we require that a relation similar to the charge formula (5.70) applies to these weak interactions:

$$Q = t_3 + \frac{y}{2}, \tag{9.2}$$

where Q is the electrical charge, t_3 is the third component of weak isospin, and y is the weak hypercharge. Then a comparison of (9.2) and (9.1) dictates the quantum number assignments in Table 9.1 (see Exercise 9.1).

Table 9.1
Weak Isospin and Weak Hypercharge Quantum Numbers for Leptons and Quarks in the Standard Electroweak Model

Particle	t	t_3	y	Q
ν_e, ν_μ	$\frac{1}{2}$	$\frac{1}{2}$	-1	0
e_L, μ_L	$\frac{1}{2}$	$-\frac{1}{2}$	-1	-1
e_R, μ_R	0	0	-2	-1
u_L, c_L	$\frac{1}{2}$	$\frac{1}{2}$	$\frac{1}{3}$	$\frac{2}{3}$
$(d_c)_L, (s_c)_L$	$\frac{1}{2}$	$-\frac{1}{2}$	$\frac{1}{3}$	$-\frac{1}{3}$
u_R, c_R	0	0	$\frac{4}{3}$	$\frac{2}{3}$
$(d_c)_R, (s_c)_R$	0	0	$-\frac{2}{3}$	$-\frac{1}{3}$

Subscripts L and R indicate left-handed and right-handed components, respectively; a subscript c denotes Cabibbo rotation (6.36). We assume neutrinos to be massless and to have only left-handed components.

Notice that the left-handed and right-handed components of particles have been assigned different quantum numbers, so they belong to different representations of the gauge group. Thus we install parity violation by hand with these choices. Furthermore, (9.2) fixes the weak hypercharge quantum number through the empirical charge Q, so the GSW theory will offer no fundamental explanation for charge quantization.

EXERCISE 9.1 Verify the weak isospin and weak hypercharge quantum number assignments in Table 9.1 and eq. (9.10).

EXERCISE 9.2 Consider the group structure of the electroweak interactions. Form weak and electric charges from the current (6.14), its Hermitian conjugate, and (1.118); show that these charges do not close under commutation. Show that two alternatives can lead to a closed algebra: (1) introduction of a fourth generator (gauge boson) with appropriate couplings to yield a closed $SU(2) \times U(1)$ algebra, or (2) addition of new (heavy) fermions to modify the currents such that the charges close under $SU(2)$.

9.2 Standard Model for Leptonic Interactions

To keep the discussion simple we first restrict the theory to leptonic weak interactions (for much of the remainder of this chapter we follow the presentations of Abers and Lee, 1973, and Quigg, 1983, Ch. 6).

9.2.1 Lagrangian Density

The gauge group is $SU(2)_{\mathrm{w}} \times U(1)_y$, requiring four gauge bosons: $\mathbf{b}_\mu \equiv (b_\mu^1, b_\mu^2, b_\mu^3)$ for $SU(2)$, and a_μ for $U(1)$. The Lagrangian density is

$$\mathcal{L} = \mathcal{L}_{\mathrm{g}} + \mathcal{L}_{\mathrm{f}} + \mathcal{L}_{\mathrm{s}} + \mathcal{L}_{\mathrm{f-s}}, \tag{9.3}$$

with $\mathcal{L}_{\mathrm{g}}$ representing the gauge fields, $\mathcal{L}_{\mathrm{f}}$ the fermion fields and their coupling to the gauge fields, $\mathcal{L}_{\mathrm{s}}$ the contribution of scalar fields, and $\mathcal{L}_{\mathrm{f-s}}$ the coupling between fermion and scalar fields. We now discuss each of these terms.

The gauge-field portion of (9.3) is

$$\mathcal{L}_{\mathrm{g}} = -\tfrac{1}{4} F_{\mu\nu}^j F_j^{\mu\nu} - \tfrac{1}{4} f_{\mu\nu} f^{\mu\nu}. \tag{9.4}$$

The Abelian field strength tensor is given by eq. (2.47):

$$f_{\mu\nu} = \partial_\mu a_\nu - \partial_\nu a_\mu, \tag{9.5}$$

and the non-Abelian field strength tensor is given by eq. (7.10b):

$$F_{\mu\nu}^j = \partial_\mu b_\nu^j - \partial_\nu b_\mu^j - g\epsilon_{jkl} b_\mu^k b_\nu^l. \tag{9.6}$$

The term in the Lagrangian density for the fermion matter fields is

$$\mathcal{L}_f = \overline{R}i\gamma^\mu \left(\partial_\mu + \frac{ig'}{2}a_\mu y \right) R + \overline{L}i\gamma^\mu \left(\partial_\mu + \frac{ig'}{2}a_\mu y + \frac{ig}{2}\boldsymbol{\tau}\cdot\mathbf{b}_\mu \right) L, \quad (9.7)$$

where right-handed and left-handed lepton fields are defined by

$$R \equiv e_R = \tfrac{1}{2}(1+\gamma_5)e \quad (9.8)$$

$$L \equiv \begin{pmatrix} \nu \\ e \end{pmatrix}_L = \tfrac{1}{2}(1-\gamma_5)\begin{pmatrix} \nu \\ e \end{pmatrix}. \quad (9.9)$$

For brevity we will write the terms in the Lagrangian density involving only the electron and its neutrino. Terms for other lepton families will have the same structure as these and can be added when required. The coupling constant associated with the weak isospin group $SU(2)_w$ is g, and the coupling constant for weak hypercharge $U(1)_y$ is $g'/2$. Two independent coupling constants are required because of the direct-product group structure $SU(2)_w \times U(1)_y$. No mass terms appear in the Lagrangian for either the lepton or gauge boson fields because they would spoil the gauge invariance; all masses must come through the Higgs mechanism (see Exercise 9.3).

This is a non-Abelian symmetry and we will eventually require that the electron and three of the gauge bosons acquire spontaneously generated masses. Therefore, more Higgs fields are needed than in the simple Abelian Higgs model considered earlier. The most economical choice consistent with phenomenology is to introduce a complex $SU(2)$ doublet of Lorentz scalar fields

$$\phi \equiv \begin{pmatrix} \phi^+ \\ \phi^0 \end{pmatrix} = \frac{1}{\sqrt{2}} \begin{pmatrix} \phi_1 + i\phi_2 \\ \phi_3 + i\phi_4 \end{pmatrix}, \quad (9.10)$$

which has $t = \frac{1}{2}$ and therefore weak hypercharge $y_\phi = 1$ [see (9.2)]. This contributes a term to the Lagrangian density of the form

$$\mathcal{L}_s = (D^\mu \phi)^\dagger (D_\mu \phi) - V(\phi^\dagger \phi), \quad (9.11)$$

where the covariant derivative is

$$D_\mu = \partial_\mu + \frac{ig'}{2}a_\mu y + \frac{ig}{2}\boldsymbol{\tau}\cdot\mathbf{b}_\mu, \quad (9.12)$$

and the potential that will be used to spontaneously break the symmetry is the familiar

$$V(\phi^\dagger \phi) = \mu^2 \phi^\dagger \phi + \lambda(\phi^\dagger \phi)^2, \quad (9.13)$$

with $\lambda > 0$. To invest the electron with mass we must couple the scalar field to the fermions and invoke spontaneous symmetry breaking. A gauge-invariant and Lorentz-invariant possibility for the fermion–scalar interaction is the *Yukawa coupling* (Exercise 9.3)

$$\mathcal{L}_{f\text{-}s} = -G_e \left[\overline{R}(\phi^\dagger L) + (\overline{L}\phi)R \right], \quad (9.14)$$

where G_e is an empirical constant that is independent of the gauge couplings g and $g'/2$.

9.2.2 Spontaneous Symmetry Breaking

Proceeding as for the examples discussed in Ch. 8, we spontaneously break the local gauge symmetry by choosing $\mu^2 < 0$ in (9.13) and select as the vacuum expectation value for the scalar field

$$\langle \phi \rangle_0 \equiv \langle 0| \, \phi \, |0 \rangle = \begin{pmatrix} 0 \\ v/\sqrt{2} \end{pmatrix}, \tag{9.15}$$

with

$$v = \sqrt{\frac{-\mu^2}{\lambda}}. \tag{9.16}$$

The generators of the $SU(2)$ algebra are $t_i \equiv \tau_i/2$, with [see (1.78)]

$$\tau_1 = \begin{pmatrix} 0 & 1 \\ 1 & 0 \end{pmatrix} \qquad \tau_2 = \begin{pmatrix} 0 & -i \\ i & 0 \end{pmatrix} \qquad \tau_3 = \begin{pmatrix} 1 & 0 \\ 0 & -1 \end{pmatrix},$$

while $U(1)$ has the single generator

$$y = \begin{pmatrix} 1 & 0 \\ 0 & 1 \end{pmatrix}.$$

From these four generators for $SU(2) \times U(1)$ it is convenient to form a new set (τ_1, τ_2, K, Q) with

$$K \equiv \frac{\tau_3 - y}{2} = \begin{pmatrix} 0 & 0 \\ 0 & -1 \end{pmatrix} \qquad Q \equiv \frac{\tau_3 + y}{2} = \begin{pmatrix} 1 & 0 \\ 0 & 0 \end{pmatrix}, \tag{9.17}$$

where Q is the charge (9.2). For every generator that does not annihilate the vacuum we expect an aspiring Goldstone boson that can be gauged from the theory to reappear as an effective longitudinal state of polarization for a gauge boson, thereby endowing it with mass. Operating with the generators on the vacuum (9.15), one finds by explicit matrix multiplication

$$\tau_1 \langle \phi \rangle_0 = \begin{pmatrix} 0 & 1 \\ 1 & 0 \end{pmatrix} \begin{pmatrix} 0 \\ v/\sqrt{2} \end{pmatrix} = \begin{pmatrix} v/\sqrt{2} \\ 0 \end{pmatrix} \neq 0$$

$$\tau_2 \langle \phi \rangle_0 = \begin{pmatrix} 0 & -i \\ i & 0 \end{pmatrix} \begin{pmatrix} 0 \\ v/\sqrt{2} \end{pmatrix} = \begin{pmatrix} -iv/\sqrt{2} \\ 0 \end{pmatrix} \neq 0$$

$$K \langle \phi \rangle_0 = \begin{pmatrix} 0 & 0 \\ 0 & -1 \end{pmatrix} \begin{pmatrix} 0 \\ v/\sqrt{2} \end{pmatrix} = \begin{pmatrix} 0 \\ -v/\sqrt{2} \end{pmatrix} \neq 0 \tag{9.18}$$

$$Q \langle \phi \rangle_0 = \begin{pmatrix} 1 & 0 \\ 0 & 0 \end{pmatrix} \begin{pmatrix} 0 \\ v/\sqrt{2} \end{pmatrix} = 0.$$

Thus three of the generators give non-zero values when applied to the vacuum, and three gauge bosons should acquire masses spontaneously. However, the charge Q annihilates the vacuum state, implying that the vacuum is invariant under the local $U(1)_{\text{QED}}$ symmetry that we wish to recover: the symmetry is spontaneously broken from $SU(2)_{\text{w}} \times U(1)_y$ down to $U(1)_{\text{QED}}$, and the electrical charge is conserved.

These results originate in the choice (9.15) for the vacuum scalar field. Any value of $\langle \phi \rangle_0$ that breaks the symmetry will generate masses for gauge bosons, but the particular choice (9.15) permits only *neutral* scalars ϕ^0 in the ground state. This breaks $SU(2)_{\text{w}}$ and $U(1)_y$ but conserves electrical charge, so $W^\pm$ and Z^0 acquire masses but the gauge symmetry $U(1)_{\text{QED}}$ survives and the photon is massless.

9.2.3 Particle Spectrum

Proceeding by analogy with the Abelian Higgs example in §8.4, we investigate the particle spectrum of the theory by examining fluctuations around the classical vacuum. It is convenient to parameterize ϕ as (Abers and Lee, 1973)

$$\phi(x) = \exp\left(i\frac{\boldsymbol{\xi} \cdot \boldsymbol{\tau}}{2v} \right) \begin{pmatrix} 0 \\ \dfrac{v + \eta(x)}{\sqrt{2}} \end{pmatrix} \equiv U^{-1}(\boldsymbol{\xi}) \begin{pmatrix} 0 \\ \dfrac{v + \eta}{\sqrt{2}} \end{pmatrix} \tag{9.19}$$

[compare (8.26)], with the four real components of (9.10) traded for a field $\eta(x)$ and the three components of a field $\boldsymbol{\xi}(x)$. The transformation to unitary gauge takes the form [compare (8.30)]

$$\phi \to \phi' = U(\boldsymbol{\xi})\phi = \frac{1}{\sqrt{2}} \begin{pmatrix} 0 \\ v + \eta \end{pmatrix}$$

$$\boldsymbol{\tau} \cdot \mathbf{b}_\mu \to \boldsymbol{\tau} \cdot \mathbf{b}'_\mu \qquad a_\mu \to a_\mu \tag{9.20}$$

$$R \to R \qquad L \to L' = U(\boldsymbol{\xi})L,$$

where $\boldsymbol{\tau} \cdot \mathbf{b}'_\mu$ is defined by the transformation (7.8) with the definition (7.7) (in the subsequent equations the primes will be omitted).

Substituting in (9.14), we find for the fermion–scalar part of the interaction (Exercise 9.3)

$$\mathcal{L}_{\text{f-s}} = -G_e \frac{v + \eta}{\sqrt{2}} (\bar{e}_R e_L + \bar{e}_L e_R)$$

$$= -\frac{G_e v}{\sqrt{2}} \bar{e}e - \frac{G_e \eta}{\sqrt{2}} \bar{e}e \quad . \tag{9.21}$$

Let us compare with the Dirac Lagrangian (2.85). The first term of (9.21) indicates that the electron now has a mass

$$m_e = \frac{G_e v}{\sqrt{2}} \tag{9.22}$$

as a consequence of the spontaneous symmetry breaking, but with our assumptions the neutrino will remain massless: it does not couple through (9.14) because it has no right-handed components. The second term represents a coupling between the electron field and the scalar field $\eta(x)$.

Substitution of (9.20) into (9.11) gives

$$
\begin{aligned}
\mathcal{L}_s =& \tfrac{1}{2}(\partial^\mu \eta)(\partial_\mu \eta) - \tfrac{1}{2}m_\eta^2 \eta^2 \\
&+ \tfrac{1}{2}m_W^2 \left(|W_\mu^+|^2 + |W_\mu^-|^2 \right) \\
&+ \tfrac{1}{2}m_Z^2 |Z_\mu^0|^2 + \text{interaction terms,}
\end{aligned} \tag{9.23}
$$

where we have defined the charged boson fields

$$W_\mu^+ \equiv \tfrac{1}{\sqrt{2}}(b_\mu^1 - ib_\mu^2) \qquad W_\mu^- \equiv \tfrac{1}{\sqrt{2}}(b_\mu^1 + ib_\mu^2), \tag{9.24}$$

and the neutral boson fields

$$Z_\mu^0 \equiv \frac{(-g'a_\mu + gb_\mu^3)}{\sqrt{g^2 + g'^2}} \qquad A_\mu \equiv \frac{(ga_\mu + g'b_\mu^3)}{\sqrt{g^2 + g'^2}}, \tag{9.25}$$

and where

$$m_\eta \equiv \sqrt{-2\mu^2} \qquad m_W = \frac{gv}{2} \qquad m_Z = m_W \sqrt{1 + \left(\frac{g'}{g}\right)^2} . \tag{9.26}$$

Comparison of (9.23) with the Lagrangian densities (2.41)–(2.49) allows the identification of

- A scalar field η with mass m_η,
- Two charged gauge boson fields $W^\pm$ with a common mass m_W,
- A massive neutral gauge boson field Z^0 with a mass m_Z,
- A gauge boson field A_μ with no mass term in the kinetic energy portion of the Lagrangian.

These particles correspond to the massive Higgs boson η, the massive intermediate vector bosons $W^\pm$ and Z^0, and the massless photon A_μ, as will be more apparent when we examine their couplings. All gauge boson masses have been acquired spontaneously through the Higgs mechanism, so we may expect that the theory is renormalizable. (To be truthful, it is not quite yet because of technicalities known as triangle anomalies, but we will remedy that shortly.) The photon remains massless on account of the residual symmetry under $\exp[iQ\theta(x)]$, where Q is the charge operator. That is, $Q\,|0\rangle = 0$ and the physical vacuum is uncharged, as anticipated in (9.18).

It is customary to introduce the *weak mixing angle* or *Weinberg angle* $\theta_{\rm w}$ through the definition

$$\tan\theta_{\rm w} = \frac{g'}{g}, \tag{9.27a}$$

implying that

$$\sqrt{g^2 + g'^2} = \frac{g}{\cos\theta_{\rm w}} = \frac{g'}{\sin\theta_{\rm w}}, \tag{9.27b}$$

so that (9.25) can be written

$$Z_\mu^0 = -a_\mu \sin\theta_{\rm w} + b_\mu^3 \cos\theta_{\rm w} \tag{9.28a}$$

$$A_\mu = a_\mu \cos\theta_{\rm w} + b_\mu^3 \sin\theta_{\rm w}. \tag{9.28b}$$

9.2.4 Interactions and Feynman Rules

From (9.7) terms may be identified that are associated with the coupling of neutral and charged gauge bosons to the lepton fields. The part representing the charged couplings may be written

$$\mathcal{L}_{\rm f}^{\rm charged} = \frac{-g}{\sqrt{2}}\left(\bar{\nu}_{\rm L}\gamma^\mu e_{\rm L} W_\mu^+ + \bar{e}_{\rm L}\gamma^\mu \nu_{\rm L} W_\mu^-\right), \tag{9.29a}$$

which coincides with the low-energy, charged weak current phenomenology of §6.2 if we set [compare (6.43) and (6.45)]

$$g^2 = \frac{8}{\sqrt{2}}Gm_W^2, \tag{9.29b}$$

where G is the Fermi coupling constant appearing in (6.11). The portion of (9.7) representing the neutral couplings may be written

$$\mathcal{L}_{\rm f}^{\rm neutral} = q\bar{e}\gamma^\mu A_\mu e - \frac{g}{2\cos\theta_{\rm w}}\left[\bar{\nu}_{\rm L}\gamma^\mu \nu_{\rm L} Z_\mu^0 + 2\sin^2\theta_{\rm w}\bar{e}_{\rm R}\gamma^\mu e_{\rm R} Z_\mu^0\right.$$
$$\left. + (2\sin^2\theta_{\rm w} - 1)\bar{e}_{\rm L}\gamma^\mu e_{\rm L} Z_\mu^0\right], \tag{9.30a}$$

provided the electronic charge q is identified with

$$q \equiv \frac{gg'}{\sqrt{g^2 + g'^2}}. \tag{9.30b}$$

From (9.30a) we recognize

- A standard QED coupling of the photons to the charged electron field;
- No coupling of the photon to the chargeless neutrinos;
- *Neutral weak currents* for both electrons and neutrinos, mediated by the uncharged gauge boson Z^0. The neutral currents couple to both left- and right-handed fermions, in contrast to the charged-current couplings.

The Feynman rules for elementary vertices in the GSW theory may be constructed from the Lagrangian density by the standard methods. We reproduce them without proof in Figs. 9.1–9.3. We will deal only with tree-level diagrams, so the Feynman rules have been given in unitary gauge and rules associated with ghost particles have been omitted. A more general set of rules for R-gauges (see §8.4), including ghost propagators and vertices, may be found in Appendix B of Cheng and Li (1984). The unitary gauge exhibits the particle spectrum clearly, but the propagators have bad high-energy behavior. For higher order calculations the R-gauges are more useful because the propagators are better behaved, but the particle spectrum in these gauges is complicated.

EXERCISE 9.3 Show that an explicit mass term in the electroweak Lagrangian violates gauge invariance. Show that a mass introduced through (9.14) is gauge invariant and obtain (9.21) from (9.14).

9.2.5 Predictions and Comparison with Data

From (9.27) and (9.30b)

$$g = \frac{q}{\sin \theta_{\rm w}} \qquad g' = \frac{q}{\cos \theta_{\rm w}}, \tag{9.31}$$

where q is the electromagnetic charge. Thus the $SU(2)_{\rm w}$ coupling constant g and the $U(1)_{\rm QED}$ coupling constant q are related through a single empirical parameter, the Weinberg angle $\theta_{\rm w}$. The mass of the intermediate vector bosons can also be given in terms of the Weinberg angle; using (9.29b) and (9.31),

$$m_W = \left(\frac{q^2}{4\sqrt{2}\, G \sin^2 \theta_{\rm w}} \right)^{1/2} = \frac{37.3}{\sin \theta_{\rm w}} \ {\rm GeV}, \tag{9.32a}$$

and from (9.26),

$$m_Z = \frac{m_W}{\cos \theta_{\rm w}}. \tag{9.32b}$$

Hence a determination of the Weinberg angle fixes the masses of the intermediate vector bosons. From (9.32a) we also see that the Fermi coupling

External Particles:

Leptons and quarks:

$$\slashed{f} = u_{\mathbf{ps}} \qquad \slashed{f} = v_{\mathbf{ps}} \qquad \slashed{f} = \bar{u}_{\mathbf{ps}} \qquad \slashed{f} = \bar{v}_{\mathbf{ps}}$$

Vector bosons: polarizations ε_μ (incoming) or ε_μ^* (outgoing)

Charged and Neutral Current Vertices:

$$A_\mu \qquad -iq\gamma_\mu$$

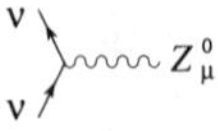

$$Z_\mu^0 \qquad \frac{-ig}{2\cos\theta_{\mathrm{w}}}\gamma_\mu\frac{1-\gamma_5}{2}$$

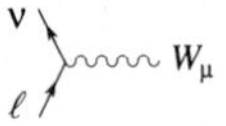

$$W_\mu \qquad \frac{-ig}{\sqrt{2}}\gamma_\mu\frac{1-\gamma_5}{2}$$

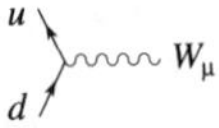

$$W_\mu \qquad \frac{-ig}{\sqrt{2}}\cos\theta_{\mathrm{c}}\gamma_\mu\frac{1-\gamma_5}{2}$$

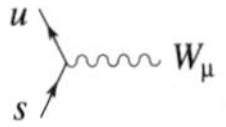

$$W_\mu \qquad \frac{-ig}{\sqrt{2}}\sin\theta_{\mathrm{c}}\gamma_\mu\frac{1-\gamma_5}{2}$$

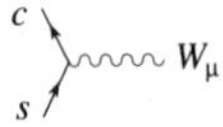

$$W_\mu \qquad \frac{-ig}{\sqrt{2}}\cos\theta_{\mathrm{c}}\gamma_{\mu}\frac{1-\gamma_5}{2}$$

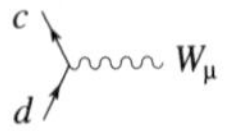

$$W_\mu \qquad \frac{-ig}{\sqrt{2}}(-\sin\theta_{\mathrm{c}})\,\gamma_\mu\frac{1-\gamma_5}{2}$$

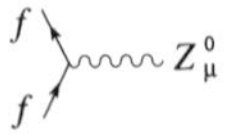

$$Z_\mu^0 \qquad \frac{-ig}{\cos\theta_{\mathrm{w}}}\gamma_\mu\left[C_{\mathrm{L}}(f)\frac{1-\gamma_5}{2}+C_{\mathrm{R}}(f)\frac{1+\gamma_5}{2}\right]$$

f	$C_{\mathrm{L}}(f)$	$C_{\mathrm{R}}(f)$
Massless ν	$\frac{1}{2}$	0
e^- , μ^-	$\sin^2\theta_{\mathrm{w}}-\frac{1}{2}$	$\sin^2\theta_{\mathrm{w}}$
u, c	$\frac{1}{2}-\frac{2}{3}\sin^2\theta_{\mathrm{w}}$	$-\frac{2}{3}\sin^2\theta_{\mathrm{w}}$
d , s	$\frac{1}{3}\sin^2\theta_{\mathrm{w}}-\frac{1}{2}$	$\frac{1}{3}\sin^2\theta_{\mathrm{w}}$

Propagators (Unitary Gauge):

Fermion

$$\frac{i(\slashed{p}+m)}{p^2-m^2+i\varepsilon}$$

Higgs scalar

$$\frac{i}{p^2-m_\eta^2+i\varepsilon}$$

Vector mesons

$$\frac{i\left(-g^{\mu\nu}+k^\mu k^\nu/m_B^2\right)}{k^2-m_B^2+i\varepsilon} \qquad (B\equiv Z \text{ or } W)$$

Fig. 9.1   Unitary gauge Feynman rules for fermion–gauge boson interactions and for propagators in the GSW electroweak theory. The inclusion of quarks in the theory is discussed in §9.3. The Weinberg angle is θ_{w}, the Cabibbo angle is θ_{c}, the symbol f stands for fermions of charge q, and ℓ denotes charged leptons. The masses m_Z, m_W, and m_η are defined in (9.26) and (9.32).

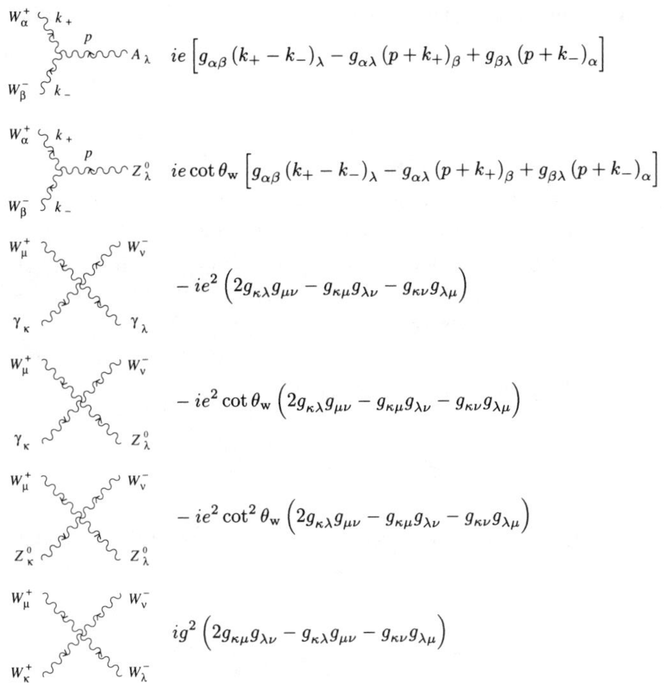

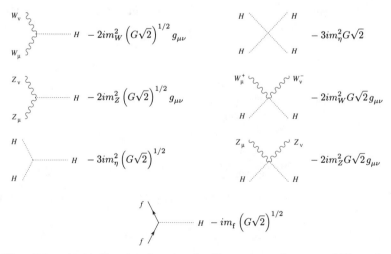

W_α^+ k_+
p
A_λ
W_β^- k_-

$ie \left[g_{\alpha\beta} (k_+ - k_-)_\lambda - g_{\alpha\lambda} (p + k_+)_\beta + g_{\beta\lambda} (p + k_-)_\alpha \right]$

W_α^+ k_+
p
Z_λ^0
W_β^- k_-

$ie \cot\theta_w \left[g_{\alpha\beta} (k_+ - k_-)_\lambda - g_{\alpha\lambda} (p + k_+)_\beta + g_{\beta\lambda} (p + k_-)_\alpha \right]$

W_μ^+ W_ν^-
γ_κ γ_λ

$- ie^2 \left(2 g_{\kappa\lambda} g_{\mu\nu} - g_{\kappa\mu} g_{\lambda\nu} - g_{\kappa\nu} g_{\lambda\mu} \right)$

W_μ^+ W_ν^-
γ_κ Z_λ^0

$- ie^2 \cot\theta_w \left(2 g_{\kappa\lambda} g_{\mu\nu} - g_{\kappa\mu} g_{\lambda\nu} - g_{\kappa\nu} g_{\lambda\mu} \right)$

W_μ^+ W_ν^-
Z_κ^0 Z_λ^0

$- ie^2 \cot^2\theta_w \left(2 g_{\kappa\lambda} g_{\mu\nu} - g_{\kappa\mu} g_{\lambda\nu} - g_{\kappa\nu} g_{\lambda\mu} \right)$

W_μ^+ W_ν^-
W_κ^+ W_λ^-

$ig^2 \left(2 g_{\kappa\mu} g_{\lambda\nu} - g_{\kappa\lambda} g_{\mu\nu} - g_{\kappa\nu} g_{\lambda\mu} \right)$

Fig. 9.2   As in Fig. 9.1, but for gauge boson self-interactions (Quigg, 1983).

W_ν
H $- 2i m_W^2 \left(G\sqrt{2} \right)^{1/2} g_{\mu\nu}$
W_μ

H H
$- 3i m_\eta^2 G\sqrt{2}$
H H

Z_ν
H $- 2i m_Z^2 \left(G\sqrt{2} \right)^{1/2} g_{\mu\nu}$
Z_μ

W_μ^+ W_ν^-
$- 2i m_W^2 G\sqrt{2}\, g_{\mu\nu}$
H H

H
H $- 3i m_\eta^2 \left(G\sqrt{2} \right)^{1/2}$
H

Z_μ Z_ν
$- 2i m_Z^2 G\sqrt{2}\, g_{\mu\nu}$
H H

f
H $- i m_f \left(G\sqrt{2} \right)^{1/2}$
f

Fig. 9.3   As in Fig. 9.1, but for the Higgs–gauge boson and Higgs–fermion interactions. The Higgs mass is m_η and m_f is the fermion mass (Quigg, 1983).

constant is inversely proportional to the square of the intermediate vector boson mass

$$G \propto \frac{q^2}{m_W^2}. \tag{9.33}$$

Thus the weakness of the weak interactions may be attributed to the large mass of the intermediate vector bosons. This observation, and the discussion of running coupling constants in §6.4.4, warn us once again that the strength of a fundamental interaction is a subtle concept that cannot be specified independent of momentum transfer and the vacuum expectation values of scalar fields that may spontaneously break the symmetry.

From (9.26), (9.15), and (9.29b) the vacuum expectation value of the scalar field is determined by the Fermi constant (6.12):

$$\langle \phi^0 \rangle_0 = \frac{v}{\sqrt{2}} = (\sqrt{8}G)^{-1/2} \simeq 174 \;\; \text{GeV}, \tag{9.34}$$

while from (9.22) and (9.34) the constant G_e that couples the electron and scalar fields and gives the electron mass is dimensionless and small:

$$G_e = \frac{\sqrt{2}m_e}{v} = 2^{\frac{3}{4}}m_e\sqrt{G} \simeq 3 \times 10^{-6}. \tag{9.35}$$

It is also arbitrary, its only justification being to give the electron a mass without doing violence to the gauge invariance.

Various experiments are sensitive to the Weinberg angle θ_w; Fig. 9.4 shows the impressive agreement among several independent determinations, with a weighted average of $\sin^2 \theta_w = 0.222$ (Marciano and Parsa, 1986). A more recent compilation of the Particle Data Group gives (Aguilar-Benitez et al., 1988)

$$\sin^2 \theta_w = 0.230 \pm 0.005. \tag{9.36}$$

With this parameter taken from experiments, (9.32) gives for the masses of the $W^\pm$ and Z^0

$$m_W = \frac{37.3}{\sin \theta_w} \simeq 78 \; \text{GeV}/c^2 \tag{9.37a}$$

$$m_Z = \frac{m_W}{\cos \theta_w} \simeq 89 \; \text{GeV}/c^2. \tag{9.37b}$$

These predictions are altered only slightly by radiative corrections. Particles within a few percent of these masses (see Table 9.2) were reported in the UA1 and UA2 experiments at the CERN $p\bar{p}$ collider in 1983, and it was quickly confirmed that they were the long-sought intermediate vector bosons (Arnison et al., 1983; Banner et al., 1983).

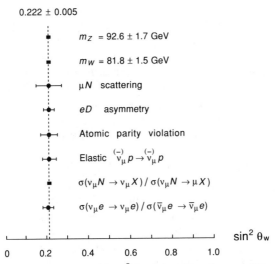

Fig. 9.4   Measured values of $\sin^2 \theta_{\mathrm{w}}$ (after Marciano and Parsa, 1986).

Because the GSW theory is a Yang–Mills theory, it has a large amount of predictive power relative to the number of phenomenological parameters. For instance, once θ_{w} has been determined a variety of decay widths may be predicted that provide stringent tests of the model. Table 9.2 compares some quantities measured in the UA1 and UA2 experiments with standard electroweak predictions. Marciano and Parsa (1986), Altarelli (1986), Bagnaia and Ellis (1988), and the more recent literature may be consulted for the status of electroweak parameters.

Table 9.2
Comparison of UA1 and UA2 Results with Standard Model Predictions[a]

Quantity	UA1	UA2	Standard Model
m_{w} (GeV)	$83.5^{+1.1}_{-1.0} \pm 2.8$	$81.2 \pm 1.1 \pm 1.3$	82.0 ± 1.3
m_{z} (GeV)	$93.0 \pm 1.4 \pm 3.2$	$92.5 \pm 1.3 \pm 1.5$	93.0 ± 1.1
$m_{\mathrm{z}} - m_{\mathrm{w}}$ (GeV)	$9.5^{+1.8}_{-1.7} \pm 0.5$	$11.3 \pm 1.7 \pm 0.2$	11.0 ± 0.2
Δr	-0.028 ± 0.12	0.08 ± 0.10	0.0696 ± 0.0020
$\sin^2 \theta_{\mathrm{w}} = \left(\frac{38.65\ \mathrm{GeV}}{m_{\mathrm{w}}} \right)^2$	0.214 ± 0.015	0.227 ± 0.010	0.222 ± 0.007
$\sin^2 \theta_{\mathrm{w}} = 1 - m_{\mathrm{w}}^2/m_{\mathrm{z}}^2$	0.194 ± 0.032	0.229 ± 0.030	0.222 ± 0.007

[a]From Marciano and Parsa (1986). The quantity Δr is a measure of radiative corrections to the vector boson masses.

We have just seen the impressive agreement of the GSW model with data, but it must also be said that the Standard Model has many parameters without a firm microscopic explanation, with the number depending on how many fermion generations are included. If we consider only electroweak interactions and only the first generation with massless neutrinos, there are still seven free parameters: the gauge couplings g and g', the scalar field parameters μ^2 and λ, and the three Yukawa couplings between the scalars and the charged fermions. In more physical terms these parameters may be traded for the electron charge e, the Weinberg angle θ_w, the IVB mass m_W, the Higgs mass m_η, the electron mass m_e, and the quark masses m_u and m_d (the inclusion of quarks is discussed in the following section). For three generations with massless neutrinos the additional fermion masses and the parameters of the quark mass mixing matrix (§9.3) increase this to 17 parameters. If the neutrinos turn out to be massive there will be additional neutrino masses and possible leptonic mixing angles in the parameter set. Nevertheless, the Standard Model correlates a huge amount of data and has far more predictive power than any previous theory of fundamental interactions.

9.3 Inclusion of Hadrons

The hadrons also undergo weak interactions, so we now incorporate them in the GSW theory. It is natural to assume that the weak interactions of the hadrons occur at the quark level, and that extension of the GSW theory to hadrons requires that quarks be included in the $SU(2)_w \times U(1)_y$ representations. This already has been anticipated in eqns. (9.1), where a parallel was conjectured between lepton doublets and singlets and the corresponding quark doublets and singlets. The relevant gauge-group quantum numbers have been assigned in Table 9.1.

Because of the close parallel between the lepton and quark doublets, the inclusion of quarks proceeds for the most part in a manner analogous to the leptons-only theory. The Feynman rules including quarks have been given already in Figs. 9.1–9.3. The primary difference between lepton–gauge boson and quark–gauge boson interactions is the Cabibbo mixing for the quarks, and we will not dwell on the details. However, the inclusion of quarks does have two far-reaching implications that are now briefly discussed. The first concerns renormalizability of the GSW theory and indications for a quark internal symmetry (*color*); the second concerns the suppression of unwanted flavor-changing currents in the theory, leading to the prediction of a fourth quark flavor (*charm*).

9.3.1 Anomalies

Renormalization in non-Abelian gauge theories is a delicate and involved task, and we will not launch into that esoteric subject here ('t Hooft, 1971; 't Hooft and Veltman, 1972; Lee and Zinn-Justin, 1972). However, it can be shown that the GSW theory as formulated to this point with leptons only is *not renormalizable*. In demonstrating renormalizability it is crucial that intricate cancellations take place in diagrams containing closed loops, and the symmetries of a theory play an essential role in guaranteeing these cancellations. However, for a field theory it may happen that a classical local conservation law derived from gauge invariance holds at the tree level, but not at the loop diagram level. Such violations of classical conservation laws in the quantized theory are called *anomalies*, and if they are not expunged they may compromise the renormalizability of the quantized theory. A famous example of a Feynman diagram producing an anomaly is shown in Fig. 9.5, where two vector currents (V) and one axial current (A) are coupled through a fermion triangle (anomalies that involve axial currents are commonly called γ^5, axial vector, or chiral anomalies).

A more descriptive name for anomalies is *quantum mechanical symmetry breaking*: an anomalous theory is one for which some symmetry of the classical fields is *necessarily* broken by the quantization process (see Jackiw, 1986). The current theoretical prejudice is that *gauge theories* with incurable anomalies are incorrect because they cannot be perturbatively renormalized. (In nongauge theories the appearance of anomalies may be more welcome. For an example, see the discussion of the $\pi^0 \to 2\gamma$ decay in §11–5–1 and §11–5–2 of Itzykson and Zuber, 1980.)

The anomalies that most concern us can occur in gauge theories with fermions and axial currents and are exemplified by diagrams such as Fig. 9.5. The standard cure is to include in the fermion multiplets of the offending theory a mix of particles with the appropriate quantum numbers to cancel exactly all anomalous currents (see Exercise 11.4 for an example). Thus the belief that fundamental theories should be perturbatively renormalizable leads to restrictions on allowed gauge group representations, and hence on allowed gauge groups. Such restrictions play an important role in limiting the possible gauge structures for grand unified theories (Ch. 11).

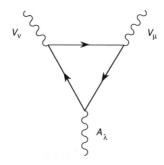

Fig. 9.5   Triangle anomaly for axial vector current.

In the GSW theory it can be demonstrated that there is only one fundamental anomaly, and that the condition for it to vanish can be formulated in
terms of the electrical charges Q as (Quigg, 1983)

$$\Delta Q = Q_{\mathrm{R}} - Q_{\mathrm{L}} \equiv \left(\sum_{\substack{\text{right-handed}\\\text{doublets}}} Q - \sum_{\substack{\text{left-handed}\\\text{doublets}}} Q \right) = 0. \qquad (9.38)$$

For the theory considered so far only left-handed particle doublets appear,
and from each lepton family in (9.1a) we find that

$$\Delta Q = -Q_{\mathrm{L}} = 1. \qquad (9.39)$$

Thus the leptonic GSW theory is anomalous and is not renormalizable! Not to
fear, however. From (9.38) we see that there are two ways to cancel this anomaly: add doublets of either right-handed fermions or left-handed fermions,
with charges selected to cancel the contributions from the left-handed particles already included. There is little evidence for right-handed lepton doublets
coupled to the weak interactions, so the second alternative appears more plausible; the obvious candidates are the left-handed quark doublets of (9.1b). A
single quark doublet such as $q_{\mathrm{L}} = \left(\begin{smallmatrix} u \\ d_c \end{smallmatrix} \right)_{\mathrm{L}}$ contributes

$$Q_{\mathrm{L}}^{\text{quarks}} = Q_u + Q_d = \tfrac{2}{3} - \tfrac{1}{3} = \tfrac{1}{3},$$

which is not sufficient to offset the lepton charge and the anomaly for the
corresponding family. However, if each flavor quark came in *three possible
internal states* the anomaly could be canceled. In fact, there is independent
evidence for just such a degree of freedom for the quarks, denoted by the *color
quantum numbers:* red, blue, and green. The existence of the color degree of
freedom is the basis of QCD, the gauge theory of the strong interactions to be
discussed in Ch. 10. If for each lepton doublet we include *three* quark doublets

$$\begin{pmatrix} u_{\text{red}} \\ d_{\text{red}}^c \end{pmatrix}_{\mathrm{L}} \qquad \begin{pmatrix} u_{\text{blue}} \\ d_{\text{blue}}^c \end{pmatrix}_{\mathrm{L}} \qquad \begin{pmatrix} u_{\text{green}} \\ d_{\text{green}}^c \end{pmatrix}_{\mathrm{L}}$$

the anomaly is canceled and the theory made renormalizable. Obviously this
can be generalized to each of the lepton–quark families, and we are led to
the remarkable conclusion that *the renormalizability of the weak interactions
requires a 3-valued internal degree of freedom for the quarks.*

9.3.2   GIM Mechanism

The formulation of the Lagrangian carries through more or less as before
when the (colored) quarks are included. The quarks may acquire masses
spontaneously by coupling to the scalar fields, as for the electron, and the

theory will be renormalizable owing to the gauge symmetry and the presence of the colored quarks to cancel anomalies, as outlined above. However, trouble still is in the offing. Because of the Cabibbo rotation (6.36) the neutral current contains terms of the form

$$Z_\mu^0 \left[\bar{d} \gamma^\mu (1 - \gamma_5) s + \bar{s} \gamma^\mu (1 - \gamma_5) d \right] \sin \theta_c \cos \theta_c, \qquad (9.40)$$

which is a potential disaster because such interactions exchange strange quarks for nonstrange quarks and lead to *flavor-changing neutral weak currents* that are systematically absent in the data. For example, the decay $K^+ \to \pi^+ \nu \bar{\nu}$ presumably is mediated by the strangeness-changing elementary transition $\bar{s} \to \bar{d} \nu \bar{\nu}$:

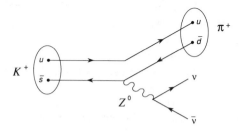

and the measured branching ratio for this transition is

$$\Gamma(K^+ \to \pi^+ \nu \bar{\nu}) / \Gamma(K^+ \to \text{all}) < 1.4 \times 10^{-7}.$$

Glashow, Iliopoulos, and Maiani (1970) proposed a solution to this problem by introducing a new flavor of quark, the *charmed quark c*. The charmed quark is assumed to be the weak isospin partner of the Cabibbo-rotated strange quark, s_c, which is defined by the combination orthogonal to d_c in (6.36):

$$s_c = s \cos_c - d \sin \theta_c. \qquad (9.41)$$

This leads to the doublet assignments in (9.1b), and it is shown in Exercise 9.6 that the dangerous term (9.40) is thereby canceled.

This method of suppressing the unwelcome strangeness-changing currents is called the Glashow–Iliopoulos–Maiani (1970) or *GIM mechanism*. Approximate properties of the putative charmed quark are implied by the GIM mechanism;[‡] the discovery of the J/ψ particle (Augustin et al., 1974; Aubert et

---

[‡]For example, it was shown that the mass of c could not exceed several GeV; otherwise the magnitude of the second-order neutral strangeness-changing currents would become too large. To set the GIM proposal in historical perspective it should be remembered that when it was offered there was no evidence for the charmed quark, nor had it even been proven that non-Abelian gauge theories were renormalizable so that it made sense to discuss perturbation theory for them!

al., 1974), and its interpretation as a bound $c\bar{c}$ state must be counted among the most impressive triumphs of the gauge theory approach to fundamental interactions.

The discovery of the Υ particle (Herb et al., 1977), its interpretation as a $b\bar{b}$ bound state, and the continued expectation that the t-quark will be found suggest that there is a third generation of quarks, with an additional left-handed quark doublet to be added to (9.1b). The most general mixing matrix extending the Cabibbo rotation to three quark flavor doublets is the Kobayashi–Maskawa (1973) matrix, which is a 3×3 unitary matrix with three empirical angles (θ_1, θ_2, and θ_3) giving all possible mixings among (u, d), (c, s), and (t, b) doublets, and an empirical phase δ leading to CP violation. Since we will not be concerned much with b and t quarks, we omit discussion of this (see Ellis et al., 1977). As we noted in Ch. 6, the absence of mixing angles in the leptonic sector is a consequence of the assumed masslessness of the neutrinos, which makes such mixings unobservable. If neutrinos have finite and nonequivalent masses, we may expect that the leptonic weak eigenstates could also be related to the mass eigenstates through a mixing matrix.

EXERCISE 9.4 (a) Use the standard electroweak theory to show that the width for decay of the W boson into a lepton and its neutrino is

$$\Gamma(W \to e\nu_e) \simeq \Gamma(W \to \mu\nu_\mu) \simeq \Gamma(W \to \tau\nu_\tau)$$

$$\simeq \frac{Gm_W^3}{6\sqrt{2}\pi}.$$

Hint: First prove that if the vertex factor for the decay amplitude of a vector boson X into two spin-$\frac{1}{2}$ particles is written in the form $-ig_X\gamma^\mu \frac{1}{2}(C_V - C_A\gamma^5)$, then the decay width is

$$\Gamma = \frac{g_X^2}{48\pi}(C_V^2 + C_A^2)m_X$$

(see Halzen and Martin, 1984, Exercise 13.2). (b) Assuming the quarks to be free within the nucleon (parton model), show that

$$\Gamma(W^+ \to u\bar{d}) = \Gamma(W^+ \to c\bar{s}) = 3\Gamma(W \to e\nu)\cos^2\theta_c$$
$$\Gamma(W^+ \to u\bar{s}) = \Gamma(W^+ \to c\bar{d}) = 3\Gamma(W \to e\nu)\sin^2\theta_c,$$

where θ_c is the Cabibbo angle.   (c) From parts (a) and (b), assuming that the decay $W \to t\bar{b}$ occurs with a width comparable with that of the decays in part (b)‡ and that no other heavy quarks or leptons exist into which W could decay, estimate the total width of the W-boson. Use this result to predict the branching ratio for leptonic decay of the W.

---

‡See the note added in proof to the solution of this exercise (Appendix D) concerning the t-quark mass.

EXERCISE 9.5 (a) Show that the width for $Z^0 \to f\bar{f}$ is

$$\Gamma_{Zf\bar{f}} = \frac{g^2}{24\pi \cos^2 \theta_{\mathrm{w}}} [C_{\mathrm{L}}(f)^2 + C_{\mathrm{R}}(f)^2] m_Z,$$

where $C_{\mathrm{L}}(f)$ and $C_{\mathrm{R}}(f)$ are defined in Fig. 9.1. *Hint*: see Exercise 9.4a.
(b) Calculate the widths for $Z^0 \to \nu\bar{\nu}$, for $Z^0 \to e^+e^-$, and for $Z^0 \to q\bar{q}$.
Assume the Z^0 to decay universally to all available fermion pairs and estimate
its total width.

EXERCISE 9.6 Demonstrate explicitly that the GIM mechanism can be
used to suppress weak, neutral, strangeness-changing currents.

9.4 Critique of the Standard Theory

The GSW electroweak theory has tightly-constrained predictive power by
virtue of the non-Abelian structure of the local gauge invariance. Ample discussion of the comparison of theoretical predictions with data may be found
in the literature [see, for example, Table 9.2 and Perkins (1987)] and we will
not dwell on that material here. Suffice it to say that the Standard Model
has met the experimental test remarkably well. Among its more significant
achievements we may include

1. A prediction of the existence and properties of weak neutral currents,
2. The partial unification of weak and electromagnetic interactions,
3. A prediction of correct masses for $W^{\pm}$ and Z^0,
4. The suggestion of a color degree of freedom for quarks,
5. The suggested existence of charmed quarks,
6. The satisfactory reproduction of QED and low-energy weak interaction
   phenomenology.

It should be appreciated that these attributes go far beyond putting lines
through data points, although the GSW theory does that extremely well too.
In addition to its triumphs, the GSW theory has some less desirable features:

1. The appearance of *two* gauge coupling constants, rather than the one
   expected in a truly unified theory,
2. The ad hoc way in which fermion masses, and parity and CP violation,
   enter the theory,
3. The absence of direct evidence for the Higgs boson.

We may hope for at least a partial resolution of the first two problems in
the grand unified theories (GUTs) to be discussed in Ch. 11. Feature (3) may
not be a serious problem because the mass of the Higgs field is not fixed by

the rest of the theory at the classical level.[‡]   Thus it can be quite massive, explaining why it has not yet been seen. Obviously the discovery of a Higgs particle, or a convincing interpretation in terms of an effective field generated dynamically at a more fundamental level (see §8.5.5), would turn liability (3) into one of the theory's greatest assets. Still, the lack of information about the Higgs field and the cavalier way that it enters the theory are cause for some unease, even though there can be little doubt that it is required phenomenologically (Chanowitz, 1988).

9.5 Background and Further Reading

Most discussions of gauge theories treat the Standard Electroweak Model in some detail. The discussion presented here has been strongly influenced by Abers and Lee (1973) and Quigg (1983), Chs. 6–7. The presentations in Ryder (1985), Cheng and Li (1984), Renton (1990), and Bég and Sirlin (1974) are also strongly recommended. Proceedings of recent conferences usually contain updates of the status of electroweak parameters and experimental tests of the model; for a recent analysis see Bagnaia and Ellis (1988). A wide-ranging discussion of the successes and shortcomings of the GSW theory, and how the shortcomings might be alleviated by embedding the Standard Model in more speculative and comprehensive theories, may be found in Chanowitz (1988).

[‡]The classical Higgs mass is unconstrained. There are some limits that come from assumptions regarding internal consistency of the quantum theory. An "upper bound" of about 1 TeV follows from partial wave unitarity, and a "lower bound" of about 7 GeV results from assumptions about the stability of the classical vacuum with respect to loop corrections.

CHAPTER 10

The Strong Interactions and Quantum Chromodynamics

In the preceding chapter a highly successful theory of the electroweak interactions based on a Yang–Mills gauge field was introduced. In this chapter we examine a Yang–Mills theory of the strong interactions, *quantum chromodynamics* (QCD). Although not yet tested as rigorously as the standard electroweak model, QCD will be seen to merit serious consideration as the microscopic theory of the strong interactions. Because QCD involves the interactions of quark and gauge fields, we must first make a detour to collect some phenomenology of the strong interactions, and to examine the role of quarks and strong interaction gauge fields in that phenomenology.

First we recall the evidence for the traditional attributes of the quarks, and then examine in more detail the hint from the electroweak theory that quarks contain a triply degenerate internal degree of freedom called color. Then the evidence for gauge quanta of the color field (gluons) will be summarized. Finally, we will proceed to a discussion of deep inelastic lepton scattering and of the theory of quantum chromodynamics.

10.1 Properties of Quarks

Some properties of the known quarks have been summarized in Table 5.3. That quarks and antiquarks have baryon number $\frac{1}{3}$ and $-\frac{1}{3}$, respectively, follows directly from the quark model $q\bar{q}$ valence structure for mesons and qqq structure for baryons, which was introduced in Ch. 5. At least three pieces of evidence support a spin-$\frac{1}{2}$ assignment for the quarks (Quigg, 1983; Close, 1979):

1. The observed hadronic spectrum is consistent with angular momentum, parity, and charge conjugation assignments, and with relative level ordering, for spin-$\frac{1}{2}$ valence constituents in the quark model. For example, combinations of spin, parity, and charge conjugation quantum numbers forbidden for spin-$\frac{1}{2}$

quark and antiquark constituents ($J^{PC} = 0^{--}$, 0^{+-}, 1^{-+}, ...) are absent from the spectrum of charmed and anticharmed quark bound states (the J/ψ family; see Fig. 12.19).

2. Objects with spin-$\frac{1}{2}$ can absorb transverse virtual photons but not longitudinal ones (angular momentum conservation); spinless objects can absorb only longitudinal virtual photons (Exercise 10.1). The deep inelastic scattering of electrons from nucleons to be discussed in §10.5.1 suggests that electrons interact with charged constituents of the nucleon by exchange of a transverse photon.

3. In high-energy e^+e^- annihilation the reaction is thought to proceed through the formation of a virtual photon, which rematerializes as a $q\bar{q}$ pair. Because of the confining property of the strong interaction, the high-energy $q\bar{q}$ pair polarizes the vacuum and creates new $q\bar{q}$ pairs (Fig. 12.13). This soup of quarks and antiquarks then combines in some (ill-understood) way to make physical hadrons, as illustrated in Fig. 10.1. The formation of hadrons from the original $q\bar{q}$ pair is called *hadronization*; such events are often characterized by a *jet structure*, where the hadronic debris is emitted in well-collimated cones. Figure 10.2 shows two- and three-jet structure measured for charged particles emitted in e^+e^- annihilation reactions. The two-jet events may be interpreted as being defined by the direction of emission for the original $q\bar{q}$ pair in Fig. 10.1 (three-jet events will be interpreted below). The angular distribution of scattered objects depends on their spins, and two-jet events are observed to have the same angular distributions as the spin-$\frac{1}{2}$ muons in $e^+e^- \rightarrow \mu^+\mu^-$. Assuming the jet angular distribution to be determined by the direction of emission of the initial $q\bar{q}$ pair, we conclude that the quarks are also spin-$\frac{1}{2}$ objects.

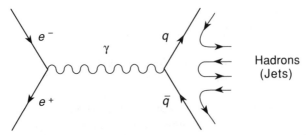

Fig. 10.1   Schematic illustration of e^+e^- annihilation into hadrons. In such diagrams particle lines are directed to the right and antiparticle lines to the left. These pictures are meant to convey the qualitative reaction mechanism; they are not Feynman diagrams and are without dynamical content. The hadronization process in the right portion of the diagram is only partially understood because it is intimately connected with the problem of confinement in QCD (see Fig. 12.13).

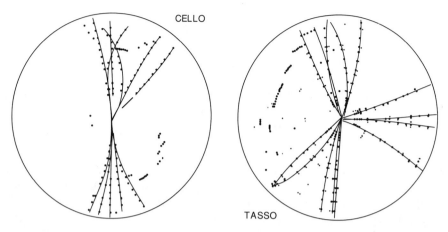

Fig. 10.2   Two- and three-jet charged particle tracks observed in e^+e^- annihilation reactions in the CELLO and TASSO detectors at PETRA. See Wu (1984) for a comprehensive survey of such experiments.

The novel third-integer charges ascribed to the quarks are also supported by several pieces of evidence (Quigg, 1983, Ch. 1):

1. The quark model charge assignments for hadrons require charges that are multiples of $\frac{1}{3}$.

2. Leptonic decays of neutral vector mesons $V \to \ell^+\ell^-$ proceed as in Fig. 10.3. The width for such a decay is given by

$$\Gamma(V \to \ell^+\ell^-) = \frac{16\pi\alpha^2 Q_q^2}{M_V^2} |\psi(0)|^2 \,, \tag{10.1}$$

where $\psi(0)$ is the amplitude for the q and $\bar{q}$ to be at the same spatial point, and M_V is the mass and Q_q^2 the mean-squared quark charge for the meson (see Close, 1979, §16.3.3; Perkins, 1987, Table 5.9). If we regard $\psi(0)$ for a series of vector mesons such as $\rho(770)$, $\omega(783)$, and $\phi(1020)$ to be approximately

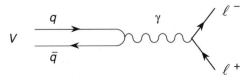

Fig. 10.3   Leptonic decay of a neutral vector meson.

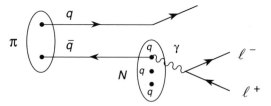

Fig. 10.4   Drell–Yan process.

equal, comparison of the leptonic decay widths in such a series provides information about quark charges. These comparisons support the usual charge assignments of the quark model (see Exercise 10.1b).

3. *Drell–Yan reactions* correspond to

$$h_1 + h_2 \to \ell^+\ell^- + \text{anything},$$

where h_1 and h_2 are hadrons and the invariant mass of the lepton pair is large compared with the mass of the nucleon. The Drell–Yan reaction $\pi N \to \ell^+\ell^- + \text{anything}$ is believed to proceed by the $\bar{q}$ in the π annihilating a q in the nucleon, with the resulting virtual photon rematerializing as lepton pairs (Fig. 10.4). Since the reaction is *inclusive* ($\to \ell^+\ell^- + \text{anything}$), presumably we may ignore details of how the q from the pion and qq remaining from the nucleon undergo hadronization. If the reaction is performed on a target with equal numbers of protons and neutrons, there should be equal numbers of up and down valence quarks. The cross section depends on the charge of the annihilated quark, and from the valence structure $\bar{u}d$ for π^- and $u\bar{d}$ for π^+ we find the prediction

$$\frac{\sigma(\pi^+ C \to \mu^+\mu^- + \text{anything})}{\sigma(\pi^- C \to \mu^+\mu^- + \text{anything})} = \frac{Q_d^2}{Q_u^2}, \qquad (10.2)$$

where C denotes an isoscalar $N = Z$ target. The measured ratio is about $\frac{1}{4}$, as expected for charges of $-\frac{1}{3}$ and $+\frac{2}{3}$ for the down and up quarks, respectively.

4. As discussed in more detail later, deep inelastic electron scattering effectively counts the number of quarks of each flavor weighted by the square of the quark charge; the data are consistent with the quark model charge assignments of Table 5.3.

EXERCISE 10.1   (a) Demonstrate in the Breit frame (Fig. 10.13) that spinless objects couple only to longitudinal photon states, but spin-$\frac{1}{2}$ objects can absorb only transverse photons. (b) Estimate the ratio of decay widths in $V \to e^+e^-$ for the ρ, ω, ϕ, and J/ψ mesons using (10.1) and the quark charges in Table 5.3.

There are several lines of inquiry suggesting that quarks possess, in addition to flavor and Lorentz degrees of freedom, an internal color degree of freedom. The three allowed values of the color quantum number will be termed *red*, *blue*, and *green* (obviously, these have nothing to do with ordinary colors). Among the arguments for the color degree of freedom we may include the following (Close, 1979; Quigg, 1983):

1. Renormalizability of the GSW electroweak theory seems to require a triply degenerate internal structure for the quarks (see §9.3).

2. Within the quark model the width for the decay $\pi^0 \to \gamma\gamma$ is calculated to be

$$\Gamma(\pi^0 \to \gamma\gamma) \propto N_c^2, \tag{10.3}$$

where N_c is the number of quark colors. The experimental value of 7.48 ± 0.32 eV compares favorably with the predicted width of 7.75 eV for $N_c = 3$, but very poorly with the prediction of 0.86 eV for $N_c = 1$.

3. Without the introduction of another quantum number the quark model comes into apparent conflict with the Pauli principle, as we have seen in Exercise 5.14a. For example, the quark model assumes Δ^{++} to have a valence quark structure uuu. This resonance is known to have spin $= \frac{3}{2}$, isospin $= \frac{3}{2}$, and the ground state is assumed to be an S-wave configuration. Now u is a spin-$\frac{1}{2}$, isospin projection $T_3 = \frac{1}{2}$ particle (Table 5.3), and the spins of all three quarks, as well as their isospins, must be aligned to give the observed spin and isospin of Δ^{++}. Unless there is an additional degree of freedom, this implies a symmetric state of identical fermions. For uuu to be properly antisymmetrized each quark flavor must exist in at least three (color) states, but more than three colors would give rise to such unobserved phenomena as more than one kind of proton. For the Δ^{++} an antisymmetric wavefunction can be written $\psi \simeq \epsilon_{\alpha\beta\gamma} u_\uparrow^\alpha u_\uparrow^\beta u_\uparrow^\gamma$, where the indices label color states, $\uparrow$ indicates spin up, and $\epsilon_{\alpha\beta\gamma}$ is antisymmetric in all indices. More generally, since the hadrons themselves are colorless a baryon state is assumed to be of the form

$$\psi_{\text{baryon}}^{ijk} \simeq \epsilon_{\alpha\beta\gamma} q_i^\alpha q_j^\beta q_k^\gamma, \tag{10.4a}$$

while a meson state is

$$\psi_{\text{meson}}^{ij} \simeq \delta_{\alpha\beta} \bar{q}_i^\alpha q_j^\beta, \tag{10.4b}$$

where i, j, and k are spin/flavor labels and α, β, γ are color labels. For example, the color/flavor wavefunction for a π^+ may be written

$$\psi(\pi^+) = \sqrt{\tfrac{1}{3}} \left(u_R \bar{d}_R + u_B \bar{d}_B + u_G \bar{d}_G \right),$$

where R, B, and G are color labels (§10.6.1).

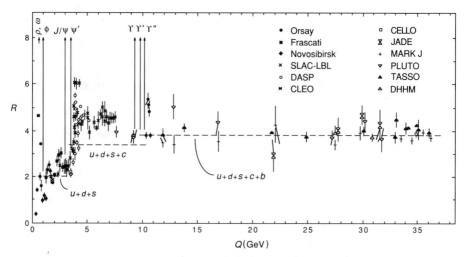

Fig. 10.5   Ratio $R \equiv \sigma(e^+e^- \to \text{hadrons})/\sigma(e^+e^- \to \mu^+\mu^-)$. After Halzen and Martin (1984); data compiled in Aguilar-Benitez et al. (1986).

4.   The cross section for the annihilation reaction $e^+e^- \to$ hadrons is sensitive to the number of quark species. The elementary process is $e^+e^- \to \gamma \to q\bar{q}$, with the $q\bar{q}$ materializing as hadronic jets (see Fig. 10.1). In Fig. 10.5 the experimental ratio

$$R \equiv \frac{\sigma(e^+e^- \to \text{hadrons})}{\sigma(e^+e^- \to \mu^+\mu^-)} \tag{10.5a}$$

is plotted as a function of center-of-mass energy. At high energies QCD predicts in leading order

$$R \simeq N_\text{c} \sum_{\text{quarks}} Q_\text{q}^2, \tag{10.5b}$$

where N_c is the number of quark colors and Q_q is the charge for quarks energetically accessible in the reaction. For CM energies between 1.5 and 3.6 GeV (above strangeness but below charm threshold)

$$R \simeq 3\left(Q_u^2 + Q_d^2 + Q_s^2\right) = 2, \tag{10.5c}$$

while for the highest energies plotted

$$R \simeq 3\left(Q_u^2 + Q_d^2 + Q_s^2 + Q_c^2 + Q_b^2\right) = \tfrac{11}{3}, \tag{10.5d}$$

where u, d, s, c, and b stand for up, down, strange, charmed, and bottom quark flavors, respectively. Both predicted ratios are in accord with the data for $N_c = 3$, but are in serious disagreement for $N_c = 1$. Higher order QCD corrections to R are small and will be discussed below [see eq. (10.43)].

5. The decay of the heavy τ lepton is sensitive to the number of quark colors through the ratio[‡]

$$\frac{\Gamma(\tau \to e\bar{\nu}_e\nu_\tau)}{\Gamma(\tau \to \text{anything})} = \frac{1}{N_c + 2}. \tag{10.6}$$

The measured value of 0.175 ± 0.004 for this ratio suggests $N_c = 3$.

These results indicate that there is indeed a color degree of freedom for the quarks. It is natural to speculate that color plays the role of a *non-Abelian charge* for the strong interactions. This will be the starting point for QCD.

10.2 Evidence for Gluons

Although less extensive than the arguments for quarks, there is strong circumstantial evidence for the existence of neutral gauge particles coupled to the color degree of freedom and playing the role of non-Abelian strong interaction "photons". The support for these *gluons* comes primarily from two experimental results (Quigg, 1983).

1. Energy–momentum sum rules in deep inelastic lepton scattering indicate that nucleon constituents interacting through weak or electromagnetic interaction carry only about one-half the nucleon momentum (§10.5.4). The remainder is apparently associated with nucleon substructure that is electrically neutral and inert to the weak interactions.
2. Three-jet events play a role in the reaction $e^+e^- \to$ hadrons above energies of about 20 GeV in the center-of-mass. Figure 10.2 shows an experimental example. The simplest interpretation of these events is in terms of *gluon bremsstrahlung*, where the q or $\bar{q}$ produced initially from the virtual photon radiates a gluon, which then hadronizes through formation of $q\bar{q}$ pairs and adds a third jet to the event (Fig. 10.6). The angular distribution of such events suggests that the gluon is spin-1.

[‡]The τ^- emits a virtual W^-, converting to a ν_τ. The W^- decays to all energetically allowed fermion–antifermion pairs with equal probability (weak universality—see Exercise 9.4): $(e\bar{\nu}_e)$, $(\mu\bar{\nu}_\mu)$, $(\bar{u}d)$, where phase-space factors, Cabibbo mixing, and QCD effects on the quarks are ignored. Thus (10.6) is $\frac{1}{3}$ if the quarks are uncolored, and $\frac{1}{5}$ if there are three colors of u and d.

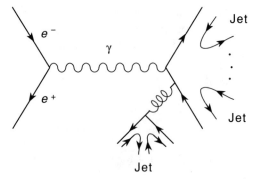

Fig. 10.6   Gluon bremsstrahlung and three-jet events in e^+e^- annihilation.

10.3 Deep Inelastic Lepton Scattering

Beginning in the late 1960s a series of experiments involving the scattering of high-energy electrons and neutrinos from hadrons gave strong impetus to the notion that the quarks were something more than just a mnemonic for flavor $SU(3)$ quantum numbers. These experiments reshaped our perception of the strong interactions. Until that time it was accepted wisdom that these interactions were terribly complicated and that there was little hope for understanding them at a microscopic level as long as perturbation theory was the standard tool of the field theorist. This indeed was the case when hadrons were scattered from hadrons. In retrospect, it appears we were posing a much too complicated question by asking hadron–hadron collisions to elucidate the basic nature of the strong interaction.

The following essential features emerged from these deep inelastic lepton scattering experiments:

1. An indication for point structure within the hadrons because of more backward scattering than would be expected for a uniform mass distribution in the proton. These were the *partons*, later to be identified with quarks and gluons; this result was reminiscent of Rutherford's discovery of the nucleus 50 years earlier when he and his colleagues observed anomalous large-angle α-scattering from atoms.
2. An indication that for high-energy leptons probing small regions of space-time the pointlike particles scattering the leptons are effectively quasi-free (*asymptotic freedom*), leading to *Bjorken scaling* of the hadronic structure functions.
3. A suggestion that the pointlike constituents of the hadrons did not exhibit the properties of any particles known to exist in the laboratory, leading

eventually to the conjecture that the partons are permanently imprisoned in the hadrons (*confinement*, or *infrared slavery*).

Thus the seeming complexity of hadron–hadron collisions could now be understood: it was as if we had tried to decipher the principles of QED by studying the collisions of molecules rather than those of more fundamental charged particles. Considerable headway could be made once attention was shifted from hadron–hadron to parton–parton interactions as the purest example of a strong interaction. Indeed, in the asymptotically free regime (small parton–parton separation) it is thought that the "strong" interaction is sufficiently meek to admit perturbative solutions.

Let us now turn to a more thorough discussion of these ideas. That hadrons should exhibit internal pointlike structure as the probe resolution is increased is not surprising, considering the history of atomism. However, that these pointlike constituents should exhibit the attributes of confinement and asymptotic freedom is altogether remarkable. Accordingly, we first examine the evidence and then ask whether any known theory can account for such exceptional behavior. The answer will surprise no one who has followed the story to this point: *the only physically acceptable theories known to exhibit both confinement and asymptotic freedom in four spacetime dimensions are non-Abelian gauge theories.*

We begin by considering the scattering of a lepton from a hadron in events for which there are many hadrons in the final state,

$$\ell + h \rightarrow \ell' + \text{anything}.$$

Figure 10.7 illustrates the likely mechanism for electrons scattered from protons. We will assume that the interaction is mediated by single-photon exchange and consider an inclusive process, since that is the easiest to measure

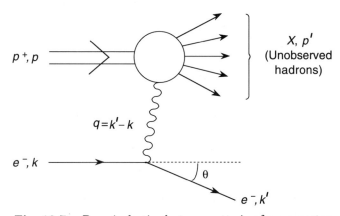

Fig. 10.7 Deep inelastic electron scattering from a proton.

experimentally. (By an inclusive process we mean one in which only the final electron is detected; no attempt is made to identify the hadronic debris.) Requiring a large momentum loss for the lepton ensures a process such as that depicted in Fig. 10.7, rather than elastic or quasielastic scattering.

Our lack of knowledge concerning the detailed hadronic structure is symbolized by the blob at the γ–proton vertex. In light of the discussion in §3.7.2, the restriction to single-photon exchange means that the unpolarized inclusive cross section must factor into a leptonic tensor contracted with a corresponding hadronic tensor,

$$d\sigma \simeq L_{\mu\nu}W^{\mu\nu}, \tag{10.7}$$

where the leptonic tensor is (3.102)

$$L_{\mu\nu} = \frac{1}{2m^2}\left(k'_\mu k_\nu + k'_\nu k_\mu + \frac{q^2}{2}g_{\mu\nu}\right), \tag{10.8}$$

with the momenta those in Fig. 10.7. Because the strong interactions are expected to modify the γ–proton vertex significantly, the form of $W_{\mu\nu}$ must be specified by general invariance arguments. We may define $W^{\mu\nu}(q,p)$ by (see the discussion in Aitchison and Hey, 1982, §3.8 and §4.1, and Cheng and Li, 1984, §7.1)

$$W^{\mu\nu}(q,p) = \frac{1}{8\pi M}\sum_{\sigma X}\langle P;p,\sigma|\,J^\mu(0)\,|X;p'\rangle$$
$$\times \langle X;p'|\,J^\nu(0)\,|P;p,\sigma\rangle\,(2\pi)^4\delta^{(4)}(p+q-p'), \tag{10.9}$$

where P denotes the proton, X the final set of undetected hadrons, the inclusive sum over X includes phase-space integrations, σ is a spin variable, $J^\mu(x)$ is the electromagnetic current, and M is the proton mass. It is convenient to introduce the independent Lorentz scalar variables Q^2 and ν:

$$Q^2 = -q^2 = 4EE'\sin^2\left(\frac{\theta}{2}\right) = 2EE'(1-\cos\theta), \tag{10.10a}$$

$$\nu = \frac{p\cdot q}{M} = E - E', \tag{10.10b}$$

which are related to the *invariant mass* W of the hadronic final states by

$$W^2 = (p+q)^2 = M^2 + 2p\cdot q + q^2$$
$$= M^2 + 2M\nu - Q^2, \tag{10.10c}$$

where we have taken the laboratory frame variables

$$p = (M,0,0,0) \qquad k = (E,\mathbf{k}) \qquad k' = (E',\mathbf{k'}), \tag{10.10d}$$

and θ is the lab scattering angle. Therefore, ν specifies the lab frame energy transfer and Q^2 is a measure of the minimum scale on which spacetime can be investigated, since a virtual photon of $(\text{mass})^2 = -Q^2$ probes the target on a characteristic length scale $1/\sqrt{Q^2}$:

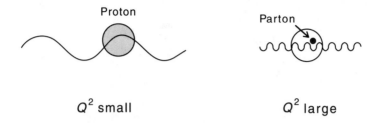

Proton

Parton

Q^2 small

Q^2 large

The hadronic tensors depend only on Q^2 and ν, and from invariance arguments take the form (see Exercise 10.3b)

$$W^{\mu\nu}(q,p) = \left(-g^{\mu\nu} + \frac{q^\mu q^\nu}{q^2}\right) W_1(Q^2,\nu)$$
$$+ \left(p^\mu - \frac{p\cdot q}{q^2}q^\mu\right)\left(p^\nu - \frac{p\cdot q}{q^2}q^\nu\right) \frac{W_2(Q^2,\nu)}{M^2}, \tag{10.11}$$

where the Lorentz-scalar *structure functions* W_1 and W_2 carry the intrinsic information about proton structure. The corresponding laboratory frame differential cross section may be written

$$\frac{d^2\sigma}{d\Omega dE'} = \frac{\alpha^2}{4E^2\sin^4\left(\frac{\theta}{2}\right)}\left[W_2\cos^2\left(\frac{\theta}{2}\right) + 2W_1\sin^2\left(\frac{\theta}{2}\right)\right], \tag{10.12}$$

where the electron mass is neglected. The appearance in these expressions of two structure functions can ultimately be related to absorption cross sections for longitudinal–scalar and transverse virtual photons in Fig. 10.7.

We may now classify lepton–hadron scattering in terms of the variables ν and Q^2, and specify precisely what is meant by deep inelastic scattering. Figure 10.8 illustrates that deep inelastic scattering occurs in the region of large Q^2 and large ν, but with Q^2/ν finite. The functions W_1 and W_2 might be expected to have a complicated dependence on ν and Q^2 since they contain all of the hadronic structure information in the approximation we are employing. Remarkably, in the deep inelastic regime the structure functions display *scaling behavior:* to a good approximation they are functions only of the *ratio,* Q^2/ν (Bjorken, 1969). Figure 10.9 exhibits scaling for a broad range of Q^2 and the quantity x defined by

$$x = \frac{Q^2}{2M\nu} = \frac{Q^2}{2p\cdot q} \qquad (0 \leqslant x \leqslant 1), \tag{10.13}$$

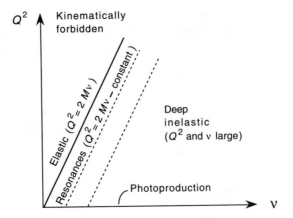

Fig. 10.8   Physical regions in the Q^2 and ν variables.

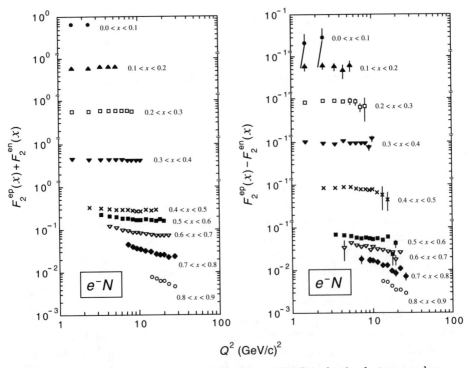

Fig. 10.9   Structure functions derived from SLAC inelastic electron–nucleon data with recoil mass greater than 2 GeV and $Q^2 > 1$ (GeV/c)^2; it was assumed that $R \equiv \sigma_L/\sigma_T = 0.21$ (see Fig. 10.12). Scaling corresponds to horizontal lines. Data compiled in Aguilar-Benitez et al. (1986).

with the structure functions scaling as

$$MW_1(Q^2, \nu) \longrightarrow F_1(x) \qquad \nu W_2(Q^2, \nu) \longrightarrow F_2(x)$$
$$Q^2 \to \infty \qquad \nu \to \infty \qquad x = \text{finite}, \tag{10.14}$$

where $F_1(x)$ and $F_2(x)$ are finite functions of only the ratio of Q^2 and ν. In Fig. 10.9 the sum and difference of experimental proton and neutron structure functions at fixed values of x are seen to be independent of Q^2, except possibly for $x \simeq 0$ and for $x \simeq 1$.

This characteristic of deep inelastic scattering is quite different from that for hadronic elastic scattering or resonance production, where the formfactors diminish rapidly at large Q^2 (see the discussion in Ch. 8 of Halzen and Martin, 1984). Such scaling behavior implies the absence of a mass scale (x is dimensionless), and is the signal to be expected for elastic scattering of the electron from pointlike constituents of the proton (Close, 1979, Chs. 9 and 11).

EXERCISE 10.2 Show that the virtual photon in deep inelastic electron scattering is spacelike.

EXERCISE 10.3 (a) Use the integral representation of the δ-function, completeness, and that translations in space and time are generated by the momentum and energy operators to prove that (10.9) can be written in the form

$$W^{\mu\nu} \simeq \sum_\sigma \int d^4x \, e^{iq \cdot x} \, \langle p\sigma | \, J^\mu(x) J^\nu(0) \, | p\sigma \rangle .$$

(b) Prove that (10.11) is the most general form allowed for the hadronic tensor in (10.7). *Hint:* recall that (10.7) is spin-averaged and that $L^{\mu\nu}$ is symmetric; impose Lorentz invariance and then charge conservation.

10.4 Partons

The physical content of Bjorken scaling can be grasped most intuitively in terms of the *parton model* (Feynman, 1972). In this model the hadrons consist of elementary pointlike constituents (partons), and deep inelastic lepton scattering from hadrons corresponds to *incoherent elastic scattering* of leptons from constituent partons through the exchange of a virtual photon (Fig. 10.10). The subsequent interactions of struck and residual partons are assumed to produce the hadronic debris.

In this picture the Bjorken scaling variable x may be interpreted as the fraction of the hadronic momentum carried by the struck parton, with $0 \leqslant x \leqslant 1$ (see Exercise 10.5b). An economical guess is to identify the partons with the valence quarks. However, the data require that hadrons contain charged "an-

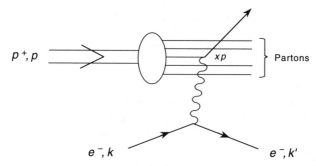

Fig. 10.10   Elastic scattering from partons as the mechanism for deep inelastic scattering of electrons from protons. The 4-momentum carried by the proton and the struck parton are p and xp, respectively.

tipartons" in addition to charged partons, and also neutral partons. The QCD gauge bosons (gluons) will be the logical candidates for the neutral partons, and this identification will lead naturally to the occurrence of antipartons through particle–hole excitation of the quark Fermi sea by gluons.

In the parton model the onset of Bjorken scaling is understood in terms of *elastic* scattering from the structureless quarks at sufficiently large Q^2 (high resolution), and deviations from scaling as Q^2 is increased further may be interpreted as either the effect of the gluons, or evidence for new internal structure of the quarks themselves. The latter possibility will be discounted in our considerations as there is no experimental evidence clearly requiring it; for an introduction to the theoretical idea that quarks (and leptons) might be composed of more fundamental *preons*, see Collins, Martin, and Squires (1989), Ch. 9.

EXERCISE 10.4   (a) Show that for the momenta defined in Fig. 10.7, $\nu \equiv p \cdot q/M$ measures the energy transfer in the lab frame and $y \equiv p \cdot q/p \cdot k$ measures the inelasticity of the reaction. (b) Show that the allowed ranges of the variables x and y in eq. (10.15) are $0 \leqslant x \leqslant 1$ and $0 \leqslant y \leqslant 1$.

EXERCISE 10.5   (a) Demonstrate that elastic scattering of electrons from hadrons and hadronic resonance production lie on the lines depicted in Fig. 10.8. Where would the scattering of real photons lie? (b) Prove that in the Bjorken limit x is the momentum fraction carried by a parton.

The parton model is basically an *impulse approximation* (the assumption of incoherence). The conditions for its validity are that[‡]

[‡]The impulse approximation is often used in intermediate energy nuclear physics for, say, the scattering of a proton from the individual nucleons of a nucleus. However, in that case the "partons" are the struck nucleons. They can recoil from the nucleus (no confinement), making the neglect of final-state interactions less problematic than for quarks.

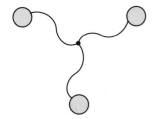

Fig. 10.11   Heuristic example of confinement.

1. The parton–parton interaction is negligible during the time of interaction between a parton and the electromagnetic current.
2. Final-state interactions associated with the confinement mechanism do not influence the interaction of the virtual photon with the parton because they act at much larger spacetime distances.

This division is technically useful but it is apparent that these conditions can be reduced to the single requirement that the photon–parton scattering not be influenced by the presence of other partons. A mechanical model for which this is satisfied is shown in Fig. 10.11, where confinement is imposed by the elastic and massless strings, but the motion is essentially free for the balls at separations less than the confinement radius. This is intended only as a suggestive model, but we will see later from the discussion of QCD, bag models, and string models (Ch. 12) that the picture of Fig. 10.11 may be at least a caricature of the truth, provided the strings are interpreted as flux tubes of the gluon fields (*chromoelectric flux*) and the balls as quarks.

> The parton model is more properly formulated in the *infinite momentum frame* using the following reasoning. The proton is assumed to exist in a virtual state of free pointlike objects. The lifetime of such a virtual state is inversely proportional to the energy difference between the virtual state and the proton ground state. This difference is large, possibly infinite, because of the confinement mechanism, but if the proton is moving arbitrarily fast with respect to the observer (the electron) this lifetime can be made arbitrarily long by time dilation. The energy transfer ν controls the characteristic time for absorption of the virtual photon, and this time is also dilated in going to the infinite momentum frame. However, consideration of the details shows that in the deep inelastic kinematic regime the parton state lifetime is much longer than the time of interaction with the virtual photon [see, for example, Bjorken and Paschos (1969); Leader and Predazzi (1982), Ch. 14.]
>
> We will proceed with discussion of the parton model in the rest frame of the proton, but it should be realized that the concepts to be employed are strictly valid only in a Lorentz frame in which the proton moves with infinite velocity with respect to the observer. Accordingly, the partons of the infinite-momentum frame need not have a simple interpretation in the rest frame of the proton. Only in the infinite-momentum frame can $q(x)dx$ be strictly interpreted as the probability of finding the quark carrying a fraction between x and $x + dx$ of the total hadronic momentum. Because

the lifetime of the virtual parton state is estimated to be much longer than the interaction time with the virtual photon, our discussion should not be far in error.

10.5 The Quark–Parton Model

If we identify the charged partons with the quarks, the *Quark–Parton Model* (QPM) results. For charged-lepton deep inelastic scattering the cross section may be written

$$\frac{d^2\sigma}{dx\,dy} = \left(\frac{2M\nu^2}{y}\right)\frac{d^2\sigma}{dQ^2 d\nu}$$

$$= \frac{4\pi\alpha^2 s}{Q^4}\left[(1-y)F_2(x) + xy^2 F_1(x)\right], \tag{10.15}$$

where a term of order M/E has been neglected, we employ the following kinematic variables

$$Q^2 = -q^2 \qquad \text{(negative square of 4-momentum transfer)}$$

$$x = \frac{Q^2}{2M\nu} \qquad \text{(scaling variable)}$$

$$y = \frac{q\cdot p}{k\cdot p} = \frac{\nu}{E} \qquad \text{(fractional energy transfer; } E \text{ is lab energy)}$$

$$\sqrt{s} = E_{\text{CM}} = \left(2ME + 2M^2\right)^{1/2}$$

$$\simeq \sqrt{2ME} \qquad \text{(center-of-mass energy)},$$

and the structure functions are [see (10.14)]

$$F_1 = MW_1 \qquad F_2 = \nu W_2.$$

10.5.1 Longitudinal and Transverse Momentum

The structure functions are related to the cross sections for longitudinal–scalar (L) and transverse (T) virtual photons by

$$\frac{\sigma_{\text{L}}}{\sigma_{\text{T}}} = \frac{F_2\left(1 + \dfrac{Q^2}{\nu^2}\right) - 2xF_1}{2xF_1}. \tag{10.16}$$

Some measured values of this ratio are shown in Fig. 10.12. Except for small

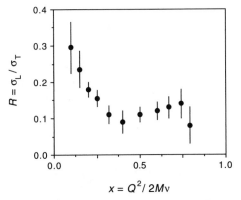

Fig. 10.12 Ratio (10.17) for longitudinal–scalar and transverse photons in deep inelastic scattering (Close, 1979).

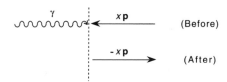

Fig. 10.13 Breit frame for collision between photon and a parton carrying momentum fraction xp.

values of x we see that $\sigma_L/\sigma_T \simeq 0.1$. Consider the *Breit frame* ("brick-wall frame") illustrated in Fig. 10.13, in which the photon and parton are collinear: the photon transfers no energy, but the parton has its 3-momentum reversed in the interaction. In this frame $\sigma_L/\sigma_T = 0$ for spin-$\frac{1}{2}$ partons (Exercise 10.1). This result survives a Lorentz boost to the infinite-momentum frame only if transverse momentum can be neglected. Therefore, it is reasonable to assume that the nonzero values of σ_L/σ_T in Fig. 10.12 result primarily from Fermi motion of the quarks.

If we allow a finite effective mass m for the parton and an initial transverse momentum k_T from Fermi motion, for spin-$\frac{1}{2}$ partons (Close, 1979)

$$\frac{\sigma_L}{\sigma_T} \simeq \frac{4\left(k_T^2 + m^2\right)}{Q^2}. \tag{10.17}$$

Neglecting the quark mass and Fermi motion and going to the deep inelastic limit we have

$$\frac{\sigma_L}{\sigma_T} \to 0 \qquad Q^2 \to \infty$$

$$\nu \to \infty \qquad \frac{Q^2}{\nu} = \text{finite},$$

and eq. (10.16) implies

$$F_2(x) = 2xF_1(x). \tag{10.18}$$

This result is called the *Callan–Gross relation* (Callan and Gross, 1969); it is a direct consequence of the spin-$\frac{1}{2}$ nature of the charged partons, and it tells us that the structure functions F_1 and F_2 are not independent if the masses and transverse momenta of the partons can be neglected in the deep inelastic regime.

10.5.2   Quark–Parton Structure Functions

In the quark–parton model the incoherence assumption allows the structure functions to be expressed in terms of summations over distribution functions for individual quarks and antiquarks. For example, letting Q_α denote the charge of a flavor α quark and $\alpha(x)$ its momentum distribution, the proton structure function may be written

$$
\begin{aligned}
F_2^{\mathrm{ep}}(x) =& \nu W_2(x) \\
=& x\left[Q_u^2\big(u(x) + \bar{u}(x)\big) + Q_d^2\big(d(x) + \bar{d}(x)\big)\right. \\
& \left. + Q_s^2\big(s(x) + \bar{s}(x)\big) + \ldots\right] \\
=& \tfrac{4}{9}x\big[u(x) + \bar{u}(x)\big] + \tfrac{1}{9}x\big[d(x) + \bar{d}(x)\big] \\
& + \tfrac{1}{9}x\big[s(x) + \bar{s}(x)\big] + \ldots,
\end{aligned} \tag{10.19a}
$$

where the ellipses stand for charmed and heavier quarks that will be ignored in the subsequent discussion. The corresponding neutron structure function may be obtained by an isospin rotation, $u \leftrightarrow d$ and $p \leftrightarrow n$:

$$
\begin{aligned}
F_2^{\mathrm{en}}(x) =& \tfrac{4}{9}x\left[d(x) + \bar{d}(x)\right] + \tfrac{1}{9}x\big[u(x) + \bar{u}(x)\big] \\
& + \tfrac{1}{9}x\big[s(x) + \bar{s}(x)\big] + \ldots \quad .
\end{aligned} \tag{10.19b}
$$

In *both* (10.19a) and (10.19b) the quantities $u(x)$, $d(x), \ldots$ mean the quark distribution functions evaluated for the *proton*.

It will sometimes prove useful to divide the quarks into valence quarks (v) and a sea (s) of $q\bar{q}$ pairs. For example,

$$
\begin{aligned}
F_2^{\mathrm{ep}}(x) =& \tfrac{4}{9}xu_{\mathrm{v}}(x) + \tfrac{1}{9}xd_{\mathrm{v}}(x) \quad\quad \text{(valence)} \\
& + \tfrac{4}{9}x\big[\bar{u}_{\mathrm{s}}(x) + u_{\mathrm{s}}(x)\big] + \tfrac{1}{9}x\left[\bar{d}_{\mathrm{s}}(x) + d_{\mathrm{s}}(x)\right] \\
& + \tfrac{1}{9}x\big[s(x) + \bar{s}(x)\big] + \ldots \quad\quad \text{(sea)},
\end{aligned} \tag{10.19c}
$$

where the valence quarks are u and d such that uud carries the proton quantum numbers, and the sea quarks come from processes such as

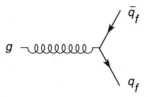

Since the gluons are assumed flavorless, we expect as many sea quarks q_f of flavor f as sea antiquarks $\bar{q}_f$. In addition, the gluons are emitted with a probability $P \simeq dx/x$ (bremsstrahlung momentum spectrum), which implies that sea quarks should be more important at small values of x. Therefore, in the partition (10.19c) we expect qualitatively that valence quarks should dominate for $x \to 1$ and sea quarks for $x \to 0$ (see Fig. 10.15).

10.5.3 Inclusion of Neutrino Data

A more complete analysis is afforded by studying deep inelastic neutrino scattering reactions such as

$$\nu_\mu + N \to \mu^- + \text{hadrons}.$$

The weak cross sections for ν and $\bar{\nu}$ scattering from isoscalar targets can be written in the convenient form

$$\frac{d^2(\sigma^\nu + \sigma^{\bar{\nu}})}{dx\,dy} = \frac{G^2 s}{\pi}\left[(1-y)F_2^\nu + xy^2 F_1^\nu\right] \tag{10.20}$$

$$\frac{d^2(\sigma^\nu - \sigma^{\bar{\nu}})}{dx\,dy} = \frac{G^2 s}{2\pi} x F_3^\nu \left(2y - y^2\right), \tag{10.21}$$

where G is the Fermi coupling constant, and we have ignored terms of order M/E and neglected Q^2 relative to the square of the weak vector boson mass. The cross sections for these charged-current processes depend on *three* structure functions: F_1^ν, F_2^ν, and F_3^ν. The additional structure function F_3^ν relative to (10.15) is a consequence of the $V-A$ neutrino current that enters $L_{\mu\nu}$ in (10.7); this allows a parity-violating term in the hadronic tensor $W_{\mu\nu}$ that is absent for electromagnetic scattering (see Exercise 10.3b). Assuming we are below charm threshold and setting the Cabibbo angle to zero, the neutrino relations corresponding to (10.19) are

$$
\begin{aligned}
F_2^{\nu p}(x) &= F_2^{\bar\nu n}(x) = 2x\big[d(x) + \bar u(x)\big] \\
F_2^{\nu n}(x) &= F_2^{\bar\nu p}(x) = 2x\big[u(x) + \bar d(x)\big] \\
F_3^{\nu p}(x) &= F_3^{\bar\nu n}(x) = 2\big[d(x) - \bar u(x)\big] \\
F_3^{\nu n}(x) &= F_3^{\bar\nu p}(x) = 2\big[u(x) - \bar d(x)\big],
\end{aligned}
\tag{10.22}
$$

where n and p denote neutrons and protons, respectively. Neutral-current processes may also be studied, but we do not reproduce the relevant formulas.

10.5.4 Significance of the Structure Functions

For lepton scattering from isoscalar targets (10.19) and (10.22) give (see Exercise 10.6)

$$
\frac{F_2^e}{F_2^\nu} \equiv \frac{\tfrac12\big(F_2^{en} + F_2^{ep}\big)}{\tfrac12\big(F_2^{\nu n} + F_2^{\nu p}\big)} = \frac{5}{18}\left[\frac{(q+\bar q) + \tfrac25(s+\bar s)}{q+\bar q}\right],
\tag{10.23}
$$

where the factor $\frac{5}{18}$ is just the average of the squared charges for the u and d quarks and

$$
\begin{aligned}
q &= q(x) \equiv u(x) + d(x) \\
\bar q &= \bar q(x) \equiv \bar u(x) + \bar d(x).
\end{aligned}
$$

The distribution function $u(x)$ is that for a u-quark in the proton or a d-quark in the neutron (isospin invariance), and $d(x)$ refers to a d-quark in the proton or a u-quark in the neutron, with analogous correspondences for the antiquarks. Thus, if strange quarks can be ignored,

$$
F_2^e = \tfrac{5}{18} F_2^\nu.
\tag{10.24}
$$

There is strong experimental support for this relationship when $x \gtrsim 0.1$, indicating that the strange-quark content of nucleons is small, except possibly near $x = 0$ where they may be produced by sea quark excitation (see Fig. 10.14). From (10.22), we see that $F_2^\nu \simeq \tfrac{18}{5} F_2^e$ measures the *sum* of quark and antiquark distributions:

$$
F_2^\nu \equiv \tfrac12\big(F_2^{\nu n} + F_2^{\nu p}\big) = x\,(q+\bar q),
\tag{10.25}
$$

and xF_3^ν measures the *difference* of quark and antiquark distributions:

$$
xF_3^\nu \equiv \tfrac12\big(xF_3^{\nu p} + xF_3^{\nu n}\big) = x\,(q-\bar q).
\tag{10.26}
$$

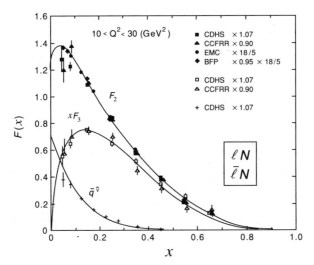

Fig. 10.14 Structure functions F_2 and xF_3^ν, and sea quark distributions $\bar{q}^\nu \equiv$ $x(\bar{u} + \bar{d} + 2\bar{s})$, from electron and neutrino data. Data are for fixed Q^2 vs. x, and $R = \sigma_L/\sigma_T = 0$ was assumed. Data compiled in Aguilar-Benitez et al. (1986).

[Recall also the Callan–Gross relation, $F_2(x) = 2xF_1(x)$; to the extent that it is satisfied, the neutrino scattering is independent of F_1^ν.] If the quarks are split into valence quarks q_v and sea quarks q_s, one finds in Exercise 10.6 that

$$F_2^\nu - xF_3^\nu = 2xq_s \qquad xF_3^\nu = xq_v. \tag{10.27}$$

That is, $F_2^\nu - xF_3^\nu$ measures the momentum distribution of the sea quarks, while xF_3^ν measures the momentum distribution of the valence quarks. Examples of the structure functions $F_2(x)$ and $F_3(x)$ are shown in Fig. 10.14, and a qualitative interpretation of the shape of $F_2(x)$ is illustrated in Fig. 10.15.

Some consequences of this quark–parton picture may readily be extracted by requiring that momentum-integrated parton distributions reproduce the proton quantum numbers. The number of valence quarks is (Flügge, 1982)

$$N_v = \int_0^1 (q - \bar{q})\,dx = \int_0^1 \frac{xF_3^\nu}{x}\,dx \simeq 3.2 \pm 0.5, \tag{10.28}$$

which compares favorably with the quark model value $N_v = 3$. The momentum fraction carried by the valence quarks is

$$x_v = \int_0^1 x(q - \bar{q})\,dx = \int_0^1 xF_3^\nu\,dx \simeq 0.32, \tag{10.29}$$

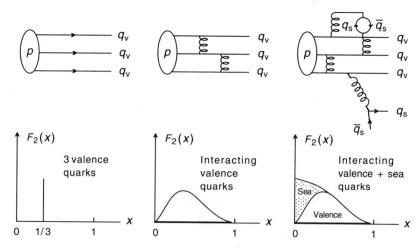

Fig. 10.15   Structure function at three levels of interaction in the quark–parton model. Non-interacting valence quarks would each carry about a third of the proton momentum (left figure). Gluon exchange between valence quarks introduces a spread in the momentum distribution of the quark interacting with the virtual photon (middle figure). Particle–hole excitation of the vacuum by virtual gluons produces sea quarks, which are favored at low x (right figure).

the momentum fraction carried by sea quarks is

$$x_s = \int_0^1 x\,(2\bar{q})\,dx = \int_0^1 (F_2 - xF_3^\nu)\,dx \simeq 0.13, \qquad (10.30)$$

and the momentum fraction carried by all quarks is

$$x_q = x_v + x_s = \int_0^1 F_2(x)\,dx \simeq 0.45. \qquad (10.31)$$

This result is the basis for the earlier contention that approximately 55% of the proton momentum is carried by uncharged partons, presumably to be identified with the gluons.

   This completes our introductory survey of strong interaction phenomenology, and of the evidence for fundamental nucleonic constituents provided by leptonic probes. We now turn to the discussion of a non-Abelian gauge field theory that appears capable of accounting for these phenomena.

EXERCISE 10.6   (a) Derive the structure function results of eqns. (10.23) and (10.25)–(10.27). (b) Show that in the quark–parton model the structure functions satisfy $\frac{1}{4} \leqslant F_2^{en}/F_2^{ep} \leqslant 4$.

10.6 Quantum Chromodynamics

Quantum chromodynamics (QCD) is based on two premises: (1) the description of strong interactions requires a *local gauge theory* (a Yang–Mills field); (2) the symmetry to be gauged is an exact $SU(3)$ symmetry based on the *color degree of freedom* carried by quarks and gluons. Where necessary to avoid confusion we will denote this symmetry by $SU(3)_c$, which should not be confused with the approximate global flavor symmetry $SU(3)_f$. The adjoint representation of $SU(3)$ is of dimension $N^2 - 1 = 8$, so we expect eight gauge vector bosons; these are the gluons. It is assumed that the gluons see color but not flavor. Thus each of the different flavor quarks experiences the same strong interactions, except for effects associated with quark mass differences.

As they are the gauge quanta of a Yang–Mills field, the gluons must be *massless*. However, no long-range fields are observed in conjunction with the strong interactions, which presents phenomenological problems. There are two possible solutions: the first takes a cue from the electroweak theory and invokes spontaneous symmetry breaking in the Higgs mode to give the gluons mass; the second conjectures that the nonlinear nature of the color gauge fields leads to a spatial confinement of the gluons and quarks. The confinement problem of QCD is not amenable to perturbative solution, but there are strong hints from non-perturbative lattice gauge calculations that local $SU(3)_c$ is confining (see Chs. 13–14). This, coupled with the absence of experimental candidates for free massive gluons and the fact that QCD with spontaneously broken symmetry would not be asymptotically free because of the self-coupling of the Higgs scalars, strongly favors the confinement alternative.

10.6.1 Lagrangian Density

The QCD Lagrangian density is

$$\mathcal{L}_{\mathrm{QCD}} = \sum_f \bar{\psi}_f \left(i\!\not{D} - m_f \right) \psi_f - \tfrac{1}{4} F^j_{\mu\nu} F^{\mu\nu}_j, \tag{10.32}$$

where the covariant derivative is (7.6)

$$D_\mu = \partial_\mu + \tfrac{i}{2} g \lambda_\ell A^\ell_\mu \qquad \not{D} \equiv \gamma^\mu D_\mu, \tag{10.33}$$

λ_ℓ denote Gell-Mann $SU(3)$ matrices [see eq. (5.63)], A^ℓ_μ represents the space-time component μ of the ℓth gauge boson vector potential ($\ell = 1, 2, 3, \ldots, 8$), f is a flavor label, and g is the gauge coupling constant. The non-Abelian gluon field tensor is defined by (7.10)

$$F^j_{\mu\nu} = \partial_\mu A^j_\nu - \partial_\nu A^j_\mu - g f_{jk\ell} A^k_\mu A^\ell_\nu, \tag{10.34}$$

where the structure constants $f_{jk\ell}$ of $SU(3)$ have been given in (5.64). The color-triplet quarks are described by a composite spinor for each flavor f,

$$\psi_f = \begin{pmatrix} q_{\text{red}} \\ q_{\text{blue}} \\ q_{\text{green}} \end{pmatrix}_f \equiv \begin{pmatrix} q_1 \\ q_2 \\ q_3 \end{pmatrix}_f \equiv \begin{pmatrix} R \\ B \\ G \end{pmatrix}_f . \tag{10.35}$$

A single gauge coupling constant g describes the coupling of the gluons to all quark flavors in (10.32), so the only breaking of flavor symmetry in the QCD Lagrangian comes through inequivalent masses m_f for different flavor quarks. We assume that these masses come from fermion coupling to the phenomenological Higgs scalar in the electroweak gauge sector (§9.2), so their origin lies outside QCD.

10.6.2 QCD Symmetries

Before proceeding we list the important symmetries of the QCD Lagrangian density (Wilczek, 1982):

1. Invariance under the discrete symmetries P, C, and T (but see §10.6.7).
2. Invariance under global phase rotations $\psi_f \to e^{i\theta_f}\psi_f$, where f denotes quark flavors. This implies conservation of the number of quarks of flavor f, in turn implying conservation of strangeness, baryon number, third component of isospin, electrical charge, charm, ... (see Exercise 5.6a).
3. Approximate flavor symmetry if the differences in quark masses m_f can be ignored; then $\mathcal{L}_{\text{QCD}}$ is invariant under transformations of the form $\psi_j = U_{jk}\psi_k$, where U_{jk} is a unitary matrix acting on the flavor indices. The existence of the relatively light quarks u and d leads to very good isospin symmetry, while the addition of the somewhat heavier strange quark s yields approximate $SU(3)_{\text{f}}$ flavor symmetries (see Exercise 12.8).
4. To the degree that the quark masses can be ignored, $\mathcal{L}_{\text{QCD}}$ may be separated into terms involving only left-handed quarks and terms involving only right-handed quarks (§6.2.2):

$$q_{\text{L}} \equiv \tfrac{1}{2}(1 - \gamma_5)q \qquad q_{\text{R}} \equiv \tfrac{1}{2}(1 + \gamma_5)q.$$

If we limit attention to the three lightest flavors, this implies an ungauged symmetry of the Lagrangian density under the direct product of global $SU(3)$ flavor rotations on the left- and right-handed fields. This invariance under $SU(3)_{\text{L}} \times SU(3)_{\text{R}}$ is called *chiral symmetry*, and will be discussed in Ch. 12. It is an important approximate symmetry of the strong interactions, and mesons like π and K are thought to be approximate Goldstone bosons resulting from the spontaneous breaking of this global symmetry.

The QCD Lagrangian density restricted to massless u, d, and s quarks actually has a larger symmetry

$$SU(3)_L \times SU(3)_R \times U(1)_V \times U(1)_A,$$

where the $U(1)$ symmetries are generated by the transformations

$$U(1)_V : \quad q_i \rightarrow e^{i\theta} q_i \qquad U(1)_A : \quad q_i \rightarrow e^{i\gamma_5 \theta} q_i.$$

The $U(1)_V$ symmetry is associated with the observed baryon current, but there is no obvious manifestation of the $U(1)_A$ symmetry in the hadronic spectrum (no parity doubling of hadronic states—see §12.3.2). If instead $U(1)_A$ is broken spontaneously it should produce a pseudoscalar Goldstone boson comparable in mass with the pion (see §12.3 and Exercise 8.3). The mesons $\eta(549)$ and $\eta'(958)$ are the obvious candidates, but they are too heavy. This is termed the "$U(1)$ problem". The existence of instantons (see §13.7) solves this problem, but the instantons bring their own difficulties (§13.7.5). A more complete discussion of the $U(1)$ problem may be found in Huang (1982), §12.5 and §12.6, and Cheng and Li (1984), §16.3.

The explicit breaking of chiral invariance in the QCD Lagrangian restricted to the lightest quarks comes through the term (Exercise 12.8)

$$\mathcal{L}' = m_u \bar{u}u + m_d \bar{d}d + m_s \bar{s}s. \tag{10.36}$$

(Recall from Exercise 6.2e that $\bar{\psi}\psi$ is not chiral invariant.) When this result is used to construct mass ratio formulas for physical particles, one finds (Gasiorowicz and Rosner, 1981)

$$m_u \simeq 4.2 \text{ MeV} \qquad m_d \simeq 7.5 \text{ MeV} \qquad m_s \simeq 150 \text{ MeV}. \tag{10.37}$$

These masses are called the *current masses* of the quarks; they are the explicit masses appearing in the Lagrangian of the field theory and are considerably smaller than the *constituent masses* for the lightest quarks

$$M_u \simeq 300 \text{ MeV} \qquad M_d \simeq 300 \text{ MeV} \qquad M_s \simeq 500 \text{ MeV}, \tag{10.38}$$

which may be derived from quantities such as magnetic moments in nonrelativistic quark models (Ch. 12).

At a qualitative level this difference should not be particularly disturbing. The *effective mass* of a particle is always determined by its interactions, and the masses of the constituent quarks presumably reflect a dressing by the confinement mechanism of the current quarks appearing in the Lagrangian. An understanding of the relationship between current masses and constituent masses awaits a first-principles solution of the QCD bound-state problem.

Empirically, the current and constituent quark masses differ by a constant of about 300 MeV for the three lightest quarks; presumably this constant shift is related in some way to the QCD scale parameter Λ discussed below.

Therefore, present theory supplies little fundamental understanding of either the origin of current masses or of how these current masses are dressed to effective constituent quark masses by the strong interactions. The former requires a microscopic understanding of the Higgs field; the latter, a full solution of non-perturbative QCD. Neither is currently available and the origin of mass remains a major unsolved problem of modern physics.

10.6.3   Feynman Rules for QCD

Feynman rules for QCD may be constructed using the standard methods, starting from the Lagrangian density (10.32). They are more complicated than for QED because of color indices, gluon self-couplings, and the necessity of ghosts in covariant gauges; however, they are simpler than those of the electroweak theory because of the absence of scalar field couplings. The basic QCD vertices and propagators are shown in Fig. 10.16.

10.6.4   Asymptotic Freedom and Confinement

The running coupling constant in QED is [eq. (6.68)]

$$\alpha(Q^2) = \frac{\alpha(m^2)}{1 - \dfrac{\alpha(m^2)}{3\pi} \log\left(\dfrac{Q^2}{m^2}\right)}, \tag{10.39}$$

where $Q^2 = -q^2 >> m^2$. The dependence on Q^2 comes from vacuum polarization graphs such as

In QCD there are similar graphs associated with gluon vacuum polarization

but because of the nonlinear gauge field there are also graphs like

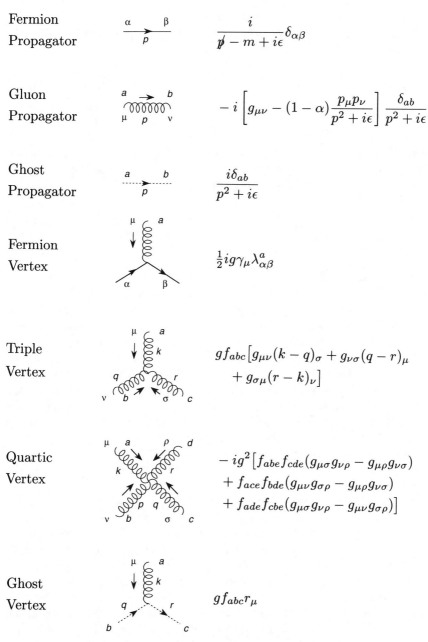

Fermion Propagator		$\dfrac{i}{\not{p} - m + i\epsilon}\delta_{\alpha\beta}$
Gluon Propagator		$-i\left[g_{\mu\nu} - (1 - \alpha)\dfrac{p_\mu p_\nu}{p^2 + i\epsilon}\right]\dfrac{\delta_{ab}}{p^2 + i\epsilon}$
Ghost Propagator		$\dfrac{i\delta_{ab}}{p^2 + i\epsilon}$
Fermion Vertex		$\tfrac{1}{2}ig\gamma_\mu\lambda^a_{\alpha\beta}$
Triple Vertex		$gf_{abc}\big[g_{\mu\nu}(k - q)_\sigma + g_{\nu\sigma}(q - r)_\mu + g_{\sigma\mu}(r - k)_\nu\big]$
Quartic Vertex		$-ig^2\big[f_{abe}f_{cde}(g_{\mu\sigma}g_{\nu\rho} - g_{\mu\rho}g_{\nu\sigma}) + f_{ace}f_{bde}(g_{\mu\nu}g_{\sigma\rho} - g_{\mu\rho}g_{\nu\sigma}) + f_{ade}f_{cbe}(g_{\mu\sigma}g_{\nu\rho} - g_{\mu\nu}g_{\sigma\rho})\big]$
Ghost Vertex		$gf_{abc}r_\mu$

Fig. 10.16 Feynman rules for QCD (Buras, 1980). The parameter α in the gluon propagator specifies the gauge; the indices α and β label quark colors; a, b, c, d, e label gluon or ghost colors; $\lambda^a_{\alpha\beta}$ is an element of the color matrix λ^a; f_{abc} is an $SU(3)$ structure constant; μ, ν, ρ, σ denote Lorentz indices; and the 4-momenta are specified by p, k, q, and r.

that contribute to the vacuum polarization and to evolution of the coupling constant with Q^2. As a consequence, the running coupling constant in QCD takes the form

$$\alpha_s\left(Q^2\right) = \frac{g_s^2\left(Q^2\right)}{4\pi} = \frac{\alpha_s\left(\mu^2\right)}{1 + \left(\dfrac{\alpha_s\left(\mu^2\right)}{12\pi}\right)\left(33 - 2N_f\right)\log\left(\dfrac{Q^2}{\mu^2}\right)}, \tag{10.40}$$

where μ^2 is a reference momentum transfer, N_f is the number of accessible quark flavors, and $g_s(Q^2)$ is the (renormalized) gauge coupling in (10.33). This can also be written (Exercise 10.7)

$$\alpha_s(Q^2) = \frac{12\pi}{\left(33 - 2N_f\right)\log\left(\dfrac{Q^2}{\Lambda^2}\right)}, \tag{10.41}$$

where the *QCD scale parameter* Λ is to be determined experimentally.

Comparing (10.40) with (10.39), we see that as long as the number of quark flavors is less than 17 the QCD coupling has a qualitatively different behavior from that for QED: $\alpha_s(Q^2)$ tends to zero for large Q^2. This is the phenomenon of *asymptotic freedom*, which provides an immediate qualitative justification for Bjorken scaling and the success of the quark–parton model (Gross and Wilczek, 1973; Politzer, 1973). The first term in the factor $(33 - 2N_f)$ comes from the gluon loops, while the second one comes from fermion loops; it is the absence of the former contribution in electrodynamics that is responsible for the opposite Q^2 dependence of the QED coupling constant. The dominance of gluon loops over quark loops for a world with only a few quark flavors is related to the structure of the gauge group: there are only three colors of $SU(3)$ quarks (fundamental representation), but there are eight gluons (adjoint representation).

Equation (10.41) implies that at large distances (small Q^2) the coupling becomes large, perturbation theory breaks down, and (10.41) is no longer reliable. Nevertheless, the trend suggested by (10.41) and (more to the point) models and non-perturbative calculations to be discussed in Chs. 12–14, indicate that the growth in $\alpha_s(Q^2)$ at large distance may prevent the separation of color sources except with the expenditure of an infinite amount of energy. This conjectured property of QCD is called *confinement*; it would reconcile the absence of free quark and gluon asymptotic states with the success of the parton model. It is remarkable that the only non-pathological theories in four spacetime dimensions that are asymptotically free are non-Abelian gauge theories, and that such theories may exhibit both confinement at large distances and asymptotic freedom at short distances.

Renormalization schemes separate divergent integrals into a piece absorbed into a renormalization constant and a finite remainder, with this partition

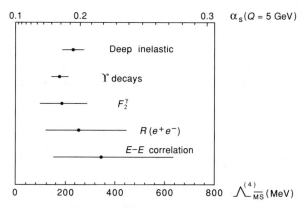

Fig. 10.17 QCD scale parameter as measured in various processes. Uncertainties are experimental only. Each point represents an average of results for the process (from the compilation by Aguilar-Benitez et al., 1988). A discussion of various determinations of Λ may be found in Duke and Roberts (1985).

scheme-dependent.[‡] The QCD scale parameter Λ depends on the finite remainder, so it depends on the scheme in use. The relation of scale parameters from different renormalization prescriptions requires a comparison of detailed calculations in each scheme (Celmaster and Gonsalves, 1979; Hasenfratz and Hasenfratz, 1980). One renormalization scheme commonly employed in phenomenological QCD is called *modified minimal subtraction* (Bardeen et al., 1978), with the corresponding scale parameter denoted by $\Lambda_{\overline{\text{MS}}}^{(n)}$, where n is the number of quark flavors included in the calculations. Figure 10.17 shows a compilation of the average values of various experimental determinations of this parameter. From this we conclude that

$$\Lambda_{\overline{\text{MS}}}^{(4)} = 200 \, {}^{+150}_{-80} \text{ MeV} \tag{10.42}$$

[or $\alpha_s(Q = 5 \text{ GeV}) \simeq 0.18$], which defines empirically the momentum scale on which the strong interactions become strong.

Thus interactions in QCD mean that the scale-invariant classical theory implied by (10.32) with massless quarks does not lead to a scale-invariant quantum theory. An unphysical dimensionless bare coupling conspires with an unphysical mass cutoff to produce a physical scale with mass dimensions. This phenomenon is called *dimensional transmutation:* particles develop a

---

[‡]A bare coupling may be separated into a finite term (renormalized coupling) and an infinite counterterm. This division is arbitrary: the physical theory is invariant under a reparameterization that transfers a finite amount between these two terms. The essence of *renormalization group invariance* is that the physics is unchanged by such redefinitions.

mass dynamically in a renormalizable theory that lacks an intrinsic mass scale at the classical level (Coleman and Weinberg, 1973). This is the physics of the parameter $\Lambda \simeq 200$ MeV appearing in (10.41), and presumably is the reason that a proton with a mass of about 1 GeV (three quarks of effective mass ~ 300 MeV) can appear in a theory with no classical mass scale.

EXERCISE 10.7 Show that the QCD running coupling (10.40) can be written in the form (10.41), which involves only one parameter.

10.6.5 Scaling Violations

Scaling is observed to be only an approximate property of deep inelastic scattering. This is expected in a field theory such as QCD where the partons are dressed by a cloud of virtual gluons and $q\bar{q}$ pairs. As Q^2 increases the probe sees the bare parton resolved more and more from its cloud of virtual particles. The virtual particles carry part of the momentum of the dressed quarks, so as Q^2 grows the structure functions no longer scale and are depleted in higher values of x. Schematically, we expect the behavior shown in Fig. 10.18, as a consequence of the processes illustrated in Fig. 10.19 (Söding and Wolf, 1981; Alterelli and Parisi, 1977; Close, 1979; Quigg, 1983).

Some low-order QCD diagrams giving deviation from scaling in deep inelastic scattering are of the form

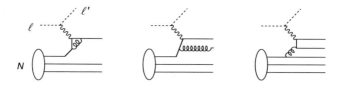

Calculation of such diagrams in perturbative QCD leads to a $\log Q^2$ leading-order deviation from scaling. This variation is slow and is obscured by other effects such as final-state interactions and heavy-quark production, but it is consistent with the experimental data shown in Fig. 10.20.

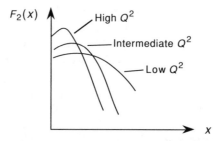

Fig. 10.18 Schematic violation of scaling in QCD. Generally, increasing Q^2 depletes high x and increases low x, with the area under the curve constant. In the limit $Q^2 \to \infty$ the structure function becomes sharply peaked at $x = 0$.

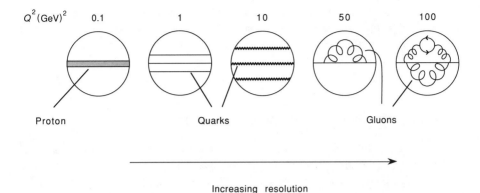

Increasing resolution

Fig. 10.19   The proton in an interacting field theory (after Quigg, 1983).

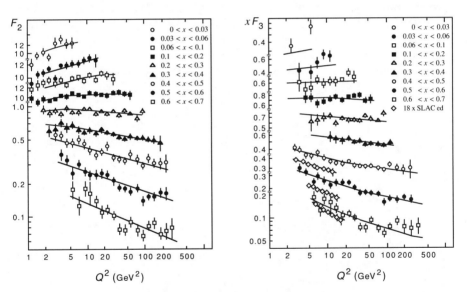

Fig. 10.20   Scaling violations in neutrino scattering. Data are measured by the CERN–Dortmund–Heidelberg–Saclay (CDHS) collaboration. The lines show a fit to the QCD evolution (Altarelli–Parisi) equations (see Altarelli and Parisi, 1977; Halzen and Martin, 1984, Ch. 10; Söding and Wolf, 1981). These equations allow an explicit computation of the Q^2 dependence of the structure functions. Scaling behavior (horizontal lines) is observed only for $x \simeq 0.2$.

In addition to these considerations, deviations from scaling behavior would be expected on a momentum scale sufficient to resolve quark substructure, if such were present. As noted previously, we assume the quarks to be structureless in all our discussions.

10.6.6   Experimental Tests of QCD

QCD is more difficult to test experimentally than the gauge theories we have discussed in preceding chapters. Nevertheless, there are some important experimental confirmations of its validity (Wilczek, 1982).

1. Two- and three-jet events in e^+e^- annihilation are described well by QCD calculations.
2. The logarithmic violation of scaling in deep inelastic lepton scattering (Fig. 10.20) is consistent with the QCD prediction.
3. Bound states of heavy quark–antiquark pairs (*quarkonium*; see §12.6) can annihilate when the q and $\bar{q}$ approach within a Compton wavelength of each other. The most important channel for states that are odd under charge conjugation is annihilation into three gluons (with subsequent hadronization), which is analogous to the three-photon decay of positronium (see §12.6.1):

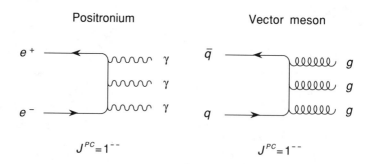

Positronium   Vector meson

$$J^{PC}=1^{--}\qquad\qquad J^{PC}=1^{--}$$

The predictions of QCD are in satisfactory agreement with data for the decays of the mesons J/ψ (charmed pair) and Υ (bottom pair).
4. The ratio R defined in eq. (10.5) for e^+e^- annihilation is modified by higher order diagrams:

In perturbative QCD the corresponding modification of (10.5b) is

$$R\left(q^2\right) = 3\sum_f Q_f^2 \left(1 + \frac{\alpha_{\mathrm{s}}\left(q^2\right)}{\pi} + \dots\right),\qquad(10.43)$$

where f is a flavor index. The data are consistent with the second term in the range of q^2 accessible to experiments.

10.6.7 QCD and CP Violation

QCD allows CP violation of undetermined magnitude in the strong interactions because of non-perturbative instanton effects (§13.7). One consequence is a predicted electric dipole moment of the neutron, which is known experimentally to have a very small upper limit. Two solutions, one passive and one active, have been proposed. The first is to ignore the problem until a more fundamental theory such as a GUT (see Ch. 11) sets the magnitude of the CP violation to be small. The second invokes a spontaneously broken global quark chiral symmetry that is also broken explicitly by instanton effects, resulting in the appearance of a nearly massless boson called the *axion* (Peccei and Quinn, 1977; Weinberg, 1978; Wilczek, 1978). Experimental searches have found no trace of the originally conjectured light boson. "Invisible axions", with smaller couplings and masses than the standard axion, are constrained but not eliminated by the data (Kim, 1979; Dine, Fischler, and Srednicki, 1981; Sikivie, 1982). The existence of axions would have important astrophysical implications (Turner, 1990); for example, there are speculations that large amounts of dark matter in the universe could be in the form of axions [see Primack, Seckel, and Sadoulet (1987) for a review]. We will return briefly to the question of strong CP violation in §13.7.5.

10.7 Background and Further Reading

A summary of quark and gluon properties may be found in Quigg (1983), Ch. 1; and in Close (1979). Deep inelastic lepton scattering is reviewed in Close (1979), Aitchison and Hey (1982), Perkins (1987), Drees and Montgomery (1983), and Quigg (1983). Our discussion of structure functions follows Flügge (1982) rather closely. The review article by Wilczek (1982) is the qualitative basis for much of the discussion of QCD; a concise explanation of QCD concepts is presented in Close (1982b). An introduction and references to the CP problem in QCD may be found in Cheng and Li (1984), Ch. 16; see also Collins, Martin, and Squires (1989), Ch. 5.

Physical processes in which the primary interaction involves QCD at large momentum transfer may be amenable to solution by perturbation theory, provided that two conditions are fulfilled: (1) the momentum transfer in the primary interaction must be sufficiently large that the interaction is asymptotically free (logarithmically decreasing coupling constant); (2) a "factorization theorem" must hold that allows hadronic amplitudes to be constructed as the convolution of a perturbatively tractable quark hard-scattering amplitude and a distribution amplitude for incoming and outgoing hadrons that incorporates non-perturbative dynamics. The calculations of Fig. 10.20 or eq. (10.43) are examples where these conditions appear to be satisfied. The status of perturbative QCD is discussed extensively in many places. See, for example, Collins and Soper (1987), Buras (1980), Muta (1987), Lepage and Brodsky (1980), Mueller (1981), Wu (1984), Halzen and Martin (1984), and Brodsky and Moniz (1986). We have deliberately paid only moderate attention to perturbative QCD for reasons of economy: most of the specific applications to be emphasized in Part III are dominated by non-perturbative effects and perturbative QCD may be of limited utility there.

CHAPTER 11

Grand Unified Theories

In the preceding chapters we have seen that the weak, electromagnetic, and strong interactions all appear to be described by local gauge theories. Neglecting gravitation, the fundamental interactions correspond to a gauge symmetry

$$SU(3)_c \times SU(2)_w \times U(1)_y \longrightarrow SU(3)_c \times U(1)_{\text{QED}},$$

where the Higgs field breaks the electroweak gauge group $SU(2)_w \times U(1)_y$ down to $U(1)_{\text{QED}}$ below a mass scale of order 100 GeV, but the color and charge symmetries remain exact. This observation leads naturally to speculation that this is just the last step in a chain of symmetry breakings originating in a deeper (gauge) symmetry that unifies all interactions at some enormous mass scale. The known or suspected elementary particles of the $SU(3)_c \times SU(2)_w \times U(1)_y$ Standard Model are listed in Table 11.1, along with some of their important characteristics. These particles and their properties must be accommodated in any deeper symmetry.

Since quantum gravitation is not well understood, we would like to exclude it from these considerations and concentrate only on the unification of the electromagnetic, weak, and strong interactions. By dimensional arguments it is expected that quantized gravity becomes important on a scale given by the Planck mass

$$M_P \simeq \sqrt{\frac{\hbar c}{G_N}} = 1.2 \times 10^{19} \text{ GeV/c}^2, \tag{11.1}$$

where G_N is the gravitational constant. Thus we ask whether it is possible that a single gauge symmetry can unify the non-gravitational interactions at energies much less than 10^{19} GeV and still be consistently broken down to $SU(3)_c \times SU(2)_w \times U(1)_y$ in the "low-energy regime" of $100 - 1000$ GeV. As we shall see, theories can be constructed that meet these criteria qualitatively; the pressing question is whether they can also be made consistent with quantitative low-energy observations, while maintaining an acceptable number of parameters that do not require unnatural fine-tuning. Theories unifying the non-gravitational interactions are called *grand unified theories* (GUTs), while those that attempt to include gravity as well are called *superunified theories*. We will discuss only grand unified theories; a simple introduction to theories

Table 11.1
Elementary Particles and Their Properties

Particle	Symbol	Charge	Mass (GeV)
Fermion Generation I			
Electron neutrino	ν_e	0	$< 1.8 \times 10^{-8}$
Electron	e	-1	0.511×10^{-3}
Up quark	u	$\frac{2}{3}$	5×10^{-3}
Down quark	d	$-\frac{1}{3}$	9×10^{-3}
Fermion Generation II			
Muon neutrino	ν_μ	0	$< 0.25 \times 10^{-3}$
Muon	μ	-1	0.106
Charm quark	c	$\frac{2}{3}$	1.25
Strange quark	s	$-\frac{1}{3}$	0.175
Fermion Generation III			
Tau neutrino	ν_τ	0	< 0.035
Tau	τ	-1	1.78
Top quark	t	$\frac{2}{3}$	> 90 (?)
Bottom quark	b	$-\frac{1}{3}$	4.5
Gauge Bosons			
Photon	γ	0	0
W boson	$W^\pm$	± 1	81.0 ± 1.3
Z boson	Z^0	0	92.4 ± 1.8
Gluon	g	0	0
Vacuum Scalars			
Higgs scalar	H	0	(?)

that aspire to incorporate gravitation in the grand synthesis may be found in Collins, Martin, and Squires (1989), Chs. 10–14.

We begin by noting that two general observations give a strong impetus to GUTs: (1) If we adopt a conservative and economical attitude and assume that the grand gauge group G operates on the familiar quarks and leptons of the low-energy world, gauge groups can readily be found that allow the particles listed in Table 11.1 to be assigned to low-dimensional representations. (2) There is evidence that the running coupling constants of the strong, weak, and electromagnetic interactions evolve with Q^2 in such a way that they will meet at an energy of approximately 10^{15} GeV.

11.1 Evolution of Coupling Constants

The evolution of the effective coupling constants with Q^2 is defined by the equations of the *renormalization group*. Denoting the running coupling constants of $U(1)_y$, $SU(2)_w$, and $SU(3)_c$, respectively, by $\alpha_1(Q^2)$, $\alpha_2(Q^2)$, and $\alpha_3(Q^2)$, we find in leading log or one-loop approximation that (see §6.4.4)

$$\frac{1}{\alpha_3(Q^2)} - \frac{1}{\alpha_3(M_x^2)} = -\frac{1}{4\pi}\left(11 - \frac{2}{3}N_f\right)\log\left(\frac{M_x^2}{Q^2}\right) \tag{11.2a}$$

$$\frac{1}{\alpha_2(Q^2)} - \frac{1}{\alpha_2(M_x^2)} = -\frac{1}{4\pi}\left(\frac{22}{3} - \frac{2}{3}N_f\right)\log\left(\frac{M_x^2}{Q^2}\right) \tag{11.2b}$$

$$\frac{1}{\alpha_1(Q^2)} - \frac{1}{\alpha_1(M_x^2)} = \frac{1}{4\pi}\left(\frac{2}{3}N_f\right)\log\left(\frac{M_x^2}{Q^2}\right), \tag{11.2c}$$

where M_x is the *unification mass*, N_f is the number of quark flavors energetically accessible at Q^2, and contributions from Higgs boson loops have been neglected. Hence, the $U(1)$ coupling increases slowly with momentum transfer (Fig. 6.5), but the $SU(2)$ and $SU(3)$ couplings decrease with increasing momentum transfer, provided N_f is not greater than 10 and 16, respectively. The factors 11 and $\frac{22}{3}$ in the $SU(3)$ and $SU(2)$ expressions result from the self-couplings of the non-Abelian gauge bosons; it is these couplings that are ultimately responsible for the asymptotic freedom of such theories. At the unification mass we require

$$\alpha_1(M_x^2) \simeq \alpha_2(M_x^2) \simeq \alpha_3(M_x^2). \tag{11.3}$$

There are some technical points concerning normalization of the $U(1)$ generator that are addressed in Exercise 11.2. We skip over them to summarize the following results for the representative GUT group $SU(5)$ (see §11.3) when the above relations are used (Georgi, Quinn, and Weinberg, 1974):

- The Weinberg angle at the unification mass is given by (Exercise 11.2)

$$\sin^2\theta_w = \tfrac{3}{8}, \tag{11.4}$$

  but being a function of coupling constants [eq. (9.27)] it decreases as Q^2 decreases; at low energies $\sin^2\theta_w \simeq 0.21$, in remarkable agreement with the measured value (9.36).
- The unification mass is found to be $M_x \simeq 10^{15}$ GeV, with the evolution of the coupling constants as in Fig. 11.1.
- $\alpha_i(M_x^2) \simeq \frac{1}{40}$.
- The lifetime of the proton is predicted to be $\tau_p \simeq 10^{30}$ years, which appears to be in conflict with the experimental limits (§11.4).

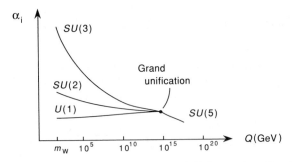

Fig. 11.1   Evolution of coupling in $SU(5)$ grand unification.

11.2 Minimal Criteria for a GUT Group

We assume, in the absence of evidence to the contrary, that the GUT gauge group G operates on the familiar quarks and leptons. These may be grouped into generations (families) such as

$$\begin{pmatrix} \nu_e \\ e \end{pmatrix}_{\mathrm{L}} \quad \begin{pmatrix} u \\ d \end{pmatrix}_{\mathrm{L}} \quad e_{\mathrm{R}} \quad u_{\mathrm{R}} \quad d_{\mathrm{R}} \qquad \text{(generation I)},$$

where d is understood to be Cabibbo rotated, L and R denote left-handed and right-handed particles, and the color index on the quarks is suppressed. Because the gauge couplings preserve chirality, it will be convenient to rewrite the generations entirely in terms of *left-handed fields* (see Exercises 6.2e and 11.1)

$$\begin{aligned} \text{I:} \quad & \left(\nu_e, e^-, e^+, u_i, d_i, \bar{u}_i, \bar{d}_i \right)_{\mathrm{L}} \\ \text{II:} \quad & \left(\nu_\mu, \mu^-, \mu^+, c_i, s_i, \bar{c}_i, \bar{s}_i \right)_{\mathrm{L}} \\ \text{III:} \quad & \left(\nu_\tau, \tau^-, \tau^+, t_i, b_i, \bar{t}_i, \bar{b}_i \right)_{\mathrm{L}}, \end{aligned} \qquad (11.5)$$

where the generations are labeled by Roman numerals, bars denote charge conjugates, $i = 1, 2, 3$ is a color index, we presume the existence of the top quark t and tau-lepton neutrino ν_τ, and the quarks are assumed mixed by the Kobayashi–Maskawa matrix (§9.3). Thus each generation consists of 15 two-component fields—three for each quark and one for each lepton. If neutrinos turn out to have a mass the addition of $\bar{\nu}_{\mathrm{L}}$ could increase this to 16, but the neutrinos will be assumed massless in our discussion.

We may expect that the grand unification gauge group should meet certain minimal criteria:

1. The group should be simple (see §5.2), so that only one gauge coupling constant appears in the theory. (Alternatively, it could be a product of

simple groups $G_1 \times G_2$, with an additional discrete symmetry ensuring that the two coupling constants are equivalent.)

2. The group must contain $SU(3)_c \times SU(2)_w \times U(1)_y$ as a subgroup; consequently, it must be at least of rank $2 + 1 + 1 = 4$.

3. The generations I, II, and III contain fields and their charge conjugates, so the group must accommodate complex representations (see the solution of Exercise 5.9 and §5.3.2).

4. The representations to which particles are assigned must satisfy technical renormalization conditions ensuring freedom from anomalies (see §9.3).

Various candidate GUT groups meeting these requirements have been suggested, with the most popular being $SU(5)$, $SO(10)$, and the exceptional group $E6$. O'Raifeartaigh (1986) may be consulted for a survey of these and other proposed unification groups.

EXERCISE 11.1 Prove that the charge conjugate of a right-handed fermion field is left-handed. Thus, in (11.5) we may take as independent fermion fields ψ_L and ψ_L^c, instead of ψ_L and ψ_R.

11.3 The $SU(5)$ Grand Unified Theory

The simplest theory meeting these conditions, and the one we will use as a prototype GUT, is the $SU(5)$ theory of Georgi and Glashow (1974). The adjoint representation of $SU(5)$ is $N^2 - 1 = 24$ dimensional, so 24 gauge bosons are required. Of these, we must assume that 12 are already familiar: eight gluons, three intermediate vector bosons, and one photon. The remaining 12 gauge bosons are denoted by X and Y, and are termed *leptoquark bosons*; it will be seen shortly that they can mediate transitions between leptons and quarks because they carry both weak isospin and color quantum numbers. Such transitions change baryon and lepton number, and the leptoquark bosons will need large masses (of order 10^{15} GeV) to avoid immediate conflict with experimental limits on baryon nonconserving processes such as proton decay.[‡]

The fundamental and conjugate representations of $SU(5)$ are the **5** and **5̄**. Each fermion generation fits nicely into a *direct sum* of $SU(5)$ representations:

[‡]This is reminiscent of the electroweak theory, where the feebleness of the weak interactions at normal energies is attributed to the large mass of the corresponding gauge bosons [see eq. (9.33)]. Notice that in a GUT we cannot change the gauge coupling strength to one particle (for example, a baryon nonconserving one) without changing the gauge coupling to all particles. To appreciate the enormity of the leptoquark boson scale, it is instructive to express it in more picturesque units: the GUT scale is comparable with the mass of a bacterium, or the kinetic energy of a charging rhinoceros (Preskill, 1984).

$\overline{5} \oplus 10$, where the **10** comes from the antisymmetric part of the product $5 \otimes 5$ [see eqns. (5.79) and (5.87)]. It is conventional to classify these $SU(5)$ multiplets according to their dimensionality under the $SU(3)_c$ and $SU(2)_w$ subgroups using the notation $\big(\text{Dim } SU(3), \text{Dim } SU(2)\big)$. Then

$$5 = (3, 1) + (1, 2) = d_R + \big(\bar{\nu}_e, e^+\big)_R$$

$$\overline{5} = (\overline{3}, 1) + (1, 2) = \bar{d}_L + \big(\nu_e, e^-\big)_L \tag{11.6}$$

$$10 = (3, 2) + (\overline{3}, 1) + (1, 1) = (u, d)_L + \bar{u}_L + e_L^+,$$

where quark color indices are suppressed. For example, the $SU(5)$ representation $\overline{5}$ consists of the left-handed components of ν_e and e^- [a weak isospin doublet, eq. (9.1a), but a color singlet since ν_e and e^- do not experience the color force] and the left-handed components of the antidown quark [a $\overline{3}$ under $SU(3)$ color, but a singlet under $SU(2)_w$ since only right-handed antiparticles partake of the charged-current weak interactions—see §9.1]. By similar reasoning the **10** of $SU(5)$ consists of the left-handed e^+, transforming as a color and weak singlet; the left-handed $\bar{u}$ quark, which is a color $\overline{3}$ and weak singlet; and the left-handed u and d quarks, which are $SU(3)_c$ triplets and constitute a weak-isospin doublet [eq. (9.1b)].

The other generations may be transcribed into $\overline{5}$ and **10** representations in analogous fashion but the $SU(5)$ GUT offers no explanation for the observed replication of quark–lepton generations. In Exercise 11.4 it is shown that grouping the fermions into $\overline{5} \oplus 10$ representations of $SU(5)$ yields an anomaly-free theory; thus it is renormalizable.

Since the $SU(5)$ representations contain both leptons and quarks, there exist operators in the theory that violate baryon number conservation and the proton can decay. On dimensional grounds the lifetime of the proton is

$$\tau_p \simeq \frac{1}{\alpha_G^2} \frac{M_x^4}{M_p^5}, \tag{11.7}$$

where α_G is the unknown $SU(5)$ coupling and M_p is the proton mass. For reasonable values of α_G the leptoquark boson masses M_x must be of order 10^{15} GeV to keep the lifetime of the proton out of gross conflict with the experiments; such masses can be acquired only through spontaneous symmetry breaking. Thus in the $SU(5)$ GUT there are two levels of symmetry breaking. At a scale $M_x \simeq 10^{15}$ GeV a 24-dimensional $SU(5)$ representation of real scalar fields is used to break the symmetry through the Higgs mechanism down to

$$SU(5) \rightarrow SU(3)_c \times SU(2)_w \times U(1)_y.$$

In this symmetry breaking the 12 X and Y bosons acquire masses of order 10^{15} GeV, but the other 12 gauge bosons remain massless. In addition,

12 Higgs particles appear in the theory with masses comparable to the lepto-quark bosons; these Higgs particles have negligible influence on the low-energy theory. The 24 gauge bosons of $SU(5)$ are classified under the $SU(3)_c$ and $SU(2)_w$ subgroups as

$$\mathbf{24} = \underset{\text{gluons}}{(\mathbf{8},\mathbf{1})} + \underset{W^{\pm}, Z^0, \gamma}{(\mathbf{1},\mathbf{3}) + (\mathbf{1},\mathbf{1})} + \underset{\overline{X}, \overline{Y}}{(\mathbf{3},\mathbf{2})} + \underset{X, Y}{(\overline{\mathbf{3}},\mathbf{2})}. \quad (11.8)$$

Thus the heavy X and Y bosons transform as color triplets and weak isodoublets, and can mediate interactions that interconvert leptons and quarks. Finally, at a mass scale of approximately 100 GeV the familiar spontaneous symmetry breaking of the GSW theory occurs:

$$SU(3)_c \times SU(2)_w \times U(1)_y \rightarrow SU(3)_c \times U(1)_{\text{QED}}.$$

In the $SU(5)$ theory this is accomplished by a set of scalar fields operating in the Higgs mode and transforming as a complex $\mathbf{5}$ under $SU(5)$. Three of 10 scalars originally in the $\mathbf{5}$ are used to give the weak vector bosons mass, leaving seven physical Higgs fields.

The disparity of the two scales for spontaneous symmetry breaking by elementary scalar fields is not easily managed in perturbation theory. This leads to the *gauge hierarchy problem:* in a quantum field theory with two vastly different mass scales, radiative corrections naturally mix the scales and equalize them unless parameters are fine-tuned. Thus, in $SU(5)$ GUTs fine-tuning is required to keep the electroweak mass scale from being forced to the leptoquark scale M_x, and in a typical calculation the square of the mass ratio between the two symmetry-breaking scales must be adjusted to disconcerting precision (about 24 decimal places) to generate reasonable and stable results. This smells of artificiality. These difficulties do not yet have a compelling solution, although there is hope that some form of supersymmetric GUT (§11.4), or the replacement of fundamental Higgs scalars by composite scalars (see §8.5.5), might supply at least a partial solution. The gauge hierarchy problem ultimately may be the most serious difficulty facing grand unified theories.

In the picture just painted there is a "desert" in the energy range $\sim 10^3 - 10^{15}$ GeV in which no new masses appear. This desert can be populated if minimal GUTs are embellished with additional assumptions. As noted below, minimal GUTs appear to be ruled out by experimental results.

The $SU(5)$ fermion assignments reproduce the factors of $\frac{1}{3}$ between the charges of the leptons and the quarks, thereby accounting for the equality of the e^+ and proton charges [experimentally,

$$\left| \frac{Q_e + Q_p}{Q_p} \right| \leqslant 10^{-21} \quad (11.9)$$

(Dylla and King, 1973)]. For example, electrical charge is a symmetry generator associated with an additive quantum number [the photon is one of the $SU(5)$ gauge bosons]; hence it is a traceless diagonal matrix acting on the $\bar{5}$ or 5 [generalize eq. (5.25) to $SU(5)$] and the sum of the charges in a multiplet must vanish. From the $\bar{5}$ for the first generation we see that

$$3Q(\bar{d}) + Q(e) + Q(\nu) = 0,$$

where $Q(\bar{d})$ denotes the charge of the antidown quark, and the factor of three is because three colors of $\bar{d}$ appear in the $\bar{5}$. Therefore, the $SU(5)$ GUT accounts for charge quantization, and suggests that the third-integer charges historically introduced into hadronic physics through approximate flavor symmetry actually arise from an exact color symmetry for the quarks.

EXERCISE 11.2 Prove that in $SU(5)$ minimal GUT the Weinberg angle at the unification mass is given by $\sin^2 \theta_w = \frac{3}{8}$. *Hint*: determine the relative normalization of the $SU(5)$ and $U(1)_y$ generators by evaluating a physical quantity such as the charge; take the $U(1)$ subgroup generator $\lambda_0/2$ of $SU(5)$ to be the diagonal matrix
$$\lambda_0 = \text{Diag } (2, 2, 2, -3, -3)/\sqrt{15};$$
assume the GSW coupling to be specified by (9.7) and the corresponding $SU(5)$ coupling to be of a similar form with y replaced by λ_0.

EXERCISE 11.3 Demonstrate that $SU(5)$ is the simplest GUT group that satisfies the minimal criteria of §11.2 for a grand unified theory.

EXERCISE 11.4 The anomaly $A(R)$ of a fermion representation R is given by (see §9.3 and Cheng and Li, 1984, §14.1)

$$\text{Tr}\left(\{T^a(R), T^b(R)\}T^c(R)\right) = \tfrac{1}{2}d^{abc}A(R),$$

where $T^a(R)$ is a representation matrix and d^{abc} are the totally symmetric constants analogous to those in eq. (5.66a). $A(R)$ is normalized to unity for the fundamental representation and is independent of the generators. Prove that the reducible $SU(5)$ representation $\bar{5} \oplus 10$ is anomaly-free because the anomalies of the $\bar{5}$ and 10 cancel each other. *Hint*: charge is a generator.

11.4 Decay of the Proton

Exotic processes are allowed in GUTs since quarks and leptons appear in the same irreducible representations. The most intriguing is proton decay, because of its implications for the long-term stability of matter. Some important mechanisms for this decay are shown in Fig. 11.2. The proton lifetime

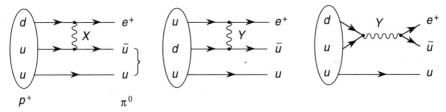

Fig. 11.2   Some diagrams mediating proton decay in minimal $SU(5)$ theory.

estimated from minimal $SU(5)$ is (Langacker, 1981)

$$\tau_{\mathrm{p}}^{SU(5)} \simeq 3.2 \times 10^{29 \pm 1.3} \text{ years}, \tag{11.10}$$

with a predicted branching ratio of 30% for $p \rightarrow e^+\pi^0$. The experimental limit (Park et al., 1985)

$$\tau_{\mathrm{p}}^{\mathrm{exp}}/B\left(p \rightarrow e^+\pi^0\right) \geqslant 2 \times 10^{32} \text{ years} \tag{11.11}$$

for $p \rightarrow e^+\pi^0$ (where B is the branching ratio) implies difficulties with minimal $SU(5)$. Extensions of minimal $SU(5)$, and GUTs based on larger groups such as $SO(10)$ [which contains $SU(5)$ as a subgroup], can push the lifetime of the proton higher, and also produce as yet unobserved phenomena like massive neutrinos and $N\bar{N}$ oscillations (see Okun, 1982).

There is also considerable theoretical interest in GUTs based on *supersymmetries,* which are symmetries with bosons and fermions appearing in the same representations; see Sohnius (1985) or Chanowitz (1988) for an introduction. A particular virtue of supersymmetric gauge theories is that they might provide help with the gauge hierarchy problem, essentially because boson and fermion partners related by supersymmetries could give canceling divergences in loop diagrams. Supersymmetries are also essential to ambitious *superstring theories* that attempt to unite gravitation with the other forces in a superunified "Theory of Everything" (Green, 1985; Schwarz, 1982). However, as yet there is no substantial evidence for the new particles required in supersymmetric theories (see Hinchcliffe, 1986, for a summary).

11.5 Background and Further Reading

Most books on gauge theories and a number of review articles give introductions to GUTs: see Quigg (1983), Chaichian and Nelipa (1984), Cheng and Li (1984), Okun (1982), Gasiorowicz and Rosner (1981), Ross (1984), or O'Raifeartaigh (1986). The $SU(5)$ model is introduced concisely in Georgi and Glashow (1974). An extensive discussion of the practical group theory required in constructing unified models is given in Slansky (1981). Perkins (1984) has reviewed proton decay experiments and Langacker (1981) has reviewed proton decay in GUTs. A short introduction to the question of neutrino masses may be found in Kuo and Panteleone (1989); see also Kayser (1985). Readable introductions to the interface between GUTs and cosmology have been given by Kolb and Turner (1983); Collins, Martin, and Squires (1989); and Kolb (1987). We shall address some of these topics in Ch. 15.

Part III

MODELS AND APPLICATIONS

CHAPTER 12

Models for QCD

The theory of QCD as outlined in Ch. 10 is a promising candidate for a microscopic description of the strong interactions. It can be demonstrated that QCD is free of ultraviolet divergences, but it has severe infrared problems. Unlike the infrared divergences of ordinary field theories, the low energy difficulties of QCD are not easily solved. They are intimately connected with the problem of confinement—although high energy probes supply abundant evidence that hadrons are composed of quarks, we do not see quarks as free asymptotic states in physical processes.

This problem of confinement in QCD is a difficult one, the recent advances in lattice gauge calculations notwithstanding (see Ch. 13). Indeed, confinement is more an extremely plausible hypothesis than a proven fact at this stage. Therefore, we are faced with the dilemma that in the regime where QCD is calculable (short distance $\rightarrow$ asymptotic freedom $\rightarrow$ perturbative QCD) only a few experimental tests are so far available, while in the regime where a plethora of data exists (large distance $\rightarrow$ confinement $\rightarrow$ non-perturbative effects) the theory admits of tangible results only grudgingly.

12.1 Confinement Phenomenology

To make headway in the interim period while large-scale non-perturbative methods are being developed, it is necessary to employ phenomenology in one form or another. In this chapter we discuss suggestive and useful models of confinement that appear in some guise in most approximations to QCD. In the first kind of model one assumes that the vacuum can have more than one phase, and that one phase of the vacuum may appear as localized regions within another; an obvious analogy is that of bubbles in a liquid. A second class of models utilizes the language of group theory and proclaims that only wavefunctions transforming as singlets under $SU(3)_c$ are permissible for real hadrons. This leads to confinement of color and of particles coupled to the color fields. The third category of phenomenological models assumes that hadrons consist of quarks moving in a confining potential, which is usually taken to be linear or quadratic in the separation of the quarks.

The phenomenological models to be discussed are at best either caricatures that emphasize particular aspects of QCD at the expense of others, or highly effective theories with only a precarious direct connection to the underlying QCD microscopy. Therefore, we should in no way confuse such models of confinement with explanations of confinement; they should rather be viewed as methods that allow us to construct portions of a house to fit a foundation before the foundation is complete.

We begin with a discussion of models that are based on a two-phase structure of the non-Abelian vacuum. These form the basis for the bag phenomenology to be introduced later in this chapter.

12.1.1 Dielectric Structure of the Vacuum

In QED we conventionally set the dielectric constant of the vacuum state to unity,

$$\kappa_{\text{vac}}^{\text{QED}} = 1. \tag{12.1}$$

The displacement vector $\mathbf{D}$, the electric field $\mathbf{E}$, and the polarization $\mathbf{P}$ are related by (Jackson, 1975, §4.3)

$$\mathbf{D} = \mathbf{E} + \mathbf{P}, \tag{12.2}$$

where $\mathbf{D}$ is the field produced by the source charges and $\mathbf{E}$ is the total field including contributions from induced charges. Under the influence of $\mathbf{E}$, atoms will have $\mathbf{P}$ in the same direction as $\mathbf{E}$ so as to produce a screening effect, and since the dielectric constant is defined by

$$\mathbf{D} = \kappa \mathbf{E}, \tag{12.3}$$

the dielectric constant for any physical medium satisfies

$$\kappa_{\text{med}}^{\text{QED}} \geqslant 1. \tag{12.4}$$

Now we hypothesize a medium for which $\kappa \ll 1$ (the medium is *antiscreening*). It is easy to demonstrate that a test charge immersed in such a medium will dig a "hole" in the medium, the hole will be stable, and inside the hole surrounding the test charge the dielectric constant κ is unity, while in the surrounding medium $\kappa \ll 1$ [see Lee (1981), Baym (1982c), and Fig. 12.1a]. On the other hand, it is not difficult to show that if a test charge is placed in a screening vacuum ($\kappa > 1$), the radius of the hole created in the medium will shrink to zero. Thus test charges can produce structures of finite spacetime extent in a vacuum, provided the vacuum is *antiscreening*.

This model can be used in conjunction with QCD to produce confinement (Lee, 1981; Baym, 1982c; Callan, Dashen, and Gross, 1979). We introduce a

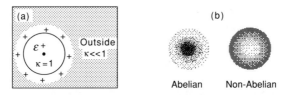

Fig. 12.1   (a) Test charge ϵ^+ in an antiscreening medium. Because the medium antiscreens the induced charge on the sphere is of the same sign as the test charge. (b) Schematic spatial distribution of effective charge in Abelian and non-Abelian field theories.

color dielectric constant κ_L associated with the QCD vacuum in a volume L^3 by the definition

$$\kappa_L \equiv \frac{g_\ell^2}{g_L^2}, \qquad (12.5)$$

where g_ℓ is the renormalized strong coupling constant on an arbitrary reference length scale ℓ, and g_L is the corresponding coupling on the length scale L. Therefore we have $\kappa_\ell \equiv 1$. Now because of asymptotic freedom we know that for non-Abelian gauge fields at short distances $g_L > g_\ell$, provided $L > \ell$ [see eq. (10.40)], from which

$$\kappa_L < \kappa_\ell \quad \text{if} \quad L > \ell. \qquad (12.6)$$

If (12.6) continues to hold in the large distance (non-perturbative) regime, then for an infinite volume,

$$\kappa_\infty \ll 1. \qquad (12.7)$$

This, in light of the previous discussion, corresponds to a medium that is *antiscreening for color charges:* with these assumptions the QCD vacuum acts as a nearly perfect *color dielectric*.

A qualitative explanation of this has already been given in §10.6.4: the gluons carry the charge of the non-Abelian field so the color charge of a quark is spread over a finite region by interaction with the virtual gluons. For the observed number of quark flavors, vacuum polarization from $q\bar{q}$ pairs is not sufficient to offset the effect of the gluons. Thus a large momentum (short distance) probe interacting with a quark sees less of the quark color charge than a low momentum probe and vacuum polarization antiscreens the quark color charge. In QED the gauge bosons are not self-coupling and the only contribution to vacuum polarization is from pair creation, which leads to a screening vacuum (§1.6.2). This dispersion of the effective charge in Abelian and non-Abelian theories is illustrated in Fig. 12.1b.

Callan, Dashen, and Gross (1979) have given plausibility arguments that the color dielectric constant $\kappa \ll 1$ in the normal vacuum (see also Baym, 1982a;

Shuryak, 1988). They first observe that by Lorentz invariance $\kappa = 1/\mu$, where μ is the permeability of the vacuum, so a determination of μ also specifies κ. The theory of QCD is known to have localized solutions in Euclidian 4-space that are called *instantons:* non-perturbative, soliton-like objects to be discussed in Ch. 13. For the argument here we note that they behave like magnetic dipoles in 4-dimensional space, causing the vacuum to exhibit color paramagnetism, $\mu > 1$, whence $\kappa < 1$.

Determining the equilibrium instanton density is nontrivial, but it appears that instantons are driven away from regions of large color displacement fields **D**, so that the color sources expel the instantons from the bag into the surrounding medium. The bags, with low instanton density, then have $\kappa \simeq 1$, but the larger concentration of instantons outside the bag produces a medium in which $\kappa \simeq 0$. The dense instantons outside the bag may also break chiral symmetry spontaneously, so that chiral symmetry is realized in the Wigner mode within the bag (massless u and d quarks) and in the Goldstone mode outside the bag (massless pions). It has been suggested that the transition from the interior region of low instanton density to the exterior region of high instanton density may take the form of a phase transition, but rigorous proof is lacking.

12.1.2 Color Singlet Bags

Whenever color charges (quarks) are immersed in the antiscreening QCD vacuum, we expect from the preceding discussion that holes develop in the vacuum around the particles; these regions are called *bags* or *domain structures*. As we will see in Ch. 13, they may correspond to *soliton solutions* of nonlinear relativistic field equations.[‡] Inside these bags $\kappa_{\mathrm{in}} = 1$, but outside $\kappa_\infty \ll 1$; for simplicity we will assume that outside the bags $\kappa_\infty = 0$. It is then possible to show that

1. If the total color inside the bag is nonzero, the mass of the bag becomes infinite as $\kappa_\infty \to 0$.
2. If the interior of the bag contains no net color (color singlet), the bag mass remains finite as $\kappa_\infty \to 0$.

EXERCISE 12.1 Convince yourself of the plausibility of the above two assertions. *Hint*: consider a test charge ϵ^+ placed in a dielectric medium ($\kappa \simeq 0$); then consider a dipole distribution $\epsilon^+ - \epsilon^-$ (net charge zero) placed in the same medium.

[‡]The bag terminology is evident. The concept of domains is borrowed from condensed matter physics where different regions of a macroscopic object may exhibit different phase structure (e.g., regions with different magnetization). Solitons will be discussed in Ch. 13; they are non-dispersive solutions of nonlinear wave equations.

This can be expressed concisely in terms of the QCD group structure. If quarks and antiquarks form the **3** and **$\bar{3}$** representations of a color $SU(3)$, the representations arising from the simplest quark and antiquark combinations are (Exercise 12.2)

$$
\begin{aligned}
q &= \mathbf{3} \qquad \bar{q} = \bar{\mathbf{3}} \\
q\bar{q} &= \mathbf{3} \otimes \bar{\mathbf{3}} = \mathbf{1} \oplus \mathbf{8} \\
qq &= \mathbf{3} \otimes \mathbf{3} = \mathbf{6} \oplus \bar{\mathbf{3}} \\
qq\bar{q} &= \mathbf{3} \otimes \mathbf{3} \otimes \bar{\mathbf{3}} = \mathbf{3} \oplus \mathbf{6} \oplus \mathbf{3} \oplus \mathbf{15} \\
qqq &= \mathbf{3} \otimes \mathbf{3} \otimes \mathbf{3} = \mathbf{1} \oplus \mathbf{8} \oplus \mathbf{8} \oplus \mathbf{10},
\end{aligned}
\tag{12.8}
$$

and quarks in the combinations qqq and $q\bar{q}$ can form color singlet states, corresponding to hadrons of finite mass. Then, if we assume all hadrons to be color singlets (10.4),

$$
\Psi^{ij}_{\text{meson}} \simeq \delta_{\alpha\beta} \bar{q}^{\alpha}_i q^{\beta}_j \qquad\qquad \Psi^{ijk}_{\text{baryon}} \simeq \epsilon_{\alpha\beta\gamma} q^{\alpha}_i q^{\beta}_j q^{\gamma}_k
$$

(where α, β, γ are color indices and i, j, k are flavor/spin indices), the confinement in hadrons and the observation of only these quark combinations is (in a sense) explained: the vacuum pushes all color electric flux into the bags, leading to confinement of finite-mass color singlet states (see Fig. 12.2). Conversely, any colored state trying to form in the vacuum produces a bag of infinite mass, which plays no role in hadronic dynamics.

Therefore, in a Yang–Mills gauge theory such as QCD the non-Abelian charges can modify the normal vacuum over a finite region of space, giving

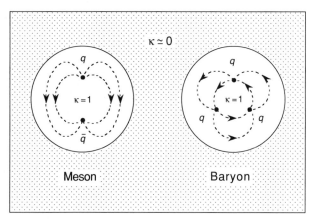

Fig. 12.2 Closed lines of chromoelectric field inside color-singlet meson and baryon bags. The bag interiors are regions of perturbative vacuum. The shaded areas are regions of normal, non-perturbative vacuum.

rise to an *abnormal vacuum* in the vicinity of the color charges. The normal vacuum is often termed the *non-perturbative vacuum*, while the abnormal vacuum in the bags is sometimes called the *perturbative vacuum*. Under ordinary conditions QCD has a lower energy ground state in the presence of non-Abelian charges if such domain structures develop.

Notice in this discussion the germ of an interesting possibility (Lee, 1975). Since hadronic matter fields modify the QCD vacuum, we might be able to "engineer the vacuum" by suitably arranging matter and energy over some region of spacetime. One possibility is that the small hadronic bags might merge if the density of color charges is made sufficiently high, generating a quark–gluon plasma deconfined over macroscopic regions of spacetime. This possibility will be discussed in Ch. 14.

EXERCISE 12.2 Verify the irreducible representations shown in eq. (12.8). What is the next simplest combination of q and $\bar{q}$ forming an $SU(3)$ color singlet? What about for a state made only of q's?

12.1.3 Analogy with Superconductivity

There is a close analogy between these effects and those observed in superconductivity, as illustrated in Fig. 12.3. The superconductor exhibits a perfect *diamagnetism*, and a magnetic field is expelled from the superconductor (Meissner effect—see the discussion in §8.5.2). Likewise, the QCD vacuum is

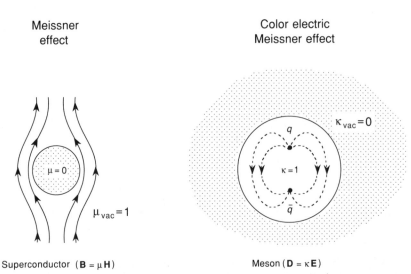

Fig. 12.3   Analogy between expulsion of the magnetic field from a superconductor and expulsion of chromoelectric field from the QCD vacuum into the bag.

a *color dielectric*; it pushes the color fields into the bags, and we may view color confinement as a kind of *color electric Meissner effect*. In both the magnetic and chromoelectric Meissner effects the different vacuum domains are separated by phase transitions, with the magnetic permeability μ and color dielectric constant κ the respective order parameters.

12.1.4 Confinement as a Boundary-Value Problem

Models such as the ones just described allow us to finesse the confinement problem in a manner similar to that employed to describe the strong short-distance repulsion (*hard-core*) acting at separations $\lesssim$ 0.5 fm in nucleon–nucleon scattering. The potential becomes large and difficult to handle there, but the very fact that the potential is so strongly repulsive means that the wavefunction tends to zero in that region. Therefore, the actual problem may be replaced with a more tractable boundary-value problem excluding the nucleons from small separations. At large separations the interaction is much weaker and standard methods may be used.

A related technique is employed in the bag models to be discussed below. The quarks are confined by boundary conditions to the interior of a particular spacetime region (the bag), which means that the mass of a quark is effectively infinite outside the bag. The strong confinement forces having been disposed of phenomenologically by the bag boundary conditions, one then hopes that the remaining interactions among the quarks and gluons inside the bag are perturbative.

12.2 Current Algebra and Chiral Symmetry

As we will see in §12.5, an important ingredient in models that confine through the imposition of a boundary is the manner in which chiral symmetry is implemented (§10.6.2). As a prelude to this discussion, we review the basic principles of current algebra and chiral symmetry.

The Cabibbo theory outlined in §6.2.3 assumes an $SU(3)$ octet of conserved vector currents (6.33) and an octet of partially conserved axial vector currents (6.34). If consideration is restricted to u and d quarks, we recover a triplet of *conserved* vector (isospin) currents

$$V_i^\alpha(x) = \bar{q}(x)\gamma^\alpha \frac{\tau_i}{2} q(x) \qquad (i = 1, 2, 3) \tag{12.9}$$

$$\partial_\alpha V_i^\alpha = 0 \qquad q = \begin{pmatrix} u \\ d \end{pmatrix}, \tag{12.10}$$

and a triplet of *partially conserved* axial-vector currents

$$A_i^\alpha(x) = \bar{q}(x)\gamma^\alpha\gamma_5\frac{\tau_i}{2}q(x) \qquad (i = 1, 2, 3) \tag{12.11}$$

$$\partial_\alpha A_i^\alpha \simeq 0. \tag{12.12}$$

To appreciate the implications of the partially conserved axial current, it is useful to consider an idealized world in which it is exactly conserved:

$$\partial_\alpha A_i^\alpha = 0. \tag{12.13}$$

The consequences of such an assumption are discussed in many places (see Campbell, 1978; Pagels, 1975; Lee, 1972). Among the most important ones are (1) the *Goldberger–Treiman relation*

$$g_A M = g f_\pi, \tag{12.14}$$

where M is the nucleon mass, $g = 13.4$ is the $\pi - N$ coupling constant, $f_\pi = 93$ MeV is the pion decay constant $(\pi^+ \to \ell^+ + \nu_\ell)$, and $g_A = 1.24$ is the nucleon axial vector coupling coefficient [eq. (6.19)], and (2) that conservation of the axial current implies the existence of a *massless particle* with the pseudoscalar, isovector quantum numbers of the pion. With the empirical values inserted, (12.14) is satisfied at the 10% level.

We may construct charges associated with the vector and axial currents. From (2.193), (12.9), and (12.11),

$$Q_i(t) \equiv \int d^3x \, V_i^0(\mathbf{x}, t) = \int d^3x \, q^\dagger(x)\frac{\tau_i}{2}q(x) \tag{12.15a}$$

$$Q_{i5}(t) \equiv \int d^3x \, A_i^0(\mathbf{x}, t) = \int d^3x \, q^\dagger(x)\gamma_5\frac{\tau_i}{2}q(x). \tag{12.15b}$$

Then the left- and right-handed charges defined by

$$\begin{aligned} Q_L^i &\equiv Q_-^i \equiv \tfrac{1}{2}(Q_i - Q_{i5}) \\ Q_R^i &\equiv Q_+^i \equiv \tfrac{1}{2}(Q_i + Q_{i5}), \end{aligned} \tag{12.16}$$

satisfy two independent commutation algebras (Exercise 12.3):

$$[Q_L^i, Q_L^j] = i\epsilon_{ijk}Q_L^k$$

$$[Q_R^i, Q_R^j] = i\epsilon_{ijk}Q_R^k \tag{12.17}$$

$$[Q_L^i, Q_R^j] = 0,$$

which we recognize as the Lie algebra of $SU(2) \times SU(2)$. Since the chirality matrix γ_5 appearing in the axial current plays a central role, this algebra is called chiral $SU(2) \times SU(2)$ or $SU(2)_{\mathrm{L}} \times SU(2)_{\mathrm{R}}$. Equations (12.17) are examples in integral form of the commutation relations for *current algebra* (Gell-Mann, 1962, 1964; Lee, 1972). The direct product structure implies that the left- and right-handed charges transform under *independent* $SU(2)$ symmetries.

Now in the real world the pion has a mass, and the *PCAC hypothesis* is that

$$\partial_\mu A_i^\mu(x) = f_\pi m_\pi^2 \pi_i(x) \neq 0, \qquad (12.18)$$

where m_π is the mass of the pion and $\pi_i(x)$ is a pseudoscalar field operator. Equation (12.18) implies that the current (12.11) is not conserved, suggesting that a realistic strong interaction Hamiltonian will be of the form

$$H = H_0 + \epsilon H_1, \qquad (12.19)$$

where H_0 is chiral invariant, H_1 is not, and ϵ is small because $\epsilon \simeq \mathcal{O}\left(m_\pi^2\right)$ and the pion mass is small compared with other hadrons (see Exercise 12.4).

EXERCISE 12.3 (a) Demonstrate that for the charges (12.15) the commutators are

$$[Q_i, Q_j] = i\epsilon_{ijk}Q_k \qquad [Q_i, Q_{j5}] = i\epsilon_{ijk}Q_{k5} \qquad [Q_{i5}, Q_{j5}] = i\epsilon_{ijk}Q_k,$$

and that these lead to the chiral $SU(2)_{\mathrm{L}} \times SU(2)_{\mathrm{R}}$ algebra.   (b) Show that the generators (12.16) of the chiral algebra (12.17) are related by parity transformation. *Hint*: under parity $q(\mathbf{x}, t) \to$ (phase) $\times \gamma^0 q(-\mathbf{x}, t)$.

12.3 Linear σ-Model

The linear σ-model (Gell-Mann and Levy, 1960) illustrates many of the ideas we have just introduced concerning PCAC and is useful for material in later chapters. The Lagrangian density is (Campbell, 1978; Dashen, 1969)

$$\mathcal{L} = \mathcal{L}_0 + \epsilon \mathcal{L}', \qquad (12.20a)$$

where

$$\begin{aligned}
\mathcal{L}_0 =& \overline{N}\left(i\slashed{\partial} + g(\sigma + i\boldsymbol{\tau} \cdot \boldsymbol{\pi}\gamma_5)\right)N \\
& + \tfrac{1}{2}\left(\partial_\mu \sigma\right)^2 + \tfrac{1}{2}\left(\partial_\mu \boldsymbol{\pi}\right)^2 - \tfrac{1}{4}\lambda(\sigma^2 + \boldsymbol{\pi} \cdot \boldsymbol{\pi} - v^2)^2
\end{aligned}$$

$$(12.20b)$$

is symmetric under the infinitesimal isospin rotations

$$\pi^i \rightarrow \pi^i + \epsilon_{ijk}\alpha^j \pi^k \qquad \sigma \rightarrow \sigma,$$

$$N \rightarrow N - i\frac{\tau^i}{2}\alpha^i N \qquad \overline{N} \rightarrow \overline{N} + i\overline{N}\frac{\tau^i}{2}\alpha^i \tag{12.21a}$$

and under the infinitesimal chiral transformations (Exercise 12.4)

$$\pi^i \rightarrow \pi^i - \alpha^i \sigma \qquad \sigma \rightarrow \sigma + \alpha^i \pi^i,$$

$$N \rightarrow N + i\frac{\tau^i}{2}\alpha^i \gamma_5 N \qquad \overline{N} \rightarrow \overline{N} + i\overline{N}\frac{\tau^i}{2}\alpha^i \gamma_5, \tag{12.21b}$$

and $\mathcal{L}'$ is an isospin invariant, chiral symmetry-breaking term, often taken to be of the form

$$\epsilon\mathcal{L}' = \epsilon\sigma. \tag{12.22}$$

This Lagrangian density contains an isodoublet of nucleons $N = (p, n)$ coupled by a pseudoscalar Yukawa interaction to an isotopic triplet of pions $\boldsymbol{\pi} = (\pi_1, \pi_2, \pi_3)$, and by a scalar interaction to an isoscalar meson σ. The mesons self-interact through the last term of (12.20b), and the scalar products such as $\boldsymbol{\pi} \cdot \boldsymbol{\pi}$ in (12.20) refer to components in the isospace:

$$\boldsymbol{\pi} \cdot \boldsymbol{\pi} = \boldsymbol{\pi}^2 = \pi_1^2 + \pi_2^2 + \pi_3^2.$$

The constants λ and ϵ will be assumed positive throughout our discussion.

EXERCISE 12.4 (a) Prove that in the Lagrangian density (12.20) of the linear σ-model, $\mathcal{L}_0$ is chirally invariant but $\mathcal{L}'$ is not. (b) For the linear σ-model, find the conserved vector isospin currents $V_\mu^i(x)$ and the conserved axial vector currents $A_\mu^i(x)$ associated with $\mathcal{L}_0$. Show that for the total Lagrangian density (12.20) the divergence of the axial current is $\partial_\mu A_i^\mu(x) = \epsilon\pi_i(x)$, so that by the PCAC hypothesis (12.18), $\epsilon = m_\pi^2 f_\pi$. *Hint*: The divergence of the current may be calculated from

$$\partial_\mu A_i^\mu = \frac{\partial(\delta\mathcal{L})}{\partial\alpha_i},$$

where $\delta\mathcal{L}$ is the change in the Lagrangian density brought about by the infinitesimal transformation defined in terms of the parameters α_i [compare (2.190) and (2.191b)].

EXERCISE 12.5 Prove that an expansion in the number of independent loops of connected Feynman diagrams is approximately an expansion in powers of $\hbar$; thus loops represent quantum corrections to the classical tree diagrams. *Hint*: reinstate the explicit $\hbar$'s that are normally suppressed in $\hbar = 1$ units, and consider the general process of quantization and perturbative expansion.

12.3.1 Classical Approximation for Meson Fields

The linear σ-model is greatly simplified if we retain only tree diagrams for the π and σ fields. Keeping only tree diagrams in a field theory is often termed the *classical approximation*, since one can show formally that tree diagrams represent lowest order contributions in $\hbar$, with loop diagrams contributing higher orders (Exercise 12.5). Therefore we treat the π and σ fields as classical, phenomenological objects, solving the model with a meson potential energy

$$V \equiv \tfrac{1}{4}\lambda(\sigma^2 + \boldsymbol{\pi} \cdot \boldsymbol{\pi} - v^2)^2 - \epsilon\sigma, \tag{12.23}$$

determined by replacements such as $\langle \sigma^4 \rangle \simeq \langle \sigma \rangle^4 \equiv \sigma^4$. Introducing μ^2 through $v^2 \equiv -\mu^2/\lambda$, the classical potential is

$$V = \tfrac{1}{4}\lambda \left(\sigma^2 + \boldsymbol{\pi} \cdot \boldsymbol{\pi}\right)^2 + \tfrac{1}{2}\mu^2 \left(\sigma^2 + \boldsymbol{\pi} \cdot \boldsymbol{\pi}\right) - \epsilon\sigma, \tag{12.24}$$

where a constant term has been discarded.

12.3.2 Particle Spectrum

We first neglect the explicit symmetry-breaking term in the meson potential by setting $\epsilon = 0$. Two cases may be distinguished, depending on the sign of μ^2 (recall that λ is always chosen positive); these are illustrated in Figs. 12.4 and 12.5.

The way to proceed is clear from our earlier discussions of symmetry and broken symmetry (Ch. 8). For $\mu^2 > 0$ the symmetry is realized in the Wigner mode, the origin is the classical vacuum ($\sigma = \boldsymbol{\pi} = 0$), and the particle spectrum for small oscillations about the origin can be read directly from the low order terms of the Lagrangian density (see §2.3):

$$\mathcal{L} \simeq \overline{N}\left(i\slashed{\partial} + g(\sigma + i\boldsymbol{\tau} \cdot \boldsymbol{\pi}\gamma_5)\right)N$$
$$+ \tfrac{1}{2}\left(\partial_\mu\sigma\right)^2 + \tfrac{1}{2}\left(\partial_\mu\boldsymbol{\pi}\right)^2 - \tfrac{1}{2}m_{\sigma,\pi}^2(\sigma^2 + \boldsymbol{\pi} \cdot \boldsymbol{\pi}). \tag{12.25}$$

This corresponds to a spectrum of massless nucleons (a nucleon mass term $m\overline{N}N$ would not be chiral invariant—see Exercise 6.2e) and massive mesons with

$$m_\pi^2 = m_\sigma^2 = \mu^2 = -\lambda v^2 \equiv m_{\sigma,\pi}^2. \tag{12.26}$$

But this is nonsense: the physical nucleons are massive, and there is no evidence in nature for a scalar partner degenerate in mass with the pion. It is easy to see that any theory with a conserved axial vector current that is implemented in the Wigner mode will produce a spectrum in which each hadron is degenerate in mass with a particle of the opposite intrinsic parity (the pseudoscalar π and the scalar σ in the present example). To prove this,

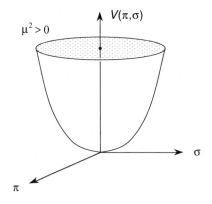

Fig. 12.4   Meson potential of the σ-model with symmetry implemented in the Wigner mode.

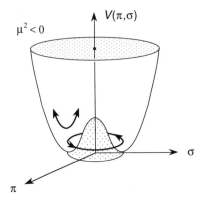

Fig. 12.5   Meson potential of the σ-model with symmetry in the Goldstone mode.

notice that if $\partial_\mu A_i^\mu = 0$, then for the axial charge (12.15b) we have (§2.8.3)

$$\frac{\partial}{\partial t} Q_{i5} = 0, \tag{12.27}$$

implying that Q_{i5} is a constant of motion,

$$[H, Q_{i5}] = 0, \tag{12.28}$$

and if an eigenstate $|\Psi_+\rangle$ of the Hamiltonian H exists with mass m, then the state

$$|\Psi_-\rangle \equiv Q_{i5} |\Psi_+\rangle \tag{12.29}$$

is also an eigenstate of H with the same mass:

$$H |\Psi_\pm\rangle = m |\Psi_\pm\rangle. \tag{12.30}$$

Furthermore, from (12.29) the states $|\Psi_-\rangle$ and $|\Psi_+\rangle$ have opposite parities (see Exercise 12.3b), so if the symmetry is implemented in the Wigner mode every hadron should have a partner of opposite intrinsic parity but of the same mass.

This unwelcome result can be evaded if the symmetry is realized in the Goldstone mode. That is, we consider an implementation of chiral symmetry such that the axial charge does not annihilate the vacuum:

$$Q_{i5} |0\rangle \neq 0. \tag{12.31}$$

As we will see, this removes the problem of the parity doublets at a cost of introducing a massless pseudoscalar Goldstone boson. This boson will be identified with the pion, which is massless in the idealized world of exact chiral symmetry.

Therefore, we choose $\mu^2 < 0$ in (12.24). To investigate the resulting particle spectrum the Lagrangian density must be expanded about a classical vacuum that is no longer at the origin. The ring of minima in Fig. 12.5 satisfies

$$\sigma^2 + \boldsymbol{\pi} \cdot \boldsymbol{\pi} = v^2, \tag{12.32}$$

and we choose the minimum on the negative σ-axis to be the classical vacuum, spontaneously breaking the symmetry.[‡] Introducing the new variable σ' by

$$\sigma' = \sigma + v, \tag{12.33}$$

the Lagrangian density becomes (Campbell, 1978)

$$\begin{aligned}
\mathcal{L} =& \overline{N}\big(i\partial\!\!\!/ - gv + g(\sigma' + i\boldsymbol{\tau} \cdot \boldsymbol{\pi}\gamma_5)\big)N \\
&+ \tfrac{1}{2}\left(\partial_\mu \sigma'\right)^2 + \tfrac{1}{2}\left(\partial_\mu \boldsymbol{\pi}\right)^2 - \tfrac{1}{4}\lambda(\sigma'^2 + \boldsymbol{\pi} \cdot \boldsymbol{\pi})^2 \\
&- \lambda v^2 \sigma'^2 + \lambda v \sigma'(\sigma'^2 + \boldsymbol{\pi} \cdot \boldsymbol{\pi}).
\end{aligned} \tag{12.34}$$

Picking out the terms quadratic in the fields, the particle spectrum now consists of (see §8.3 and §9.2)

- A nucleon isodoublet with mass $m_N = gv$,
- An isoscalar σ-meson of mass $m_\sigma = \sqrt{2\lambda}v = \sqrt{-2\mu^2}$,
- An isotriplet π-meson with $m_\pi = 0$.

The σ corresponds to radial oscillations of the field, while the π corresponds to a massless excitation (no restoring force) running around the bottom of the valley in Fig. 12.5; this should be compared with Fig. 8.5.

Things now look somewhat more realistic: the nucleons are massive and the

---

[‡] We choose the vacuum to lie on the σ-axis so that the vacuum state does not violate parity (σ is a Lorentz scalar but π is a pseudoscalar).

degeneracy of the parity doublets has disappeared; the interpretation of the massive scalar is more problematic, but at least the mass of σ is independent of the masses of N and π and can be adjusted through the parameter λ. The pion is massless and is the Goldstone boson associated with the spontaneous breaking of chiral $SU(2) \times SU(2)$ symmetry. In real life $m_\pi \simeq 140$ MeV, but this is small on the scale of elementary particle masses and in many instances the pion mass has negligible effect on physical processes. However, we can do even better since we still have at our disposal in (12.20) the linear term $\epsilon\sigma(x)$ that explicitly breaks chiral symmetry.

12.3.3 Explicit Symmetry Breaking

The linear term tilts the potential of Fig. 12.5, resulting in the potential of Fig. 12.6 (see Exercise 8.3). Unlike the preceding example, which had an infinite number of classical vacuum states, the ground state is now unique because of the explicit symmetry-breaking term $\epsilon\sigma$. If we assume ϵ to be small the new minimum is on the σ-axis at

$$f \simeq v + \frac{\epsilon}{2\lambda v^2}.$$

If the Lagrangian density is expanded around this minimum, the resulting mass spectrum is (Campbell, 1978, and Exercise 12.6)

$$m_N = gf \simeq g\left(v + \frac{\epsilon}{2\lambda v^2} + \dots\right) \tag{12.35a}$$

$$m_\sigma^2 = \lambda(3f^2 - v^2) \simeq \lambda\left(2v^2 - \frac{3\epsilon}{\lambda v} + \dots\right) \tag{12.35b}$$

$$m_\pi^2 = \lambda(f^2 - v^2) = \frac{\epsilon}{f} \simeq \frac{\epsilon}{v} + \dots \quad . \tag{12.35c}$$

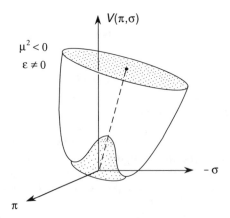

Fig. 12.6   Meson potential of the σ-model with spontaneous symmetry breaking and an explicit symmetry-breaking term $\epsilon\sigma$.

In addition, since (Exercise 12.4b)

$$\partial_\mu A_i^\mu = \epsilon \pi_i, \qquad (12.36)$$

we see from (12.18) and (12.35c) that f is just the pion decay constant: $f = f_\pi$.

Thus, for spontaneous symmetry breaking in the presence of a small explicit symmetry breaking the pion acquires a small mass, the axial current is only partially conserved, and the nucleon and σ are shifted from their unperturbed masses by $\mathcal{O}(\epsilon)$. Furthermore, since $f = f_\pi$, eq. (12.35a) is the Goldberger–Treiman relation (12.14) with $g_A = 1$, rather than its physical value of 1.24,

$$m_N = g f_\pi. \qquad (12.37)$$

This is consistent for the tree level at which we are working since it is the higher-order diagrams that renormalize g_A from the bare value of unity to its physical value.

EXERCISE 12.6 Verify the mass equations (12.35) for the linear σ-model with explicit symmetry breaking of the form (12.22).

EXERCISE 12.7 Show that the vector and axial charges of the linear σ-model generate the $SU(2)_L \times SU(2)_R$ algebra (12.17). *Hint*: the currents are derived in Exercise 12.4.

12.4 Chiral $SU(3) \times SU(3)$ Symmetry

We have illustrated the idea of chiral symmetry for u and d quarks, leading to chiral $SU(2) \times SU(2)$ symmetry, but the same procedure can be applied for u, d, and s quarks. This generates a chiral $SU(3) \times SU(3)$ symmetry, sometimes denoted $SU(3)_L \times SU(3)_R$, associated with the Cabibbo vector and axial vector current octets (Dashen, 1969).

The symmetry pattern of the $SU(3)$ flavor model can be explained in terms of a hierarchy of symmetry breakings (Wigner and Goldstone modes), with the explicit breakings controlled by the mass terms of the QCD Lagrangian (10.32). With three *massless* quarks (u, d, s) the QCD Lagrangian is invariant under a global chiral $SU(3)_L \times SU(3)_R$ symmetry. Physical particles do not occur in parity doublets so it is assumed that the vacuum symmetry is spontaneously broken, producing an octet of massless Goldstone bosons (the mesons π, η, K, and $\overline{K}$, with the first an isotriplet, η a singlet, and the last two isodoublets—see Fig. 5.7). If the strange quark is then given a mass

$$m_s \neq 0 \qquad m_u = m_d = 0,$$

the symmetry is explicitly broken to $SU(2)_\mathrm{L} \times SU(2)_\mathrm{R}$, all mesons except the π acquire a mass, and the hadronic masses are split in the hypercharge direction. If instead we set

$$m_u = m_d = m_s \neq 0,$$

the chiral $SU(3)_\mathrm{L} \times SU(3)_\mathrm{R}$ symmetry is broken explicitly to flavor $SU(3)$, and all octet mesons acquire a mass. If we then set

$$m_s \neq 0 \qquad m_s \neq m_u \qquad m_u = m_d \neq 0$$

the $SU(3)$ symmetry is explicitly broken to $SU(2)$ isospin. Finally, if $m_u \neq m_d$ the isospin symmetry is broken explicitly to $U(1)$ charge conservation and the isospin multiplets are split. Because the approximation $m_u = m_d \simeq 0$ appears to be much better than $m_u = m_d = m_s \simeq 0$, isotopic spin is a better symmetry than flavor $SU(3)$, and $SU(2)_\mathrm{L} \times SU(2)_\mathrm{R}$ is better than $SU(3)_\mathrm{L} \times SU(3)_\mathrm{R}$. These assertions become more tangible in Exercise 12.8.

Statements like $m_u \simeq 0$ require the specification of a scale: "small relative to what?" The scale of the strong interactions is that of the QCD parameter Λ (§10.6.4), which is of the order of several hundred MeV. Therefore, the precise meaning of the foregoing statements is that approximate chiral, flavor, and isospin symmetries exist because the characteristic scale of the strong interactions is large compared with the current quark masses and their differences. The quark masses originate in the phenomenological Higgs sector of the electroweak theory [from terms analogous to eq. (9.22)] and Λ is a phenomenological QCD number (Fig. 10.17). Thus the preceding arguments are empirically correct, but fall short of a microscopic explanation for these symmetries.

EXERCISE 12.8 Show that for the QCD Lagrangian density (10.32) restricted to u, d, and s quarks the (current) quark mass terms can be written $M_q = M^0 + M' + M''$, where

$$M^0 = \tfrac{1}{3}(m_u + m_d + m_s)\left(\overline{u}u + \overline{d}d + \overline{s}s\right)$$

$$M' = \tfrac{1}{6}(2m_s - m_u - m_d)\left(2\overline{s}s - \overline{u}u - \overline{d}d\right)$$

$$M'' = \tfrac{1}{2}(m_u - m_d)\left(\overline{u}u - \overline{d}d\right).$$

Show that M^0 breaks chiral $SU(3) \times SU(3)$ down to flavor $SU(3)$, that M' breaks flavor $SU(3)$ to isospin $SU(2)$, and that M'' breaks isospin $SU(2)$ down to electromagnetic $U(1)$. Show that if all strange quarks are neglected, M^0 breaks chiral $SU(2) \times SU(2)$ to isospin $SU(2)$. These symmetry breakings are explicit. Now use these results, the current quark masses given in (10.37), and spontaneous symmetry breaking in the near-Goldstone mode (§12.3.3) to explain the mass spectrum for hadrons containing only u, d, or s quarks.

12.5 Bag Models

Having been motivated in §12.1 by a general discussion of the QCD vacuum and the phenomenology of confinement expressed in terms of domain structure in an antiscreening medium, we now investigate bag models of hadrons. Since these phenomenological models represent approximations to QCD, we would like them to incorporate the characteristic features of that theory. These are taken to be (1) confinement, (2) asymptotic freedom, and (3) approximate chiral symmetry. The first requires that the bag boundary conditions confine color so that only color-singlet hadronic states are observed. The second means that within a bag the quarks must move relatively freely, in order to reproduce the successes of the quark–parton model in deep inelastic scattering. The third implies that one of two features must characterize the particle spectrum: (1) if chiral symmetry is implemented in the Wigner mode, the hadrons must enter the theory as degenerate parity doublets; (2) if chiral symmetry is realized in the Goldstone mode, there will be massless pseudoscalar particles in the theory (see §12.3).

We will see that the simplest bag models incorporate confinement and asymptotic freedom in a relatively satisfactory way, but there is a basic difficulty in implementing chiral symmetry for any system that reflects quarks at a boundary. A possible solution will be seen to lie in the chiral bag models, where chiral symmetry is restored through a pion field.

12.5.1 Basic Bags

As a starting point we take seriously the hints of preceding discussions that a hadron looks like a bubble immersed in a complicated medium that is the true QCD vacuum. To make the simplest possible model, let us assume that the interior and exterior of the bag are in two different phases—the effective color dielectric constant κ can serve as a phenomenological order parameter—and that the boundary between the phases is sharp. No one has proved that QCD leads to such a structure, but there are strong reasons to believe that it does, as we have anticipated in §12.1.

Although realistic bags may be deformed (see §12.5.6), we will assume them to be spherical for now. The bag model problem then consists of (1) finding the equilibrium radius for the bag, (2) solving the Dirac equation for free quarks confined to this spherical cavity, and (3) calculating residual gluon exchange interactions among the confined quarks by perturbation theory starting from the solutions in (2).

Vacuum Bubbles. If we consider the hadron to be a bubble of the perturbative vacuum surrounded by non-perturbative vacuum, the hadronic energy is (Rho, 1982)

$$E_{\text{bag}} = E - \epsilon_{\text{vac}}V, \tag{12.38}$$

where E is the total energy (bubble + surrounding medium), ϵ_{vac} is the energy density of the normal vacuum, and V is the total volume of the system. If V_{bag} is the volume of the bubble and E_0 the energy of the fields inside the bubble,

$$E = E_0 + \epsilon_{\text{vac}}(V - V_{\text{bag}}), \tag{12.39}$$

from which

$$E_{\text{bag}} = E_0 - \epsilon_{\text{vac}}V_{\text{bag}}. \tag{12.40}$$

To mesh with later notation we may define a bag constant B by

$$B \equiv -\epsilon_{\text{vac}} \tag{12.41}$$

so that

$$E_{\text{bag}} = E_0 + BV_{\text{bag}}. \tag{12.42}$$

If B were negative the system could minimize its energy by allowing V_{bag} to increase without bound, so finite bag size requires $B \geqslant 0$. This implies that the energy density of the normal vacuum is lower than that of the bubble (taken to be zero by convention): it costs an energy BV_{bag} to make a bubble in the vacuum in which the quarks can move relatively freely.

The energy E_0 consists of kinetic energies associated with the confined quarks and the field energies for gluons in the bag. The former contribution dominates for hadrons composed of light (relativistic) quarks; the latter assumes increasing importance for heavy-quark systems, or for bags stretched into strings (§12.5.5).

Bag Equations. The MIT bag model can be constructed by heuristic arguments starting from the preceding observations, but for later purposes it will be useful to derive it from a Lagrangian density

$$\mathcal{L}_{\text{bag}} = \left(\mathcal{L}_{\text{QCD}} - B\right)\theta_{\text{V}} - \tfrac{1}{2}\bar{q}(x)q(x)\Delta_{\text{S}}, \tag{12.43}$$

where $\mathcal{L}_{\text{QCD}}$ is the QCD Lagrangian (10.32), the step function θ_{V} is unity inside the spacetime region occupied by the bag and zero on the surface and outside, B is the bag constant (which will be interpreted as the energy density of the bag), and Δ_{S} is a function that is unity on the bag surface and zero otherwise (surface δ-function). The last term may be viewed as a Lagrange multiplier imposing the confinement condition on the quark fields (see Exercise 12.9a). For static, spherical bags we may take $\theta_{\text{V}}(x) = \theta(R - r)$ and $\Delta_{\text{S}}(x) = \delta(r - R)$.

Neglecting the gluons and applying Hamilton's principle, the action

$$S = \int dt \int_V d^3x\, \mathcal{L}_{\text{bag}}$$

exhibits an extremum under variations of the quark fields q and the bag volume

V provided that (DeTar and Donoghue, 1983; Miller, 1984; Thomas, 1983, 1984; Johnson, 1978; Heller, 1982)

$$\text{Dirac:} \qquad (i\partial\!\!\!/ - m)q = 0 \qquad \text{(in } V) \qquad (12.44)$$

$$\text{linear condition:} \qquad in^\mu\gamma_\mu q = q \qquad \text{(on } S) \qquad (12.45)$$

$$\text{nonlinear condition:} \qquad \tfrac{1}{2}n^\mu\partial_\mu(\bar{q}q) = -B \qquad \text{(on } S), \qquad (12.46)$$

where n^μ is an outward normal to the surface S enclosing the spherical volume V [for a static spherical bag we may take $n^\mu = (0,\hat{\mathbf{r}})$]. The first two equations describe quarks moving in a volume V with a *linear boundary condition* requiring the normal component of the vector current to vanish at the surface (Exercise 12.9a); thus the vector current is conserved:

$$n^\mu J_\mu|_S = n^\mu(\bar{q}\gamma_\mu q)|_S = 0. \qquad (12.47)$$

Because the quarks are described by a linear Dirac equation, the confining boundary condition is not that the quark density $\bar{q}\gamma^0 q = q^\dagger q$ vanish at the confinement radius R (see Fig. 12.7a). Instead, confinement requires that the color current normal to the surface at R vanish. The third equation (12.46) requires that the outward pressure of the quark fields balances the bag pressure B coming from the normal vacuum surrounding the bag. This is called the *nonlinear boundary condition*, because it depends on $\bar{q}q$.

Chiral Symmetry Breaking. The bag is stable because of the nonlinear boundary condition; it incorporates confinement by fiat, because of the linear boundary condition; the quarks will move relatively freely within the bag if we choose to allow only weak residual interactions among them, so the property of asymptotic freedom is approximately included. However, the bag described

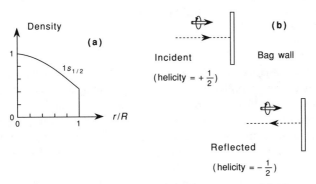

Fig. 12.7   (a) Quark density for $1s_{1/2}$ bag model orbit. (b) Non-conservation of helicity: reflection from the bag wall changes the quark momentum direction but not the spin direction.

by the Lagrangian density (12.43) violates chiral symmetry, even for massless quarks. The last term is not chiral invariant because it has the form of a Dirac mass term, which we already know to be intolerable for chiral symmetry. Since this term is associated with confinement, a relationship is suggested between confinement and chiral symmetry breaking. This connection can be seen heuristically without explicit appeal to the Lagrangian density. The chiral direct-product symmetry $SU(2)_L \times SU(2)_R$ means that no part of the Lagrangian mixes right-handed and left-handed fields [see eq. (12.17)]. Bag models that confine by scalar potentials violate this symmetry because a reflection at the boundary alters the momentum direction but not the spin of a quark, as shown in Fig. 12.7. This changes the helicity of a fermion, mixes right-handed and left-handed fields, and breaks chiral symmetry (see also Exercise 6.2e).

As we might guess from prior discussions of PCAC, this breakdown of chiral symmetry means that the bag fails to conserve the axial vector current, even for massless quarks. The axial current within the bag is carried by the quarks

$$\mathbf{A}^\mu(x) = \bar{q}(x)\gamma^\mu\gamma_5\frac{\boldsymbol{\tau}}{2}q(x)\theta_V \qquad \text{(interior)}, \qquad (12.48)$$

and at the bag surface (Exercise 12.9b)

$$\partial_\mu \mathbf{A}^\mu = i\bar{q}\gamma_5\frac{\boldsymbol{\tau}}{2}q\Delta_S \neq 0, \qquad \text{(surface)}, \qquad (12.49)$$

so the vector current is conserved (12.47) but the axial vector current is not. For now we ignore this failure of the bag model to respect a basic QCD symmetry; later, we will see how to restore it through the introduction of additional fields.

EXERCISE 12.9 (a) Prove that the linear boundary condition (12.45) implies that $\bar{q}q = 0$ on S and therefore leads to (12.47). (b) Use the hint given in Exercise 12.4b to show that the divergence of the axial current at the bag surface is given by (12.49).

Static Cavity Approximation. In the interest of tractable equations, and in the spirit of most applications of the bag model, we will restrict consideration to static bags fixed in space (*cavity approximation*). The implications of this symmetry violation will be considered below. To implement the cavity approximation, the bag is frozen in a shape and size that agrees with the nonlinear boundary condition in an average sense (DeTar and Donoghue, 1983)

$$\left\langle \tfrac{1}{2}n^\mu \partial_\mu(\bar{q}q) \right\rangle = -B. \qquad (12.50)$$

For static bags this requires the bag energy to be minimized as a function of size and shape; if we restrict consideration to spherical bags of radius R, (12.50) is equivalent to the condition

$$\frac{\partial E}{\partial R} = 0. \tag{12.51}$$

The Dirac equation (12.44) is now solved subject to the boundary conditions. The two lowest spin-$\frac{1}{2}$ solutions Ψ_κ are characterized by the Dirac quantum number

$$\kappa = \pm \left(j + \tfrac{1}{2} \right) = \pm 1,$$

and are given by (Close, 1979)

$$q_{s_{1/2}}(r) \equiv \Psi_{\kappa=-1} = N \begin{pmatrix} \sqrt{\dfrac{\omega+m}{\omega}}\, i j_0 \left(\dfrac{rx}{R} \right) \chi \\[3mm] -\sqrt{\dfrac{\omega-m}{\omega}}\, j_1 \left(\dfrac{rx}{R} \right) \boldsymbol{\sigma} \cdot \hat{\mathbf{r}} \chi \end{pmatrix} \tag{12.52a}$$

$$q_{p_{1/2}}(r) \equiv \Psi_{\kappa=+1} = N' \begin{pmatrix} \sqrt{\dfrac{\omega+m}{\omega}}\, i j_1 \left(\dfrac{rx}{R} \right) \boldsymbol{\sigma} \cdot \hat{\mathbf{r}} \chi \\[3mm] \sqrt{\dfrac{\omega-m}{\omega}}\, j_0 \left(\dfrac{rx}{R} \right) \chi \end{pmatrix}, \tag{12.52b}$$

where N and N' are normalizations, j_0 and j_1 are spherical Bessel functions, χ is a 2-component Pauli spinor, and R is the bag radius. The energies are determined by solutions x of the transcendental equation

$$\tan x = \frac{x}{1 - mR \pm \omega R}, \tag{12.53a}$$

where the $\pm$ sign is for the $\kappa = \pm 1$ solution and

$$E = \omega = \sqrt{\frac{(mR)^2 + x^2}{R^2}}. \tag{12.53b}$$

These solutions are labeled by orbital angular momentum l, but they are relativistic so $\boldsymbol{j} = \boldsymbol{l} + \boldsymbol{s}$ is conserved, but neither l nor s is separately a good quantum number (see Exercise 12.10).

In the limit of massless quarks the square root factors of (12.52) are unity, $\omega R = x$, equation (12.53a) reduces to

$$\tan x = \frac{x}{1 \pm x}, \tag{12.54}$$

and the energies are $E = x/R$:

$$E_{s_{1/2}} = \frac{2.043}{R} \qquad E_{p_{1/2}} = \frac{3.812}{R}. \qquad (12.55)$$

The Dirac fields within the bag may be expanded in terms of such *cavity eigenfunction* basis states. A more extensive discussion of the Dirac equation in spherical cavities may be found in Thomas (1984) and in Bhaduri (1988), Ch. 2.

If we neglect the gluon fields and the quark masses, the Hamiltonian following from (12.43) is

$$H = \int d^3x \left(\mathcal{H}_{\text{Dirac}} + B \right) + E_0, \qquad (12.56)$$

where E_0 is the finite part of the zero-point energy associated with the fermion fields.[‡] The value of E_0 is controversial, but it is often assumed that

$$E_0 = -\frac{Z_0}{R}, \qquad (12.57)$$

with Z_0 to be determined phenomenologically; such fits favor $Z_0 \simeq 1 - 1.8$. The parameter Z_0 may also contain corrections for spurious translational bag modes (center-of-mass correction; see below) and quark self-energies. Most theoretical attempts to account for the empirical value of Z_0 have had limited success. For an example of a calculation of Z_0, see Milton (1983).

With this prescription, if we assume that n massless quarks occupy the lowest cavity eigenstate the bag mass is

$$E = \langle H \rangle = \frac{2.043n - Z_0}{R} + \tfrac{4}{3}\pi R^3 B. \qquad (12.58)$$

The nonlinear boundary condition requires that for stability $\partial E/\partial R = 0$, so that

$$R^4 = \frac{2.043n - Z_0}{4\pi B}, \qquad (12.59)$$

and after substituting back in (12.58)

$$E = \tfrac{4}{3}(2.043n - Z_0)^{3/4}(4\pi B)^{1/4}. \qquad (12.60)$$

[‡]A delicate point, hinging on the fact that we are working with fields in a *finite* volume. In field quantization we typically must deal with infinite zero-point energies (§2.4.4). However, for quantization in a cavity of finite volume there will also be finite zero-point contributions that are dependent on cavity size. This is related to the Casimir effect in QED (for example, see Plunien, Müller, and Greiner, 1986; Itzykson and Zuber, 1980, §3–2–4).

The parameters B and Z_0 may be determined by fits to known hadronic masses. In the minimal implementation of the bag model it is assumed that they are universal constants; once they are fixed the bag radius R follows from (12.59). Taking the average mass of the nucleon and the Δ-resonance to fix the baryonic mass scale for hadrons composed of light (we assume massless) u and d quarks, and ignoring Z_0 for now, one finds as a crude initial estimate

$$B^{1/4} \simeq 110 \text{ MeV} \qquad R \simeq 1.5 \text{ fm} \qquad \text{(baryons)}. \qquad (12.61)$$

In this exercise we have neglected the residual interactions of the quarks within the bag that cause the split of approximately 300 MeV between the nucleon and Δ masses. Residual one-gluon effects will be introduced shortly. In addition, we will see below that in hadrons containing strange and heavier quarks the neglect of quark masses may no longer be justified, and that in some cases we cannot ignore the energy of the gluon fields in the bag.

If the bag constant B is assumed to be universal and we neglect Z_0, eqns. (12.59) and (12.60) fix the meson masses in terms of the baryon masses

$$\frac{M_{\text{meson}}}{M_{\text{baryon}}} \simeq \left(\frac{2}{3}\right)^{3/4} \simeq 0.74, \qquad (12.62)$$

and the meson radii in terms of the baryon radii

$$\frac{R_{\text{meson}}}{R_{\text{baryon}}} = \left(\frac{2}{3}\right)^{1/4} \simeq 0.90. \qquad (12.63)$$

12.5.2 Refinements: Quark Masses and Gluons

The preceding discussion introduces bag models in their most rudimentary form, but there must be more to the story than this. Figure 12.8 displays the masses of some light hadrons. The simple considerations to this point would suggest that all baryons have the same mass and that all mesons have a mass of approximately $\frac{3}{4}M_{\text{baryon}}$. This is a clear oversimplification: from Fig. 12.8 the replacement of a u or d quark by an s-quark increases the mass of a hadron by about 150 MeV, and there seems to be a spin dependence in the masses (for example, the $J = \frac{1}{2}$ baryons lie several hundred MeV lower in energy than the $J = \frac{3}{2}$ baryons).

We may expect that the excess mass of the hadrons containing strange quarks results from a finite effective mass for the strange quark within the bag. Therefore, we introduce the strange quark mass m_s as a parameter. To include additional mass splittings it is necessary to provide for residual interactions between the quarks in the bag. It is usually assumed that this residual interaction can be calculated perturbatively in terms of one-gluon exchange

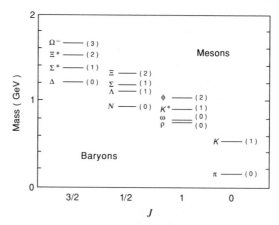

Fig. 12.8   Mass spectrum for some light hadrons. Numbers in parentheses give the number of strange quarks or antiquarks in the valence configuration. The η and η' have been omitted from the $J = 0$ states [see the "$U(1)$ problem"; §10.6.2].

between the quarks in the bag. A formal justification of this assumption is tenuous, but the expected nearly free nature of the quarks within the bag makes it plausible.

For a cavity containing only gluons (*pure glue*) the equations of motion following from (12.43) are (DeTar and Donoghue, 1983)

$$\partial^\mu F^a_{\mu\nu} = 0 \qquad \text{(in } V) \qquad\qquad (12.64a)$$

$$n^\mu F^a_{\mu\nu} = 0 \qquad \text{(on } S) \qquad\qquad (12.64b)$$

$$-\tfrac{1}{4} F^a_{\mu\nu} F^{\mu\nu}_a = B \qquad \text{(on } S), \qquad\qquad (12.64c)$$

where $F^a_{\mu\nu}$ is the QCD non-Abelian field strength tensor (10.34). The first equation is the QCD analog of Maxwell's equations in the bag; the condition $n^\mu F^a_{\mu\nu} = 0$ requires that the chromoelectric flux be tangent to the surface in the rest frame of the surface point; the last equation imposes energy balance.

If consideration is restricted to lowest order in the color interaction $\alpha_s = g^2/4\pi$, the problem is greatly simplified: the gluon self-couplings make no contribution, the non-Abelian fields reduce to eight independent gluon fields, the non-Abelian field strength tensor reverts to the Abelian Maxwell form $F^a_{\mu\nu} = \partial_\mu A^a_\nu - \partial_\nu A^a_\mu$, and eqns. (12.64) have the demeanor of eight independent electromagnetic fields confined to a spherical cavity. The lowest energy cavity normal modes E_{J^P} for the gluon field are (DeTar and Donoghue, 1983)

$$E_{1+} = \frac{2.74}{R} \qquad E_{2-} = \frac{3.87}{R} \qquad E_{1-} = \frac{4.49}{R}. \qquad\qquad (12.65)$$

The boundary conditions (12.64) require that the total color charge of such states vanish.

In the hydrogen atom the Fermi–Breit interaction associated with transverse one-photon exchange splits the degenerate levels according to whether the spins are parallel or antiparallel, with the $^3S_1$ state pushed up in energy and the $^1S_0$ state pushed down. (The electromagnetic transition between these states gives the 21 cm spin-flip line of radio astronomy fame.) In one-gluon exchange QCD the gluon fields act as eight independent Abelian gauge fields (color photons). This leads to a "color magnetic" interaction of the Fermi–Breit form [see (12.98) for the full Fermi–Breit Hamiltonian]

$$\Delta H_{\sigma\sigma} \simeq \sum_{i \neq j} C_{ij} \frac{(\mathbf{F}_i \cdot \mathbf{F}_j)\,(\mathbf{S}_i \cdot \mathbf{S}_j)}{m_i m_j}, \tag{12.66}$$

where i and j label quarks, $\mathbf{F} = \boldsymbol{\lambda}/2$ denotes color $SU(3)$ generators (5.67), $\mathbf{S} = \boldsymbol{\sigma}/2$ denotes spin $SU(2)$ generators (1.78), and m_i is the effective mass of quark i (Close, 1979).

By analogy with the hyperfine splitting of the hydrogen atom, this interaction will push vector meson $q\bar{q}$ states up and pseudoscalar meson states down in energy (Fig. 12.8). That is, if we assume the members of the π and ρ octets to have their quarks in the lowest ($s_{1/2}$) bag states, the difference in angular momentum between the pseudoscalar and vector multiplets comes from the former being $^1S_0$ (antiparallel spins) and the latter being $^3S_1$ (parallel spins). Likewise, the $J = \frac{3}{2}$ baryons may be expected to lie higher in energy than the $J = \frac{1}{2}$ baryons. On comparing the meson and baryon spectra in Fig. 12.8, we see that such a spin-dependent interaction is strongly indicated by the data. The contribution to the bag energy from one-gluon exchange for color singlet hadrons is (Thomas, 1984)

$$\Delta E_{\sigma\sigma} = \epsilon \frac{\alpha_s}{R} \sum_{i<j} f(m_i, m_j, R)\boldsymbol{\sigma}_i \cdot \boldsymbol{\sigma}_j, \tag{12.67}$$

where $\epsilon = 1$ for baryons and $\epsilon = 2$ for mesons (this factor arises from the color matrices—see Exercise 12.11b), and $f(m_i, m_j, R)$ is a known function coming from radial integration of quark wavefunctions (DeGrand et al., 1975). These hadronic spin–spin "hyperfine" splittings have typical magnitudes of several hundred MeV.

EXERCISE 12.10 Show that solutions of the Dirac equation conserve the total angular momentum

$$\mathbf{J} = \mathbf{L} + \tfrac{1}{2}\boldsymbol{\Sigma} \qquad \mathbf{L} = \mathbf{r} \times \mathbf{p} \qquad \boldsymbol{\Sigma} = \begin{pmatrix} \boldsymbol{\sigma} & 0 \\ 0 & \boldsymbol{\sigma} \end{pmatrix},$$

but not the orbital ($\mathbf{L}$) or spin ($\boldsymbol{\Sigma}$) angular momenta separately.

EXERCISE 12.11   (a) The baryons Σ^0 and Λ have a uds valence quark content (Figs. 5.7, 5.9). Assume these hadrons to result from placing quarks in the lowest available bag states and consider a generalized exclusion principle in the spin, flavor, spatial, and color degrees of freedom. Show that if the baryons are color singlets the Σ^0 angular momentum gets contributions from all three valence quarks, but the spin of the Λ is carried entirely by the strange quark. *Hint*: use the results of Exercise 12.14 and consider the symmetries of the ud pair first.   (b) Show that in eq. (12.66), $\langle \mathbf{F}_i \cdot \mathbf{F}_j \rangle = -\frac{4}{3}$ for the q and $\bar{q}$ in a color singlet meson and $-\frac{2}{3}$ for two quarks in a color singlet baryon, so that $\epsilon_{\text{meson}}/\epsilon_{\text{baryon}} = 2$ in (12.67). *Hint*: use

$$\langle \mathbf{F}_1 \cdot \mathbf{F}_2 \rangle = \tfrac{1}{2} \left(\langle \mathbf{F}^2 \rangle - \langle \mathbf{F}_1^2 \rangle - \langle \mathbf{F}_2^2 \rangle \right)$$

with $\mathbf{F} = \mathbf{F}_1 + \mathbf{F}_2$, use the results of Exercise 12.14 for the baryon, and use Table 5.4.

12.5.3   Spurious Center-of-Mass Motion

The bag model is a kind of relativistic shell model for a few-body system. By their very nature shell models violate translational invariance, and this violation is more severe the fewer the number of particles (Ring and Schuck, 1980). While this difficulty causes mass shifts in ground states, it is even more important for excited states because it produces a set of spurious states associated with motion of the center-of-mass (CM).

The CM problem may be the most serious deficiency of the bag model. The standard shell-model techniques for projecting states of good momentum cannot be used directly in the cavity approximation because they require knowledge of the underlying translationally invariant Hamiltonian. This is not available because the Hamiltonian is defined only for a particular cavity— there is no superposition principle for different cavities.

One empirical procedure to correct for CM motion assumes that the bag energy for a hadron is (DeTar and Donoghue, 1983)

$$E_{\text{bag}} = \langle H \rangle = \left\langle \sqrt{\mathbf{p}^2 + m^2} \right\rangle, \tag{12.68}$$

where the bag wavefunction is assumed to be a wavepacket of nucleon momentum states with $\langle \mathbf{p} \rangle = 0$, but $\langle \mathbf{p}^2 \rangle \neq 0$. If $m \gg 1/R$, then $m^2 \gg \mathbf{p}^2$ and

$$E_{\text{bag}} \simeq m + \frac{\langle \mathbf{p}^2 \rangle}{2m}. \tag{12.69}$$

A simple estimate for the correction term is

$$\langle \mathbf{p}^2 \rangle \simeq \frac{n x^2}{R^2}, \tag{12.70}$$

where $x = 2.043$ is the dimensionless wave number for a massless $s_{1/2}$ quark. For a nucleon $n = 3$, and taking $R = 1.5$ fm gives $\langle \mathbf{p}^2 \rangle /2m_n \simeq 115$ MeV for the difference between the bag energy and nucleon mass, a 12% correction. For the pion the problem becomes acute: the preceding estimate would give a correction much larger than the pion mass. Of course the assumption that $m \gg 1/R$ leading to the above equations is not satisfied for this case, but the results indicate serious problems.

The discussion of more sophisticated attempts to correct for CM motion is technical and we will not consider it further. Let us simply note that the correction term in (12.69) can be written $\langle \mathbf{p}^2 \rangle /2m \propto 1/R$, which is the same form as the zero-point energy contribution (12.57). Therefore, one empirical approach is to keep a term such as (12.57) in the bag energy and interpret the constant Z_0 as measuring both zero-point energy and center-of-mass energy (and possible self-energy contributions). From the preceding discussion, if we take $Z_0 = 1.8$ and $R \simeq 1.5$ fm in (12.57), then $Z_0/R \simeq 240$ MeV and for the nucleon we estimate that about half of the phenomenological energy Z_0/R is actually CM energy and not zero-point energy.

12.5.4 Hadrons Containing u, d, and s Quarks

Our basic bag now embellished with gluons, massive quarks, and approximate corrections for CM energy and zero-point motion, we confront the data. For light-quark bags we ignore contributions from the gluon field except for one-gluon exchange (for heavy quarks, or high angular momentum states corresponding to elongated bags, this may not be a good approximation).

Masses. The simple energy expression (12.58) now takes the form

$$E(R) = \sum_i \omega_i + \frac{4}{3}\pi B R^3 + \Delta E_{\sigma\sigma} - \frac{Z_0}{R}, \qquad (12.71)$$

where the first term is a sum of contributions (12.53b) for the quarks (possibly massive), the constant B is the bag pressure, $\Delta E_{\sigma\sigma}$ is the spin–spin splitting coming from one-gluon exchange (12.67) and depending on the coupling constant $\alpha_s = g_s^2/4\pi$, and the last term accounts empirically for omitted effects such as zero-point and CM motion, and quark self-energies. The strong coupling g_s varies with momentum transfer [eq. (10.41)], so the parameter α_s entering (12.71) must be understood as an effective color coupling operating on a length scale characteristic of the bag. If we assume the u and d quarks to be massless there are four parameters: the strange quark mass m_s, the bag constant B, the effective bag color coupling constant α_s, and Z_0. Application of (12.51) to (12.71) determines the stability condition, as for the basic bag. In Fig. 12.9 we show a calculation of hadronic masses using the MIT bag

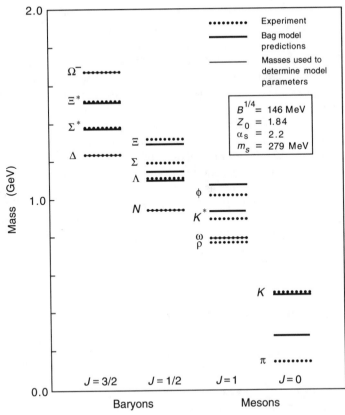

Fig. 12.9 Hadronic masses in the MIT bag model (DeGrand et al., 1975).

model (DeGrand et al., 1975). The masses of the N, Δ, ω, and Ω were used to fix the four parameters:

$$B^{1/4} = 146 \text{ MeV} \qquad m_s = 279 \text{ MeV}$$
$$Z_0 = 1.84 \qquad \alpha_s = 2.2; \tag{12.72}$$

all other masses are predictions. The calculated masses are all rather good except for the pion, which is expected to have severe CM motion problems.

Radii. Once the bag parameters have been determined by fitting the hadronic mass spectrum, a variety of other hadronic properties can be calculated. The value of R for a baryon is, from (12.59) and the parameters of Fig. 12.9,

$$R_{\text{baryon}} = 5.2 \text{ GeV}^{-1} \simeq 1 \text{ fm},$$

while for a meson

$$R_{\text{meson}} = 4.5 \text{ GeV}^{-1} \simeq 0.9 \text{ fm}.$$

[In using (12.59) we are neglecting contributions to the bag energy from one-gluon exchange and finite quark masses, insofar as these have a bearing on the equilibrium radius.] Thus, a more careful treatment of the bag yields a smaller radius than our original simple estimate in (12.61).

The matter density for a Dirac particle confined in the bag is given by the $\mu = 0$ component of the Dirac vector current

$$J_i^0(\mathbf{r}) = \bar{q}_i(\mathbf{r})\gamma^0 q_i(\mathbf{r})\theta_V = q_i^\dagger(\mathbf{r})q_i(\mathbf{r})\theta_V, \tag{12.73}$$

where i labels the quarks in the bag. The mean-square charge radius for a hadron is then given by

$$\langle r^2 \rangle_{\text{charge}} = \sum_i Q_i \int_{\text{bag}} r^2 d\mathbf{r} \, q_i^\dagger(\mathbf{r}) q_i(\mathbf{r}), \tag{12.74}$$

where Q_i is the charge of the ith quark. Evaluating this using the bag wavefunctions (12.52) gives $\langle r^2 \rangle \propto R^2$, and for the proton (DeGrand et al., 1975)

$$\langle r^2 \rangle_{\text{proton}}^{1/2} = 0.73 \text{ fm}, \tag{12.75}$$

whereas the experimental value is $\langle r^2 \rangle^{1/2} \simeq 0.88 \pm 0.03$ fm. The corresponding bag prediction for the neutron is zero, which may be compared with the experimental value of $\langle r^2 \rangle = -0.12 \pm 0.01$ fm^2.

Magnetic Moments. The magnetic moment is (Thomas, 1984)

$$\boldsymbol{\mu} = \tfrac{1}{2} \int (\mathbf{r} \times \mathbf{j}_{\text{em}}) d\mathbf{r}$$

$$= \tfrac{1}{2} \int_{\text{bag}} d\mathbf{r} \, \mathbf{r} \times \sum_i (\bar{q}_i(\mathbf{r})\boldsymbol{\gamma}_i q_i(\mathbf{r})) Q_i, \tag{12.76}$$

where a sum over Dirac currents for quarks of charge Q_i has been substituted, and $\gamma \equiv (\gamma_0, \boldsymbol{\gamma})$. Inserting the $1s_{1/2}$ wavefunctions (12.52) into (12.76) and employing some spin algebra yields formulas for hadronic magnetic moments that have no free parameters once the four bag parameters are fixed by the hadronic masses. On an absolute scale the predicted moments are not very good. For example, the best-fit parameters of DeGrand et al. (1975) yield a proton magnetic moment

$$\mu_p^{\text{bag}} = 1.9 \text{ n.m.}, \tag{12.77}$$

Table 12.1
Baryon Magnetic Moments in Units of the Proton Moment

Baryon	Experiment	MIT Bag Model	Quark Model
n	−0.685	−0.67	$-\frac{2}{3}$
Λ	−0.220	−0.26	$-\frac{1}{3}$
Σ^-	−0.414	−0.36	$-\frac{1}{3}$
Σ^+	+0.866	+0.97	+1
Ξ^0	−0.450	−0.56	$-\frac{2}{3}$
Ξ^-	−0.247	−0.23	$-\frac{1}{3}$

(n.m. = nuclear magnetons), which compares poorly with the measured

$$\mu_p^{\text{exp}} = 2.792 \text{ n.m.} \tag{12.78}$$

However, the ratios of baryon magnetic moments fare much better. Table 12.1 shows the bag model predictions for some nucleon octet magnetic moments normalized to the proton moment. Also shown in Table 12.1 are the ratios for the constituent quark model (see Exercise 12.12). The bag model is seen to do as well, or somewhat better, in each case.

EXERCISE 12.12 The nonrelativistic magnetic moment of a point quark is $\mu_i = q_i/2m_i$, where q_i is the quark charge and m_i the effective mass. Define the magnetic moment of the proton as a sum over quark contributions,

$$\mu_p = \sum_{i=1}^{3} \langle p_\uparrow | \mu_i \sigma_3^i | p_\uparrow \rangle,$$

where $|p_\uparrow\rangle$ denotes a proton in the $M = J = \frac{1}{2}$ angular momentum state, and σ_3^i is a Pauli matrix operating on the spin wavefunction of the ith quark. Use the proton wavefunction of Exercise 5.14b to calculate μ_p in terms of the quark moments μ_u and μ_d. Construct a quark wavefunction for the neutron and show that $\mu_p/\mu_n = -\frac{3}{2}$, if we assume $m_u = m_d$.

Axial-Vector Coupling Coefficient. One of the liabilities of the $SU(6)$ nonrelativistic quark model is the prediction (see Close, 1979, Ch. 6)

$$\left(\frac{g_A}{g_V}\right)_{SU(6)} = \frac{5}{3} \tag{12.79}$$

for the ratio of the axial vector to vector weak coupling coefficients. Experimentally the ratio is

$$\left(\frac{g_A}{g_V}\right)_{\text{exp}} = 1.26 \quad . \tag{12.80}$$

This ratio can be calculated for the bag model using the cavity wavefunctions, with details to be found in Thomas (1984). The result with massless quarks (DeGrand et al., 1975),

$$\left(\frac{g_A}{g_V}\right)_{\text{bag}} = 1.09, \tag{12.81}$$

is in much better agreement with experiment than the corresponding prediction of the $SU(6)$ model [empirical center-of-mass corrections and finite quark masses typically increase (12.81) to near the value of (12.80)]. Explicit relativistic effects associated with the lower components of the Dirac wavefunction cause the difference between (12.81) and the nonrelativistic ratio (12.79). Since the lower components are negligible in the nonrelativistic limit (Exercise 1.9d), the value of g_A/g_V is often cited as evidence supporting the formal necessity of a relativistic description of hadrons containing light quarks.

12.5.5 A Variety of QCD States

We have thus far considered only the simplest bag states involving three quarks, or a quark and antiquark, coupled to a color singlet in the lowest cavity eigenstates of a spherical bag. However, a variety of other states appear to be compatible with QCD in general, and with bag models and nonrelativistic potential models in particular. In this section some of those possibilities will be summarized (DeTar and Donoghue, 1983).

Glueballs. Since QCD is non-Abelian, the gauge quanta themselves might bind into multiparticle states, either for pure gluon fields or in the presence of quarks. States of pure glue are called *gluonium* or *glueballs*, while states involving the binding of gluons to quarks are called *hybrids*. Given that the quarks transform as triplets under color $SU(3)$ and the gluons as an **8**, the simplest combinations yielding color singlets are (Exercise 12.13 and Close, 1982a)

$$\begin{aligned}
(GG)_{\mathbf{1}} &: (\mathbf{8} \otimes \mathbf{8})_{\mathbf{1}} \\
(GGG)_{\mathbf{1}} &: (\mathbf{8} \otimes \mathbf{8} \otimes \mathbf{8})_{\mathbf{1}} \\
(Gq\bar{q})_{\mathbf{1}} &: [\mathbf{8} \otimes (\mathbf{3} \otimes \bar{\mathbf{3}})_{\mathbf{8}}]_{\mathbf{1}} \\
(Gqqq)_{\mathbf{1}} &: [\mathbf{8} \otimes (\mathbf{3} \otimes \mathbf{3} \otimes \mathbf{3})_{\mathbf{8}}]_{\mathbf{1}},
\end{aligned} \tag{12.82}$$

where the subscripts **1** and **8** denote $SU(3)$ coupling to singlet and octet representations, respectively.

EXERCISE 12.13 Prove that (12.82) gives the simplest gluon states exhibiting $SU(3)$ singlet structure.

The possible existence of glueballs is more general than their occurrence in bag models, but bags provide one convenient framework for estimates (DeTar and Donoghue, 1983). From the cavity gluon modes (12.65) we may construct multigluon states. The lowest two-gluon states have $J^{PC} = 0^{++}$ and 2^{++}, with first excited states $J^{PC} = 0^{-+}$ and 2^{-+}, and the lowest three-gluon states have $J^{PC} = 0^{++}, 1^{+-}$, and 3^{+-}. The energies of these states may be calculated in bag models, lattice gauge theory, and nonrelativistic quark models. Typical mass estimates for the lowest states are in the range 1–2 GeV. There are candidate experimental states for glueballs (Burnett and Sharpe, 1990), but a better theoretical understanding seems necessary to establish this interpretation. Obviously the discovery of glueballs or hybrids would be a major triumph for QCD; it is less clear whether a failure to find direct evidence for such states would constitute a serious problem for the theory.

Exotic and Cryptoexotic Hadrons. Although $q\bar{q}$ and qqq constitute the simplest quark combinations with a singlet $SU(3)_c$ representation, there are other more complicated assemblages of quarks and gluons that can have a singlet structure and therefore could form bag states of finite mass [see eq. (12.82) and Exercises 12.2 and 12.13]. Some possibilities are

$$q\bar{q}q\bar{q} \qquad q\bar{q}G \qquad qqqq\bar{q} \qquad qqqG \qquad qqqqqq.$$

Among such singlet states, some have quantum numbers that cannot occur for the normal hadronic states qqq and $q\bar{q}$; these are called *exotics*. Others have states with the same quantum numbers as qqq and $q\bar{q}$; these are called *cryptoexotics*. As in the case of glueballs, the energies of these states can be estimated in the bag model, in potential models, and in lattice calculations. However, there are many theoretical uncertainties, and although various candidates have been proposed there are few data clearly demanding exotics or cryptoexotics for an interpretation. References and a compressed review of candidates for glueballs and non-$q\bar{q}$ mesons may be found in Hernandez et al. (1990) and in Burnett and Sharpe (1990).

String Models and Regge Trajectories. It has been known for some time that there is a striking relationship between the mass and angular momentum of many hadronic states. This is illustrated in Figs. 12.10 and 12.11, where one sees that many hadrons can be grouped on curves

$$J = \alpha_0 + \alpha' M^2, \tag{12.83}$$

with $\alpha' \simeq 1$ GeV^{-2}.

The lines on which the hadrons lie in the (J, M^2) plane are called *Regge tra-*

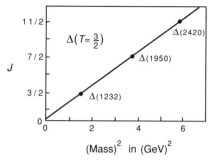

Fig. 12.10 A Regge trajectory for baryons.

jectories. They are reminiscent of the rotational bands familiar from nuclear and molecular physics, but the relationship between the energy and angular momentum is not the familiar $E \propto J(J+1)$ of the elementary nonrelativistic quantum rotor. The assorted particles lying on a Regge trajectory may be interpreted as *recurrences* of the same basic particle with differing amounts of angular momentum, similar to a rotational band in nuclear physics where a deformed intrinsic state generates a sequence of physical states by rotating collectively at different frequencies while maintaining the same internal structure (see §8.5.4).

Such trajectories have been explained in terms of *string models*, where the hadrons are described by a string rotating such that its ends travel at light velocity. It is possible to show that the relation (12.83) follows from such a model. However, we may view these string models as just special cases of bag models in which the bag has become elongated at it rotates. This is illustrated in Fig. 12.12, which should be compared with Fig. 12.2. In these figures the dashed lines indicate the color electric field, and the notation $(qq)_{\bar{3}}$ means that the qq pair must be in an $SU(3)$ state $\bar{3}$ to yield a color singlet when coupled to the other quark. In the color space we have a generalization of the rule for

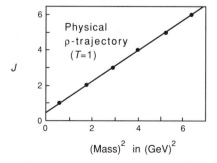

Fig. 12.11 A Regge trajectory for mesons (after Dalitz, 1982).

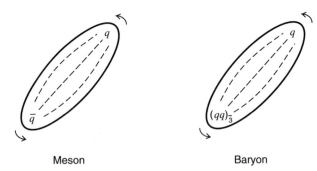

Meson   Baryon

Fig. 12.12   Rotating meson and baryon bags that deform to approximate rotating strings. Such rotating bags can generate the Regge trajectories shown in Figs. 12.10 and 12.11.

simpler Abelian electrical charges: like charges repel, unlike charges attract. From the valence quark model we conclude (qualitatively) that colors attract anticolors, and that two particles with color repel each other, because of the existence of stable mesons ($q\bar{q}$), but not stable qq states. Since the quark model assigns a qqq structure to baryons, in some sense the colors of two of the quarks in a baryon must combine such that they look like an antiquark to the third quark, thereby yielding an attraction as for the mesons. The $SU(3)$ algebra has this property since

$$qq = \mathbf{3} \otimes \mathbf{3} = \mathbf{6} \oplus \bar{\mathbf{3}} \qquad q = \mathbf{3} \qquad \bar{q} = \bar{\mathbf{3}},$$

so the $\bar{\mathbf{3}}$ occurring in qq looks like a $\bar{q}$ to the third quark (see Exercises 12.11 and 12.14).

By neglecting the quark kinetic energy within the deformed bag and balancing the pressure of the confined color fields against the bag pressure, Johnson and Thorn (1976) demonstrated that the bag mass could be written in the form (12.83), with α' determined by the bag parameters B and α_s:

$$\alpha' = \left(\frac{3}{128\pi^3 \alpha_s B}\right)^{1/2} \simeq 0.88 \text{ GeV}^{-2}, \tag{12.84}$$

which is in reasonable agreement with the observed trajectories. The linear energy density (also called the *string tension*) of the string corresponding to the deformed bag is (Exercise 12.15)

$$\sigma = \frac{1}{2\pi\alpha'} \simeq 0.18 \text{ GeV}^2 \simeq 900 \text{ MeV/fm}. \tag{12.85}$$

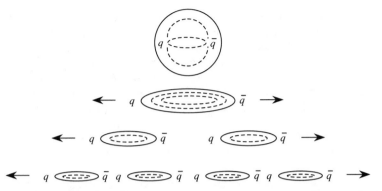

Fig. 12.13   Schematic illustration of an attempt to separate the q and $\bar{q}$ in a meson. As the bag deforms it becomes energetically feasible to create additional $q\bar{q}$ pairs from the vacuum and the strings break into smaller strings, with each still a color singlet. Thus the QCD analog of ionization is colorless meson production.

Thus the bag model can produce a string model.[‡]   The lines of force that connect the quarks in Fig. 12.12 become parallel for sufficient elongation of the bags, basically in response to the constrictive effect of the non-perturbative vacuum outside the bag (more microscopically: because of the self-coupling of the gluons); this produces confinement because the force needed to separate the quarks is constant, implying that an infinite amount of energy is required to separate them by an infinite distance. In practice, as the color sources are separated it becomes advantageous to create a $q\bar{q}$ pair from the vacuum at some intermediate point of the string and the string breaks into smaller (colorless) strings, as illustrated in Fig. 12.13. Presumably this figure is a caricature of the hadronization process taking place when one of the quarks is given a large recoil momentum in a deep inelastic process (see Fig. 10.1).

EXERCISE 12.14   Show that for color $SU(3)$ with colors $R, B,$ and G [see (10.35)], two-quark states are of the form $qq = \mathbf{3} \otimes \mathbf{3} = \mathbf{6} \oplus \bar{\mathbf{3}}$, with

$$\bar{\mathbf{3}}\,(\text{antisymmetric}) = (RG - GR, RB - BR, BG - GB)$$
$$\mathbf{6}\,(\text{symmetric}) = (RG + GR, RB + BR, GB + BG, RR, GG, BB),$$

so to have three quarks in a color singlet each pair of quarks belongs to the

---

[‡]Although the bag model can lead to a string model, quarks generally play a different role in bag and string models. In the spherical bag the quarks are *dynamical* objects obeying Dirac equations. In the simplest string version of the deformed bag the quarks are *static* objects that serve only to terminate the color field lines: the energy and angular momentum reside primarily in the color fields of the rotating bag.

$\bar{\mathbf{3}}$ of $\mathbf{3} \otimes \mathbf{3}$ and is antisymmetric in the color indices. Thus the noncolor part of any two-quark wavefunction in a baryon must be symmetric. *Hint*: the wavefunctions for this simple two-particle case may be guessed by forming all possible antisymmetric and symmetric combinations. More systematically, the wavefunctions may be constructed by finding all *standard arrangements for Young diagrams*. Lichtenberg (1978), Ch. 4, describes how to do this. A brief account may be found in the solution of this Exercise (Appendix D).

EXERCISE 12.15 Show that for a rotating string of energy density σ, with its endpoints moving at the velocity of light,

$$J = \alpha' M^2 \qquad \alpha' = \frac{1}{2\pi\sigma},$$

where J is the angular momentum and M the mass of the string.

Born–Oppenheimer Approximation. For bound states involving heavy quarks such as $b\bar{b}$ or $c\bar{c}$ the quark masses are sufficiently large that nonrelativistic quantum mechanics and potential descriptions may be formally justified (§12.6). For a bag description of such states a *Born–Oppenheimer or adiabatic approximation* may be appropriate. Recall that the Born–Oppenheimer approximation in molecular physics assumes that the electrons move much more rapidly than the nuclei, so that in molecules the electrons follow the motion of the nuclei adiabatically. If it is assumed that the gluon fields and bag boundaries move quickly relative to the motion of the heavy quarks, the bag problem can be solved in two steps: (1) the quarks are fixed in space at positions $\mathbf{r}_i$ and the gluon field and bag boundary resulting from the fixed sources are calculated, yielding the energy as a function of quark positions $E(\mathbf{r}_1, \mathbf{r}_2, \dots)$; (2) the energy calculated in the first step is used as a potential energy in a Schrödinger equation for the quark motion (see, e.g., Heller, 1982).

In this approximation the bag develops naturally around the quarks, the potential depends only on the relative coordinates $\mathbf{r}_i - \mathbf{r}_j$, and we do no violence to the translational invariance. This is not true for the bag model with light quarks, where one expects the quark motion to be too fast for the Born–Oppenheimer approximation. There the bag is first specified, then the quarks are fed into it; this guarantees a violation of translational invariance, with attendant spurious modes associated with motion of the center-of-mass.

The heavy-quark bound states are well described by nonrelativistic potentials and they will be discussed later in that context (§12.6). However, we observe here that in a bag description the short-range potential is Coulombic $(1/r)$, while at larger separations the bag forms flux tubes that generate a linear potential (constant force). This was anticipated in our previous discussion.

P-Wave and D-Wave Hadrons; Radial Excitations. So far we have concentrated on the lighter hadrons, which can be described by placing quarks in the $1s_{1/2}$ cavity eigenstates, where $n = 1$ is a radial quantum number. These are predominantly S-wave ($L = 0$) states, in the nonrelativistic limit. There are also many states of higher angular momentum in the hadronic spectrum that can be interpreted as orbital angular momentum excitations of quarks in the bag. In addition, there are excited states that correspond to radial excitations ($n > 1$) of the quarks (see Figs. 12.22 and 12.19). In the nonrelativistic limit these states may be classified according to an $SU(6) \times O(3)$ symmetry, where the $SU(6)$ is just the (flavor) $\times$ (spin) symmetry of §5.3.6, and $O(3)$ is the group associated with orbital angular momentum (Close, 1979, Ch. 5).

Such excited states can be investigated in the bag model, but there are severe difficulties because of spurious states associated with the center-of-mass motion. The nonrelativistic potential models handle the center-of-mass problem better than a bag does, and orbitally excited states are more easily discussed in terms of those models.

12.5.6 Chiral Bag Models

The bags that we have been considering violate a rudimentary QCD symmetry: for massless quarks the QCD Lagrangian density (10.32) is chiral invariant, but the Lagrangian density (12.43) of the MIT bag model is not. We have seen that this is an inherent property of the confinement itself (Fig. 12.7b), and that it leads to a nonconserved axial current at the bag surface [(12.48)–(12.49)]. The cure is to introduce new fields that couple to the surface of the bag in such a way that the axial current is continuous at the surface, with the axial current flowing from the bag being carried into the non-perturbative vacuum by the new fields. The new fields must be free to propagate in the normal QCD vacuum, so they must be color singlets. The pion field is well-suited to this task, and the *chiral bag models*[‡] restore chiral invariance by coupling a phenomenological pion field to the bag so that the axial current is conserved (Chodos and Thorn, 1975).

Two basic philosophies have been followed in implementing such hybrid models. The first assumes that chiral symmetry is present in two phases: the Wigner mode in the bag interior and the Goldstone mode in the exterior, with the bag surface defining a sharp boundary between the two phases (Brown and Rho, 1979). This hypothesis rigorously excludes pions from the bag, since they are assumed to appear only when the chiral symmetry is spontaneously broken. In this model the pions exert anisotropic pressure on the surface of the bag, which tends to shrink the bag and deform it; for this reason, it is often called the *little bag model*. The reduction in size of the bags would presumably suppress explicit quark effects in nuclei, bringing one back to the

---

[‡]These are also called hybrid bag models. In the discussion to follow we will concentrate on those modifications of basic bags that are required by chiral symmetry. Features such as perturbative one-gluon exchange and correction for zero-point motion may also be included in realistic hybrid bag calculations, even if we do not display such terms explicitly.

Yukawa picture of nucleons interacting through pion clouds for nuclei under ordinary conditions.

The second class of chiral bag models allows the pion fields to exist both inside and outside the bag; these are called *cloudy bag models* (Thomas, 1983, 1984). Not surprisingly, the exterior pion clouds for such models are much less important than those in the little bag models, and the cloudy bag radius is comparable with the MIT bag radius. This suggests that explicit quark effects might be more important for nuclear matter under normal conditions in the cloudy bag models than in the little bag models (but the situation may be more subtle than this; see §13.8).

Nonlinear σ-Model. Chiral bags basically correspond to an extension of the σ-model of §12.3 to include the bag as a finite-size source for the pion field. In this extension the pion is treated in a long-wavelength approximation—as a phenomenological field with its internal quark structure ignored. This evades the important question of how the pion, in its role as a Goldstone boson born of spontaneous chiral symmetry breaking, is related to the $q\bar{q}$ pion of the quark model. Such an approximation can only be expected to describe phenomena on a length scale where the internal quark structure of the pion is immaterial.

A bothersome conceptual problem for the linear σ-model is an interpretation for the scalar field σ. The lightest scalar resonances seen in $\pi\pi$-scattering are the $f_0(975)$ [formerly $S(975)$] with a width of 34 MeV, and the $f_0(1400)$ [formerly $\epsilon(1300)$] with a width of 150–400 MeV; there is little evidence that either should be identified with fluctuations of the scalar field that appears in the linear σ-model.

It is possible to eliminate the σ-field altogether, at a price of working with a nonlinear representation of $SU(2) \times SU(2)$ (Gell-Mann and Levy, 1960; Dashen, 1969). For example, suppose the mass of the σ field in (12.20) is made infinite by allowing $\lambda \to \infty$. From eq. (12.32)

$$\sigma^2 = f_\pi^2 - \boldsymbol{\pi}^2, \tag{12.86}$$

where the Goldberger–Treiman relation (12.14) at tree level ($g_A = 1$) and the relation $M = m_N \simeq gv$ following from (12.35) have been used to set $v^2 = f_\pi^2$. If this is substituted back into (12.20)–(12.22) we obtain the Lagrangian density of the *nonlinear σ-model*, in which no explicit σ field appears. The axial current of the linear σ-model without fermions is (Exercise 12.4)

$$\mathbf{A}_\mu = \sigma\partial_\mu\boldsymbol{\pi} - \boldsymbol{\pi}\partial_\mu\sigma. \tag{12.87}$$

If we introduce the *chiral angle* θ by (Brown, 1982a)

$$\sigma \equiv f_\pi \cos\theta \qquad \boldsymbol{\pi} \equiv f_\pi \hat{\boldsymbol{\pi}} \sin\theta, \tag{12.88}$$

where $\hat{\boldsymbol{\pi}}$ is a unit vector, (12.86) is satisfied automatically, and (12.87) may be written (Exercise 12.16)

$$\mathbf{A}_\mu = f_\pi D_\mu\Phi, \tag{12.89}$$

where the pion field Φ is defined by

$$\Phi \equiv f_\pi \hat{\boldsymbol{\pi}} \tan\theta, \tag{12.90}$$

and the *nonlinear derivative* D_μ is

$$D_\mu \equiv \left(1 + \frac{\Phi^2}{f_\pi^2}\right)^{-1} \partial_\mu. \tag{12.91}$$

EXERCISE 12.16 Prove that the expression (12.89) for the axial current is equivalent to the relation (12.87).

Little Bag Model. In the little bag model (LBM) the chiral symmetry is restored by coupling the pion field (12.90) to the surface of the bag. In the bag interior the axial current is (12.48)

$$\mathbf{A}_\mu = \bar{q}\gamma^\mu \gamma_5 \frac{\boldsymbol{\tau}}{2} q, \tag{12.92}$$

and the imposition of chiral symmetry requires $\partial_\mu A_j^\mu = 0$. In the LBM the pion field exists only outside the bag (Goldstone phase) and the current (12.92) exists only inside the bag (Wigner phase), so chiral symmetry is restored by making (12.89) and (12.92) continuous at the bag boundary. That is, $\partial_\mu A_j^\mu = 0$ is an operator equation that must be satisfied in both phases; the pion field, which carries no color and therefore can propagate in the non-perturbative vacuum, carries off the axial current lost at the surface of the bag because of reflection of the Dirac particles. This restores chiral invariance for massless quarks.

For processes in which chiral symmetry is important the fields are usually asymptotically weak and the nonlinear derivative D_μ can be approximated by the linear one:

$$D_\mu \to \partial_\mu \qquad (\Phi^2 \ll f_\pi^2). \tag{12.93}$$

Then the Lagrangian density of the basic little bag model is equivalent to the chiral invariant form

$$\begin{aligned}
\mathcal{L}_{\mathrm{LBM}} =&(i\bar{q}\slashed{\partial}q - B)\theta_{\mathrm{V}} - \frac{1}{2f_\pi}\bar{q}(\sigma + i\boldsymbol{\tau}\cdot\boldsymbol{\pi}\gamma_5)q\Delta_{\mathrm{S}} \\
&+ \tfrac{1}{2}(\partial_\mu\sigma)^2\bar{\theta}_{\mathrm{V}} + \tfrac{1}{2}(\partial_\mu\boldsymbol{\pi})^2\bar{\theta}_{\mathrm{V}},
\end{aligned} \tag{12.94}$$

where the σ field is normally eliminated using transformations similar to (12.88) that satisfy (12.86), $\theta_{\mathrm{V}} = 1$ inside the bag and zero otherwise, $\bar{\theta}_{\mathrm{V}} = 1$ *outside* the bag and zero otherwise, and $\Delta_{\mathrm{S}} = 1$ on the bag surface and zero otherwise. The exact solution of the resulting field theory is difficult because the pion field is found to exert a strong pressure on the surface of those bags that are strongly coupled to pions—a decidedly non-perturbative effect. Solutions are possible for a hypothetical baryon called the *hedgehog* that is an eigenstate neither of angular momentum nor of isospin (Vento et

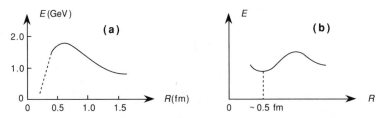

Fig. 12.14   (a) A hedgehog solution for the little bag model. (b) Conjectured stabilization of a bag by coupling to other mesons.

al., 1980).[‡]   The bag energy vs. bag radius for a hedgehog solution is sketched in Fig. 12.14a.

The picture suggests that the hedgehog bag is unstable against collapse to a small radius $\leqslant 0.3$ fm because of the pressure of the pion field on the bag surface. The hedgehog model is rather unrealistic, but it illustrates a basic feature of the little bag model. It has been argued that in a more realistic case the additional coupling of other mesons to the surface will stabilize it at $R \simeq 0.5$ fm; this is illustrated schematically in Fig. 12.14b.

Thus the little bag model predicts smaller bags than the MIT model for hadrons strongly coupled to pions, and the little bags are surrounded by meson clouds. Hadrons containing multiple strange quarks such as Ξ^0 or Ξ^- do not couple strongly to pions and the radius could be larger, approaching the MIT value of about 1 fm. Since the bags of nonstrange quarks are small (and heavy vector meson exchange at short distance causes them to repel each other strongly), the quark cores of nucleons remain far apart in nuclei and the traditional picture of nucleons interacting by boson exchange is recovered, as illustrated in Fig. 12.15. In this picture two and three pion exchanges are replaced by effective vector and scalar boson exchanges:

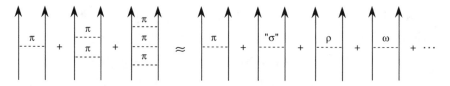

where "σ" stands for the fictitious scalar boson described in the caption of

---

[‡]The pion couples to the nucleon spin through a term $\sigma \cdot \mathbf{k}_\pi$, where $\mathbf{k}_\pi$ is the momentum of the pion and σ_i is a Pauli matrix. Thus in the LBM the pion field pressure is largest at the poles of the nucleon (defined by direction of the nucleon spin) and smallest at the equator, and the nucleon is deformed, even in the ground state. The bag can be forced to remain spherical if the pion pressure is made uniform with angle by choosing the expectation value of the spin operator σ to always be in the radial direction. The resulting hadron is called a "hedgehog" (Brown, 1982a; Bhaduri, 1988, §4.4).

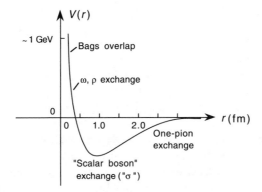

Fig. 12.15   Boson exchange description of phenomenological nucleon–nucleon interaction. The scalar boson exchange pieces are viewed as phenomenology describing aspects of multipion resonances not accounted for by ω and ρ; they should not be identified with the scalar field of the σ-model.

Fig. 12.15. This is made plausible by the large width for ω decay to three pions and for ρ to two pions.

Properties such as magnetic moments come out well, but the LBM magnetic moment is attributed partially to the contribution of constituent quarks and partially to the meson cloud; for the nucleon, calculations indicate that about half comes from each (Brown, 1982b).

A serious problem for the LBM is that in its simplest form it predicts a value of g_A/g_V that is too large [see (12.80)]. The value is about $\frac{5}{3}$ in perturbation theory, as in the $SU(6)$ model, but grows to greater than 2 with inclusion of higher orders (Thomas, 1984). The increase of g_A over the MIT bag value is a consequence of the restriction of the pion field to the bag exterior. It has been proposed that the tensor coupling of the π field to the nucleon generates large deformed components (40% D-state admixture in the ground state is required), and that this decreases g_A/g_V to near the experimental value (Brown, 1982a; Glashow, 1979; Vento, Baym, and Jackson, 1981). This conjecture is not easy to test since it is based on a non-perturbative effect that is difficult to calculate. In §13.8 a possible resolution of these problems through topological effects will be introduced.

Cloudy-Bag Model. The essential feature of the cloudy-bag model (CBM) is that the chiral symmetry is restored by coupling to a pion field at the bag surface, but with the pion field not restricted to the bag exterior. Because the pion is allowed to roam inside the bag, it is not surprising to find that the modification of the MIT bag is small in the CBM, unlike the non-perturbative effects associated with the pion field in the LBM.

The CBM Lagrangian density in the chiral symmetry limit is (Thomas, 1984)

$$\mathcal{L}_{\text{CBM}} = (i\bar{q}\not{\partial}q - B)\theta_V - \tfrac{1}{2}\bar{q}e^{i\tau\cdot\Phi\gamma_5/f_\pi}q\Delta_S + \tfrac{1}{2}(D_\mu\Phi)(D^\mu\Phi), \qquad (12.95)$$

which is essentially (12.94) with the $\bar{\theta}_V$ factor restricting the pion to the bag exterior discarded, and with the σ-field eliminated using parameterizations of the form (12.86)–(12.91). Notice that the cloudy-bag pion is allowed to penetrate the bag volume, but it does not couple to the quarks in the bag interior. The qualitative argument given for omitting $\bar{\theta}_V$ is that a static bag surface with pions rigorously excluded from the interior by a sharp phase transition is unrealistic phenomenology: pions interacting with the bag surface will alter it, leading effectively to pions inside the bag.

In practical applications of the cloudy bag it is assumed that the internal structure of the pion can be ignored and the pion field treated in terms of small fluctuations about $\Phi = 0$. Then the nonlinear derivative may be replaced by the usual derivative, the exponential in (12.95) may be expanded, and the linearized CBM Lagrangian density

$$\mathcal{L}_{\text{CBM}} \simeq (i\bar{q}\slashed{\partial}q - B)\theta_V - \tfrac{1}{2}\bar{q}q\Delta_S$$

$$+ \tfrac{1}{2}(\partial_\mu\Phi)^2 - \tfrac{1}{2}m_\pi^2\Phi^2 - \frac{i}{2f_\pi}\bar{q}\gamma_5\Phi\cdot\tau q\Delta_S \qquad (12.96)$$

is obtained.

The CBM, with the addition of standard bag improvements such as perturbative one-gluon exchange and center-of-mass corrections, does a good job of describing the data to which bags normally are applied (Thomas, 1983, 1984). Notably, because the pion field is not restricted to the bag exterior the MIT result $g_A \simeq 1.09$ is recovered without the introduction of bag deformations.

Cloudy Bags or Little Bags? The cloudy bag and little bag models address basically the same physics; however, the philosophies that they represent and their implications for relating QCD to nuclear structure are superficially very different. The root issue is the role of pions and vector mesons in nuclear physics. The little bag model assays to salvage a long history of phenomenological meson physics by relegating the quarks to the small cores of nucleon bags. These bags repel each other strongly at short distances so that at normal nuclear densities the quarks in the bags never see each other and effective meson physics is all that remains. Of course, this sidesteps the important issue of the *structure* of the meson field—the mesons too are made of quarks. However, for nuclear physics the detailed structure of the pion could be irrelevant.

Conversely, the cloudy bag model finds that the pion distribution surrounding the nucleon is more a halo than a cloud (Thomas, 1984, estimates $\langle n \rangle \simeq \tfrac{1}{2}$ for the number of pions around a nucleon bag in the CBM). In the CBM pionic effects are but small perturbations on the MIT bag and it is argued that heavy vector-meson exchange has no place in describing the nucleon–nucleon interaction: the pion field may be relevant for the long-range (one-pion exchange) part of the interaction, but all shorter-range components in the N–N

interaction should be treated as explicit bag or quark effects. In particular, since the vector mesons have radii of about 0.9 fm in the MIT bag model and must by virtue of their masses mediate exchange interactions over ranges of a few tenths of a fermi, it is argued that it makes no sense to neglect the internal structure of a ρ or ω when it is exchanged between nucleons. They should be treated as sea quark excitations (Thomas, 1983, 1984).

In addition to the implications of this debate for fundamental nucleonic and nuclear structure, these differing points of view could have important consequences in astrophysics. For example, the collapse of supernovae cores and the structure of neutron stars depend crucially on the behavior of matter at very high density. Naively, the MIT bag and little bag models might be expected to give rather different qualitative predictions for this behavior (but see §13.8).

Thus, the restoration of chiral bag symmetry through the LBM or the CBM has generated an intense controversy that is difficult to resolve because the primary disagreement can be addressed only through non-perturbative strong interaction calculations. In §13.8 we will discuss topological developments suggesting that these seemingly divergent points of view may be less contradictory than one might suppose. These developments indicate that the chiral bag radius R, which is a primary bone of contention in the controversy between the little bag and cloudy-bag models, may not play as direct a physical role as had been assumed.

12.6 Nonrelativistic Potential Models

Considerable attention has been devoted to the study of hadronic structure through models that use a nonrelativistic potential to impose confinement. We will discuss such models both for heavy quark systems, where the neglect of relativity is formally justified, and for light quark systems where it is not but the potential description often seems to work anyway. Our discussion will be brief because the nonrelativistic potential models are conceptually simple, and because they employ many features already introduced in our discussion of bag models.

12.6.1 Charmonium and Upsilonium

The near simultaneous discovery of the J/ψ (3097) resonance in e^+e^- annihilation at SLAC, and in the reaction $p\text{Be} \to e^+e^-\text{X}$ at Brookhaven (Aubert et al., 1974; Augustin et al., 1974) was a dramatic event with far-reaching implications. The most startling property of the J/ψ is its narrow width ($\Gamma = 68$ keV), which is about 1000 times smaller than a typical hadronic width. The J/ψ is produced copiously from the virtual photons in e^+e^- reactions,

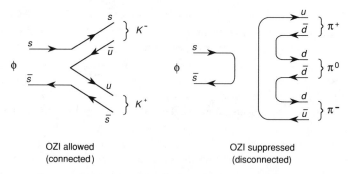

Fig. 12.16   Some quark diagrams illustrating OZI allowed and suppressed transitions. Disconnected diagrams, such as for $\phi \to 3\pi$, are found to be strongly suppressed.

which suggests that it carries the photon quantum numbers, $J^{PC} = 1^{--}$. An early explanation of the new vector resonance, which proved to be correct, was as a meson state $c\bar{c}$ formed from quarks and antiquarks carrying a new flavor quantum number called *charm* (c). We may recall that the existence of a fourth quark flavor had been predicted, in a sense, by the GIM mechanism for suppressing strangeness-changing weak neutral currents in the standard electroweak theory (§9.3.2). A discussion of charmed particles can be found in Gaillard, Lee, and Rosner (1975) and Trilling (1981).

OZI Suppression. The narrow width of J/ψ can be "explained" by appeal to a phenomenological device called the *OZI selection rule*, which had been observed previously to hold in the decay of vector mesons.[‡]   For example, consider the decay of the $\phi(1020)$ meson, which also has a narrow width, $\Gamma = 4.4$ MeV. The decay $\phi \to 3\pi$ is highly suppressed relative to $\phi \to K^+K^-$, even though the former channel is favored strongly by phase space. In addition, one finds surprises like

$$\Gamma(\phi \to 3\pi) << \Gamma(\omega \to 3\pi),$$

even though phase space again favors the former reaction. The OZI rule accounts for these observations by assuming that *disconnected quark diagrams are suppressed relative to connected ones*. Valence quark structure diagrams for the decays $\phi \to 3\pi$ and $\phi \to K^+K^-$ are shown in Fig. 12.16. The small width of the ϕ results then because $\phi \to 3\pi$ is OZI suppressed, and OZI-allowed $\phi \to K^+K^-$ has little phase space ($m_\phi = 1019$ MeV and $2m_K = 988$ MeV; see Appendix B).

By analogy with the ϕ decay, the decay of J/ψ can proceed through processes such as those shown in Fig. 12.17. We would then expect that the

---

[‡]OZI = Okubo, Zweig, Iizuka (see Zweig, 1964; Okubo, 1963; Iizuka, 1966); it is often just called the *Zweig rule*.

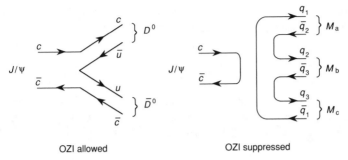

Fig. 12.17 Strong decay modes of J/ψ. The symbol M_i stands for a meson and q_i stands for any u, d, or s quark.

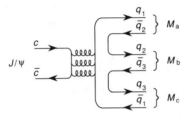

Fig. 12.18 Schematic quark diagram illustrating the role of gluons in OZI suppressed decay modes for J/ψ.

strong decay of J/ψ would occur primarily by decay to pairs of charmed–anticharmed mesons. However, the lightest charmed meson is $D^0(1865)$, so J/ψ (3097) is below the energy threshold for decay into $D^0\overline{D}^0$ and can decay strongly only through OZI-forbidden processes.

The empirical OZI rule is now understood to have its foundation in QCD. For OZI-suppressed diagrams with their quarkless intermediate states the communication between reactants and products must be through gluons. Single-gluon exchange violates color conservation because the initial and final states are colorless hadrons, but the gluon is colored. Two gluons can couple to a color singlet, but a $J^{PC} = 1^{--}$ vector meson state cannot couple to the two gluons because a two-gluon state must be even under charge conjugation—this is the same reason that two-photon decay for orthopositronium is forbidden. Thus, the simplest coupling involves three gluons (Fig. 12.18), just as orthopositronium decays by three-photon emission (see Exercise 12.17).

The diagrams in Figs. 12.16–12.18 are not Feynman diagrams and have no dynamical content. However, from them we can see qualitatively how the OZI rule arises in QCD. At the three-gluon vertex the emitted gluons must carry all the mass–energy of the $c\bar{c}$ pair and thus are "hard". Because of asymptotic freedom, perturbative QCD should be approximately valid to describe this vertex and the rate for emission of n hard gluons is

$$R \simeq \left(\alpha_{\rm s}(q^2)\right)^n,$$

where $\alpha_{\rm s}(q^2)$ is the running strong coupling constant. Thus, if $\alpha_{\rm s}(q^2)$ is small on the distance scale appropriate to J/ψ (it seems to be, see §10.6.4), the OZI-suppressed diagrams contribute small rates for the decay of J/ψ.

Quarkonia and Quark–Antiquark Potentials. Since the discovery of J/ψ a detailed spectroscopy of *charmonium* (bound $c\bar{c}$) states has been established. In addition, the upsilon resonance $\Upsilon(9.46\ \text{GeV})$ (Herb et al., 1977) has been interpreted as a vector meson $b\bar{b}$ bound state (with $b = bottom$ or *beauty*), and a spectroscopy of excited $b\bar{b}$ states (*upsilonium*) has emerged (Franzini and Lee–Franzini, 1983). The $c\bar{c}$ and $b\bar{b}$ spectra are shown in Fig. 12.19. It is expected that there is a sixth quark flavor ($t = top$ or *truth*) with the promise of $t\bar{t}$ states, but it has not appeared below a mass of about 90 GeV. Collectively, the heavy $q\bar{q}$ bound states are termed *quarkonium*.

The J/ψ resonance is the lowest $^3S_1$ state, but it is not the ground state of the $c\bar{c}$ system. This is because of the spin–spin splitting already discussed in connection with the bag model that pushes the triplet states up relative to the singlet states. However, $^3S_1$ states carry the appropriate quantum numbers to be strongly populated in e^+e^- reactions. The others are populated by radiative decay, as indicated by dashed lines in Fig. 12.19.

EXERCISE 12.17 Compare the widths of the states $\eta_c(2980)$ (*paracharmonium*) and $J/\psi(3097)$ (*orthocharmonium*) shown in Fig. 12.19 [see Aguilar-Benitez et al. (1988) for the experimental widths]. Give a plausible explanation for the difference. *Hint*: compare with the decay of *ortho* and *para* positronium.

The higher lying states of Fig. 12.19 are interpreted as orbital and radial quark excitations within a potential model. Thus we may use the quarkonium spectra to investigate the effective potential acting between a q and $\bar{q}$ on the scale of a J/ψ or Υ system ($0.2\ \text{fm} \leqslant R \leqslant 1\ \text{fm}$). These potentials are determined by fitting the eigenvalues of a Schrödinger equation to the energy levels. In addition, expressions such as (10.1) for the leptonic decay widths can be used to constrain the zero-distance wavefunction. The Hamiltonian is normally taken to be of the form (De Rújula, Georgi, and Glashow; 1975)

$$H = \sum_i \left(m_i + \frac{\mathbf{p}_i^2}{2m_i}\right) + \sum_{i>j}(\alpha Q_i Q_j + \kappa \alpha_{\rm s})\, S_{ij} + V_{\rm c}, \qquad (12.97)$$

where α is the electromagnetic fine structure constant, $\alpha_{\rm s}$ is a corresponding strong coupling constant, Q_i, $\mathbf{r}_i$, m_i, and $\mathbf{p}_i$ are the electric charge, position,

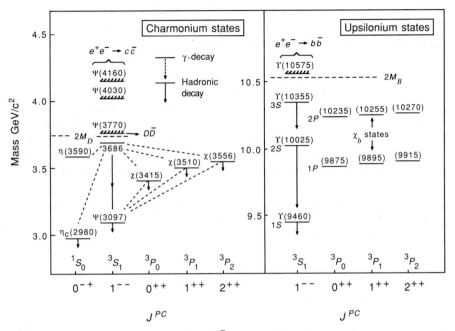

Fig. 12.19   Spectra of the $c\bar{c}$ and $b\bar{b}$ systems (Perkins, 1987). Dashed lines indicate radiative transitions. States above the thresholds $2M_D$ and $2M_B$ are broad because they can decay by OZI-allowed diagrams. A comprehensive review of J/ψ decays may be found in Köpke and Wermes (1989).

mass, and momentum of quark i, the color factor κ is

$$\kappa = \langle \mathbf{F}_i \cdot \mathbf{F}_j \rangle = \begin{cases} -\dfrac{4}{3} & (q\bar{q}) \\ -\dfrac{2}{3} & (qqq) \end{cases}$$

(see Exercise 12.11b), V_c is the long-range confining part of the potential, and S_{ij} is of the Fermi–Breit form

$$\begin{aligned}
S_{ij} =& \frac{1}{r} - \frac{1}{2m_i m_j}\left(\frac{\mathbf{p}_i \cdot \mathbf{p}_j}{r} + \frac{\mathbf{r}\cdot(\mathbf{r}\cdot\mathbf{p}_i)\mathbf{p}_j}{r^3}\right) \\
& -\frac{\pi}{2}\delta^{(3)}(\mathbf{r})\left(\frac{1}{m_i^2} + \frac{1}{m_j^2} + \frac{16\mathbf{s}_i\cdot\mathbf{s}_j}{3m_i m_j}\right) \\
& -\frac{1}{2r^3}\left[\frac{1}{m_i^2}\mathbf{r}\times\mathbf{p}_i\cdot\mathbf{s}_i - \frac{1}{m_j^2}\mathbf{r}\times\mathbf{p}_j\cdot\mathbf{s}_j\right. \\
& \left. +\frac{1}{m_i m_j}\left(2\mathbf{r}\times\mathbf{p}_i\cdot\mathbf{s}_j - 2\mathbf{r}\times\mathbf{p}_j\cdot\mathbf{s}_i - 2\mathbf{s}_i\cdot\mathbf{s}_j + \frac{6(\mathbf{s}_i\cdot\mathbf{r})(\mathbf{s}_j\cdot\mathbf{r})}{r^2}\right)\right],
\end{aligned}$$

$$(12.98)$$

where $\mathbf{s}_i$ is a quark spin operator, $\mathbf{r} \equiv \mathbf{r}_i - \mathbf{r}_j$ and $r = |\mathbf{r}|$. Often in practical calculations only a part of (12.98) is used; for example, see (12.102) below. Notice that in the nonrelativistic limit the spin–orbit and tensor interactions (the terms within square brackets) do not contribute for S-wave two-quark states, while for P-wave and D-wave states the factor $\delta^3(\mathbf{r})$ is zero and the third term (*Fermi contact term*) vanishes. The motivation for the Fermi–Breit form lies in the assumption that the bulk of the strong interaction coupling has been exhausted in the phenomenological confining potential, leaving only weak short-range interactions among the confined quarks. If we keep only lowest order gluon exchange the $SU(3)$ gauge fields look like eight $U(1)$ gauge fields, except for a different coupling constant and factors coming from color matrices (see §12.5.2).

If we assume that the confining potential depends only on r, the QCD portion of the potential appearing in (12.97) may be written

$$V = V_{\text{radial}} + V', \tag{12.99}$$

where

$$V(r) \equiv V_{\text{radial}} = \frac{\kappa \alpha_{\text{s}}}{r} + V_{\text{c}}(r), \tag{12.100}$$

and V' is everything else. Many forms have been proposed for $V_{\text{c}}(r)$. The string model discussed in §12.5.5 motivates a linear potential and one often takes

$$V(r) \simeq -\frac{4}{3}\frac{\alpha_{\text{s}}}{r} + \sigma r \tag{12.101}$$

for a meson system; this is of the form shown in Fig. 12.20. Notice that the Coulomb term alone is not sufficient because it would allow free quarks to ionize from the system; in contrast, the potential (12.101) confines and the

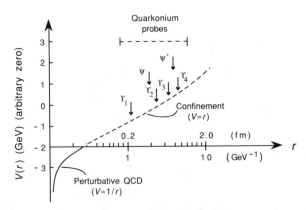

Fig. 12.20 Approximate QCD interquark potential. The average radii probed by some J/ψ and Υ states are indicated.

corresponding Schrödinger equation has no continuum states.[‡] In the string model σ is called the string tension.

The range of the potential to which the Υ and J/ψ states are sensitive in Fig. 12.20 is about 1–5 GeV^{-1}, corresponding to a distance scale of about 0.2–1.0 fm. Thus charmonium and upsilonium levels yield information only on the linear part of (12.101). We may also note that the same potentials appear to be consistent for both upsilonium ($b\bar{b}$) and charmonium ($c\bar{c}$) states (Kwong, Rosner, and Quigg; 1987). An approximate flavor independence of the interquark potential would be expected from fundamental considerations, since flavor enters QCD only through the differing current masses for the quarks in (10.32). The discovery of the hypothesized top quark and a study of $t\bar{t}$ states might allow the potential in Fig. 12.20 to be probed at distances of less than 1 GeV^{-1}. In particular, some $t\bar{t}$ states could be sensitive to the Coulomb (asymptotically free) region of (12.101). However, recent lower limits in excess of 90 GeV for the mass of the t quark could modify these conjectures.

12.6.2 Nonrelativistic Models for Light Quarks

By uncertainty principle arguments the momentum of a quark confined to a volume of radius one fermi is

$$\langle \mathbf{p} \rangle \simeq r^{-1} \simeq 200 \text{ MeV}.$$

For the charmonium or upsilonium systems the constituent quark masses are about 1.6 GeV and 5 GeV, respectively, and since $m_0 c^2 \gg pc$ a nonrelativistic approximation is justified. For u or d quarks, however, we have $m_0 c^2 \simeq pc$ [eq. (10.38)] and a nonrelativistic approximation is questionable; even for strange quarks relativity effects should be significant. Nevertheless, nonrelativistic models of quark structure for hadrons have been found to work surprisingly well, even for light hadrons. An early example is the $SU(6)$ quark model (§5.3.6), which does a credible job on quantities such as magnetic moments despite its separation of orbital and spin angular momenta, which is untenable in a covariant theory.

More recent interest in nonrelativistic quark models issues largely from an influential paper of De Rújula, Georgi, and Glashow (1975), in which it was shown that a nonrelativistic Hamiltonian consisting of a confining potential and a Fermi–Breit interaction (12.98) yields various hadronic mass formulas that are well-satisfied by the data. (It is also one of the few papers in the physics literature that begins "Once upon a time ... ".) Such calculations have

---

[‡]This picture will be altered by meson production at large string elongation, as we discuss in §14.3

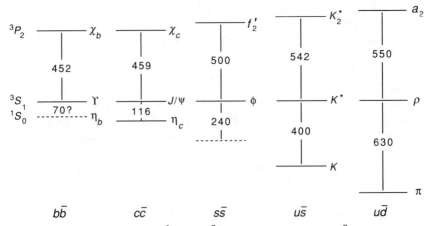

Fig. 12.21 Evolution of the $^1S_0$ and $^3P_2$ levels relative to the $^3S_1$ level in the transition from heavy to light $q\bar{q}$ systems. Dashed levels have been estimated theoretically (after Isgur and Karl, 1983).

been expanded and applied to a variety of both heavy and light hadronic data in recent years. The most extensive work is probably that based on the papers of Isgur and Karl (1979a,b) and references therein.

In these systematic calculations one observes a fundamental continuity between the properties of the heavy quark and light quark systems. For example, in Fig. 12.21 the smooth evolution of the $^1S_0$ and $^3P_2$ levels relative to the $^3S_1$ level is shown for a range of $q\bar{q}$ systems. Therefore, even though a nonrelativistic calculation is formally suspect it is argued that the continuity of basic features as the quark flavors (and hence masses) are varied justifies the potential model as a useful phenomenological device, with the formal shortcomings of the model absorbed into a smooth modification of parameters.

The potential could be any confining form (linear, log, power law, ...), but in many practical applications a harmonic oscillator potential yields spectra not very different from those found for potentials such as Coulomb + linear that QCD prejudice would favor. For example, Fig. 12.22 shows the difference between a linear, linear + Coulomb, and harmonic potential in describing $c\bar{c}$ states. Since harmonic oscillator models have nice mathematical properties, that form has often been employed for the confining potential. Isgur and Karl (1978, 1979a,b) have used such two-body confining potentials, keeping only the spin–spin and tensor components of the Fermi–Breit interaction (12.98) between quarks i and j,

$$H_{\text{hyperfine}}^{ij} = -\kappa \frac{\alpha_s}{m_i m_j} \left[\frac{8\pi}{3} \delta^{(3)}(\mathbf{r}) \mathbf{s}_i \cdot \mathbf{s}_j \right.$$

$$\left. + \frac{1}{r^3} \left(\frac{3(\mathbf{s}_i \cdot \mathbf{r})(\mathbf{s}_j \cdot \mathbf{r})}{r^2} - \mathbf{s}_i \cdot \mathbf{s}_j \right) \right]$$

$$(12.102)$$

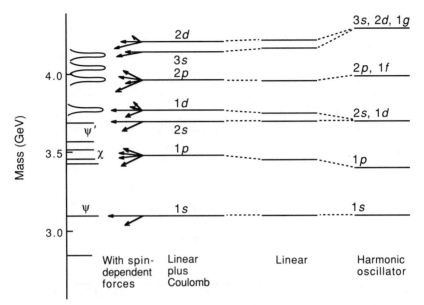

Fig. 12.22 Comparison of linear, linear + Coulomb, and harmonic potential levels with the charmonium spectrum (Close, 1979).

($\mathbf{r} \equiv \mathbf{r}_i - \mathbf{r}_j$ and $r = |\mathbf{r}|$), with α_s considered an adjustable parameter, to describe a broad range of hadronic ground states, and excited states of both negative and positive parity resulting from the promotion of quarks from the ground-state orbits.[‡]

Two more recent examples of systematic nonrelativistic calculations that employ a linear confining potential with one-gluon exchange and approximate relativistic corrections are shown in Figs. 12.23 and 12.24. Here one sees a detailed example of the phenomenological continuity suggested by Fig. 12.21: mesons with light quarks (Fig. 12.23) and heavy quarks (Fig. 12.24) are described well by the same theory when empirical relativistic corrections are included.

12.6.3 Attenuation of Spin–Orbit Coupling

We make several remarks about the interaction (12.102). First, the tensor component averages to zero when evaluated in a spherically symmetric state.

[‡]In the usual implementations there is a substantial difference in philosophy between the bag model approach to residual interactions and that of the nonrelativistic potential model. In typical potential model calculations the one-gluon exchange interaction (or a part of it) is *diagonalized* in a limited basis; in bag models the one-gluon exchange is normally treated as a *perturbation* on the free cavity states. Nevertheless, these approaches give qualitatively similar descriptions of the data.

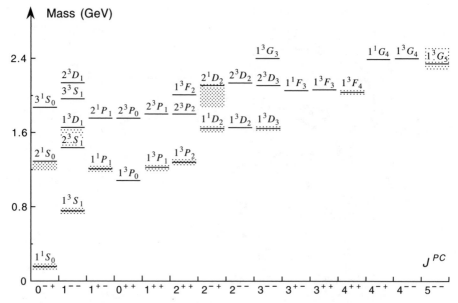

Fig. 12.23 Spectrum of isovector mesons. The lines are the calculations of Godfrey and Isgur (1985); shaded areas represent experimental masses. The horizontal axis is J^{PC} and each state is labeled by the predicted energy and the dominant spectroscopic component $n^{2S+1}L_J$ of the $q\bar{q}$ wavefunction.

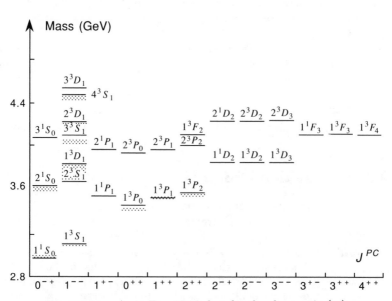

Fig. 12.24 As in Fig. 12.23, but for the charmonia ($c\bar{c}$).

To the degree that the hadronic ground states are S-wave the tensor force is unimportant, but it can be significant for excited states. Second, we note the conspicuous absence of spin–orbit terms in (12.102), which vanish for S-waves but might be expected to play a role in excited states.

In hydrogen atoms the $\mathbf{s} \cdot \mathbf{s}$ interaction of (12.98) is "hyperfine" because it is suppressed by the large mass m_j of the proton; it manifests itself in the ground state only because it has no competition. In excited states the spin–orbit terms

$$V_{\mathrm{so}} \simeq \mathbf{l} \cdot \mathbf{s} = (\mathbf{r} \times \mathbf{p}) \cdot \mathbf{s} \tag{12.103}$$

have pieces that are proportional to m_i^{-2}, where m_i is the electron mass, so they are of order $(m_{\mathrm{hydrogen}}/m_{\mathrm{electron}})$ times the size of the spin–spin terms; this is the fine structure. For hadrons made of light quarks (likewise for positronium) $m_i \simeq m_j$, and the $\mathbf{l} \cdot \mathbf{s}$ terms should be about equal in importance with the $\mathbf{s} \cdot \mathbf{s}$ terms. Actually, inclusion of strong $\mathbf{l} \cdot \mathbf{s}$ interactions in (12.102) would greatly worsen the description of the light hadronic data—*the nonrelativistic quark model neither requires, nor desires, spin–orbit interactions of significant magnitude.*

One standard explanation for this, which is on firmer ground for heavy quarkonia than for light hadrons, is that there is a contribution from the confining potential that largely cancels the spin–orbit term coming from the one-gluon exchange potential (see Isgur and Karl, 1978). However a convincing justification for the effective suppression of the spin–orbit interaction in the nonrelativistic quark model remains to be given. References to a more extensive discussion of the issue may be found in Close (1982a) and in Myhrer and Wroldsen (1988).

12.6.4 Potential Models or Bags?

In spite of the tenuous formal justification of the nonrelativistic quark model for light hadrons, there are two features of these models that must be taken seriously by practitioners of alternative phenomenologies:

1. Empirically the nonrelativistic models do well for light hadrons—at least comparable with bag models in most cases, somewhat better in many cases. This is particularly true in questions of detailed spectroscopy.
2. The nonrelativistic models, especially those employing harmonic oscillator potentials, are better equipped than bag models to isolate the center-of-mass motion. This is important for excited states because of the danger of spurious solutions if the translational invariance is violated.

With respect to this last point, Isgur and Karl (1983) have stated the nonrelativistic quark model position eloquently: "It is our prejudice that in spectroscopy it is better to have the right number of moving parts, moving at the wrong speed, than to have the wrong number of moving parts moving at the right speed." A full understanding of why the nonrelativistic models work—how much of this success is fundamental and how much fortuitous— will require QCD solutions in the non-perturbative regime. This is the subject of the next chapter.

12.7 Background and Further Reading

For a good qualitative introduction to the issues considered in this chapter, see the articles by Isgur and Karl (1983) and Brown and Rho (1983), and the review by Wong (1986). The phenomenology of confinement is discussed in Lee (1981), Ch. 17; and Baym (1982a,c). Because of Lorentz invariance, the picture of the QCD vacuum as a color dielectric may also be discussed in terms of *color paramagnetism* (§12.1.1); see Nielsen (1981) and Bhaduri (1988). Readable introductions to current algebra and to the σ-model may be found in Campbell (1978); Lee (1972); Dashen (1969); Lee (1981), Ch. 24; Thomas (1984); Pagels (1975); and Cheng and Li (1984), Ch. 5. General treatments of bag models include DeTar and Donoghue (1983), Thomas (1983, 1984), Heller (1982), and DeGrand et al. (1975). Chiral bags are covered in Thomas (1983, 1984), Miller (1984), Myhrer (1984), and Brown (1982a). For nonrelativistic models see Isgur and Karl (1978; 1979a,b; 1983); Godfrey and Isgur (1985); De Rújula, Georgi, and Glashow (1975); Capstick and Isgur (1986); Capstick et al. (1986); Close (1979); Kwong, Rosner, and Quigg (1987); Bhaduri (1988); and Isgur (1980).

Our discussion in this chapter has concentrated on hadronic structure. The important question of a microscopic description of the nucleon–nucleon force and nucleon–nucleon scattering has been omitted. Myhrer and Wroldsen (1988) and Oka and Yazaki (1984) present reviews with extensive references that can set this omission right.

CHAPTER 13

Non-Perturbative Methods In Gauge Field Theories

The standard methodology of the relativistic quantum field theorist is diagrammatic perturbation theory. This results from two circumstances: many problems of interest can be treated as perturbations of a simple problem, and until recently perturbation theory was the only approach technically feasible for most problems. However, there are many indications that it is possible for a gauge field theory to support a rich non-perturbative structure. Until this structure is isolated and understood, we must face the unpleasant possibility that the nature of such a field theory may be very different from expectations based on perturbative investigations.

A case in point is confinement in QCD. Although plausible arguments suggest such a property, only non-perturbative solutions of the relevant field equations can provide a definitive answer. Because its non-perturbative aspects remain rather mysterious, the point is sometimes made that we may not even know how many free parameters are present in QCD. For example, the study of instantons reveals a CP-violating parameter θ that does not appear to any order in perturbative QCD (§13.7). Are there other effective parameters hidden in the unexplored jungles of non-perturbative field theories? The complete answer to this question can only come when reliable non-perturbative solutions are available for gauge fields.

In this chapter we examine two general approaches to this problem. The first pursues numerical solution of the field theory in its full non-perturbative glory; the second looks for classical bound-state solutions of the field equations (solitons) as a starting point for introducing non-perturbative effects semiclassically.

13.1 Lattice Gauge Theory

Although QCD is well studied at short distances and asymptotic freedom can be taken as proven, the same certainty does not prevail for the large distance (low momentum transfer) regime. Perturbation theory appears to be valid in the asymptotically free region, but perturbation theory to any finite order in QCD gives only quarks and gluons as the particle spectrum. This is in stark contrast to the actual situation where we find a complex set of multiquark bound states, but no free quarks or gluons. To understand this confinement mechanism—indeed, to determine whether confinement actually is a property of QCD—a non-perturbative calculation is required. Lattice gauge theories provide a systematic numerical method by which such calculations may be implemented.

13.1.1 Introduction

Practical calculations in any field theory employ a *regularization procedure* that sets a length scale below which processes have no influence on the theory. For example, we saw that in calculating QED loop diagrams the divergent integrals could be cut off at some momentum k, rendering them finite (§6.4). In effect, the cutoff momentum defines a length scale $L \simeq \mathcal{O}\left(k^{-1}\right)$ below which phenomena make no contribution to the field theory. The lattice spacing plays a similar role since space and time are defined only at discrete lattice points. Like the momentum cutoffs for loop integrals, this length scale must disappear in the final theory: the lattice spacing is only an artifact to facilitate calculations; in a correct theory, no physical quantity may depend on it. Thus, a lattice is intended to play the same role that a discrete mesh plays on a computer in solving differential equations. However, the requirements of Lorentz invariance, gauge invariance, the Pauli principle for fermions, and consistency with perturbatively renormalized solutions make a lattice gauge calculation much more difficult than an ordinary numerical calculation.

A lattice theory is well defined as long as one is on the lattice: there is no infinite renormalization on a finite lattice because the lattice spacing provides an ultraviolet momentum cutoff. It is only when we attempt to recover the physical continuum limit that the question of a smooth joining to (renormalized) perturbative results becomes important. In addition, we will see that it is possible to formulate gauge theories on a lattice without the gauge-fixing requirements of the continuum path integral theory, basically because gauge elements on a lattice vary over a compact group manifold (Wilson, 1974).

There are two general methods for implementing gauge theories on lattices. In the *Lagrangian formulation* the theory is constructed on a 4-dimensional spacetime lattice (Wilson, 1974). This approach is also often termed Euclidian, because as a practical matter the actual calculations are done in Euclidian 4-space after an analytical continuation from Minkowski space. In an alternative approach only space is discretized, time is treated as a continuous variable, and there is no rotation to Euclidian space. This is called the *Hamiltonian formulation* of lattice gauge theory (Kogut and Susskind, 1975). For much of our discussion we will focus on the Lagrangian method, primarily because the most powerful numerical method used in lattice gauge theory, the Monte Carlo simulation, is formulated in those terms (Rebbi, 1983). In the continuum limit the lattice spacing tends to zero and the Lagrangian and Hamiltonian formulations should give the same results (see below the discussion of universality in critical phenomena). However, for finite lattice spacing they need not represent equivalent approximations.

13.1.2 Lattice Concepts

Quarks and Antiquarks. We will consider (hyper) cubic lattices, although equal spacings are not essential to the method. A two-dimensional set of lattice points is shown in Fig. 13.1. A gauge theory may then be constructed on the lattice by the following prescription (Wilson; 1974, 1976, 1977). The quarks and antiquarks are confined to the lattice sites. A quark or antiquark state can be specified by $|n, \alpha, i, \sigma\rangle$, where n is a lattice vector with integer components that gives the location of a site in the lattice ($x^\mu = an^\mu$, where a is the lattice spacing), α is a flavor index (up, down, strange, ...), i is a color index (red, blue, green), and σ is a Dirac spinor index. Multiple quark states may be formed by putting more than one quark on the lattice, there can be a configuration with no quarks at all (quark vacuum), and if it is useful we may construct linear combinations of quark states. The number of quarks that can exist on a single lattice site is limited only by the Pauli principle; for example, if we restrict consideration to u, d, and s flavors, as many as 18 quarks and 18 antiquarks could occupy a single lattice site. The Dirac operators $\overline{\psi}_{n\alpha i\sigma}$ and $\psi_{n\alpha i\sigma}$ that create and destroy quarks are like those of a continuum theory, except they act only at the lattice sites (however, see the discussion below of fermion doubling on the lattice).

Gauge Fields. The segments joining adjacent sites on a lattice are called (directed) *string bits;* an example is shown in Fig. 13.1. In the simplest implementation of a lattice gauge theory the string bits connect nearest-neighbor

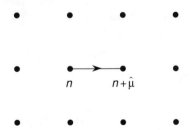

Fig. 13.1   A plane section of a cubic lattice with a directed string bit.

sites and are assumed rigid. The string bits are significant because the gauge fields reside on the links between the lattice sites. This follows because the links transport color information between lattice sites: the string bits are lattice analogs of the gluon operators in the continuum theory (§7.3). Each string bit has a color at one end and an anticolor at the other end (see Exercise 13.1). A single string bit state can be written $|n\hat{\mu}kld\rangle$ where n and $n + \hat{\mu}$ denote the nearest-neighbor lattice sites that the bit connects (Fig. 13.1), k is the anticolor index, l is the color index, and d indicates the direction of the bit. Only the colored string ends, which are attached to the lattice sites, will be relevant for the quantum theory since the string bits are assumed to have no internal degrees of freedom. The string bits will generally have a mass, which will be related to the slopes for hadronic Regge trajectories.

The lattice gauge theory of strong interactions represents a solution of QCD, so in the continuum limit it must be locally $SU(3)_c$ invariant. It is useful to build this property into the lattice theory even for finite lattice spacings. This means that a quark on a lattice without gauge fields (links) cannot move in spacetime from one lattice site to another. For example, suppose a red quark at site A moves to site B:

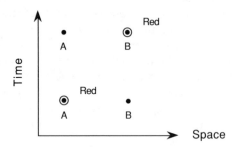

This violates local color conservation because site A changes from red to colorless, and B changes from colorless to red (color is conserved globally in the transition, but the whole point of QCD is *local* color conservation). However, the color gauge field acts like a string with color at one end and anticolor at the other, and a string can combine with two colored quarks to form a colorless object (color scalar or color singlet). For example

is a color scalar, where r denotes red and $\bar{r}$ antired. A linear combination of this and the other color combinations that yield a singlet for a quark and an antiquark joined by a string bit constitutes a lattice gauge model of a meson.

This object is free to propagate in spacetime without violating local color symmetry because it is colorless.

In the simplest static picture the hadronic states are built by combining quarks, antiquarks, and string bits subject to the requirement of local color neutrality. For example, the meson ground states consist of a $\bar{q}$ and q at the same lattice site, the simplest excited states consist of a $\bar{q}$ and q, and a string bit in a locally color neutral combination, and so on. Of course, as long as the lattice spacing is finite such a picture is but a caricature of a real meson. For example, the discrete lattice does not possess rotational invariance, so lattice meson states do not conserve angular momentum (see Hasenfratz and Hasenfratz, 1985, §3.2). A similar statement applies to the failure of translational invariance and the related difficulty of separating relative motion and internal degrees of freedom.

Gauge Invariant Operators. It is useful at this point to recall the geometrical meaning of a gauge field (§7.3). If we have a local symmetry a gauge field is required. If we wish to compare the direction of a vector in an internal space at one spacetime point with that at another spacetime point, one vector must first be translated to the location of the other vector; the gauge field specifies how this parallel transportation is to take place. For an infinitesimal displacement from x to $x + dx$, the required non-Abelian transport operator is of the form

$$U(\delta x) = \exp\left(igA_\mu^i(x)\tau_i dx^\mu\right) \equiv \exp\left(igA_\mu dx^\mu\right), \tag{13.1}$$

where g is the gauge coupling constant, $A_\mu^i(x)$ is the gauge vector potential (μ = spacetime, i = gauge group internal index), and τ_i is an infinitesimal generator of the gauge group.

In order to construct gauge invariant operators in continuum QCD it is useful to define a *string operator* or *transport operator* (7.18) as an exponentiated line integral of the gauge field

$$U_\gamma = \text{P} \exp\left(ig \int_\gamma A_\mu^i(x)\tau_i dx^\mu\right), \tag{13.2}$$

where γ denotes a path for the line integral and the symbol P indicates that the integral is to be *path ordered*. That is, eq. (13.2) represents a generalization of the infinitesimal displacement (13.1) to a path of finite length, with proper care taken for quantum-mechanical commutation relations. The matrix U_γ is a path-dependent representation of a gauge-group element. It can then be demonstrated in the continuum theory that two *gauge invariant operators* are (Bander, 1981)

$$M(x, y, \gamma) = \bar{q}(y)U_\gamma(x, y)q(x) = \bar{q}^k(y)U_\gamma^{k\ell}(x, y)q^\ell(x) \tag{13.3}$$

$$W(\gamma) = \text{Tr}\, U_\gamma(x, x), \tag{13.4}$$

where $U_\gamma(x,y)$ denotes a string operator evaluated on the path γ with endpoints x and y, and k, ℓ are anticolor and color indices. The first of these can be interpreted as a component of a meson creation operator, while the last will be associated with the action of a pure gauge field. Notice from (13.4) that $W(\gamma)$ depends only on the closed path γ and not on the endpoints. These results may be transcribed to the lattice, subject to the restriction that transport be limited to discrete jumps between lattice sites. The main criteria in constructing lattice operators are that the local gauge symmetry and as many global symmetries as possible be respected, and that the ordinary Yang–Mills action with its attendant symmetries emerge from the continuum limit of the fields defined on the lattice.

Link Operators. Consider a cubic spacetime lattice of spacing a. The lattice string operator corresponding to (13.2) for an Abelian field $A_{n\mu}$ on the link between lattice site n and $n + \hat{\mu}$ is

$$U_{n\mu} \equiv U(n, n + \hat{\mu}) = \exp\left(igaA_{n\mu}\right). \tag{13.5}$$

For a non-Abelian color field this generalizes to

$$U_{n\mu} = \exp\left(igaA_{n\mu}^i F_i\right), \tag{13.6}$$

where F_i are the eight 3×3 Hermitian generators of $SU(3)$ given in (5.67). Therefore, $U_{n\mu}$ is a unitary matrix with components $(U_{n\mu})_{kl}$, where k is the anticolor index and l is the color index. Each component of this matrix is an operator that creates or destroys a string bit. For example, $U_{n\mu}$ creates a forward string bit or destroys a reverse string bit, and the Hermitian conjugate operator $U_{n\mu}^\dagger$ destroys a forward string bit or creates a reverse string bit:

By unitarity the link operators satisfy

$$(U_{n\mu})_{k\ell}(U_{n\mu}^\dagger)_{\ell q} = \delta_{kq}. \tag{13.7}$$

To simplify expressions we will often drop the vector notation and just use U_{ij} to mean the link

In this notation the two indices label adjacent sites of the lattice; they should not be confused with the color indices of the matrix U, which will not be displayed henceforth except where necessary. The relation

$$U_{ji} = (U_{ij})^{-1} = U_{ij}^{\dagger}$$

is understood to mean the inverse defined in the group sense (conjugate rotation in the gauge group).

The physical meaning of the matrix U_{ij} is that the wavefunction of a particle transported between sites i and j undergoes an internal symmetry rotation in the gauge group defined by the link U_{ij}. Thus in a continuum theory we work with the matrix vector potential $A_{\mu}^{i}\tau_{i}$ of eq. (7.7), which takes values in the *Lie algebra* of the gauge group (Exercise 7.4). In a lattice theory we work instead with the variables U, which are elements of the gauge group itself [see eqns. (5.8) and (5.9)].[‡] One important consequence is that lattice calculations generally do not require gauge fixing. Another is that lattice methods allow the study of local gauge theories defined for discrete groups that have no Lie algebra. Indeed, the first use of lattice gauge methods was for spin systems with discrete symmetries (Wegner, 1971).

On a lattice the continuum equation (13.2) becomes

$$U_{\gamma} = U_{i_N i_{N-1}} \cdots U_{i_3 i_2} U_{i_2 i_1}, \tag{13.8}$$

for a path through a sequence of neighboring sites $i_1, i_2, \ldots, i_N$. Of particular importance are those transport operators associated with a closed path [see (13.4)]; the simplest example is a rectangle of four sites and four links, which is called a *plaquette* (Fig. 13.2). For a closed path c the trace of the transport operator

$$W(c) = \operatorname{Tr} U_c \equiv \operatorname{Tr} \prod_{\substack{\text{closed} \\ \text{path } c}} U_{ij} \tag{13.9}$$

is called a *Wilson loop*; we will see that quantities such as (13.9) define the structure of the pure gauge field. Under the gauge group the link variables transform as (see Exercise 13.1)

$$U_{ij} \rightarrow G_i^{-1} U_{ij} G_j, \tag{13.10}$$

where the elements of the gauge group G are defined locally at the sites i and j. From (13.10) the Wilson loop (13.9) is gauge invariant since class functions of matrix representations such as traces are invariant under similarity transformations:

$$\operatorname{Tr}(G_i^{-1} U_c G_i) = \operatorname{Tr} U_c. \tag{13.11}$$

[‡]The gauge-group elements U constitute the primary variables of a lattice theory, but vector potentials may still be defined through (13.6) when that is convenient.

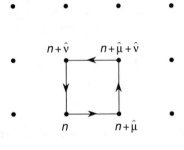

Fig. 13.2   A lattice plaquette.

EXERCISE 13.1 Define a string operator (13.2) for a continuous Abelian gauge theory with coupling constant g. Show that under gauge transformations this operator behaves as if it has charge $+g$ on one end of the string and $-g$ on the other end. The generalization of this to non-Abelian $SU(3)_c$ is the source of the assertion that the string bits in a lattice gauge theory act as if they have color on one end and anticolor on the other.

13.1.3 Gauge Invariant States

Static Configurations. In a static picture the total energy of a field theory on a lattice is the sum of the quark masses on the lattice sites and the masses of the string bits, which are proportional to their lengths. Neglecting kinetic energy terms for now, the simple static gauge invariant states are (Wilson, 1976)

- *Ground-state mesons, constructed from a quark and an antiquark at the same lattice site.* These contain no string bits, and the mass is a sum of quark and antiquark masses. The quark and antiquark must be in a color singlet state, and for u, d, s quarks with $m_u = m_d \neq m_s$ there are 36 such states (§5.3.6), corresponding to the $SU(6)$ meson multiplet $(\pi, \eta, \rho, \omega, K, K^*, \phi, \eta')$, with

$$m_\pi = m_\rho = m_\omega = m_\eta = 2m_u$$

$$m_K = m_{K^*} = m_u + m_s \qquad m_\phi = m_{\eta'} = 2m_s.$$

- *Ground-state baryons, constructed from three quarks at the same lattice site.* There are no string bits, and the quarks must be in a color singlet state. This leads to the standard $SU(6)$ multiplet **56** containing the octet (spin-$\frac{1}{2}$) and decuplet (spin-$\frac{3}{2}$). As in the case of the ground-state mesons, the masses at this level of approximation are given by a sum of quark masses; for example, $m_p = m_n = 3m_u$ and $m_\Omega = 3m_s$.

- *Excited meson and baryon states, constructed from quarks and string bits:*

Thus, for a string energy density σ and lattice spacing a the lowest excited state lies an energy σa above the ground state, in static approximation. As for the ground states, only color singlets are allowed.

- *Glueballs, built from string bits arranged in closed loops.* The simplest involves a closed loop around four nearest-neighbor sites with no quarks at the sites:

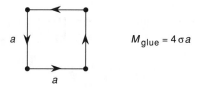

Such states are color scalars.

- *Exotic states, constructed from four (or more) valence quarks.* The simplest ground states will involve four quarks on the same site in a color singlet state, no string bits, and a minimum energy of four times the quark mass.

Kinetic Energies. Gauge invariant kinetic energies must be added to the static energies of these states to construct the total masses of a realistic theory. When kinetic terms are added many of the states described above become unstable and acquire a finite width; detailed calculations then are required to determine if they remain narrow enough to have physical relevance.

Kinetic energies are described by lattice operators that cause quarks or string bits to move on the lattice, by destroying them at one spacetime point or link and creating them at another. An example of a fundamental gauge invariant quark kinetic energy operator is [see (13.3)]

$$\Gamma_q = \overline{\psi}_n U_{n\mu} \psi_{n+\hat{\mu}}, \tag{13.12}$$

This operator can destroy a quark at lattice site $n+\hat{\mu}$ and create one at lattice site n. At the same time, to preserve gauge invariance a string bit is created by $U_{n\mu}$ between the sites n and $n + \hat{\mu}$. We also require lattice kinetic terms for the gauge fields. The simplest gauge invariant kinetic energy operator for the string bits is (see Fig. 13.2)

$$\Gamma_g = \text{Tr}\,(U_{n\mu}U_{n+\mu,\nu}U_{n+\nu,\mu}^{\dagger}U_{n\nu}^{\dagger}). \tag{13.13}$$

13.1.4 Transcription of a Field Theory to a Lattice

Having had an introduction to lattice concepts, we now consider realistic field theories on a lattice (Rebbi, 1983). From the general discussion of path integral quantization in Ch. 4, a convenient method of formulating a quantum field theory consists of the following steps:

1. Define a suitable action functional S.
2. Perform a Wick rotation $t \to -i\tau$ to Euclidian space to obtain better behaved integrals.
3. Construct the Euclidian quantum expectation value of operators Q as an average over field configurations [see (13.19)], using a measure e^{-S}:

$$\langle Q \rangle = \frac{\int [D\phi]Q[\phi]e^{-S}}{\int [D\phi]e^{-S}},$$

   with a suitable regularization procedure employed to give the Euclidian integrals exact meaning.
4. Analytically continue $\langle Q \rangle$ back to real time to recover the physical (Minkowski) expectation values.

Adapting this procedure to a lattice is formally rather simple. The Wick rotation is usually assumed to be no problem, and the functional integrals are replaced by sums or by ordinary integrals, with the discrete lattice providing an automatic momentum regularization. The crucial points are the construction of a suitable lattice action and the recovery of appropriate continuum limits.

The Euclidian action in the continuum limit for a pure non-Abelian gauge field is

$$S = \tfrac{1}{4} \int d^4x\, F_{\mu\nu}^{j}F_{j}^{\mu\nu}. \tag{13.14}$$

A corresponding lattice action for $SU(N)$ gauge groups is (Rebbi, 1983)

$$S = \beta \sum_{p} E_p, \tag{13.15}$$

where the sum is over plaquettes p (see Fig. 13.2),

$$E_p = 1 - \frac{1}{2N} \, \text{Tr} \left(U_p + U_p^\dagger \right) \qquad (13.16a)$$

$$U_p \equiv U_{i_1 i_4} U_{i_4 i_3} U_{i_3 i_2} U_{i_2 i_1} \qquad (13.16b)$$

$$\beta = \frac{2N}{g^2}, \qquad (13.17)$$

and $i_1 \ldots i_4$ label adjacent sites around a plaquette. This form of the lattice gauge action is called the *Wilson action*; it is constructed subject to the requirement that in the continuum limit the action (13.14) be recovered. Since that only places constraints on the behavior of E_p for U_p near the identity matrix, this choice of the lattice action is not unique and there are other forms with the same continuum limit. In Exercise 13.2 it is shown that the Wilson lattice action is equivalent to the continuum action up to terms of order a^2:

$$S_{\text{Wilson}} \xrightarrow[a \to 0]{} \tfrac{1}{4} \int d^4x \, F_{\mu\nu}^j F_j^{\mu\nu} + \mathcal{O}\left(a^2\right). \qquad (13.18)$$

EXERCISE 13.2 Prove that in the continuum limit eq. (13.15) yields (13.14) to order a^2. *Hint*: write (13.16b) for the elementary plaquette in Fig. 13.2, commute the middle two U factors, expand all U's [see (13.6)], replace finite differences with derivatives, and use (13.15), (13.16a), and (7.10b).

13.1.5 Euclidian Lattices and Statistical Mechanics

In a lattice regularized quantum field theory the expectation value of an observable Q is given by (Rebbi, 1983)

$$\langle Q \rangle = \frac{1}{Z} \int \left(\prod_{\{ij\}} dU_{ij} \right) Q(U_{ij}) e^{-S(U)} \qquad (13.19)$$

$$Z = \int \left(\prod_{\{ij\}} dU_{ij} \right) e^{-S(U)}, \qquad (13.20)$$

where we assume for notational simplicity that the gauge fields defined on the links U_{ij} are the only dynamical variables (dynamical fermions will be introduced in §13.1.10). The integrals are sums over group elements for discrete groups, or invariant integrals over (compact) group manifolds for continuous groups. For a finite lattice these integrals are well-defined ordinary multiple integrals; unlike the continuum theory there is no need to fix the gauge

Table 13.1

Formal Phrasebook Connecting Euclidian Field Theory and 4-Dimensional Statistical Mechanics ($\beta \equiv 1/kT$)

Statistical Physics	Euclidian Field Theory
Sum over configurations	Path integral
Energy of configuration $\times \beta$	Classical action $\times \hbar^{-1}$
Free energy density	Vacuum energy density
Correlation length	Inverse mass gap
Temperature	Coupling strength
Low temperature expansions	Weak coupling expansions
High temperature expansions	Strong coupling expansions
Averages	Expectation values
Thermal fluctuations	Quantum fluctuations
Spin variables	Gauge variables
Magnetic field	External source
Correlation functions	Time-ordered field products

(see §7.4), and a lattice allows the formulation of explicitly gauge invariant quantum averages.

Now (13.19) and (13.20) resemble a thermodynamical average; this formal analogy with statistical mechanics becomes manifest if we set

$$\beta \equiv \frac{1}{kT} \qquad S = \beta \sum E_p, \qquad (13.21)$$

where k is the Boltzmann constant, T is the "temperature", and $\sum E_p$ is the "internal energy" of the system. Then Z is the "partition function", and the sums over U_{ij} in eqns. (13.15) and (13.16) are analogous to sums over statistical variables (e.g., spins). Table 13.1 illustrates the analogies between a Euclidian field theory and 4-dimensional statistical mechanics at a temperature T defined through $\beta = (kT)^{-1}$. The relativistic field theory looks like statistical mechanics in the Euclidian space, even if it corresponds to zero temperature in the physical Minkowski space. It is important to remember this, for later we will discuss field theories at finite *physical* temperatures (Chs. 14 and 15): a Minkowski space field theory and the corresponding Euclidian theory obtained by Wick rotation describe two different physical systems; only on continuing the Euclidian results back to Minkowski space do we obtain quantities relevant to the original Minkowski problem. (However, many quantities of interest are not altered by the Wick rotation; for these we may use the Euclidian result directly.) A thorough introduction to the statistical physics heritage of lattice gauge theories can be found in Kogut (1979).

13.1.6 Strong Coupling and Monte Carlo Methods

Strong Coupling Expansions. Weak coupling expansions correspond to normal perturbation theory. On a lattice these are complicated by problems such as the loss of Lorentz invariance, and a lattice offers few advantages as a regularization for weak coupling. Lattice theories are most useful for strong coupling situations. A *strong coupling approximation* may be implemented by expanding the measure $\exp(-\beta \sum_p E_p)$ appearing in a Euclidian path integral in powers of $\beta \simeq 1/g^2$, which is a small number if the coupling is large.

Strong and weak coupling expansions produce useful results for their respective domains, but it is frequently necessary to determine how these domains are related. For example, the question of confinement in QCD turns on whether the field theory exhibits a phase transition between the strong and weak coupling regimes. In the strong coupling approximation one may take the expansion to high order, in the hope of providing a reliable extrapolation to the weak coupling domain. However, the most powerful method for investigating the intermediate range of β has been the numerical Monte Carlo technique (Rebbi, 1983; Creutz, 1985).

Lattice Monte Carlo Procedure. In a lattice theory the observables are given by the integrals (13.19) and (13.20). If the chosen group G is finite (feasible for a theory defined on a lattice) and the system is finite in volume, the integrals are ordinary sums with a finite number of terms. However, even for the simplest group (the additive group of integers Z_2, with elements $+1$ and -1) and a $4 \times 4 \times 4 \times 4$ lattice the sums contain approximately 10^{310} terms (Rebbi, 1983). The only practical calculations resort to *importance sampling:* a few configurations are chosen as representative and the exact averages are replaced by averages over these configurations. More precisely, importance sampling calculates a sum over a set of configurations with the probability that a given configuration C is included in the set given by a Boltzmann weight,

$$P_C \simeq \exp\left[-S(C)\right],$$

where S is the action. A brief introduction to specific algorithms used to accomplish this is given below. A more complete discussion may be found in Kogut (1983), Hasenfratz and Hasenfratz (1985), or Creutz (1985), Chs. 18–19. In the following sections some examples of calculations using strong coupling expansions will be presented for pedagogical reasons, but the important quantitative results will be obtained using Monte Carlo procedures.

Importance Sampling. In Monte Carlo importance sampling one replaces the expression (13.19), which involves multiple integrals or sums of unmanageably high dimensionalities, by an average over gauge configurations $\{U\}_k$:

$$\langle Q \rangle \simeq \frac{1}{\ell} \sum_{k=1}^{\ell} Q\left(\{U\}_k\right). \tag{13.22}$$

In this expression $\{U\}_k$ labels a configuration k corresponding to a specified set of matrices U_{ij} on all links of the lattice, and $Q(\{U\}_k)$ represents the observable Q calculated in the configuration $\{U\}_k$. The probability that the configuration $\{U\}_k$ occurs is given by a Boltzmann factor

$$P_k = \exp[-S(\{U\}_k)],$$

where S is the action associated with the configuration (Kogut, 1983).

Thus we require an ensemble of configurations distributed according to a Boltzmann weight in order to approximate (13.19) by (13.22). There are various systematic procedures available to do this; two widely used ones are the *heat-bath algorithm* and the *Metropolis algorithm* (see Creutz, 1985, Ch. 18). In each of these one begins with some configuration $\{U\}$ and generates from this a new configuration $\{U'\}$ by varying a single link variable U_{ij}. If the conditions of the algorithm are satisfied the new configuration $\{U'\}$ replaces the old one; otherwise the old one is kept. This is done successively for each link in the lattice, and when all links have been sampled one says that a "sweep" or "iteration" of the lattice has been completed.

The algorithms are designed to bring a lattice to approximate thermal equilibrium after a finite number of iterations. Thus the configuration $\{U\}_1$ obtained from the requisite number of sweeps is assumed to be an element of a Boltzmann distribution. This configuration is used to calculate a contribution to eq. (13.22). Then, the algorithm is applied many times to $\{U\}_1$ to generate a new element $\{U\}_2$ of a Boltzmann distribution, its contribution to (13.22) is recorded, a new configuration is determined, and so on (Kogut, 1983).

Metropolis and Heat-Bath Algorithms. We complete our brief discussion of the Monte Carlo procedure by outlining the Metropolis algorithm (Metropolis et al., 1953), which is the most popular method for bringing gauge lattices to thermal equilibrium.

1. From a configuration $\{U\}$, generate a new configuration $\{U'\}$ by changing a link variable with a suitably chosen procedure.

2. Compute the change in action, $\Delta S = S' - S$.

3. If $\Delta S < 0$, accept the new link variable.

4. If $\Delta S > 0$, accept the new link variable with a conditional probability determined by the following procedure: pick a random number $0 < x < 1$; if $\exp(-\Delta S) > x$, accept the change; if $\exp(-\Delta S) \leqslant x$, reject the change.

Kogut (1983) may be consulted for a proof that this procedure applied to all links and iterated enough times will bring a lattice to thermal equilibrium.

An alternative to the Metropolis procedure is the heat-bath algorithm, which consists of successively placing each link of the lattice in contact with a "heat-bath" (the computer) that selects a new link variable stochastically with a probability given by a Boltzmann factor. In this algorithm the previous value of the link variable plays no direct role in the choice of a successor. The heat-bath algorithm typically brings a lattice to equilibrium in fewer iterations than the Metropolis algorithm, but the required computations within each iteration are usually slower. Therefore, for lattice gauge calculations with uncomplicated groups a heat-bath method may be preferable, but lattice Monte Carlo simulations with more complex groups commonly use some variant of the Metropolis algorithm.

13.1.7 Wilson Loops and Confinement

For gauge theories the physically significant matrix elements must be gauge invariant. This means that standard candidates for order parameters are trivially zero, a phenomenon well known in the spin systems of condensed matter physics (see Exercise 13.3a). As an example, we take the Ising model, which is often used to describe ferromagnetic behavior. It corresponds to a lattice with electron spins that can be up or down at each lattice site, with nearest-neighbor coupling between the spins.[‡] Consider an Ising approximation of a magnet. As it is cooled below the Curie point the global rotational symmetry is spontaneously broken and we find $\langle U_{ij} \rangle \neq 0$, where U_{ij} are spin variables located on the bonds of the crystal; thus, the magnetization order parameter is finite. It must choose a direction to point (a vanishingly small field will accomplish this at the transition point). Once a direction has been selected and the system cooled below the Curie point, the magnetization direction is stable because to alter it a macroscopic number of spins must change their direction coherently, and thermal fluctuations cannot bring this about in a large crystal. This is spontaneous (global) symmetry breaking.[‡‡]

But now consider a lattice gauge theory; because of the local gauge symmetry the action is unchanged by an arbitrary rotation of the direction of U_{ij}

[‡]Lattice spin models in which a spin can be described by a single number (up or down) are called *Ising models;* those for which the spins can be oriented anywhere in a plane and require two numbers to specify them are called *X–Y models;* for *Heisenberg models* the spins can point in any spatial direction and require three numbers for their specification. Ising, *X–Y*, and Heisenberg models are useful approximations for uniaxial, planar, and isotropic ferromagnets, respectively.

[‡‡]This analysis assumes that we deal with an effectively infinite number of quantum-mechanical degrees of freedom. For a finite system the ground state is unique because degenerate minima would differ by finite action and be mixed by tunneling (see §8.5.1). Strictly speaking, only infinite systems can exhibit spontaneous symmetry breaking; finite systems have unique ground states that display the same symmetries as the Lagrangian.

through the substitution (13.10)

$$U_{ij} \rightarrow G_i^{-1} U_{ij} G_j,$$

where G_i is an element of the gauge group defined locally at the site i. As a consequence of this *local* symmetry, it can be shown that two macroscopic spin configurations differ only by a *finite action* (Kogut, 1979, §V.C). Therefore, the situation resembles tunneling in a quantum mechanical system with only a few degrees of freedom: there is a finite barrier between the possible degenerate ground states, they restore the symmetry by tunneling into each other through thermal fluctuations, and there is no spontaneous magnetization at any T. This is a specific example of *Elitzur's theorem* (Elitzur, 1975): theories with continuous local symmetries cannot have a spontaneous magnetization.

The lesson we retrieve from this discussion is that for local gauge symmetries a useful order parameter should respect the *local* symmetry. A gauge invariant and nontrivial candidate for such an order parameter is the expectation value of the Wilson loop operator, which is given in a lattice theory by (13.9); schematically

$$\langle W(c) \rangle = \left\langle \mathrm{Tr} \prod_c U_{ij} \right\rangle, \qquad (13.23)$$

where c denotes a closed path consisting of directed links U_{ij}. Physically, the Wilson loop measures the response of the gauge fields to a color source that moves around its perimeter adiabatically . The dependence of the expectation value of $W(c)$ on the path c for large loops can be used to specify the phase structure of the theory (Exercise 13.3b). If for large loops

$$\langle W(c) \rangle \simeq \exp\left[-\mathrm{constant} \times (\text{area enclosed by } c)\right], \qquad (13.24)$$

the phase is considered *disordered;* if instead

$$\langle W(c) \rangle \simeq \exp\left[-\mathrm{constant} \times (\text{perimeter traced by } c)\right], \qquad (13.25)$$

the phase is termed *ordered.* The ordered phase corresponds to low T (weak coupling) in the statistical mechanics analog problem, while the disordered phase corresponds to high T (strong coupling) in the analog problem, as we would expect.

For a color $SU(3)$ theory the disordered phase (area law for Wilson loop) corresponds to *quark confinement,* while the ordered phase (perimeter law for Wilson loop) corresponds to free quarks. This follows because if we define a rectangular contour c of temporal length T and spatial length R, the static potential between a quark and an antiquark a distance R apart is (Wilson, 1974; Kogut, 1979)

$$V(R) = -\lim_{T \to \infty} \frac{1}{T} \log \langle W(c) \rangle. \qquad (13.26)$$

Therefore, the disordered phase with the area law loop correlation function (13.24) gives a confining linear potential

$$V(R) \simeq \sigma R \qquad \text{(area law)}, \qquad (13.27)$$

where σ is the string tension [see also (12.101) and (12.85)], while for perimeter law behavior we obtain the nonconfining potential

$$V(R) \simeq \text{constant} \qquad \text{(perimeter law)}. \qquad (13.28)$$

This condition for confinement is called the *Wilson criterion*.

The string tension σ in a confining model is the negative of the coefficient of the area A appearing in the exponent of a large Wilson loop evaluated in the confining phase,

$$\langle W(c) \rangle \simeq e^{-\sigma A(c)}. \qquad (13.29)$$

In principle, the string tension in the confined phase can be inferred from the decay of $\langle W(c) \rangle$ as the contour c is increased in size (Exercise 13.3b). In practice, the determination of σ is more complicated than this because of effects that may obscure the area law behavior (Creutz, 1980a,b).

This discussion of the Wilson condition for confinement assumes *pure gauge fields*. The area law criterion is less useful when dynamical quarks are placed on the lattice (§13.1.10). Then widely separated color sources can create quark pairs from the vacuum to screen the long-range gauge fields (§14.3), and a large Wilson loop measures an effective meson–meson potential instead of a quark–quark potential (see Fig. 12.13). The utility of the Wilson loop in the presence of dynamical quarks will be investigated further in §14.5.2.

EXERCISE 13.3 (a) Demonstrate that a conventional order parameter such as the magnetization always vanishes in a lattice gauge theory, necessitating the definition of a gauge invariant order parameter. *Hint*: show that configurations with two different values of the order parameter differ by finite action on a lattice with local symmetry. (b) For strong coupling, show that the QCD Wilson loop obeys an area law. Show that $\langle W(c) \rangle$ vanishes as $g \to \infty$, and that the strong-coupling string tension is

$$\sigma \simeq \frac{1}{a^2} \log g^2.$$

Hint: use (13.19) to write $\langle W(c) \rangle$ as an integral, make a strong-coupling expansion, and employ the following properties of invariant $SU(N)$ group integration:

$$\int [dU]\, U_{ij} = 0 \qquad \int [dU]\, U_{ij} U_{kl}^{\dagger} = \frac{1}{N} \delta_{il} \delta_{jk}.$$

(c) In perturbation theory the $q\bar{q}$ potential for an $SU(3)$ gauge theory is [Kogut,

1983, eq. (2.42)]

$$V(R, a, g) = -\frac{\text{Tr}\,(F_k F_k)}{4\pi R}\left[g^2 + \frac{22g^4}{16\pi^2}\,\log\left(\frac{R}{a}\right) + \cdots\right]$$

$$\equiv -\frac{\text{Tr}\,(F_k F_k)}{4\pi R}g^2(R),$$

where the $SU(3)$ matrix F_k is defined in (5.67), R is the separation, $g = g(a)$ is the coupling constant for lattice spacing a, the factor $g(R)$ is the effective "running coupling constant" on a scale R, and $1/a$ is the momentum scale cutoff. Impose the condition that $V(R, a, g)$ at fixed R be independent of a and use the resulting equation to obtain (13.38).   (d) Show that eq. (13.37) is equivalent to

$$B(g) = \frac{2\partial g}{\partial\big(\log(Q^2/\mu^2)\big)}$$

and obtain (13.38) from (10.40) for QCD without quarks. Obtain the result analogous to (13.38) for QED starting from (10.39).

13.1.8   The Continuum Limit and Scaling Behavior

A lattice is a regularization method allowing non-perturbative calculations to be done in systematic and gauge invariant fashion. However, a lattice is unphysical for a relativistic quantum field theory—our only real interest is in the continuum limit of such a theory. Once a lattice theory has been formulated in Euclidian space the problem is similar to a statistical mechanics problem (Table 13.1). The general procedure to be followed in establishing the corresponding continuum limit is well known from statistical physics (Kogut, 1979): (1) map out the phase diagram of the lattice theory; (2) locate the critical points of continuous (second-order) phase transition; (3) approach the critical points carefully to recover a continuum limit for the theory.

The first step is usually the least difficult; standard methods for establishing the phase structure in statistical physics include mean field theory, high and low temperature expansions, and numerical calculations using the renormalization group. The phase structure may be characterized by the values of suitably chosen order parameters; for a gauge field without dynamical fermions (13.23) can serve that purpose.

Once the phase structure has been established the essential task is to find the *critical points* for the theory—values of the coupling constants where the correlation length, measured in units of the lattice spacing, diverges (see Wilson and Kogut, 1974; Kogut, 1979). The mass gap m of a field theory is related to the correlation length ξ (measured in lattice units a) of the analogous statistical mechanics problem by

$$m = \frac{1}{\xi a}. \tag{13.30}$$

To make a field theory with finite masses the corresponding statistical mechanics must be nearly critical, since in the continuum limit of (13.30) a finite mass is possible only if $\xi \to \infty$ as $a \to 0$. The critical points must be points of *second-order phase transition*, because these are characterized by a divergent correlation length (ξ remains finite in a first-order transition).

Near a critical point the theory loses memory of the lattice as the correlation length tends to infinity, the granularity of the lattice is no longer relevant, and continuous spacetime symmetries are restored. In that region various thermodynamical functions of the system become singular; the degree of their singularity is recorded in their *critical exponents*, which can be used to define the nature of the phase transition.[‡]   The singular behavior in the critical region is associated with long-range correlations and large fluctuations: even though the underlying model action has only short-range forces, correlations can appear over infinite distances near a critical point.

The recovery of a continuum limit imposes stringent renormalization conditions on the couplings of a lattice theory (Rebbi, 1983). Consider a physical observable ℓ with the dimension of length. A lattice regularized quantum

---

[‡]In the physics of critical behavior it is commonly found that near critical points the macroscopic variables depend on a power of $T - T_c$, where T is the temperature and T_c is the critical temperature, while order parameters depend on some power of the applied fields. If Q is a macroscopic observable, $\mathcal{M}$ an order parameter, and $\mathcal{H}$ an applied field, a typical behavior is

$$Q(\mathcal{H} = 0) \simeq (T - T_c)^a \qquad \mathcal{M}(T = T_c) \simeq \mathcal{H}^b,$$

where a and b are called *critical exponents*. For example, in ferromagnets near the Curie point the magnetic susceptibility $\chi = (\partial M / \partial H)_T$, where M is the magnetization and H is the applied magnetic field, is given for $H = 0$ by

$$\chi \propto (T - T_c)^{-\gamma} \quad (T > T_c) \qquad\qquad \chi \propto (T_c - T)^{-\gamma'} \quad (T < T_c),$$

the specific heat C at $H = 0$ by

$$C \propto (T - T_c)^{-\alpha} \quad (T > T_c) \qquad\qquad C \propto (T_c - T)^{-\alpha'} \quad (T < T_c),$$

and the magnetization M by

$$M \propto H^{1/\delta} \quad (T = T_c) \qquad\qquad M \propto (T_c - T)^{\beta} \quad (H = 0).$$

Experimentally

$$\gamma \simeq \gamma' \simeq 1.33 \qquad \alpha \simeq \alpha' \simeq -0.11 \qquad \delta \simeq 4.1 \qquad \beta \simeq 0.34,$$

for a variety of ferromagnetic materials (Ma, 1976, Ch. 1; Kogut, 1979).

theory such as QCD with massless quarks has no intrinsic scale. Therefore, ℓ will be a product of the lattice spacing a and some function $\lambda(g)$ of the dimensionless coupling constant g

$$\ell = a\lambda(g), \tag{13.31}$$

since a carries the only dimensions available. (We assume that the gauge theory is described by a single g; the generalization to a theory with more than one coupling constant is not difficult conceptually.) Now ℓ is a physical observable, and it must remain finite in the continuum limit $a \to 0$. This requires the existence of at least one critical value g_0 of the coupling constant so that

$$\lambda(g) \xrightarrow[g \to g_0]{} \infty \tag{13.32}$$

in just such a way that

$$\lim_{\substack{g \to g_0 \\ a \to 0}} \ell = \text{finite constant}. \tag{13.33}$$

The requirement that the observable ℓ be independent of the lattice spacing as $a \to 0$ induces a relationship between the coupling constant and the lattice spacing,

$$g = g(a). \tag{13.34}$$

If the lattice theory is to yield an acceptable continuum limit, the same rate of approach $g \to g_0$ that makes ℓ constant as $a \to 0$ must simultaneously make *all* observables independent of the lattice spacing as $a \to 0$. This property is called *scaling*, and if it is realized g_0 is called a *scaling critical point*. For non-Abelian gauge theories $g_0 = 0$ has the properties of a scaling critical point. In the short-distance regime of these asymptotically free theories the dependence of g on the lattice momentum cutoff is calculable from perturbation theory [see eq. (13.37) and the discussion that follows it]; the appearance of this known perturbative behavior as $a \to 0$ is called *asymptotic scaling*. We may also note that for theories exhibiting scaling there generally is a *scaling window* on a finite lattice. Below a certain lattice spacing scaling is expected, but as $a \to 0$ the correlation lengths grow until they are larger than the size of the lattice. Then the system no longer scales.

Although many lattice theories can be defined, only those exhibiting critical points with acceptable scaling behavior are legitimate candidates for an approximation to a continuum quantum field theory. It is apparent that obtaining such behavior is not a trivial matter.

> The preceding discussion of critical phenomena and the continuum limit may be summarized in the *correlation length scaling hypothesis*: the long-range correlation of fluctuations near a critical point is responsible for all

singular behavior (Fisher, 1967; Kogut, 1979). In particular, the scaling hypothesis asserts that as far as the singular behavior is concerned the correlation length ξ is the *only relevant length*. In the language of spin correlations, ξ is the size of the largest spin patch in which an appreciable number of spins point in the same direction. *Universality* is attained in critical phenomena because the detailed microstructure of the sample is much smaller than the relevant scale for the phenomena. At the critical point $\xi \rightarrow \infty$, the system has fluctuations on all scales, and it is scale invariant (looks the same on all length scales). Such universal behavior is common in statistical physics; for example, the critical behavior of ferromagnets is largely independent of the crystal structure of the ferromagnet, as we saw in the preceding footnote concerning critical exponents.

According to the *universality hypothesis*, the critical behavior of many physical systems is determined by only three things: the local symmetry of the coupling between variables (*gauge group*), the *dimensionality d of space*, and the *dimensionality n of the order parameter* (Kadanoff, 1976; Kogut, 1979). Systems with the same values of d and n, and the same local symmetry, have equivalent critical exponents and are said to belong to the same *universality class;* thus, they are expected to exhibit similar critical behavior. For example, a 3-dimensional Ising model of a uniaxial ferromagnet near the Curie point and a 3-dimensional fluid near its critical point exhibit equivalent critical behavior.[‡] The order parameter of the Ising model is the magnetization (number of up spins minus number of down spins), and n is the number of components required to define a spin vector. Only one number is needed to specify an Ising spin, so $n = 1$ and $d = 3$ for the 3-dimensional Ising model. For a fluid the order parameter is the difference in density between the liquid and vapor phases. Just as the magnetization is zero at the Curie point, the difference in vapor and liquid densities vanishes at the critical point of the fluid. The density is a scalar quantity, so for the 3-dimensional fluid $n = 1$ and $d = 3$. Thus the fluid and the uniaxial ferromagnet are quite distinct physical systems, but they exhibit similar critical behavior because they belong to the same universality class (Wilson, 1979; Fisher, 1982).

More than one action can be defined in lattice theories as a consequence of this concept of universality in critical phenomena, which is assumed to hold in Euclidian lattice gauge theories because of their mathematical affinity with statistical mechanics. Loosely stated, if theories formulated on the lattice have the same local gauge symmetry and spacetime dimensionality they belong to

---

[‡]A uniaxial ferromagnet can be easily magnetized only along one axis; the Curie point is the temperature of spontaneous magnetization. The fluid critical point is that combination of temperature and pressure beyond which the distinction between liquid and gas disappears. For example, the critical point for water is $T = 647$ K at a pressure of 218 atmospheres. If the pressure is increased from the critical point, there is no further distinction between gas and liquid and only a single undifferentiated "fluid phase" exists.

the same universality class, and are expected to yield the same continuum theories. For example, actions can be defined on the lattice that differ from the continuum action by terms of order a^4, instead of a^2 as for the Wilson action. One consequence of this change is that the Wilson and improved actions have the same continuum limits, but scaling over larger ranges of $g(a)$ may be obtained for the improved actions (see Kogut, 1984).

13.1.9   Confining Properties of Gauge Theories

Lattice gauge theories are primarily of interest for strong-coupling calculations. The holy grail of QCD is the proof that a color $SU(3)$ gauge theory confines in the non-perturbative regime. This is not difficult to show for lattices with large spacing; unfortunately, such a demonstration does not constitute a proof of QCD confinement: to do that we must also demonstrate that the same theory that confines at large lattice spacing (strong coupling) has a continuum limit (weak coupling) that is consistent with the asymptotically free short-distance behavior of QCD.

An indication that a naive proof of confinement at large lattice spacing is inadequate is provided by the observation that most lattice theories confine under such conditions, even ones that should not do so in the continuum limit. For example, the Abelian $U(1)$ model exhibits confinement for strong coupling, but in continuous spacetime the $U(1)$ photons are known to be free; consistency of the lattice formulation requires the existence in $U(1)$ lattice theory of at least one critical point separating a strong-coupling phase from a "spin-wave" or free photon phase (Rebbi, 1983). In the event of such a two-phase structure the continuum limit may be recovered in the spin-wave phase, with the spin-wave excitations corresponding to the photons. Monte Carlo simulations have demonstrated that for $U(1)$ there is such a two-phase structure with free photons in the weakly coupled phase, as required by the data (Creutz, Jacobs, and Rebbi, 1979; Lautrup and Nauenberg, 1980).

Phase Structure for Lattice QCD. Thus the lattice formulation passes a consistency test for QED. Now we investigate whether QCD on a lattice can reproduce the salient properties of the strong interactions. To do so requires examination of a local $SU(3)$ gauge theory; however, many early lattice studies of QCD properties (confinement in particular) used the simpler group $SU(2)$ rather than the relevant group $SU(3)$. The general argument is that the confining phase is a disordered phase, and if $SU(2)$ is shown to confine then $SU(3)$ (with more degrees of freedom and more disorder) must also yield a confining phase (Creutz, 1979). However, we may expect that details such as the nature of the transition from confining to deconfining phases at finite physical temperature may depend on the gauge group (see Ch. 14).

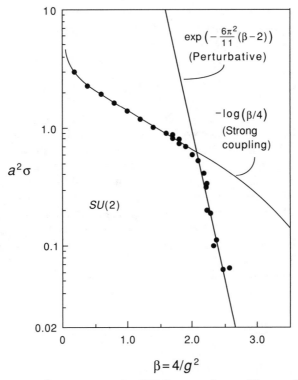

Fig. 13.3 String tension for $SU(2)$ gauge theory (Creutz, 1980a).

For large β (small coupling) in $SU(2)$ the renormalization group yields

$$a^2\sigma \simeq \exp\left(\frac{-6\pi^2\beta}{11}\right) \tag{13.35}$$

for the string tension σ, while for small β the leading order of the Wilson strong-coupling expansion gives (Creutz, 1980a)

$$a^2\sigma \simeq -\log\left(\frac{\beta}{4}\right). \tag{13.36}$$

In Fig. 13.3 a Monte Carlo calculation of the $SU(2)$ string tension is compared with these weak and strong coupling limits. The calculation exhibits a smooth and rapid transition between the two limits near $\beta \simeq 2$ (corresponding to $\alpha_s \equiv g^2/4\pi \simeq 0.16$), suggesting that the strong and weak coupling domains are in the same phase.

Renormalization Group and Callan–Symanzik Function. Further insight into the phase structure of QCD may be gained by considering the *Callan–Symanzik* (or *Gell-Mann, Low*) function

$$B(g) \equiv -a\frac{\partial g}{\partial a}, \qquad (13.37)$$

where g is the coupling and a is the lattice spacing. [In perturbative field theory $B(g)$ is commonly denoted $\beta(g)$, and is called the β-*function*.] Explicit equations for $B(g)$ follow from the renormalization properties of the field theory, and the zeros of $B(g)$ map the phase structure of the model. In particular, the function $B(g)$ may be found using a differential equation that specifies how the parameters of a theory must change with length scale to hold physical observables invariant against scaling of the lattice (see Exercise 13.3c). The points where $B(g) = 0$ are called *fixed points*, because from (13.37) the coupling constant does not vary with lattice spacing (or equivalently, momentum scale) at a fixed point.

In weak coupling the β-function for $SU(3)$ can be evaluated by perturbation theory. To one-loop level the result is (Exercise 13.3c)

$$\frac{-B(g)}{g} = \frac{11}{16\pi^2}g^2 + \dots \qquad \text{(weak coupling)}, \qquad (13.38)$$

so that the dependence of g on the cutoff is

$$\frac{\partial g}{\partial \log a} = \frac{11}{16\pi^2}g^3 + \dots,$$

which can be integrated to give the form (10.40) for no quarks ($N_{\mathrm{f}} = 0$), if we assume $Q \simeq 1/a$ (Exercise 13.3d). For strong coupling the β-function can be evaluated in powers of g^{-2}, with the result (Kogut, Pearson, and Shigemitsu, 1981)

$$\frac{-B(g)}{g} = 1 - 0.2156863\left(\frac{2}{g^4}\right)^2 - 0.1681985\left(\frac{2}{g^4}\right)^3$$

$$- 0.04137858\left(\frac{2}{g^4}\right)^4 + \dots \qquad \text{(strong coupling)}. \qquad (13.39)$$

β-Function at Intermediate Coupling. In Fig. 13.4 the $SU(3)$ β-function for strong and weak coupling is indicated. The nature of the lattice $SU(3)$ continuum limit depends crucially on the behavior of $B(g)$ at intermediate values of g. Two possible scenarios are sketched. If the upper curve applies, there are no zeros of $B(g)$ between the strong and weak coupling regimes and the continuum limit of the strong-coupling $SU(3)$ lattice theory can be ob-

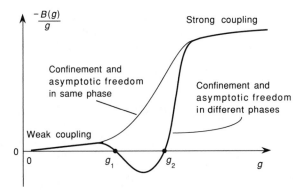

Fig. 13.4 Possible behavior of the β-function for QCD.

tained as $g \to 0$. Then the corresponding continuum theory can exhibit within a single phase both asymptotic freedom at small coupling and confinement at large coupling. This could be QCD. If instead the situation resembles the lower curve, the continuum limit of the strong-coupled theory cannot be recovered as $g \to 0$ and the confining phase observed for strong coupling cannot be related to the weak coupling limit. In this case confinement and asymptotic freedom lie in different phases, and the continuum limit of the $SU(3)$ lattice theory is presumably not QCD.

In Fig. 13.5 the $SU(3)$ β-function is shown for weak coupling (13.38) and for the strong-coupling expansion (13.39) carried to sixth order (order g^{-24}). These results are not conclusive, but they suggest that the strong- and weak-coupling regimes probably join smoothly near $g = 1.0$, and that there are no zeros for $B(g)$ except at the origin. Thus Fig. 13.5 is consistent with the strong and weak-coupling regimes of the $SU(3)$ gauge theory existing in the same

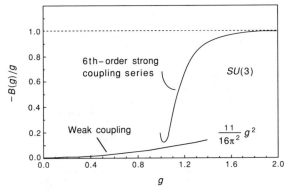

Fig. 13.5 Sixth-order strong coupling series for the $SU(3)$ β-function (Kogut, 1980).

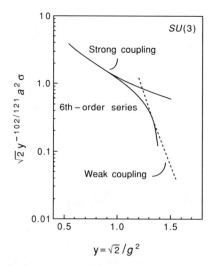

Fig. 13.6   Sixth-order strong coupling series for the $SU(3)$ string tension σ. The dashed line is the asymptotic freedom scaling law; the strong coupling limit is the first term in the strong coupling series (Kogut, 1980).

phase, as expected of a theory required to exhibit both asymptotic freedom and confinement. Figure 13.6 illustrates the rapid and smooth transition of the $SU(3)$ string tension between the weak and strong coupling regions, again as determined from a strong-coupling expansion (Kogut, 1980).

These examples suggest a smooth evolution from strong coupling to weak coupling, with the transition occurring over a narrow range of coupling constants. This behavior means that quarks at small separation move rather freely with flux spread out smoothly, but as the quarks begin to separate the flux between them rapidly collimates into a tube, generating a linearly rising potential:

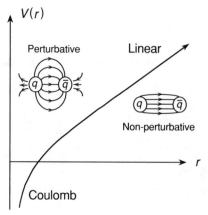

Qualitatively, such a picture makes the successes of bag and string models of hadrons easier to understand (compare Fig. 12.20). This is also consistent with the experimental results from deep inelastic lepton scattering (§10.3), which suggest that for momentum transfers $Q^2 \simeq 2$ GeV^2 the reaction changes character from the excitation of hadronic resonances (implying an important role for the interquark potential in the corresponding quasibound states) to the Bjorken scaling regime where the partons seem to be almost free. This transition appears to occur in a narrow range of Q^2, signaling a rapid change in the nature of the quark–quark interaction over a small change in distance scale.

Monte Carlo Thermal Cycles. Detailed Monte Carlo calculations provide more substantial evidence that the $SU(2)$ and $SU(3)$ theories in four dimensions exhibit no phase transition as the coupling β is varied between the strong-coupling and weak-coupling limits. A standard procedure to search for phase transitions that is called a *thermal cycle* may be borrowed from statistical physics. The inverse coupling β is set to some value above a suspected phase transition, and an order parameter such as the average action per plaquette is calculated starting from a completely ordered lattice configuration (e.g., all $U_{ij} = 1$); then β is decremented by a small amount and another calculation is performed. The procedure is repeated until some value of β below the suspected phase transition is reached; then the process is reversed, with β incremented in small steps until reaching the starting value.

This is called a thermal cycle because varying β is equivalent to varying the inverse temperature in the corresponding statistical mechanics problem. The procedure is thus a *computer experiment* analogous to the gradual cycling of a physical system in temperature. If the increments in β are small enough, the system is only slightly out of equilibrium unless there is a phase transition point in the temperature range of the cycle. Near a phase transition the relaxation time of the system increases and hysteresis effects may be expected in the thermal cycle.

Figure 13.7 shows thermal cycles for lattice Monte Carlo calculations in three different gauge theories (Creutz, 1979). The solid lines indicate strong and weak coupling limits, pluses represent the Monte Carlo results on the heating cycle, and circles the Monte Carlo results on the cooling cycle. There is a clear indication of hysteresis (and a phase transition) for the $SU(2)$ theory in five dimensions and the $SO(2)$ theory [locally isomorphic to $U(1)$] in four dimensions, but the $SU(2)$ theory in four dimensions exhibits a smooth crossover from strong to weak coupling with no evidence of a phase transition.

Once a thermal cycle has established that a phase transition exists, the order may be investigated by comparing repeated iterations of the lattice starting with ordered and random configurations for β fixed near the phase transition. For a first-order phase transition the original hysteresis gap is expected to remain after repeated iteration because the system can exist in

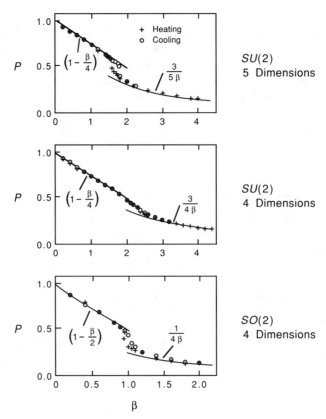

Fig. 13.7   The average plaquette action P as a function of β. (a) $SU(2)$ in five dimensions, (b) $SU(2)$ in four dimensions, and (c) $SO(2)$ in four dimensions. Pluses denote heating cycles and circles denote cooling cycles (Creutz, 1979).

two distinct stable phases at the transition point, and in metastable phases for a range of temperatures on either side of the transition point (superheating and supercooling). For higher order phase transitions the same calculations show a tendency to fluctuate and slowly converge under repeated iteration, while for no phase transition there is rapid convergence (Creutz, Jacobs, and Rebbi, 1979). Figure 13.8 illustrates this procedure for the theories considered in Fig. 13.7. The conclusions from Figs. 13.7 and 13.8 are that $SU(2)$ exhibits no phase transition in four dimensions and a first-order phase transition in five dimensions, while $SO(2)$ exhibits a second-order phase transition in four dimensions.

Thus, Figs. 13.7 and 13.8 suggest that the $SU(2)$ gauge theory in four spacetime dimensions has no phase transition between the strong-coupling and weak-coupling limits. Similar results are found for $SU(3)$ in four dimensions, and it appears plausible that these non-Abelian theories can exhibit

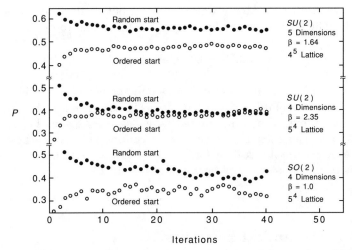

Fig. 13.8   The average plaquette action as a function of the number of iterations at a fixed β near the critical point for the cases in Fig. 13.7 (Creutz, 1979).

asymptotic freedom and confinement within the same phase, and that the continuum limit of the lattice gauge theory can be recovered in this phase.

Lattice Quark-Antiquark Potential. As an example of evaluating physical quantities, we show in Fig. 13.9 the $q\bar{q}$ potential calculated by several groups for lattice QCD. The potential V is in units of $\sqrt{\sigma} \simeq 0.42$ GeV and r is in units of $1/\sqrt{\sigma} \simeq 0.48$ fm, where σ is the empirical string tension (12.85). These calculations were done on lattices ranging in size from $12^3 \times 16$

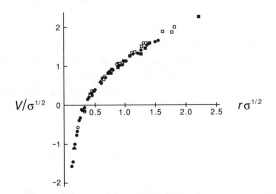

Fig. 13.9   Quark–antiquark potential calculated on $SU(3)$ lattices by several groups for 10 different values of β in the range $5.7-7.2$. The zero has been chosen arbitrarily, but the shape and scale are fixed by the calculations. Results are taken from Hasenfratz and Hasenfratz (1985), where the original references may be found.

to $24^3 \times 48$, with high Monte Carlo statistics but no dynamical quarks (see §13.1.10). The lattice spacing for $\beta = 6/g^2 = 6$ is found to be

$$a(\beta = 6) = 0.53 \pm 0.05 \ \text{GeV}^{-1} = 0.11 \pm 0.01 \ \text{fm},$$

so that a 16^4 lattice at $\beta = 6$ represents a periodic box of total width 1.76 fm, with a distance resolution of about 0.1 fm. The agreement of the calculations for different values of β is interpreted as evidence for scaling behavior when $\beta \geqslant 5.7$, and hence for the recovery of a continuum theory; there is some evidence of asymptotic scaling for $\beta \geqslant 6$. These results may be compared with the phenomenological $q\bar{q}$ potential of Fig. 12.20.

13.1.10   Dynamical Fermions

The inclusion of fermions in lattice gauge calculations presents thorny problems at both fundamental and technical levels. The fundamental difficulty has to do with the very definition of a lattice fermion theory. This problem is called *fermion doubling*, and it can be shown to be inevitable for any lattice formulation that attempts to preserve the chiral symmetry of the massless continuum action. It produces unphysical fermion degeneracy on the lattice. The two standard methods for dealing with this problem employ *Wilson fermions* and *Kogut–Susskind staggered fermions*, respectively. The first defines a lattice regularization that is not chiral symmetric, but the chiral breaking is proportional to the lattice spacing so it should disappear in the continuum limit; the second accepts the doubling but preserves a vestige of chiral symmetry; neither is completely satisfactory. It can be shown on general grounds that it is impossible to formulate a lattice theory with fermions that retains all the formal properties of a continuum theory—something must be sacrificed. We will not pursue this point, but a simple introduction may be found in Hasenfratz and Hasenfratz (1985), a more detailed one in Kogut (1983).

The technical difficulty with fermions concerns the implementation of lattice Monte Carlo algorithms in the presence of dynamical fermion fields. A schematized lattice action including Wilson fermions is of the form (Kogut, 1984; Satz, 1985a)

$$S = S_{\text{G}} + \sum \overline{\psi}(1 - \kappa M)\psi = S_{\text{G}} + S_{\text{F}}, \tag{13.40}$$

where the *hopping parameter* κ is approximately the reciprocal of the quark mass, the sum is over quark flavors, the single-site *hopping matrix* M depends on the gauge fields and plays the role of a covariant derivative of the continuum theory, the action S_{G} is the contribution from the pure gauge field, and $\overline{\psi}$ and ψ are independent elements of a Grassmann algebra (§4.5). The Euclidian

partition function is

$$Z = \int [dU][d\psi][d\overline{\psi}] e^{-S_G(U)} e^{-\sum \overline{\psi}(1-\kappa M)\psi}, \tag{13.41}$$

where U denotes the gauge variables. Because the fermions enter the action in bilinear form, they can be integrated out using (4.71) for Euclidian integrals, leading to a *fermion determinant* in the partition function:

$$Z = \int [dU] \operatorname{Det} (1 - \kappa M)^{n_f} e^{-S_G} = \int [dU] e^{-S_{\text{eff}}} \tag{13.42a}$$

for n_f quark flavors. The effective action is

$$\begin{aligned} S_{\text{eff}} &= S_G - n_f \log \left[\operatorname{Det} (1 - \kappa M) \right] \\ &= S_G - n_f \operatorname{Tr} \left[\log (1 - \kappa M) \right], \end{aligned} \tag{13.42b}$$

where the second step follows because if a matrix A can be diagonalized with eigenvalues λ_i, then

$$\operatorname{Det} A = \prod_i \lambda_i = \exp \left(\sum_i \log \lambda_i \right) = e^{\operatorname{Tr} (\log A)}. \tag{13.43}$$

As it stands this result is intractable because of the determinant of the hopping matrix, which represents the effects of virtual quark loops. This enormously increases the required computational time in standard Monte Carlo algorithms because the determinant is that of a very large matrix (typically many-thousand dimensional), and in the absence of clever tricks it must be evaluated at each gauge-link update. This is technically impossible and various approximations have been employed to circumvent the full inclusion of the fermion determinant.

In the *quenched approximation* $\operatorname{Det} (1 - \kappa M)$ is set to unity. This extreme assumption ignores the effect of virtual quark–antiquark excitations on the lattice, and fermion matrix elements are computed as averages over gauge field variables:

Included in quenched approximation

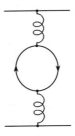

Excluded from quenched approximation

Thus in this approximation the gauge fields affect the quarks, but the quarks have no dynamical influence on the gauge fields; it may be viewed as the QCD heavy-quark limit, or the limit of a vanishing number of quark flavors. Until recently most QCD calculations on large lattices with good Monte Carlo statistics have been performed in the quenched approximation. The overriding reason is computational simplicity but there are qualitative arguments based on the Zweig rule (§12.6.1), the success of the naive valence quark model, and the dominance of gluon loops over quark loops in QCD with a large number of colors (§13.8.1) that provide some rationalization of the approximation.

Several methods have been proposed to include dynamical fermions on a lattice without making the quenched approximation (see Kogut, 1984; Fucito et al., 1981; Polonyi and Wyld, 1984; Duke and Owens, 1985; Cleymans, Gavai, and Suhonen, 1986). We will discuss some of these calculations in Ch. 14 when the deconfining and chiral restoration transitions of QCD are investigated.

13.1.11 Lattice Prospects

To recover continuum symmetries while avoiding unphysical finite-size effects, a realistic lattice calculation should satisfy

$$a << \xi << Na,$$

where a is the lattice spacing, N is the number of lattice units in a particular direction, and ξ is a typical correlation length. Wilson (1983b) estimates that a lattice spacing of 0.01 fm would give a reliable approximation to QCD. In a 10-fm-wide cube with this spacing there are 10^9 sites; taking the same number of time steps gives 10^{12} spacetime sites. In a Lagrangian formulation of QCD eight $SU(3)$ angle integrations are required for each nearest-neighbor link, so 32×10^{12} simultaneous integrations must be performed. The quarks in such a calculation are described by a $(72 \times 10^{12}) \times (72 \times 10^{12})$ complex matrix, and this matrix must be inverted and its determinant taken to compute the integrand for the 32×10^{12} integrations. Standard integration methods are out of the question; even with the Monte Carlo procedure some $10^{60} - 10^{65}$ computations are required to do the lattice calculation with 10^{12} sites. The present state of the art uses fewer than 10^6 lattice sites and employs the quenched approximation for the largest lattices.

Advances in lattice calculations have been impressive, and many would argue that these estimates are too pessimistic. Nevertheless, a conservative evaluation suggests that we are still far from a completely realistic computation of most physical quantities. Although lattice gauge calculations represent our best current understanding of non-perturbative QCD and must be taken seriously, it seems prudent to maintain a healthy skepticism about the details of such calculations until larger computers and better algorithms bring uncertainties such as continuum-limit scaling, finite lattice size effects, and the influence of dynamical fermions under full control.

13.2 Topology of Group Manifolds

Monte Carlo lattice gauge calculations have proven to be the most powerful method of studying non-perturbative aspects of gauge theories. However, such calculations are numerically complex and computationally intensive; they are in reality large computer experiments. There is an alternative to the direct numerical assault of lattice gauge methods on non-perturbative field theories that may be formulated in terms of soliton-like solutions to nonlinear field equations. These approaches have not been quantitatively as successful as lattice calculations, but they are more analytical and provide valuable insight into the non-perturbative aspects of quantum field theory. An important feature of such theories is the topological structure of the relevant group. Accordingly, we begin our discussion of these methods with a brief review of group manifolds and their topological properties.

13.2.1 Homotopies

Two paths with the same endpoints are topologically equivalent if there is a function that can continuously deform one into the other, as illustrated in Fig. 13.10a. If two paths possess this property they are said to be *homotopic*, and the continuous function that deforms one path into the other is called a *homotopy;* thus, homotopy is an equivalence relation for a collection of paths with a specified common beginning and ending. In Fig. 13.10b paths Ψ and Φ belong to different homotopy classes. The number of classes of paths on this surface is infinite; for example, paths that wrap n times around the ring cannot be continuously distorted into those that wrap m times if $m \neq n$.

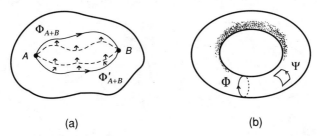

(a)   (b)

Fig. 13.10   (a) Two paths that can be continuously deformed one into the other. (b) The paths Φ and Ψ belong to different homotopy classes since they cannot be continuously deformed into each other.

13.2.2 Connectedness

In a matrix representation of a group G the angle variables θ_a parameterize the group, $U = \exp(-i\theta_a T_a)$. The space $\{\theta_a\}$ over which the θ_a vary is called the *group manifold* of G; this space must be suitably restricted if there is to be a one-to-one correspondence between group elements and $\{\theta_a\}$. Two important properties of a group manifold are its compactness and its connectedness (Huang, 1982, §4.2). If a group manifold is compact the group is termed *compact* (recall that in Ch. 5 we loosely defined compact to mean a parameter space with finite volume). If every closed path in the manifold of a connected group can be deformed continuously to a point, the group is *simply connected;* if not, the group is *multiply connected* (see §5.3).

The idea of connectedness can be illustrated by compact manifolds defined on spheres. Generally, a sphere may be denoted by the symbol S^n, where n is the number of dimensions on the surface of the sphere; thus S^1 is a circle, S^2 is the usual sphere of 3-dimensional space, and S^3 is a sphere in 4-dimensional space. The circle S^1 is *infinitely connected* (the number of homotopic classes for the paths is infinite), because paths that wind around the circle an integer k times cannot be continuously deformed into paths that go around m times if $k \neq m$. The different homotopic sectors for the paths on S^1 may be characterized by an integer *winding number* giving the number of loops the path makes around a circle (with the sign of the winding number specifying the direction), as illustrated in Fig. 13.11.

On the other hand, any sphere S^n with $n \geqslant 2$ is *simply connected;* this is readily seen for S^2 because any loop on the surface of a two-dimensional sphere may be contracted to a point. This theorem is sometimes stated in more picturesque terms: "you can't lasso a basketball".

In general, more than one Lie group shares the same Lie algebra. For a given algebra only one of these groups will be simply connected, and it is called the *covering group* of the algebra. As an example, we may consider the Lie algebra

$$[T_a, T_b] = i\epsilon_{abc} T_c, \tag{13.44}$$

which is shared by the groups $SU(2)$ and $SO(3)$. The group $SO(3)$ of real,

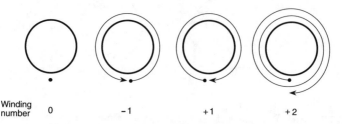

| Winding number | 0 | -1 | $+1$ | $+2$ |

Fig. 13.11   Schematic illustration of different homotopic classes for paths on the circle S^1, as labeled by the winding number n.

orthogonal, 3×3 matrices of unit determinant and the group $SU(2)$ of 2×2 unitary matrices with unit determinant are homomorphic. The manifold of $SO(3)$ may be represented geometrically by a vector that points along the axis of rotation, with the angle of rotation equal to the length of the vector. Thus, the manifold will be the volume of a sphere of radius π, but with diametric points on the surface of the sphere identified because a rotation of π about an axis is equivalent to a rotation of $-\pi$ about the same axis. This parameterization associates two kinds of closed paths with the manifold of $SO(3)$: (I) loops within the sphere, and (II) paths connecting a point on the surface of the sphere with the (identical) point on the opposite side of the sphere:

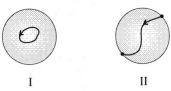

I   II

The paths I can be continuously deformed to a point but the paths II cannot; thus $SO(3)$ is a *doubly connected group*.

The situation is different for $SU(2)$, where the matrices may be parameterized (see Exercise 5.6b)

$$U = \begin{pmatrix} b_1 + ib_2 & b_3 + ib_4 \\ -b_3 + ib_4 & b_1 - ib_2 \end{pmatrix}, \tag{13.45a}$$

subject to the requirement

$$\sum_{i=1}^{4} b_i^2 = 1. \tag{13.45b}$$

This is just the equation for S^3, the unit sphere in 4-dimensional Euclidian space. On the surface of a sphere there is but one kind of closed path ("you can't lasso a basketball!"): $SU(2)$ is *simply connected* and is, in fact, the covering group for the Lie algebra (13.44).

In group theoretical jargon the group $SU(2)$ contains a *center Z_2*, which is the (discrete) subgroup commuting with all group elements [it consists of the two unit matrices ± 1; see the footnote preceding eq. (5.58)]. The group $SO(3)$ is the *factor group*, with Z_2 as the identity element (Wybourne, 1974, Ch. 16; Huang, 1982, §4.2):

$$SO(3) = SU(2)/Z_2. \tag{13.46}$$

In geometrical language, a rotation of 4π is required in $SU(2)$ to return to the starting point instead of the 2π required in $SO(3)$, as Fig. 13.12 illustrates.

A well-known consequence of these topological differences is that $SU(2)$ has single-valued integer or half-integer angular momentum representations, but $SO(3)$ admits only integer single-valued representations.

Another group with topological properties relevant to our considerations is $U(1)$, with elements of the form

$$U = e^{i\theta}. \tag{13.47}$$

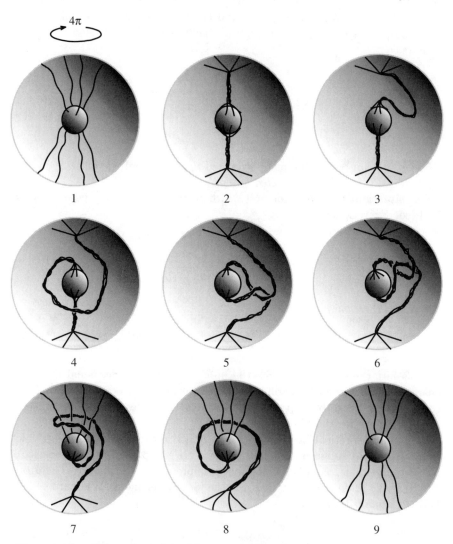

Fig. 13.12   Illustration of $SU(2)$ topology. After a rotation of 4π the strings can be disentangled by steps 3–8. This cannot be done after a rotation of 2π (Misner, Thorne, and Wheeler, 1973, §41.5).

The manifold of $U(1)$ [or of $SO(2)$] is the unit circle S^1, which is compact but not simply connected. There are infinitely many classes of closed paths that cannot be continuously deformed into each other. Each class is characterized by a winding number n (Fig. 13.11). The covering group is the group of real numbers under addition, which is simply connected; however, the covering group manifold is the real number line, which is not compact.

This concludes our brief introduction to the topology of group manifolds. We now turn to a discussion of topological solitons. The initial examples will be simple, and for them it will not be necessary to invoke explicitly the preceding ideas; in the more complicated illustrations to be discussed later in the chapter these considerations will prove extremely useful.

13.3 Solitons in (1+1) Dimensions

The wave equation of a massless real scalar field in one space plus one time (1+1) dimension,

$$\Box \phi(x,t) = \left(\frac{\partial^2}{\partial t^2} - \frac{\partial^2}{\partial x^2} \right) \phi(x,t) = 0, \tag{13.48}$$

is the simplest of relativistic wave equations and has two noteworthy properties (Rajaraman, 1982, Ch. 2): (1) it is *dispersionless*, meaning that all of its plane wave components have the same wave velocity; (2) it is *linear*, so the sum of any two solutions is also a solution. Because of these properties it is easy to demonstrate that a wavepacket solution of (13.48) can be formed that travels with uniform velocity and no distortion in shape, and that two such wave packets can collide and then return to the shapes and velocities that they had before the collision.

However, the disposition of wave packets satisfying more complicated wave equations is not so simple. For example, the Klein–Gordon equation in (1+1) dimensions,

$$(\Box + m^2)\phi(x,t) = 0, \tag{13.49}$$

has the dispersion relation $\omega^2 = k^2 + m^2$; different wavelengths travel at different velocities $\omega(k)/k$, and a localized wave packet formed by superposing solutions at time $t = 0$ will spread as time goes on. Nonlinearities will also cause wave packets to disperse; for example, adding a cubic term to (13.48),

$$\Box \phi(x,t) + \phi^3(x,t) = 0, \tag{13.50}$$

leads to wavepackets that spread in time.

Remarkably, there are certain wave equations where dispersive and nonlinear effects can compensate and wave packets that do not spread with time can be formed. A solution to a wave equation that is nondispersive is called a *solitary wave*. For a certain subset of these cases two solitary waves may collide and recede from each other, exhibiting their original shapes and velocities after separation. In mathematics this special kind of solitary wave is called a soliton; in physics it is common to gloss over this distinction and to classify any nondispersive wave solution as a soliton. We will follow the physics convention in our discussion and define a soliton as any extended but spatially localized and nondispersive solution of a nonlinear wave equation.[‡]

Equations of Motion. Let us first consider classical boson-field solitons. It is clear from the preceding remarks that wave equations must be nonlinear to harbor soliton solutions. A simple Lagrangian density yielding nonlinear classical wave equations in (1+1) dimensions is

$$\mathcal{L} = \frac{1}{2}(\partial_\mu \phi)(\partial^\mu \phi) - U(\phi)$$

$$= \frac{1}{2}\left(\frac{\partial \phi}{\partial t}\right)^2 - \frac{1}{2}\left(\frac{\partial \phi}{\partial x}\right)^2 - U(\phi), \tag{13.51}$$

where $x^\mu \equiv (t, x)$ and the form and strength of $U(\phi)$ characterizes the nonlinearity of the field. From Hamilton's principle the corresponding equation of motion is [see eq. (2.5)]

$$\frac{d^2\phi}{d^2x} = \frac{dU}{d\phi}, \tag{13.52}$$

where we restrict consideration to time-independent solutions. The energy density follows from a Legendre transformation [see (2.58)] on the Lagrangian density for time-independent fields

$$\mathcal{E}(\phi) = \frac{1}{2}\left(\frac{d\phi}{dx}\right)^2 + U(\phi), \tag{13.53}$$

and the total energy is

$$E(\phi) = \int_{-\infty}^{+\infty} dx\, \mathcal{E}\big(\phi(x)\big). \tag{13.54}$$

We will assume that $U(\phi)$ has one or more absolute minima, and that a suitable constant has been added so that the minima are at $U = 0$. A classical

---

[‡]The "solitons" of interest in high energy physics usually are not solitons in the strict mathematical sense. Sometimes in the physics literature solitary waves are termed *solitons* and true solitons are called *indestructible solitons*. Coleman (1977) calls non-dispersive solutions *lumps* to avoid this terminology problem; this leads inexorably to the designation *quivering lumps* for excited states of solitary waves. We also note that solitons are nonsingular and should be distinguished from shock waves and related phenomena.

ground state is one where ϕ is independent of space or time and equal to one of the minima (zeros) of $U(\phi)$; if $U(\phi)$ has more than one minimum there will be more than one classical ground state (spontaneous symmetry breaking).

Soliton Boundary Conditions. If soliton solutions of (13.52) exist they must have finite energy and hence a localized energy density. It is apparent from (13.53) and (13.54) that this will only be the case if $\phi(x)$ tends to one of the zeros of $U(\phi)$ as $x \to \pm\infty$, so we must solve (13.52) subject to this boundary condition. Finding solutions of (13.52) is simplified conceptually by the realization that our field theory problem is mathematically identical to the motion of a classical particle of unit mass in a potential $-U(x)$, with x in the field theory playing the role of time in the particle problem and ϕ in the field theory corresponding to a position coordinate in the particle problem (Coleman, 1977; Lee, 1981). That is, Newton's second law for $m = 1$ is

$$\frac{d^2 x}{dt^2} = -\frac{\partial U}{\partial x};$$

comparing with (13.52) and remembering that a field amplitude ϕ plays the role of a coordinate in a field theory, the identification is immediate.

We first consider the situation where $U(\phi)$ has a unique minimum. This is illustrated in Fig. 13.13, along with the potential $-U(\phi)$ for the classical particle analog problem. By considering the analog problem it is apparent that there can be no solitons in this case. There is only one zero of $U(\phi)$, and the boundary conditions necessary for the energy (13.54) to converge require a particle trajectory beginning and ending at the unique minimum $\phi = \phi_1$. From the shape of the potential in Fig. 13.13b it is obvious that this can never happen, apart from the trivial solution where the particle is motionless at ϕ_1: if we start the particle with a gentle push at ϕ_1 for $x = -\infty$, it cannot return to ϕ_1 at $x = +\infty$.

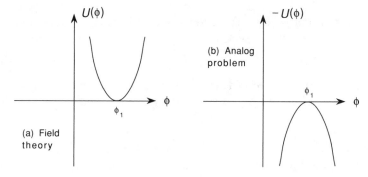

Fig. 13.13 Potentials $U(\phi)$ for a field theory and $-U(\phi)$ for the classical particle analog problem when the classical ground state is nondegenerate.

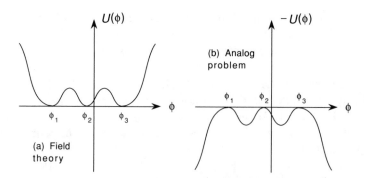

Fig. 13.14   As Fig. 13.13, but for a triply degenerate classical ground state.

Next we consider a case where $U(\phi)$ exhibits spontaneous symmetry breaking. To be definite let us consider the example shown in Fig. 13.14, where there are three degenerate minima for the field theory problem and three degenerate maxima for the particle analog problem.

Now the soliton boundary conditions can be satisfied nontrivially. For the energy to be finite in (13.54) the field must approach any one of the zeros of $U(\phi)$ as $x \to \pm\infty$. In the analog problem we could start the particle for $x = -\infty$ at ϕ_1 with a gentle push to the right, such that at $x = +\infty$ it arrives at ϕ_2. It is easy to verify that the particle moves from ϕ_1 and approaches ϕ_2 asymptotically—it can neither return to ϕ_1 nor go past ϕ_2 as $x \to +\infty$. There are also three other distinct possibilities satisfying the soliton boundary conditions: start the particle with a push to the left at ϕ_2 so that as $x \to +\infty$ the particle approaches ϕ_1; start the particle with a push to the right at ϕ_2 so that as $x \to +\infty$ the particle approaches ϕ_3; start the particle with a push to the left at ϕ_3 so that as $x \to +\infty$ the particle approaches ϕ_2.

This kind of soliton is called a *topological soliton;* it is stable because the boundary conditions at infinity are topologically distinct from those for the vacuum state. The preceding discussion indicates that a necessary condition for the existence of topological solitons in a field theory is vacuum degeneracy. Therefore, we may expect an intimate relationship between topological solitons and spontaneous symmetry breaking.

Topological Charge. Topological solitons can be assigned to a *topological class.* A simple criterion requires specifying the values of ϕ that the field takes at spatial infinity. In the example just discussed the solutions are characterized by the values of $\big(\phi(-\infty), \phi(+\infty)\big)$ shown in Table 13.2. These are in distinct topological classes because they cannot be continuously distorted one into another except at the expense of an infinite amount of energy. This is apparent from Fig. 13.15: to convert one solution into another requires a finite change $\Delta\phi$ in the field over an infinite expanse of space.

Table 13.2
Classification of Solitons in Fig. 13.15 by
Boundary Conditions at Spatial Infinity

Solution	Boundary Conditions
1	(ϕ_1, ϕ_2)
2	(ϕ_2, ϕ_1)
3	(ϕ_2, ϕ_3)
4	(ϕ_3, ϕ_2)

We will see later that in most physical problems the relevant topological quantity is the *difference* between the vacuum configurations at spatial infinity. Up to constants, this difference is called the *topological charge Q*. For topological solitons $Q \neq 0$; there can also exist *non-topological solitons*, for which $Q = 0$ (see §13.3.3). In this language the stability of topological solitons may be attributed to the conservation of topological charge. Conserved topological charges have quite a different origin from those conserved quantities such as energy or momentum that follow from Noether's theorem applied to a continuous Lagrangian symmetry (§2.8). The conservation of topological charge is associated with *boundary conditions*, not equations of motion: topological charges are conserved *identically*, rather than as a consequence of dynamics. The difference between topological and non-topological solitons is illustrated in Fig. 13.16.

Although *classical* wave equations are being discussed at this point, notice that the concept of an extended particle-like object already exists. This is in contrast to ordinary field theory, where the classical field exhibits wave properties and particle aspects are made manifest only by second-quantization of the field (the quantization of classical solitons will be discussed in §13.4).

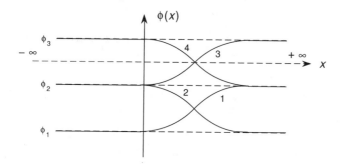

Fig. 13.15 Schematic illustration of the solitons in Table 13.2.

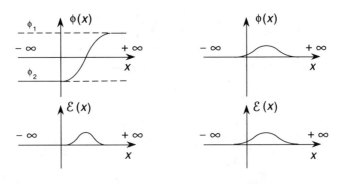

Topological   Non-topological

Fig. 13.16   Examples of field and energy density variations for topological and non-topological boson solitons in one spatial dimension. The topological soliton is stable because the boundary conditions at infinity are topologically distinct from those of the vacuum state ($\phi_{\text{vac}} = $ constant for all x); its existence requires degenerate vacua. A non-topological soliton (see §13.3.3) has the same boundary conditions as the vacuum and can occur in the absence of degenerate vacua; it does require an additive conservation law (Lee, 1981).

Notice that the concept of particle–antiparticle conjugation exists already at the classical level for solitons. By convention, solutions for which ϕ increases as we go from $x = -\infty$ to $x = +\infty$ are called solitons, while those that slope downward as we go to positive x are called antisolitons. Thus 1 and 3 are solitons and 2 and 4 are the corresponding antisolitons in Fig. 13.15.

Soliton Solutions. Equation (13.52) has an immediate solution. Multiplying both sides by $d\phi/dx$ and integrating over dx gives

$$\int \left(\frac{d^2\phi}{dx^2}\right) \frac{d\phi}{dx}\, dx = \int \frac{dU}{d\phi}\frac{d\phi}{dx}\, dx$$

$$\frac{1}{2}\left(\frac{d\phi}{dx}\right)^2 = U(\phi), \qquad (13.55)$$

where the integration constant is zero because both $d\phi/dx$ and $U(\phi)$ vanish at spatial infinity. Therefore

$$\pm\frac{d\phi}{\sqrt{2U(\phi)}} = dx,$$

and integration gives

$$x = x_0 \pm \int_{\phi(x_0)}^{\phi(x)} \frac{d\phi}{\sqrt{2U(\phi)}}, \qquad (13.56)$$

where x_0 is arbitrary, reflecting the translational invariance of the solution.

13.3.1 The Kink

A simple example of a topological soliton is the *kink* solution for a ϕ^4 field theory in (1+1) dimensions. The Lagrangian density is that of eq. (13.51) with (Rajaraman, 1982, Ch. 2)

$$U(\phi) = \frac{1}{4}\lambda \left(\phi^2 - \frac{m^2}{\lambda}\right)^2, \qquad (13.57)$$

where λ and m^2 are positive constants. The degenerate vacuum states occur at

$$\phi_0 = \pm\frac{m}{\sqrt{\lambda}}, \qquad (13.58)$$

as illustrated in Fig. 13.17. From previous considerations, two nontrivial solutions are consistent with the boundary conditions:

$$(\phi(-\infty), \phi(+\infty)) = \begin{cases} \left(-m/\sqrt{\lambda}, +m/\sqrt{\lambda}\right) \\ \left(+m/\sqrt{\lambda}, -m/\sqrt{\lambda}\right) \end{cases}.$$

These are called the kink and antikink, respectively. From (13.56) the kink solution is

$$x - x_0 = \pm\sqrt{\frac{2}{\lambda}} \int_{\phi(x_0)}^{\phi(x)} \left(\frac{d\phi}{\phi^2 - m^2/\lambda}\right).$$

The integral is standard; integrating over ϕ, taking $\phi(x_0) = 0$, and inverting we find

$$\phi(x) = \pm\frac{m}{\sqrt{\lambda}} \tanh\left[\frac{m}{\sqrt{2}}(x - x_0)\right], \qquad (13.59)$$

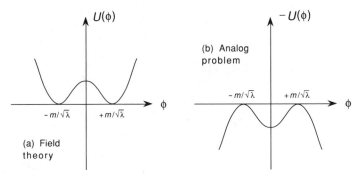

Fig. 13.17 Classical vacuum states for the kink in ϕ^4 theory, and the analog mechanical problem.

where the solution with the positive sign is the kink and the one with the negative sign the antikink. From (13.53) and (13.55) the energy density for time-independent solutions is

$$\mathcal{E}(x) = \frac{1}{2}\left(\frac{d\phi}{dx}\right)^2 + U(\phi) = 2U(\phi)$$

$$= \frac{m^4}{2\lambda}\operatorname{sech}^4\left[\frac{m}{\sqrt{2}}(x - x_0)\right], \tag{13.60}$$

where (13.57) and (13.59) have been used. The total energy is

$$E = \int_{-\infty}^{+\infty} \mathcal{E}(x)\,dx = \frac{2\sqrt{2}}{3}\frac{m^3}{\lambda} \tag{13.61}$$

for both the kink and antikink (see Exercise 13.5b). The solutions and the associated energy density are plotted in Fig. 13.18, where we see that the integration constant x_0 defines the spatial location of the soliton and $1/m$ determines its extension. In the strict terminology mentioned earlier, the kink and antikink turn out to be solitary waves but not solitons. We will call them solitons, however.

In this example the solutions and energy become *singular* as $\lambda \to 0$ [eqns. (13.59) and (13.61)], which illustrates a general feature of solitons: they are non-perturbative. This implies that *solitons can exist even for very small non-linearities* (characterized by λ in the kink problem), and that solitons cannot be obtained by perturbative expansion about the solutions of the linearized wave equation.

The topological charge of the kink can be defined by

$$Q = \frac{\sqrt{\lambda}}{m}\left[\phi(x = +\infty) - \phi(x = -\infty)\right], \tag{13.62}$$

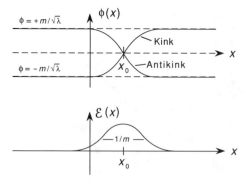

Fig. 13.18   Kink and antikink solutions and the associated energy density for ϕ^4 theory in $(1+1)$ dimensions.

and the kink and antikink are topological solitons since $Q \neq 0$. Conversely, the trivial $Q = 0$ solutions $\phi(x) = \pm m/\sqrt{\lambda}$ are non-topological.

EXERCISE 13.4 Stationary solitons may be converted to moving solitons by a Lorentz boost. Convert the static kink solution (13.59) to a moving kink solution, and demonstrate that the moving kink behaves as a relativistic particle with respect to the dependence of its spatial extension and mass on v/c.

13.3.2 Sine–Gordon Solitons

As a second example of solitons for a scalar field in (1+1) dimensions we consider the Sine–Gordon Lagrangian density (Rajaraman, 1982)[‡]

$$\mathcal{L}(x', t') = \frac{1}{2}(\partial'_\mu \phi')(\partial'^\mu \phi') + \frac{m^4}{\lambda}\left(\cos\frac{\sqrt{\lambda}}{m}\phi' - 1\right), \tag{13.63}$$

which leads to the $(1+1)$-dimensional field equation

$$\Box'\phi' + \frac{m^3}{\sqrt{\lambda}}\sin\left(\frac{\sqrt{\lambda}}{m}\phi'\right) = 0. \tag{13.64}$$

Making the change of variables

$$x = mx' \qquad t = mt' \qquad \phi = \frac{\sqrt{\lambda}}{m}\phi',$$

the equation of motion becomes

$$\frac{d^2\phi}{dt^2} - \frac{d^2\phi}{dx^2} + \sin\phi = 0 \tag{13.65}$$

and the energy is

$$E = \frac{m^3}{\lambda}\int dx \left[\frac{1}{2}\left(\frac{d\phi}{dt}\right)^2 + \frac{1}{2}\left(\frac{d\phi}{dx}\right)^2 + (1 - \cos\phi)\right]. \tag{13.66}$$

On identifying a potential

$$U(\phi) = 1 - \cos\phi, \tag{13.67}$$

[‡] The name Sine–Gordon is a play on words inspired by the similarity of the field equation and the Klein–Gordon equation: if (13.64) is expanded in powers of λ the Klein–Gordon equation is recovered as $\lambda \to 0$. Our discussion will instead concern *non-perturbative* solutions of (13.64); these are singular as $\lambda \to 0$.

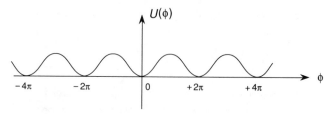

Fig. 13.19   Potential for the Sine–Gordon Lagrangian.

we see that the classical ground states occur at

$$\phi(x,t) = 2\pi N, \tag{13.68}$$

where N is an integer (Fig. 13.19). From the kink discussion we anticipate the possibility of topological solitons by virtue of the degenerate classical vacuum. All finite energy solutions can be classified by the asymptotic values $2\pi N_1$ and $2\pi N_2$ that the fields approach as x tends to $-\infty$ and $+\infty$, respectively. There will be an infinite number of such distinct topological sectors. In problems of physical interest the variable ϕ is normally an angle and only $\phi \bmod 2\pi$ is meaningful. In that case, the topological sectors are characterized only by the difference in N_1 and N_2, which is the topological charge Q

$$Q \equiv N_1 - N_2 = \frac{1}{2\pi} \int_{-\infty}^{+\infty} dx \, \frac{d\phi}{dx}. \tag{13.69}$$

The discussion of the kink indicates that for the static problem only solutions connecting neighboring minima are possible ($Q = \pm 1$). Two explicit solutions may be obtained from the general formula (13.56):

$$\phi_1(x) = 4\tan^{-1}\left(e^{x-x_0}\right) \qquad \phi_{-1}(x) = -4\tan^{-1}\left(e^{x-x_0}\right), \tag{13.70a}$$

where we term the solution with $Q = +1$ the soliton and the one with $Q = -1$ the antisoliton. The corresponding energy is easily found from (13.66)

$$E = \frac{8m^3}{\lambda} \tag{13.70b}$$

(see Exercise 13.5b), and the solutions are sketched in Fig. 13.20.

As before, moving solitons may be constructed from the static solutions by a Lorentz boost [replace $x - x_0$ with $(x - x_0 - \beta t)/\sqrt{1-\beta^2}$]. If that is done and we scatter the moving solitons or antisolitons from each other, it is found that these objects meet the rigid requirements of the mathematician for solitons: they collide and reconstitute themselves after the collision. So the Sine–Gordon solitons are true "indestructible" solitons, not just solitary waves as in the kink solution. This remarkable perseverance is related to the

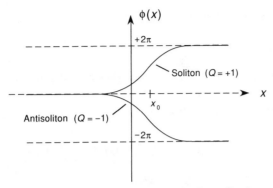

Fig. 13.20   Soliton and antisoliton for the Sine–Gordon equation.

existence of an infinite number of conserved quantities associated with each solution.

There are other solutions of the Sine–Gordon equation corresponding to multisoliton (or antisoliton) states. A particularly interesting one is the *doublet* or *breather* solution

$$\phi_u(x,t) = 4\tan^{-1}\left[\frac{\sin\left(ut/\sqrt{1+u^2}\,\right)}{u\cosh\left(x/\sqrt{1+u^2}\,\right)}\right], \tag{13.71}$$

where u is a velocity parameter. This solution may be interpreted as a bound soliton and antisoliton that oscillate with respect to each other with a period

$$\tau = \frac{2\pi}{u}\sqrt{1+u^2}. \tag{13.72}$$

On the other hand, there are no bound states of two solitons or two antisolitons for the Sine–Gordon system. This is confirmed by theoretical scattering experiments performed by Lorentz boosting solitons and allowing them to collide: two solitons or two antisolitons bounce off each other, suggesting a repulsive force, but a soliton and an antisoliton pass through each other with a time delay, suggesting an attraction.

EXERCISE 13.5 (a) Prove that stationary solitons do not exist in scalar field theories of more than one space dimension ($n = 1$). *Hint:* one way to do this is to show that for $n > 1$ the soliton is unstable against scaling and finds it energetically advantageous to shrink to a point, which is no longer an extended object. (b) Derive the expressions (13.61) and (13.70b) for the energies of the kink and the Sine–Gordon solitons.

EXERCISE 13.6 Show that eq. (13.52) has soliton solutions for a potential of the form

$$U = \tfrac{1}{2}\phi^2\left(\phi^2 - 1\right)^2.$$

13.3.3   Non-topological Soliton Bags

The bag model discussed in Ch. 12 bears a resemblance to a soliton since each is assumed to be a localized finite-energy solution of a relativistic wave equation. Lee (1981) has given an extensive discussion of *non-topological soliton bag models* consisting of scalar and quark fields, and gauge fields that are neglected in lowest order. A phenomenological Lagrangian density is taken of the form

$$\mathcal{L} = -\tfrac{1}{4}\kappa \mathbf{F}_{\mu\nu} \cdot \mathbf{F}^{\mu\nu} + \overline{\psi}(i\not{D} - f\sigma - m_0)\psi$$

$$+ \tfrac{1}{2}(\partial^\mu \sigma)(\partial_\mu \sigma) - U(\sigma), \tag{13.73}$$

where $F^a_{\mu\nu}$ and D^μ are, respectively, the field tensor and covariant derivative for a color gauge field, and the coupling g entering the QCD terms through (10.33) is an effective coupling appropriate to the interior of a hadron.

The coefficient κ is the dielectric constant introduced previously in the phenomenological discussion of confinement (§12.1). The field $\sigma = \sigma(\kappa)$ is a classical scalar field; thus we consider only long wavelength effects and neglect loop diagrams for the σ-field. Inside the hadron $\kappa = 1$ and $\sigma(\kappa) = 0$, while outside in the normal vacuum $\kappa = 0$ and $\sigma(\kappa) = \sigma_{\text{vac}} \neq 0$. The energy density is of the form

$$U(\sigma) = \alpha\sigma^2 + \beta\sigma^3 + \gamma\sigma^4 + B,$$

which has the shape displayed in Fig. 13.21. The constant energy density B plays the same role as the corresponding parameter in the MIT bag (§12.5).

The Yukawa term $f\overline{\psi}\sigma\psi$ couples the quarks directly to the scalar field, and confinement is imposed by the *ansatz*

$$f\sigma \xrightarrow[\sigma \to \sigma_{\text{vac}}]{} \infty. \tag{13.74}$$

The effective fermion mass in the Lagrangian density (13.73) is $m = f\sigma + m_0$ [compare (2.146)]. Inside the hadron $\sigma = 0$ and $m = m_0$, where m_0 is the mass of the quark in the hadron; outside the hadron $\sigma = \sigma_{\text{vac}}$, and

$$m = f\sigma_{\text{vac}} + m_0 \longrightarrow \infty.$$

Thus the quarks acquire a large mass outside the hadron and are effectively confined by the requirement (13.74).

Since confinement has been imposed through the σ-field, we assume that in the hadron the explicit gauge fields provide only perturbative effects. We may neglect their influence in lowest order, which is analogous to omitting gluon exchange in the MIT bag model. The phenomenological σ-field represents the confining effect generated by nonlinear interactions of the color fields. The

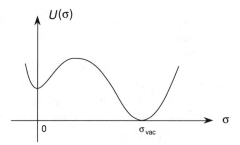

Fig. 13.21   Energy density used in the Lagrangian density (13.73).

color fields are also included in the first term of (13.73), so the Lagrangian may overcount QCD degrees of freedom and must be used with caution. Because the σ-field is a color scalar, these problems do not occur if only one-gluon exchange is considered (see Wilets, 1987). Lee (1981) shows that with the Lagrangian density (13.73) the σ-field has a soliton-like solution and the quark wavefunctions in the hadronic ground state are solutions of a Dirac equation

$$\psi_{j=\frac{1}{2}} = \begin{pmatrix} u(r) \\ i\boldsymbol{\sigma}\cdot\hat{\mathbf{r}}v(r) \end{pmatrix} \chi, \tag{13.75}$$

where χ is a spinor, σ_i is a Pauli matrix, and the radial wavefunctions $u(r)$ and $v(r)$ obey a pair of coupled differential equations with a family of solutions characterized by a single parameter n. The MIT model, which neglects the surface energy of the bag, corresponds to the limit $n \to 0$.

Thus, by viewing hadrons as non-topological solitons immersed in the QCD vacuum one obtains a systematic account of possible bag models that is pedagogically useful and provides dynamical flexibility (Lee, 1981, Ch. 20). However, the basic property of any bag model is confinement, and that is put by hand in the soliton bag as surely as in any other bag. A recent discussion of the non-topological soliton bag model may be found in Wilets (1987).

13.4 Quantization of Solitons

Solitons possess many of the features of quantum-mechanical particles, but we must remember that they are only solutions to the *classical* field theories; they must be quantized before much headway can be made in the application of solitons to high energy physics. That task will not be undertaken here, given the scope of this book. The interested reader should consult Lee (1981), Coleman (1977), and Rajaraman (1982) for an introduction to quantum solitons and references to the technical literature on this subject. We limit ourselves to a few qualitative remarks.

Quantum solitons are similar in many respects to the usual bound states of composite systems. However, ordinary bound states are approximate eigenstates of particle number in the weak coupling limit. For example, in lowest order positronium is a bound state of one electron and one positron, since many-electron, many-positron configurations are represented by higher-order terms. Therefore, for weak coupling the masses of ordinary bound states tend to the threshold masses for some multiparticle configurations (twice the e^- mass for positronium). The quantum soliton is an approximate eigenstate of a field operator that does not commute with particle number. As a consequence, we find that as the coupling vanishes the mass of a soliton goes, not to a multiparticle threshold, but to *infinity* (Coleman, 1977).

The classical solitons are harbingers of (heavy) coherent bound states in the full quantum theory. Typically the classical soliton energy appears as a *weak coupling* approximation to the quantum soliton energy, and the Fourier transform of the classical soliton field gives approximations to the quantum matrix elements between one-soliton momentum states. Thus, the most direct way to construct quantum solitons is by an expansion in powers of the coupling constant, with the leading term related to the classical soliton and higher terms giving quantum corrections (Coleman, 1977). The corresponding solutions are valid only for *weak coupling*, but still are *non-perturbative:* the quantum corrections are incorporated perturbatively in λ but the leading term is the classical term, which is non-perturbative [it typically diverges with λ; see (13.59) and (13.61)].

13.5 More Dimensions and Derrick's Theorem

We are now ready to consider solitons in a more realistic number of dimensions. Unfortunately, it can be shown that for scalar field theories in more than one spatial dimension there are no time-independent soliton solutions. Put more precisely (Coleman, 1977), if Φ defines a vector of scalar fields and the Lagrangian density is of the form

$$\mathcal{L} = \tfrac{1}{2}(\partial_\mu \Phi) \cdot (\partial^\mu \Phi) - U(\Phi),$$

with U non-negative and equal to zero for the ground states of the theory, then for the number of (space + time) dimensions $d > 2$ there are no nonsingular, time-independent solutions of finite energy other than the ground states (see Exercise 13.5a).

This discouraging state of affairs constitutes the celebrated *Derrick Theorem* (Derrick, 1964). Thus, if we wish to find solitons in $d > 2$ (space + time) dimensions it will be necessary to work harder than in the examples discussed so far. There are two options:

1. Search for multidimensional, *time-dependent* soliton solutions.
2. Include fields of nonzero spin.

The first category of solutions is discussed by Lee (1981); these solitons are *non-topological* because their stability is not associated with boundary conditions. The most extensively discussed examples of the second category involve gauge fields and are *topological* in character; well-known cases include the two-dimensional superconducting vortex solution of Nielsen and Olesen (1973) and the three-dimensional magnetic monopole solution of 't Hooft (1974a) and Polyakov (1974). We shall consider only topological solitons in the subsequent discussion.

13.6 Topological Conservation Laws in (2+1) and (3+1) Dimensions

As in the case of the scalar theories discussed in (1+1) dimensions, the existence of finite-energy solutions in higher dimensions requires that the fields approach a zero of the potential U at spatial infinity. Provided U has more than one zero, nontrivial mappings are possible. In a realistic (3+1) dimensions, for theories with scalar, vector, and possibly fermion fields, the topological classification of nondissipative solutions will be more complicated than in the simple kink or Sine–Gordon examples.[‡] Nevertheless, the basic question is the same in the simple and complex cases—how is spatial infinity mapped into the zeros of the potential? This question is intimately related to the connectedness of the spaces involved. To answer it for these more complicated systems it will prove useful to appeal to the theory of *homotopy groups*, which is a standard technique for analyzing the implications of connectedness in topological spaces (Wybourne, 1974, Ch. 16).

To illustrate the procedure we consider theories having gauge and scalar fields. The ground states will have vanishing gauge fields, with the scalar fields constant and corresponding to a zero of U. The gauge group G will be assumed simple, compact, and connected, and we will require that all the zeros of U are of the form $g\phi_0$, where ϕ_0 is any one of the zeros and g is an element of G (this latter condition simplifies the discussion by excluding accidental degeneracy and nongauge internal symmetries). In the language of group theory these assumptions mean that the zeros of U are identified with the *coset space G/H*, where G is the gauge group and H is the unbroken

---

[‡]For the examples presented here, fermion fields and the associated complications arising from the Grassmann algebra (see §4.5) will be omitted.

subgroup of G (see §8.3). As Coleman (1977) emphasizes, it is sufficient for our purposes to view G/H as eccentric shorthand for the set of zeros for U.

With these assumptions and definitions the search for topological conservation laws becomes a question of how spatial infinity is mapped into the zeros G/H of U. This mapping will often assume the form

$$S^n(\text{ordinary space}) \to G/H\ (\text{field space}), \qquad (13.76)$$

with $n = 1$ for two spatial dimensions and $n = 2$ for three spatial dimensions.

The existence of nontrivial homotopic classes, and hence topological solutions, depends on whether all mappings $S^n \to G/H$ can be continuously deformed into each other. If they cannot, the necessary condition for topological solitons exists. This has one immediate consequence: if G/H consists of one element (U has but one zero) all mappings belong to the same homotopy class, there are no nontrivial topological conservation laws, and there are no solitons. This is just a fancy restatement of our previous assertion that topological solitons require degenerate classical ground states.

For the examples we will discuss only two results from the theory of homotopy groups will be required:

$$\begin{aligned} \pi_n\left(S^n\right) &= Z \\ \pi_n\left(S^m\right) &= 0 \qquad (n < m), \end{aligned} \qquad (13.77)$$

where the homotopy group for the mapping $S^n \to S^m$ is designated by $\pi_n\left(S^m\right)$, Z is the additive group of integers, and a zero on the right side means that all mappings can be deformed into a single mapping. Thus for the mapping of spheres to spheres ($S^n \to S^m$) the topology is trivial if $n < m$. If $n = m$ the mapping is nontrivial and a topological invariant can be defined that is an integer winding number specifying how many times one sphere is "wrapped around" the other in the mapping (see Fig. 13.11).

We now give a few illustrations taken from Coleman (1977) of how all this works. The gauge-invariant Lagrangian density for these examples is assumed to be

$$\mathcal{L} = -\tfrac{1}{4}F_{\mu\nu}^a F_a^{\mu\nu} + (D_\mu\phi)^\dagger(D^\mu\phi) - U(\phi), \qquad (13.78)$$

with D_μ the covariant derivative and $F_{\mu\nu}^a$ the non-Abelian field tensor; we will assume that $U = 0$ for the classical ground states of the theory and that it is positive otherwise.

1. The number of spatial dimensions is two, spatial infinity is S^1, the gauge group is $U(1)$, ϕ is a single complex scalar field, and

$$U = \tfrac{1}{2}\lambda\big(\phi^*\phi - a^2\big)^2,$$

where λ and a are positive constants. The zeros G/H of U occur for $\phi = ae^{i\sigma}$, with σ a real number. This is the equation of a circle, S^1, and the relevant homotopy classes are those of the mapping

$$S^1(\text{ordinary space}) \;\to\; S^1(\text{field space}).$$

The mapping of a circle to a circle breaks into homotopy classes characterized by an integer winding number n [see (13.77) and Fig. 13.11], and any solution with $n \neq 0$ can be shown to be nondissipative. This model is mathematically identical to the Landau–Ginzburg theory of Type II superconductors, and the resulting solitons are two-dimensional cross sections of the magnetic flux tubes penetrating a Type II superconductor (this is discussed nicely in Ryder, 1985, §10.2). Of course, the solitons have not yet been found; we have only made a topological classification of possible solutions. This example illustrates concisely the power, and the limitation, of topological argument: topology tells us what is permitted; the equations of motion tell us what is realized. For example, topological existence arguments do not guarantee corresponding solutions that are time independent (the vortex solutions for this particular example are time independent, however).

2. The number of spatial dimensions is two, spatial infinity is S^1, the group is $SO(3)$, the scalar fields constitute an isovector Φ, and

$$U = \tfrac{1}{2}\lambda\big(\Phi^2 - a^2\big)^2.$$

Since Φ has three components the topology of the zeros G/H of U is that of an ordinary sphere, S^2. The relevant mapping is $S^1(\text{ordinary space}) \to S^2(\text{field space})$, for which every mapping is continuously deformable to the trivial mapping (13.77): there is one homotopy class, no topological conservation law, no soliton.

3. The number of spatial dimensions is three, spatial infinity is S^2, the group is $SO(3)$, the scalar fields comprise an isovector, and once again $U = \lambda(\Phi^2 - a^2)^2/2$. The relevant mapping is now

$$S^2(\text{ordinary space}) \;\to\; S^2(\text{field space}).$$

This mapping of a sphere into a sphere admits homotopic classes characterized by a generalized winding number called the *Pontryagin index*, which is a topological invariant taking integer values labeling the number of times one sphere is wrapped around the other (13.77). The nontrivial homotopic structure[‡] of this example admits the solution, found independently by Polyakov

---

[‡]Coleman (1977) gives a simple intuitive argument that at least two homotopy classes exist for $S^2 \to S^2$: "you can't peel an orange without breaking the skin." The skin against the orange is the identity map. This cannot be distorted to a point (the trivial map) and lifted off the orange without breaking the skin. Thus at least two mappings exist that cannot be distorted into each other.

and 't Hooft, called the *magnetic monopole* (this solution too is discussed nicely in Ryder, 1985, §10.4).

The interested reader should consult Coleman (1977) for further examples of searching for topological conservation laws in gauge field theories. These will be sufficient illustration for our purposes.

13.7 Yang–Mills Instantons

A question of considerable interest is whether Yang–Mills fields can have topological soliton solutions. The answer may be summarized as follows: a pure Yang–Mills theory does not possess topological solitons in $(3+1)$ dimensional Minkowski space, but it can have static solitons in four *Euclidian* dimensions; a specific example to be discussed below is the instanton. Generally, for $n \neq 4$ spatial dimensions there are no static solitons for pure Yang–Mills fields. However, if there are matter fields present in addition to Yang–Mills fields there may exist solitons in dimension $n \neq 4$; a specific example is afforded by the 't Hooft–Polyakov monopole in three spatial dimensions, which was mentioned in the preceding section.

We consider here the significance of soliton-like solutions of the classical Euclidian equations of motion for non-Abelian gauge fields that are localized in time as well as space (Polyakov, 1975; Belavin et al., 1975). Because they are localized in time they have been dubbed *instantons* (they have also been called *pseudoparticles*). We will see that these phenomena are of order $\exp(-8\pi^2/g^2)$, where g is the gauge coupling constant; therefore, they are *non-perturbative*. However, instanton solutions can serve as the starting point for a perturbative expansion that is "improved", in the sense that the vacuum state about which we calculate fluctuations is not the normal vacuum, but one that already incorporates non-perturbative effects (see §8.5.4 and Bander, 1981).

13.7.1 Euclidian Space and Vacuum Tunneling

The occurrence of these solutions in 4-dimensional Euclidian space suggests that they represent a tunneling effect in the physical $(3 + 1)$-dimensional Minkowski space. It is well known in ordinary semiclassical mechanics that tunneling through barriers is associated with *complex classical trajectories*—with evolution in imaginary time of the classical trajectory during the barrier penetration. We may anticipate this by noting that the Euclidian form of Newton's second law is given by

$$ m\frac{d^2x}{dt^2} = -\frac{\partial V}{\partial x} \qquad \xrightarrow[t \to -i\tau]{} \qquad m\frac{d^2x}{d\tau^2} = +\frac{\partial V}{\partial x}. \qquad (13.79) $$

Minkowski
space

Euclidian
space

Fig. 13.22 Effect of Wick rotation on classical barriers.

Therefore, the replacement $t \to -i\tau$ effectively changes the sign of the potential. This illustrates clearly the relation between a Wick rotation of the classical equations of motion and quantum-mechanical barrier penetration: *the Wick rotation inverts the barrier* and motion that is classically forbidden in real time becomes quite acceptable in imaginary time, as illustrated in Fig. 13.22. There is no free lunch, however; the motion becomes classically allowed in Euclidian space, but it occurs with exponentially damped probability, as befits a quantum barrier penetration process (see Polyakov, 1977, and Exercise 13.7c).

A specific example of barrier penetration by complex classical trajectories is given in Fig. 13.23. The potential is of the form

$$V(x) = \frac{V_0}{\cosh^2(x/a)},$$

where a is a parameter. Four different paths through the branch-cut structure of the complex time plane are shown, along with the associated paths in the barrier (Massmann, Ring, and Rasmussen, 1975; Miller and George, 1972). If the classical equations of motion are integrated along the real time path

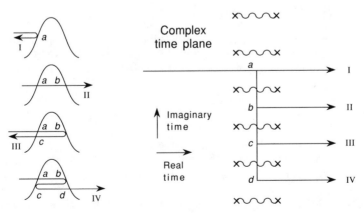

Fig. 13.23 An example of barrier penetration in semiclassical mechanics by evolution of the classical equations of motion in complex time.

I the particle is reflected from the barrier at point a. If instead at point a the time integration path is turned in the imaginary direction the effect is to invert the barrier (see Fig. 13.22) and the particle propagates to b. If at b the time path is rotated back parallel to the real axis (path II) the barrier is effectively reinverted and the particle continues to the right; this is a leading-order barrier penetration process. If instead the time path is continued in the imaginary direction after point b the particle is reflected at the right side of the barrier (as long as the time increment is purely imaginary the particle may be viewed as oscillating between the turning points of the potential well formed by the inverted barrier; see Fig. 13.22). At point c the particle will either exit the barrier traveling to the left, or once again be reflected to the right, depending on whether the time path III or IV is followed.

By this process a multiple-reflection series for higher order contributions to the transmission probability can be established. Each of the paths I, II, III, and IV has associated with it a distinct classical action. While the particle is propagating in the barrier (imaginary time increments), the classical action acquires imaginary components. This leads to an exponential damping of the probability for barrier penetration, and to the dominance of paths I and II for energies well below the top of the barrier.

Thus instanton solutions represent quantum-mechanical tunneling of field configurations in ordinary 3-dimensional space between classical vacuum minima; this is somewhat reminiscent of conduction band structure in the solid state, which develops because of electron tunneling between adjacent potential wells (see §13.7.5 and Kittel, 1976, Ch. 7). As we shall see, these solutions correspond to interpolations between classical vacua with distinct topological properties: they connect vacua that are characterized by different winding numbers.

13.7.2 Topology of the Yang–Mills Vacuum

With these qualitative remarks as introduction, we now turn to the Yang–Mills instantons. These are finite-action solutions of the Euclidian Yang–Mills field equations, and the steps in obtaining them are similar to those for finding any topological soliton:

1. Identify the boundary conditions satisfied by finite-action field configurations.
2. Use the boundary conditions to make a homotopy classification.
3. Look for explicit solutions of the Yang–Mills field equations within a given homotopic sector of finite-action configurations.

We begin with boundary conditions. A gauge transformation on the vector potentials takes the form [see eq. (7.8)]

$$B_\mu \to U B_\mu U^{-1} - U \partial_\mu U^{-1}, \tag{13.80}$$

where $B_\mu \equiv (g/i)A_\mu = (g/i)\tau_a A_\mu^a$ with τ_a a gauge group generator, and where U is a matrix representation of the gauge group. A configuration obtained from $B_\mu = 0$ by a gauge transformation,

$$B_\mu(x) = -U(x)\partial_\mu U^{-1}(x), \qquad (13.81)$$

is called a *pure gauge*. For a pure gauge the Euclidian Yang–Mills action

$$S = -\frac{1}{2g^2}\int d^4x \, \mathrm{Tr}\,(G_{\mu\nu}G_{\mu\nu}) \qquad (13.82)$$

vanishes because (Exercise 13.7b)

$$G_{\mu\nu} \equiv \frac{g}{i}\tau_a F_{\mu\nu}^a = 0 \qquad \text{(pure gauge)}. \qquad (13.83)$$

Conversely, $G_{\mu\nu} = 0$ everywhere implies that $B_\mu(x)$ is a pure gauge.

EXERCISE 13.7 (a) Show that the second form of (7.8) and eq. (13.80) follow from the first form of (7.8). (b) Prove that the classical Euclidian action vanishes if the instanton vector potential B_μ is pure gauge. (c) Consider a particle in a double-well potential

$$V(x) = \left(x^2 - x_0^2\right)^2.$$

Transform to imaginary time, $t \to -i\tau$, and show that the Euclidian classical trajectory satisfies

$$x(\tau) = \pm x_0 \tanh\left(\sqrt{2}x_0\tau\right).$$

Show that the semiclassical tunneling amplitude is $T \simeq \exp\left(-S_{\mathrm{E}}\right)$, where the Euclidian action from one minimum to the other on this trajectory is given by $S_{\mathrm{E}} = \frac{4}{3}\sqrt{2}x_0^3$.

Now consider finite-action configurations. The boundary of Euclidian 4-space can be taken as the 3-dimensional spherical surface at $r = \infty$, with

$$r = \sqrt{x_1^2 + x_2^2 + x_3^2 + x_4^2}. \qquad (13.84)$$

Let us call this boundary S^3(space). The condition of finite action imposed on (13.82) requires that $G_{\mu\nu}$ vanish faster than $1/r^2$ at the boundary,

$$G_{\mu\nu} \xrightarrow[r\to\infty]{} 0 + \mathcal{O}\left(r^{-3}\right), \qquad (13.85a)$$

which implies that the vector potentials must tend to pure gauges at spatial infinity faster than $1/r$,

$$B_\mu(x) \xrightarrow[r\to\infty]{} -U\partial_\mu U^{-1} + \mathcal{O}\left(r^{-2}\right). \qquad (13.85b)$$

The matrix functions U define a mapping of S^3(space) into the gauge group manifold. For the normal ground state we may take U as the identity matrix and $B_\mu(x) = 0$ over all space, which clearly gives finite action. But (13.85b) implies that finite action solutions topologically distinct from the trivial vacuum are possible if there are mappings U that cannot be continuously deformed into $U = 1$. Therefore, we are led to investigate the mapping of S^3(space) into the manifold of the gauge group G.

There is a famous theorem that any continuous mapping of S^3 into a simple Lie group G can be continuously deformed to a mapping into $SU(2)$ subgroups of G (Bott, 1956). It follows that for the homotopy classification in a Euclidian Yang–Mills theory with a simple gauge group it is only necessary to consider the mapping $S^3 \rightarrow$ manifold of $SU(2)$. The $SU(2)$ manifold is S^3 so the relevant mapping is of the form

$$S^3(\text{space}) \rightarrow S^3(\text{field}).$$

The surface S^3(space) may be parameterized by the angles $(\theta_1, \theta_2, \theta_3)$, and the boundary condition necessary for finite action is

$$\left(B_\mu(x)\right)_{S^3(\text{space})} = -U(\theta_1, \theta_2, \theta_3)\partial_\mu U^{-1}(\theta_1, \theta_2, \theta_3), \tag{13.86}$$

where $U(\theta_1, \theta_2, \theta_3)$ need only be defined on S^3(space) and B_μ has been gauge-transformed to remove its radial component (Rajaraman, 1982, §4.2).

Thus the functions $U(\theta_1, \theta_2, \theta_3)$ represent a mapping of S^3(space) into the manifold of $SU(2)$, which topologically is the sphere S^3(field). The mapping $S^3 \rightarrow S^3$ is nontrivial (13.77), admitting a discrete infinity of homotopy classes labeled by a topological index Q that specifies the number of times the sphere S^3(field) is wrapped around S^3(space) at infinity. The topological index is given by

$$Q = -\frac{1}{16\pi^2}\int d^4x \ \text{Tr}\left(\tilde{G}_{\mu\nu}G_{\mu\nu}\right), \tag{13.87}$$

where the *Euclidian dual* $\tilde{G}_{\mu\nu}$ is defined by

$$\tilde{G}_{\mu\nu} = \tfrac{1}{2}\epsilon_{\mu\nu\rho\sigma}G_{\rho\sigma}, \tag{13.88}$$

with $\epsilon_{\mu\nu\rho\sigma}$ antisymmetric and $\epsilon_{1234} = 1$.

13.7.3 The Instanton Solution

We have identified nontrivial topological properties of the Euclidian Yang–Mills system. The next step is to determine whether there are any solutions of the equations of motion in the nontrivial ($Q \neq 0$) homotopy sectors. Such solutions can be found: for $Q = 1(-1)$ they are called the instanton (anti-instanton); multi-instanton (or anti-instanton) solutions can also be found with $|Q| > 1$, but they will not concern us here. The instantons are selfdual or anti-selfdual solutions of the Yang–Mills field equations:

$$\tilde{G}_{\mu\nu} = \pm G_{\mu\nu}, \tag{13.89}$$

since it is easily shown that such configurations extremize the classical action (Exercise 13.8). This underscores the role of the Euclidian metric in constructing instanton solutions; as shown in Exercise 13.9, eq. (13.89) cannot be satisfied in a Minkowski metric.

For an $SU(2)$ Yang–Mills field an explicit solution is (Rajaraman, 1982)

$$B_\mu(x) = -2i\Omega_{\mu\nu}\frac{(x-b)_\nu}{|x-b|^2 + \lambda^2} = U\left(\partial_\mu U^{-1}\right)\frac{y^2}{y^2 + \lambda^2} \tag{13.90a}$$

or

$$G_{\mu\nu} = 4i\Omega_{\mu\nu}\frac{\lambda^2}{\left(|x-b|^2 + \lambda^2\right)^2}, \tag{13.90b}$$

where $y_\mu \equiv x_\mu - b_\mu$ and

$$\Omega_{\mu\nu} = \begin{cases} \epsilon^{i\mu\nu}\dfrac{\sigma_i}{2} & (\mu, \nu = 1, 2, 3) \\[2mm] \delta^{i\mu}\dfrac{\sigma_i}{2} & (\nu = 4) \end{cases} \tag{13.91a}$$

$$U(x) \equiv \frac{x_4 + i\mathbf{x}\cdot\boldsymbol{\sigma}}{r} \tag{13.91b}$$

$$U(y)\frac{\partial}{\partial y_\mu}U^{-1}(y) = -2i\,\Omega_{\mu\nu}\frac{y_\nu}{y^2}, \tag{13.91c}$$

σ_i is an $SU(2)$ Pauli matrix, and b_μ and λ are parameters. It is readily verified that

- The field tensor $G_{\mu\nu}$ is selfdual and hence satisfies the Euclidian Yang–Mills equations of motion.
- In accord with the boundary condition required for finite action, the field (13.90) is pure gauge at infinity.
- The conserved topological charge is $Q = +1$.
- The action associated with the $Q = +1$ solution is

$$S = -\frac{1}{2g^2}\int d^4x\,\mathrm{Tr}\,(G_{\mu\nu}G_{\mu\nu}) = \frac{8\pi^2}{g^2}. \tag{13.92}$$

- A corresponding anti-instanton may be found that is anti-selfdual and has $Q = -1$.

- The 4-vector b_μ represents the location, and the constant λ the spatial extent, of the action density associated with the instanton.

The arbitrariness of b_μ reflects the translational invariance of the Yang–Mills equations. The arbitrary size λ (but $\lambda \neq 0$) is a consequence of the *scale invariance* of the classical Yang–Mills fields, which contain no parameter with length units: if $B_\mu(x)$ is a solution of the Yang–Mills equations, so is $\ell B_\mu(\ell x)$.

EXERCISE 13.8 Prove that selfdual fields satisfy the Yang–Mills equations within a given homotopic sector, and that the corresponding solutions have action

$$S = \frac{8\pi^2}{g^2} Q,$$

where Q is the topological winding number. *Hint*: use the Euclidian relation $\epsilon_{\mu\nu\rho\sigma}\epsilon_{\mu\nu\kappa\lambda} = 2(\delta_{\rho\kappa}\delta_{\sigma\lambda} - \delta_{\rho\lambda}\delta_{\sigma\kappa})$, the identity

$$-\int d^4x \, \text{Tr} \left[\left(G_{\mu\nu} \pm \tilde{G}_{\mu\nu} \right)^2 \right] \geq 0,$$

and eqns. (13.82) and (13.87).

13.7.4 Physical Interpretation of the Instanton

A schematic picture of the instanton is shown in Fig. 13.24a. The condition of finite action, which is the analog of finite energy for a Minkowski soliton, requires that $G_{\mu\nu} = 0$ on the boundary of Euclidian 4-space. The normal ground state satisfies this with a vector potential that vanishes over all of space. However, this condition is more restrictive than necessary; the requirement that $G_{\mu\nu}$ vanish at the boundary is also met if B_μ is pure gauge on S^3 [that is, if B_μ can be obtained from $B_\mu = 0$ by a continuous local gauge transformation—see (13.80) and (13.81)]. For this case a nontrivial topology ensues if $G_{\mu\nu} = 0$ on S^3 but $G_{\mu\nu} \neq 0$ inside; the instanton is an example of such a configuration.

Additional intuition is afforded by distorting the instanton boundary in Fig. 13.24a to that shown in Fig. 13.24b. Then the instanton may be viewed as a solution of the Euclidian gauge field equations in which a vacuum at $x_4 = -\infty$ evolves by propagation in imaginary time to a different vacuum (belonging to a different homotopy class) at $x_4 = +\infty$. This is the generalization of the boundary condition illustrated in Fig. 13.15 and Table 13.2 for topological solitons in (1+1)-dimensional Minkowski space. Between these vacua is a region that has positive field energy (a barrier), because $G_{\mu\nu}$ is non-vanishing there. A comparison with Figs. 13.22 and 13.23 suggests an identification of the instanton with tunneling between classical vacuum states in the Minkowski space.

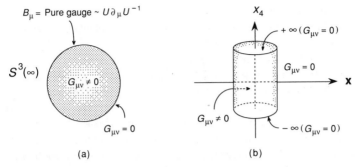

Fig. 13.24 (a) The instanton. (b) Distortion of the Euclidian-space boundaries in (a). The flat surfaces correspond to all of 3-dimensional space ($\mathbf{x}$) at $x_4 = \pm\infty$; the curved surface represents all of time (x_4) at spatial infinity (after Ryder, 1985; Rajaraman, 1982).

13.7.5 The θ-Vacuum and CP Violation

By analogy with related tunneling problems in ordinary quantum mechanics (see Kittel, 1976, Ch. 7), the existence of instantons and vacuum tunneling means that the true QCD vacuum is a coherent superposition of winding number vacua (Callan, Dashen, and Gross, 1976; Jackiw and Rebbi, 1976; Gross, 1979),

$$|0\rangle_\theta = \sum_{n=-\infty}^{+\infty} e^{in\theta}|n\rangle, \tag{13.93}$$

where n labels winding number vacua and the parameter θ with $|\theta| \leqslant \pi$ labels independent sectors of QCD that no gauge-invariant operator can connect. The superposition (13.93) is often called the $\theta-vacuum$. Gauge transformations may be labeled by a topological index giving the difference in winding number between the states connected by the gauge transformation. Transformations that do not change the winding number are called *small gauge transformations;* those that change the winding number are called *large gauge transformations.* The state (13.93) is an eigenstate of both large and small gauge transformations.

The presence of instantons in the QCD ground state implies that if we use a simple vacuum of fixed n instead of (13.93), there is an additional term in the effective QCD Lagrangian density that depends on θ:

$$\mathcal{L}_{QCD}^{eff} = \mathcal{L}_{QCD} + \frac{\theta}{16\pi^2}\,\mathrm{Tr}\left(G_{\mu\nu}\tilde{G}_{\mu\nu}\right),$$

where $\mathcal{L}_{QCD}$ is given by (10.32) and θ plays a role analogous to a coupling constant—different values of θ correspond to different Hilbert spaces and thus to different theories. Normally the θ-term could be discarded because it can be expressed as a surface term in the action. However, this term may survive in non-Abelian gauge theories because of nontrivial instanton configurations that do not vanish sufficiently fast at infinity.

We are forced by the experimental limits on a neutron dipole moment to choose $|\theta| \leqslant 10^{-9}$, because P and CP are not symmetries of the state (13.93) unless $\theta = 0$. The attempt to provide an understanding of such a small number constitutes the *strong CP problem* (Wilczek, 1978). We will not pursue this, but conclude by observing that the appearance of the parameter θ is a non-perturbative QCD effect illustrating the assertion made in the introduction to this chapter: even the effective number of fundamental parameters in a theory may be uncertain if non-perturbative effects are not properly analyzed.

> **EXERCISE 13.9** Use the properties of the appropriate metric to show that selfdual (or anti-selfdual) field tensors are possible in a Euclidian space but not a Minkowski space. *Hint*: take the dual of the dual and use the Euclidian relation $\epsilon_{\mu\nu\rho\sigma}\epsilon_{\rho\sigma\alpha\beta} = 2\left(\delta_{\mu\alpha}\delta_{\nu\beta} - \delta_{\mu\beta}\delta_{\nu\alpha}\right).$

13.8 Skyrmions

Instantons represent topological solutions for fundamental gauge fields. It is interesting to ask whether solitons may also play a role in various effective field theories that approximate QCD. The light meson fields have traditionally represented such effective degrees of freedom mediating the strong interaction. In this section we consider the possibility of topological solitons for meson fields. We will see that such solutions exist and have quite a remarkable interpretation: the solitons of nonlinear meson fields will prove to be fermions in disguise; in fact, they will turn out to be the baryons!

13.8.1 Effective QCD Lagrangians for Large N_c

The QCD gauge coupling α_s cannot serve as the basis for a perturbative expansion at large distance. However, QCD is believed to possess a hidden "weak-coupling" parameter N_c, where N_c is the number of colors entering the strong interaction gauge theory ('t Hooft, 1974b; Witten, 1979). Therefore, it has been proposed that QCD be investigated as an expansion in $1/N_c$. The large N_c limit emphasizes the role of gluons at the expense of quarks since the quark fields [fundamental representation of the color group $SU(N_c)$] have N_c components, but the gauge quanta (adjoint representation) correspond to $N_c^2 - 1 \simeq N_c^2$ gluon fields. It has been shown that QCD in the limit $N_c \to \infty$ reduces to a field theory of weakly interacting meson-like objects, with the dynamics described by a local, nonlinear, effective Lagrangian. It is conjectured that the physical situation $N_c = 3$ retains many of the characteristics of the $N_c \to \infty$ limit. As we shall see, there is some empirical evidence in support of this hypothesis.

Since the $N_c \to \infty$ limit of QCD is a theory of meson fields, a question arises as to the origin of the baryons if such a theory is relevant to the real ($N_c = 3$) world. As we have seen, a nonlinear field theory can support topological solitons even if it is effectively weakly interacting [see the comments following eq. (13.61)]. Therefore, an interesting possibility is that the baryons are topological creations of an effective meson Lagrangian that approximates QCD in the low-energy world. This conjecture antedates QCD (Skyrme, 1961, 1962), but it is only recently that interest has revived in such topological miracles. The basic idea of the Skyrme model is that the nucleon may be viewed as a soliton of the nonlinear σ-model in which a fourth-order term is added to circumvent Derrick's Theorem and stabilize the soliton against scaling. Such objects are now commonly termed Skyrme solitons, or *Skyrmions*. Our discussion of Skyrmions follows that of Balachandran (1986) and Rho (1984).

In §12.4 we discussed the chiral algebra

$$G \equiv SU(N_f)_L \times SU(N_f)_R,$$

where N_f is the effective number of flavors for the strongly interacting particles. The empirical evidence is that the group G is spontaneously broken to the subgroup (the little group—see §8.3) $H = SU(N_f)$ that transforms the right-handed and left-handed fields equivalently. Since $G \to H$ spontaneously the Goldstone theorem requires $N_f^2 - 1$ massless bosons. For $N_f = 2$ the subgroup is $H = SU(2)_{isospin}$ and the Goldstone bosons are associated with the three pions; for $N_f = 3$ we have $H = SU(3)_{flavor}$ and the eight Goldstone bosons are identified with the pseudoscalar meson octet (Fig. 5.7 and §12.4). If the quarks were massless these mesons would be massless; when the interactions are turned on, G will be *explicitly broken* by quark masses (Higgs couplings) and the mesons will acquire a small mass.[‡] By an *effective Lagrangian density* $\tilde{\mathcal{L}}$ for QCD we shall mean one that models the dynamics of the Goldstone modes in the presence of soft explicit symmetry breaking.

If the masses of the quarks are neglected, such an effective Lagrangian density may be expected to exhibit three fundamental properties (Balachandran, 1986):

1. $\tilde{\mathcal{L}}$ must be invariant under $G = SU(N_f)_L \times SU(N_f)_R$.
2. For a minimal theory the multicomponent field Φ transformed by G must have $N_f^2 - 1$ degrees of freedom for each spacetime point.
3. The subgroup of G leaving any value of the field Φ invariant is isomorphic to precisely H.

[‡]See §12.3.3 and Exercises 8.3 and 12.8. It is common to refer to explicit symmetry breaking by terms in the Lagrangian of order $(mass)^1$ or $(mass)^2$ as "soft" breaking.

These requirements on G and H limit the values of the field Φ to the *coset space* G/H, which was introduced in §13.6. In the example relevant to this discussion the manifold of G/H is equivalent to the manifold of $SU(N_f)$. Thus the field Φ can be identified with a field $U(x)$ that is an element of $SU(N_f)$ and is of dimension $N_f^2 - 1$.

13.8.2   The Skyrme Lagrangian

A Lagrangian density fulfilling requirements (1)–(3) was originally proposed by Skyrme (1961, 1962),

$$\tilde{\mathcal{L}} = \mathcal{L}_0 + \mathcal{L}_1 + \ldots$$

$$= -\frac{f_\pi^2}{4}\,\mathrm{Tr}\,(L_\mu L^\mu) + \frac{\varepsilon^2}{4}\,\mathrm{Tr}\,\left([L_\mu, L_\nu]^2\right) + \ldots, \tag{13.94}$$

where

$$L_\mu = U^\dagger \partial_\mu U \qquad U = \frac{1}{f_\pi}(\sigma + i\boldsymbol{\tau}\cdot\boldsymbol{\pi}) \qquad U^\dagger U = UU^\dagger = 1,$$

$f_\pi = 93$ MeV is the pion decay constant, σ is a scalar field, $\boldsymbol{\tau}$ represents Pauli matrices, $\boldsymbol{\pi}$ is the pion field, and ε is a dimensionless constant. The second term $\mathcal{L}_1$ of this effective Lagrangian is called the *Skyrme term*.

The most remarkable property of $\tilde{\mathcal{L}}$ for the present discussion is that it has topological soliton solutions. Let us consider this for $N_f = 2$, corresponding on the one hand to QCD with u and d quarks, and on the other to an effective pion field theory based on spontaneously broken $SU(2)_L \times SU(2)_R$ symmetry in the $N_c \to \infty$ limit.

> **EXERCISE 13.10** Prove that the Skyrme Lagrangian (13.94) cannot support stable solutions of finite energy and size without the second term. Show that in the presence of the Skyrme term there is a lower bound to the radius of the soliton. *Hint:* see Exercise 13.5a.

As discussed in §13.3 and §13.6, the requirement that a soliton exhibit finite energy means that the field $U(\mathbf{x}, t)$ must approach a constant matrix U_0 as $r = |\mathbf{x}| \to \infty$. The Lagrangian density (13.94) is invariant under global chiral transformations, so the constant U_0 can be reduced to the unit matrix by a chiral rotation without affecting $\tilde{\mathcal{L}}$; therefore, without loss of information we take for the boundary condition $U \to 1$ as $r \to \infty$. As Balachandran (1986) points out, this is analogous to the choice of a particular direction to align the spins in the ground state of the Heisenberg ferromagnet (see Ch. 8). Now $U \to 1$ and not an angle-dependent limit as $r \to \infty$, so all points at spatial infinity may be identified as a single point, and for our purposes this converts the Euclidian space R^3 of the coordinates $\mathbf{x}$ at constant time to a 3-sphere (see Balachandran, 1986). We have already established that for spontaneous breaking of $SU(2)_L \times SU(2)_R$

$$G/H = \text{Manifold of } SU(2), \tag{13.95}$$

and that the topology of the $SU(2)$ manifold is S^3 (see §13.2). Therefore, the relevant mapping is $S^3(\text{space}) \rightarrow S^3(\text{field})$, and from (13.77)

$$\pi_3(S^3) = Z, \tag{13.96}$$

where Z is the additive group of integers. Therefore, we may expect the possibility of nontrivial topologies characterized by an integer winding number, topologically conserved currents and charges, and soliton solutions to the equations of motion for the field.

The conserved topological charge is

$$B = \frac{1}{24\pi^2} \epsilon_{ijk} \int d^3x \, \text{Tr} \, (L_i L_j L_k), \tag{13.97}$$

and is an integer with this normalization. Skyrme originally surmised that the topological charge B was the *baryon number* of the system, and this has subsequently been confirmed (Balachandran et al., 1982, 1983; Witten, 1983). The fields $U^{(B)}$ may be classified by the topological index B. The field $U^{(0)}$ maps all $\mathbf{x}$ to the identity of $SU(2)$ and defines the $B = 0$ sector consisting of all maps that can be deformed continuously to $U^{(0)}$. It follows that the trivial $B = 0$ sector has no solitons and baryon number zero; it corresponds to the normal solutions of meson physics in the nonlinear σ-model (§12.5.6).

For $B = 1$, solutions have been found numerically in the *hedgehog form*

$$U^{(1)}(r) = e^{i\tau \cdot \hat{\mathbf{r}}\theta(r)}. \tag{13.98}$$

An example is shown in Fig. 13.25. The boundary conditions require that $\theta(r)$ monotonically decrease from π at $r = 0$ to zero at $r = \infty$; as $\mathbf{x}$ ranges over all its values the manifold of $SU(2)$ is covered exactly once under the map (13.98). This is a typical unit winding number map, and all maps homotopic to this map constitute the $B = 1$ sector of the Skyrme model. These solutions are solitons, and since B is to be regarded as the baryon number they presumably may be identified with baryons such as N and Δ made from u and d valence quarks. (The $B = -1$ solutions are topologically distinct from those for $B = 1$, so baryons and antibaryons correspond to distinguishable solitons.) For this solution the topological charge (13.97) is

$$B = \frac{1}{\pi} \left[\theta(0) - \theta(\infty) - \frac{1}{2} \left(\sin 2\theta(0) - \sin 2\theta(\infty) \right) \right], \tag{13.99}$$

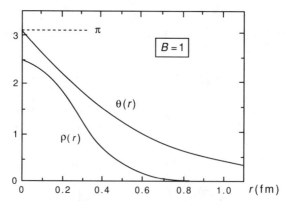

Fig. 13.25   Skyrmion hedgehog solution $\theta(r)$ and baryon charge density $\rho(r)$ in arbitrary units (Jackson and Rho, 1983).

which shows explicitly that with the boundary conditions $\theta(0) = \pi$ and $\theta(\infty) = 0$ the baryon number is $B = 1$. Topology has indeed conjured a fermion from the bose fields!

Adiabatic rotation of the Skyrmion in the internal symmetry space generates a rotational band (Skyrme, 1961; Adkins, Nappi, and Witten, 1983)

$$E_{J=T} = M_0 + \frac{1}{2\mathcal{J}}J(J+1), \tag{13.100}$$

where M_0 is the classical soliton mass, $\mathcal{J}$ is a moment of inertia parameter, J is the angular momentum, and T is the isospin with the restriction

$$J = T = \frac{1}{2}, \frac{3}{2}, \frac{5}{2}, \ldots, \frac{N_c}{2}. \tag{13.101}$$

Therefore, for a realistic $N_c = 3$ the $B = 1$ topological sector delivers up the N and the Δ. The first term in (13.100) is of order N_c and the second of order $1/N_c$, so the rotational angular momentum projection in this case corresponds to the leading order quantum correction to the classical soliton energy. Calculations for the static properties of the N and Δ are found to agree with observation to about 30% (Adkins, Nappi, and Witten, 1983; Adkins and Nappi, 1984).

13.8.3   Bags as Defects in the Skyrme Field

The calculations for static N and Δ properties using the Skyrme model are comparable in accuracy with those using bag models or nonrelativistic quark models. At first sight these approaches appear to be unrelated, but each is an approximation to the same physics. Therefore, we are led to ask how the Skyrme solitons are related to bags. It has been proposed that the bag and Skyrme approaches can be unified if bags are viewed as defects in the background Skyrme field (Rho, Goldhaber, and Brown; 1983). Then, through the topological interpretation of the baryonic charge it is possible to move continuously from one description to the other.

From the chiral bag model it is found that the baryonic charge inside a spherical bag of radius R is (Rho, 1984)

$$B_{\rm in} = 1 - \frac{1}{\pi} \left[\theta(R) - \frac{1}{2} \sin 2\theta(R) \right], \qquad (13.102)$$

where θ is the chiral angle introduced in (12.88); thus, the baryonic charge leaks from the chiral bag.[‡] On the other hand, if we view the bag as a defect excluding the Skyrme field from the region $r \leqslant R$ and replace the lower limit of (13.99) with R instead of 0, the baryonic charge in the region outside the bag is

$$B_{\rm out} = \frac{1}{\pi} \left[\theta(R) - \frac{1}{2} \sin 2\theta(R) \right] \qquad (13.103)$$

and the total baryonic charge is conserved,

$$B_{\rm in} + B_{\rm out} = 1, \qquad (13.104)$$

but is shared between the quarks in the interior of the bag and the exterior region where the Skyrme soliton lives. Likewise, it is found that other physical quantities such as the axial charge g_A and the magnetic moment receive contributions from both the exterior Skyrmion and the bag interior.

These considerations underlie the *Cheshire Cat Model* of the baryons (see Brown et al., 1984; Rho, 1985; Brown and Rho, 1986; Vepstas and Jackson, 1990; the allusion is to the cat in *Alice in Wonderland* that could disappear, leaving behind only its smile). The nucleon may be described either as a Skyrmion with a point bag defect, or as a larger bag surrounded by a small pion cloud, and these descriptions are but two sides of the same coin. If this conjecture is correct, it suggests that much of the controversy in Ch. 12 concerning the radii of bags is pointless: the bag size becomes a question not of correctness, but of utility, because the chiral bag radius R is not directly a physical quantity. The relevant expansion parameter in the bag

---

[‡]This may be understood in terms of a polarization of the bag negative-energy Dirac sea by the Skyrmion, because the boundary conditions destroy CP symmetry between positive- and negative-energy quarks (Brown and Rho, 1986). This "leakage" of baryon charge (color is confined by the bag; baryon number is not) is one example of a general phenomenon of nonlinear fields called *fermion number fractionation*, which is also observed in various condensed matter systems (see Niemi and Semenoff, 1986, for a review).

(asymptotically free quarks) is the effective color coupling α_s (§12.5.4), but the corresponding expansion parameter in the Skyrmion domain is $1/N_c$. The appropriate bag radius is a function of the phenomena being studied—which value of R strikes a balance between control of the higher order terms in the bag interior expansion in α_s and the Skyrmion exterior expansion in terms of $1/N_c$? It has been suggested that for low energy and momentum transfer at normal densities $R \simeq 0.4 - 0.5$ fm is appropriate, and that at higher energies or densities the "optimal radius" may grow, reflecting the increased importance of explicit QCD degrees of freedom (Brown et al., 1984).

13.8.4   Applications of Skyrmions

An extensive literature now exists on the use of Skyrmions in nuclear and particle physics. A reprint collection (Liu, 1987) and several reviews have appeared (Rho, 1984; Balachandran, 1986; Meissner and Zahed, 1986; Zahed and Brown, 1986; Dothan and Biedenharn, 1987), so we will omit discussion of specific applications. However, to whet the appetite a few claims that have been made in the literature are listed. These are taken from Rho (1985) and Meissner and Zahed (1986), where the original references may be found.

(1) The standard results of current algebra and soft-meson theorems can be reproduced. (2) Just as approximate quantization for $SU(2) \times SU(2)$ was seen to yield the N and Δ, the corresponding procedure for $SU(3) \times SU(3)$ produces the $J = \frac{1}{2}$ octet and $J = \frac{3}{2}$ decuplet. (3) Dibaryon resonances and the deuteron have been studied in the $B = 2$ sector. (4) Excited states of the N and Δ have been described through rotation–vibration coupling of the Skyrmion. (5) Many basic features of phenomenological nucleon–nucleon interactions (Fig. 12.15) are reproduced by Skyrmion–Skyrmion interactions. In particular, one recovers the one-pion exchange potential (OPEP) at large distance and a finite repulsion of order one baryon mass at short distance;[‡] the shallow intermediate-range attraction is found to be more problematic. (6) The phenomenology of the Skyrme term $\mathcal{L}_1$ in (13.94) is now believed to represent a stabilization of the nonlinear σ-model by heavier mesons such as the ρ, thereby establishing a connection with one-boson exchange potentials (Fig. 12.15).

The reader is warned that these and other claims discussed in the cited references are attended by a degree of controversy; for example, some Skyrmion successes may be more a consequence of the current algebra embodied in the nonlinear σ-model than of the Skyrme term and topological magic. As a second example, there are fundamental questions of why the Skyrme term, and not some other nonlinear contribution to the Lagrangian, plays such an im-

---

[‡]Although the meson fields are assumed weakly interacting [strength $\simeq \mathcal{O}(1/N_c)$], the baryonic solitons are strongly interacting [strength $\simeq \mathcal{O}(N_c)$]; the essential short-range nucleon–nucleon repulsion is *non-perturbative* in the Skyrme model.

portant role. Finally, the assumptions and approximations used in deriving these results reflect the extreme difficulty of dealing with nonlinear theories, and their validity has been challenged in specific cases. These issues are addressed at great length in the review articles cited above; as our intentions are pedagogical, we send the interested reader to the literature for the details and bring to a close our introductory discussion of Skyrme solitons.

13.9 Background and Further Reading

A useful resource for material in this chapter is the collection of reprints edited by Gervais and Jacob (1983); the introduction to lattice concepts is based on Wilson (1976, 1977); the general material on lattice gauge methods may be found in Creutz (1985), Rebbi (1983), Bander (1981), Hasenfratz and Hasenfratz (1985), and Kogut (1983). The discussion of critical phenomena and the relation of lattice QCD to methods of statistical physics owes much to Wilson (1983a, 1979), Ma (1973, 1976), Wilson and Kogut (1974), and Kogut (1979). The discussion of solitons and instantons was derived from Rajaraman (1982), Coleman (1977), Lee (1981, Chs. 7, 20), Huang (1982, Chs. 4, 5), Ryder (1985, Ch. 10), and 't Hooft (1978); see also Shuryak (1984, 1988), Coleman (1985), and Bhaduri (1988). A readable introduction to instantons as a basis for nonperturbative strong interaction theory may be found in Gross (1979). We discussed magnetic monopoles only in passing; they are extremely interesting and an overview may be found in Preskill (1984). For the status of including dynamical fermions in lattice calculations see Duke and Owens (1985), Toussaint (1987), Hasenfratz (1987), Karsch (1988), Ukawa (1989), Gottlieb et al. (1989), and Fukugita (1988, 1989). Several reviews have appeared on Skyrmions: Rho (1984), Meissner and Zahed (1986), Balachandran (1986), Zahed and Brown (1986), and Dothan and Biedenharn (1987); see also the reprint collection Liu (1987). Coleman (1985, Ch. 8) explains $1/N$ expansions. A discussion of the "Cheshire Cat Principle" may be found in Vepstas and Jackson (1990) and Brown and Rho (1986). A general discussion of the merits and limitations of nonlinear chiral field theories and their soliton solutions has been given by Weise (1989). Finally, an interesting panel discussion addressing many of the issues in this chapter and the previous one appears in *Nucl. Phys.* **A434**, 629–722 (1985).

CHAPTER 14

Deconfined Quarks and Gluons

In the preceding chapter we saw that a considerable effort has been expended to demonstrate that QCD confines color. Although a constructive proof is lacking, the calculations discussed there make this hypothesis extremely plausible for matter under normal conditions. However, those same calculations also suggest that hadronic matter at high density or temperature may undergo a phase transition to a new state in which quarks and gluons interact weakly in a deconfined macroscopic plasma. This hypothesized transition to a *quark–gluon plasma* is of interest because of the fundamental information it would provide on the confinement mechanism, because its production may be feasible in existing or proposed accelerators, and because the early universe may have existed for a short time in such a phase.

14.1 Deconfining and Chiral Phase Transitions

The hypothesized transition from normal hadronic matter to a quark–gluon plasma has been described both as a phase transition from confinement to deconfinement, and as a phase transition that alters the chiral properties of the QCD vacuum.

14.1.1 Confinement

We have already discussed phenomenological models of confinement in terms of a two-phase structure for the QCD vacuum. In this picture quarks and gluons modify the vacuum in their vicinity, carving out regions of perturbative vacuum (bags) immersed in the normal non-perturbative vacuum: the quarks and gluons propagate freely in the bags, but are strongly repelled by the non-perturbative vacuum. The bag constant B measures the excess energy density of the perturbative vacuum, and hence is a measure of the pressure of the normal vacuum on the bubble of perturbative vacuum. This bubbly picture of the QCD vacuum represents a medium with a dielectric constant $\kappa \simeq 1$ in the perturbative vacuum and $\kappa \simeq 0$ in the surrounding non-perturbative vacuum (§12.1), and may be likened to a chromoelectric Meissner effect (Fig. 12.3).

14.1.2 Chiral Symmetry

The preceding description of the ground state in terms of dielectric properties is intuitively appealing. However, the usual method of studying dielectric properties is to measure the force between test charges placed in the medium. This test is ambiguous for QCD: previous considerations suggest that a pair of color charges at large separation will produce $q\bar{q}$ excitations between them so as to screen the charges (Fig. 12.13).

An alternative characterization of the QCD vacuum is the manner in which the chiral $SU(2) \times SU(2)$ symmetry is implemented (Baym, 1982b). In the normal vacuum global chiral symmetry is broken spontaneously, with the pion appearing as the associated Goldstone excitation (§12.3), but in the perturbative vacuum the symmetry is realized in the Wigner mode. Thus instead of a color dielectric constant, an alternative order parameter characterizing the vacuum is $\langle \bar{q}q \rangle = \langle 0| \bar{q}q |0\rangle$, where q is an operator for light (u or d) quarks:

$$\langle \bar{q}q \rangle = 0 \qquad \text{(perturbative vacuum)} \qquad (14.1)$$

$$\langle \bar{q}q \rangle \neq 0 \qquad \text{(normal vacuum)}. \qquad (14.2)$$

In the normal vacuum this expectation value is not zero because of the presence of coherent $q\bar{q}$ pairs (pions) in the ground state. This is analogous to the order parameter $\langle \psi_\uparrow(\mathbf{x})\psi_\downarrow(\mathbf{x}) \rangle$ for a BCS superconductor, which measures the amplitude for removing a pair of electrons (one with spin up, one with spin down) from the ground state and returning to the ground state. This matrix element vanishes for the normal phase, but it is finite for the superconductor because of the macroscopic pair condensate.

Thus there are two ways to characterize the phase structure of the QCD vacuum: in terms of the effective color dielectric constant of the medium, or in terms of the order parameter $\langle \bar{q}q \rangle$ specifying the manner in which chiral symmetry is implemented. Although the two are closely related, they need not be identical. We may appeal to our old pedagogical friend, the superconductor, for a related illustration: near the surface of a superconductor the magnetic permeability falls to zero exponentially over a penetration depth from the surface, but the material in that region is superconducting.

14.1.3 Quark–Gluon Transition

Now suppose the density of bags is increased, either by compressing hadronic matter or by raising the temperature of a hadronic system so that it is energetically favorable to create mesons from the normal QCD vacuum between the hadrons. (Notice that these two possibilities may be associated with very different baryon number densities.) Initially a region of space is characterized by isolated pockets of perturbative vacuum (hadrons) surrounded

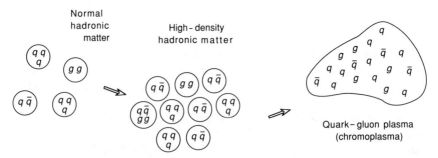

Fig. 14.1   Schematic illustration of the transition from ordinary matter to a quark–gluon plasma as the density or the temperature is increased.

by non-perturbative vacuum; as the density is increased the non-perturbative vacuum is squeezed from between the bags; finally, the system may find it favorable to coalesce all the hadronic bags into a large region filled with perturbative vacuum, as illustrated in Fig. 14.1. We call this state of matter a *quark–gluon plasma*, or a *chromoplasma*. In such a plasma the quarks and gluons are free to propagate over macroscopic distances.

This transition from the dominance of non-perturbative vacuum to that of perturbative vacuum is called *deconfinement* if the phase structure is characterized by a color dielectric constant or equivalent parameter, and *chiral restoration* if $\langle \bar{q}q \rangle$ serves as an order parameter. We may obtain a simple estimate of the critical density for these transitions by observing that nucleons begin to touch at a nucleon density

$$n_{\rm c} \simeq \left(\tfrac{4}{3}\pi R^3\right)^{-1}, \tag{14.3}$$

where R is the effective nucleon radius. If we denote the normal nucleon density in nuclear matter by $n_0 = 0.17/\,{\rm fm}^3$, then for $R = 1$ fm we have $n_{\rm c} \simeq 1.4 n_0$, while $R = 0.4$ fm gives $n_{\rm c} \simeq 22 n_0$.

EXERCISE 14.1 Make a crude estimate of the deconfining phase transition temperature by calculating the equilibrium temperature at which pions would fill all of space (see Fig. 14.1). *Hint:* see §15.1.2 and assume that the pions are relativistic.

The simplest possibility is that the deconfinement temperature $T_{\rm c}$ and the chiral restoration temperature $T_{\rm ch}$ coincide: $T_{\rm ch} = T_{\rm c}$. However, nature might not be simple. It seems unlikely that chiral restoration would precede the deconfining transition (Shuryak, 1981; Pisarski, 1982), but it is conceivable that $T_{\rm c} < T_{\rm ch}$. Then deconfinement would precede chiral restoration and the intermediate phase would consist of deconfined quarks and gluons, but the quarks would carry masses. At present the relationship between the chiral

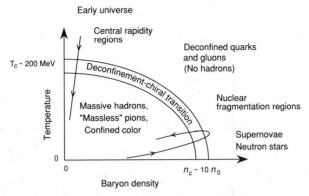

Fig. 14.2   Phase diagram for extended and equilibrated nuclear matter. Central rapidity and fragmentation regions refer to the ultrarelativistic heavy ion collisions discussed in §14.7. The density of normal nuclear matter is n_0 (after Baym, 1984).

transition and deconfining transition is only partially understood; we will address the question more quantitatively in §14.5.

The preceding arguments survive further scrutiny, and more sophisticated calculations discussed below suggest the phase diagram shown in Fig. 14.2. Thus, high-temperature or high-density nuclear matter is expected to exist in a deconfined, chiral-symmetric phase; a hadron dropped into such matter would be melted into deconfined quarks and gluons. It is this unusual state of matter that we wish to discuss in this chapter.

14.1.4 Phase Transition in the Early Universe

The schematic evolution of the universe is illustrated in Fig. 14.3, where we see that early times were characterized by very high temperatures. Thus the universe is likely to have passed through a quark–gluon plasma phase shortly after its inception, as we have anticipated in Fig. 14.2. In the early universe the mean baryon density was much less than the density of quanta produced by thermal excitation, so the primordial quark–gluon plasma would have existed at high temperature but low baryon density. By solving the equations of motion for the early universe the temperature is found to evolve in time as [eq. (15.19)]

$$T(\text{MeV}) \simeq \frac{0.50}{\sqrt{t_{\text{sec}}}}. \tag{14.4}$$

Assuming a critical temperature of 200 MeV (corresponding to 2.4×10^{12} K), the time after the big bang until confinement sets in is

$$t_{\text{c}} \simeq 6 \times 10^{-6} \text{ sec}, \tag{14.5}$$

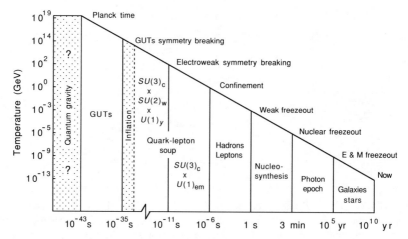

Fig. 14.3   Evolution of the universe (see Ch. 15). The scale is highly nonlinear and 1 GeV $\simeq 1.2 \times 10^{13}$ K (after D. Schramm).

at which time the event horizon is of planetary scale. Several possible cosmological implications for this phase transition are discussed in Olinto (1988) and references cited therein.

14.2 Simple Thermodynamical Considerations

Because of asymptotic freedom, as the quark density or temperature becomes large the quark–quark interaction becomes weak. To be more precise, the characteristic momentum p entering the QCD running coupling constant equation (10.41)

$$\alpha_{\mathrm{s}}(p^2) = \frac{12\pi}{(33 - 2N_{\mathrm{f}}) \log\left(\dfrac{p^2}{\Lambda^2}\right)} \qquad (14.6)$$

grows as $n_{\mathrm{B}}^{1/3}$, where n_{B} is the baryon density, or as the temperature T. Thus the strength of the quark–quark interaction decreases logarithmically with increasing temperature or density. Further, the long-range color force is screened in the chromoplasma (§14.3), just as the long-range Coulomb force is screened for an equilibrated, globally neutral electromagnetic plasma. Under these conditions quark matter in an overall color neutral state may be approximated as an ideal relativistic gas (Collins and Perry, 1975; Baym, 1982c; see Karsch, 1988 for a discussion of deviations from ideal gas behavior that are indicated by lattice gauge calculations).

Let us illustrate by analyzing a simple two-phase ideal gas picture of strongly interacting matter (Satz, 1983; Cleymans, Gavai, and Suhonen, 1986). At low density QCD must lead to weakly interacting hadrons since the color forces are essentially saturated, leaving only small residual interactions between color neutral objects. (This may be likened somewhat to the weak residual interaction between molecules that remains after electromagnetic forces have been nearly saturated by the atoms.) At high density we expect a plasma of quarks and gluons that is weakly interacting because of asymptotic freedom. The thermodynamic variables for our model are the temperature T, baryonic chemical potential μ, and volume V. Consideration will be limited to u and d quarks, nucleons, and pions, and we will adopt the equation of state of an ideal ultrarelativistic gas.

First, consider the case of zero baryon number ($\mu = 0$). At low T the system may be approximated by a noninteracting pion gas. The pressure is

$$P(T, \mu = 0) = \frac{\pi^2}{90} g(T) T^4,$$ (14.7)

where $g(T)$ is a degeneracy factor. Since the pion has three charge states but no spin, the degeneracy factor is 3 and the pressure for the hadronic gas is

$$P_{\text{H}}(T, \mu = 0) = \frac{\pi^2}{90} \times 3 \times T^4.$$ (14.8)

The equation of state for an ultrarelativistic ideal gas is $P = \varepsilon/3$, where ε is the energy density, so eq. (14.7) implies the Stefan–Boltzmann energy density

$$\varepsilon_{\text{H}}(T, \mu = 0) = \frac{\pi^2}{30} g(T) T^4$$

$$= \frac{\pi^2}{30} \times 3 \times T^4.$$ (14.9)

For a noninteracting ultrarelativistic quark–gluon plasma the pressure and energy density are related by $P = (\varepsilon - 4B)/3$ and

$$P_{\text{Q}}(T, \mu = 0) = \frac{\pi^2}{90} (2 \times 8 + \tfrac{7}{8} \times 2 \times 2 \times 2 \times 3) T^4 - B$$ (14.10)

$$\varepsilon_{\text{Q}}(T, \mu = 0) = \frac{\pi^2}{30} (2 \times 8 + \tfrac{7}{8} \times 2 \times 2 \times 2 \times 3) T^4 + B.$$ (14.11)

In eqns. (14.10) and (14.11) the first term in parentheses is the gluon degeneracy (two helicity states and eight color states), the second term is the quark degeneracy (two helicity states, two flavors, quark and antiquark, and three colors; the factor $\tfrac{7}{8}$ accounts for the difference between Fermi and Bose statistics in the Boltzmann factor). The positive constant B is the pressure exerted by the non-perturbative vacuum on the perturbative vacuum (see §12.5.1). It

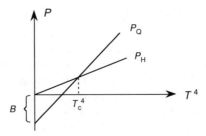

Fig. 14.4 Pressure as a function of temperature for a two-phase ideal gas model of strongly interacting matter at zero baryon density (Satz, 1983).

may be interpreted as the energy required to convert a unit volume of non-perturbative vacuum into perturbative vacuum, and represents a ground-state energy shift in the plasma relative to the hadronic phase.

To minimize the free energy the system must be in the state of higher pressure [lower thermodynamic potential; see (14.27)], so the plasma is thermodynamically favored at high temperature and the hadronic phase is favored at low temperature, as illustrated in Fig. 14.4. The phase transition temperature in our simple model comes from setting $P_H(T_c, \mu = 0) = P_Q(T_c, \mu = 0)$, and is given by

$$T_c = \left(\frac{45B}{17\pi^2} \right)^{1/4} \simeq 0.72 B^{1/4} \simeq 105 \text{ MeV},\qquad(14.12)$$

where $B^{1/4} = 146$ MeV has been used (see Fig. 12.9; however, the phenomenological B used here need not have the MIT bag value). The latent heat per unit volume for the phase transition is

$$L = \varepsilon_Q(T_c, \mu = 0) - \varepsilon_H(T_c, \mu = 0) = 4B.\qquad(14.13)$$

EXERCISE 14.2 Verify eqns. (14.12) and (14.13) for the transition temperature and latent heat per unit volume of the hadron to plasma phase transition.

Now let us consider *cold baryonic matter* as a function of density. In the hadronic phase the pressure and energy density of a degenerate, relativistic Fermi gas of nucleons are given by

$$P_H(T = 0, \mu) = \frac{\mu^4}{24\pi^2} \times 2 \times 2\qquad(14.14)$$

$$\varepsilon_H(T = 0, \mu) = \frac{\mu^4}{8\pi^2} \times 2 \times 2,\qquad(14.15)$$

where the degeneracy 2×2 is for neutrons and protons, each with two possible spins, and the baryon number density is

$$n_B = \frac{2\mu^3}{3\pi^2}. \tag{14.16}$$

For a high density, cold plasma (include quarks but neglect gluons)

$$P_Q(T = 0, \mu) = \frac{\mu_Q^4}{24\pi^2} \times 2 \times 2 \times 3 - B \tag{14.17}$$

$$\varepsilon_Q(T = 0, \mu) = \frac{\mu_Q^4}{8\pi^2} \times 2 \times 2 \times 3 + B, \tag{14.18}$$

where the quark density n_Q is related to the quark chemical potential μ_Q by

$$n_Q = \frac{2\mu_Q^3}{\pi^2}, \tag{14.19}$$

from which, utilizing $n_B = n_Q/3$ and (14.16),

$$\mu = \mu_Q. \tag{14.20}$$

Once again the minimal thermodynamic potential requires the higher pressure, so the plasma phase is favored at large baryon density (large chemical potential) and the hadronic phase at low density, as shown in Fig. 14.5. The critical density is determined by setting $P_H(T = 0, \mu_c) = P_Q(T = 0, \mu_c)$ and using the relation (14.16) between density and chemical potential,

$$n_c = \frac{2}{(3\pi^2)^{1/4}} B^{3/4} \simeq 2n_0, \tag{14.21}$$

where $n_0 = 0.17 \text{ fm}^{-3}$ and $B^{1/4} = 0.146$ GeV have been used. As before, the latent heat per unit volume is

$$\varepsilon_Q(T = 0, \mu_c) - \varepsilon_H(T = 0, \mu_c) = 4B. \tag{14.22}$$

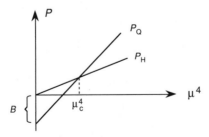

Fig. 14.5   Pressure as a function of baryon chemical potential for a two-phase ideal gas model of cold, strongly interacting matter (Satz, 1983).

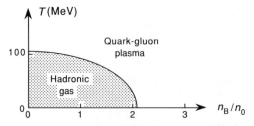

Fig. 14.6   Phase diagram for a simple two-phase ideal gas model of strongly interacting matter.

Such considerations lead to the qualitative phase diagram shown in Fig. 14.6 (compare Fig. 14.2). The details of Fig. 14.6 cannot be taken too seriously,[‡] but it should readily be appreciated from the simple examples in Figs. 14.4 and 14.5 that the essential factors are the bag constant and the large degeneracy of the quark–gluon plasma relative to the hadronic phase. The constant B is a ground state energy shift that favors the hadronic phase at low temperature; the plasma phase dominates at high temperature because it has an order-of-magnitude more degrees of freedom than the hadronic phase (a degeneracy factor of 37 compared with 3 for the case in Fig. 14.4). We may expect these qualitative features to persist in the more sophisticated calculations to be discussed below.

14.3 Debye Screening in the Plasma

Screening Lengths.   In the Debye–Hückel theory of conductance each ion is surrounded by ions of opposite charge that screen the central ion. The effective potential associated with the central ion is spherically symmetric and given by (Landau and Lifshitz, 1958)

$$\phi(r) = \frac{Ze}{r}e^{-r/r_d}, \tag{14.23}$$

where the length r_d measures the radius of the surrounding cloud of screening ions. The small and large distance behavior of ϕ is

$$\phi \xrightarrow[r/r_d \to 0]{} \frac{Ze}{r} - \frac{Ze}{r_d} + \dots \tag{14.24a}$$

$$\phi \xrightarrow[r/r_d \to \infty]{} 0. \tag{14.24b}$$

[‡]The critical temperatures and densities in Fig. 14.6 are a factor 2–3 lower than those of the more realistic Fig. 14.2. Much of this discrepancy could be removed by using a larger empirical constant $B^{1/4} \simeq 200$ MeV in Fig. 14.6.

Thus at very short distances $\phi \simeq Ze/r$, by virtue of the bare central ion (the second term, Ze/r_d, is due to the field of the surrounding cloud of opposite charge). At radii much larger than r_d the field of the central ion is completely shielded, while at intermediate distances it is partially shielded. This phenomenon is called *Debye screening,* and the characteristic distance r_d is called the *Debye screening length.* Generally,

$$r_d \simeq n^{-1/3},$$

where n is the charge density.

The screening length can be used to characterize the transition from insulator to conductor in the solid state (*Mott transition*). An example is the hypothesized conversion of solid hydrogen to a metallic conducting state under very high pressure.[‡] Under ordinary pressure solid hydrogen is an insulator: the electrons are bound to individual protons and they experience little screening because the atomic spacing is large. However, as the density is increased the electron of each atom begins to be Debye screened by the presence of other nearby electrons and protons. As the density continues to increase the Debye radius decreases ($r_d \simeq n^{-1/3}$), and the electron is increasingly screened from its proton. Finally, when the Debye radius is comparable with the electron orbital radius the electron is unable to see clearly the charge on "its proton" anymore, the electrons are liberated from the atoms to propagate in the lattice, and a metallic conducting state is established.

QCD Mott Transition. In an $SU(N)$ Yang–Mills theory there should be an analogous transition from the normal QCD confining phase (color insulator) to a plasma phase (color conductor) brought about by a "Debye screening" of the color charges. Satz (1984) has estimated the temperature for such a *QCD Mott transition:* taking the Debye length to be the typical hadronic scale of ~ 1 fm, the critical temperature is found to be $T_c \simeq 175$ MeV for QCD with three colors and three quark flavors. Lattice gauge determinations of various screening lengths in the plasma are discussed in Karsch (1988).

We can take the preceding analogy further; in insulating solids the electrical conductivity σ_e is exponentially small, but not zero:

$$\sigma_e \simeq e^{-\Delta E/T},$$

where ΔE is the ionization energy and T the temperature. Even above the Mott transition temperature thermal ionization can produce a few conducting electrons. A corresponding effect occurs in QCD: any attempt to remove a quark from a hadron will eventually materialize a $q\bar{q}$ pair from the vacuum

---

[‡]The complex magnetic field of Jupiter is thought to result from electrical currents in a large, rapidly spinning (1 Jovian day $\simeq 10$ hours), metallic hydrogen core.

because of the confining potential (see Fig. 12.13). The $\bar{q}$ can then neutralize the q we were trying to remove, forming a colorless meson that can "ionize" from the system. Thus *the QCD analog of ionization is meson production*, and the color conductivity σ_c is not zero, even in the confinement region:

$$\sigma_c \simeq e^{-m_{\mathrm{H}}/2T},$$

where m_{H} is the mass of the lowest $q\bar{q}$ state in the system. Only if $m_{\mathrm{H}} \to \infty$ does σ_c vanish identically in the confinement regime. This corresponds to a gauge field with static quark sources in which the quarks play no dynamical role (see §13.1.10), and σ_c serves as an order parameter distinguishing the confined ($\sigma_c = 0$) and deconfined ($\sigma_c > 0$) phases (Satz, 1985a). We will return to the influence of dynamical quarks on confinement order parameters in §14.5.

14.4 Finite Temperature Yang–Mills Theory

The simple two-phase ideal gas model and the QCD Mott transition model provide useful phenomenology, but a quantitative description of the transition to a quark–gluon plasma requires that we investigate QCD at finite physical temperatures. In this section we consider finite temperature field theory by expressing the partition function in terms of lattice approximations to Euclidian path integrals.

Partition Function. The *grand partition function* for an ensemble at temperature T and chemical potential μ is (Fetter and Walecka, 1971; Morse, 1969; Cleymans, Gavai, and Suhonen, 1986)

$$Z = \mathrm{Tr}\, e^{-\beta(H-\mu N)}, \tag{14.25}$$

where N is the particle number operator, H is the Hamiltonian, $\beta \equiv 1/T$, and $\hbar = c = k_{\mathrm{B}} = 1$ units are used (Appendix A). The *thermodynamic potential* Ω is related to the partition function by

$$\Omega = -\frac{1}{\beta} \log Z, \tag{14.26}$$

and satisfies

$$\Omega = F - \mu N = -PV, \tag{14.27}$$

where the Helmholtz free energy is

$$F = E - TS, \tag{14.28}$$

with T the temperature, S the entropy, μ the chemical potential, and N the particle number. From (14.27) and the first law of thermodynamics,

$$dE = TdS - PdV + \mu dN, \tag{14.29}$$

we find that

$$d\Omega = -SdT - PdV - Nd\mu \tag{14.30}$$

$$S = -\left(\frac{\partial\Omega}{\partial T}\right)_{V,\mu} \tag{14.31}$$

$$P = -\left(\frac{\partial\Omega}{\partial V}\right)_{T,\mu} \tag{14.32}$$

$$N = -\left(\frac{\partial\Omega}{\partial\mu}\right)_{T,V} \tag{14.33}$$

Thus a determination of the partition function yields, by differentiation of the resulting thermodynamic potential, the equilibrium thermodynamical properties of the system. Generally, for any observable Q the *ensemble average* may be calculated from

$$\langle Q \rangle = \frac{1}{Z} \operatorname{Tr} e^{-\beta(H-\mu N)} Q. \tag{14.34}$$

Euclidian Path Integral. Now we come to the crucial point for evaluation of thermodynamical properties: *the partition function for a field theory may be replaced by a Euclidian path integral in which the fields ϕ are periodic in the temperature (imaginary time) direction for bosons and antiperiodic for fermions:*

$$\phi(\mathbf{x},\beta) = +\phi(\mathbf{x},0) \qquad \text{(bosons)} \tag{14.35}$$

$$\phi(\mathbf{x},\beta) = -\phi(\mathbf{x},0) \qquad \text{(fermions)}. \tag{14.36}$$

These boundary conditions are consequences of the trace in eq. (14.25), as may be appreciated from Exercise 14.4. Thus the spatial integration is over the total volume V of the system, but the imaginary time integration is over a *finite slice* determined by the physical temperature $T = 1/\beta$. For QCD one finds (Satz, 1985a; Kapusta, 1989)

$$Z = \int [dA][d\psi][d\overline{\psi}]e^{-S[A,\psi,\overline{\psi}]} \tag{14.37}$$

$$S[A,\psi,\overline{\psi}] = \int_0^\beta d\tau \int_V d^3x \left(\mathcal{L}\left(A,\psi,\overline{\psi}\right) + \sum_f \mu_f N_f \right), \tag{14.38}$$

where N_f and μ_f are, respectively, the number density and chemical potential of quarks with flavor f, the Euclidian ($\tau \equiv it$) QCD Lagrangian density is $\mathcal{L}(A,\psi,\overline{\psi})$, and the boundary conditions are

$$\overline{\psi}(\mathbf{x},\beta) = -\overline{\psi}(\mathbf{x},0) \qquad \psi(\mathbf{x},\beta) = -\psi(\mathbf{x},0)$$
$$A_\mu^a(\mathbf{x},\beta) = A_\mu^a(\mathbf{x},0). \tag{14.39}$$

These equations may be used to derive finite temperature Feynman rules in a way similar to that outlined for zero temperature in Ch. 4 (Bernard, 1974; Shuryak, 1980; Kapusta, 1989). However, we anticipate that perturbation theory will be of limited utility as a tool for studying the deconfining phase transition: the high physical temperature phase may be perturbative, but the low temperature confined phase surely is not.

Lattices at Finite Temperature. We require a non-perturbative evaluation of statistical QCD, and the obvious candidate is a generalization to finite physical temperature of the lattice regularization introduced in §13.1. The transcription of (14.37) to a lattice closely parallels that of the zero temperature theory. The primary difference is that finite temperature simulations employ asymmetric lattices with the number of temporal links N_τ, temporal lattice spacing a_τ, and physical temperature T related by

$$a_\tau T = \frac{1}{N_\tau}, \qquad (14.40)$$

with periodic boundary conditions in τ for the gauge fields and antiperiodic boundary conditions for the fermion fields (14.39). (In practical calculations periodic boundary conditions are often taken for the spatial directions as well, to suppress finite size effects.)

Thus, if the lattice spacings in the temporal and space directions are taken to be equivalent and equal to a, calculations on an $N_\tau \times N_s^3$ lattice approximate a system at a temperature T and volume V given by

$$T = \frac{1}{aN_\tau} \qquad V = (aN_s)^3,$$

measured in lattice units. (The relation of such quantities to physical units is discussed in §13.1.8.) The physical temperature may be changed by varying the coupling $g^2(a)$, since this changes the length scale in the lattice calculation; alternatively, the number of temporal links N_τ may be varied at fixed g^2. We omit details (see Satz, 1985a) and proceed directly to summarize the important results.[‡]

EXERCISE 14.3 Show that the boundary conditions (14.35) and (14.36) follow from the trace appearing in (14.25).

EXERCISE 14.4 Derive a path integral formula for the partition function of a one-dimensional harmonic oscillator. *Hint:* use the definition (14.25) and evaluate $Z = \int dx \, \langle x| \, e^{-\beta H} \, |x\rangle$, where $|x\rangle$ are coordinate eigenstates, by the path integral approximation of §4.1.

[‡]There are unresolved difficulties in simulating finite baryon number on a lattice (Barbour et al., 1986). Therefore, in all subsequent discussions it will be assumed that $\mu = 0$ in the partition function (14.25).

14.4.1 $SU(2)$ Yang–Mills Theory

In Fig. 14.7 the dependence of the energy density on temperature is shown for a pure $SU(2)$ Yang–Mills system with no dynamical fermions. The large increase of ε near $T/\Lambda_L = 45$ is suggestive of a phase transition. To investigate this further it is useful to identify an order parameter. For an $SU(N)$ gauge symmetry the lattice average $\langle L \rangle$ of a thermal Wilson loop (the trace of the product of U's along a time-oriented string running the temporal length of the lattice at fixed $\mathbf{x}$),

$$L(\mathbf{x}) \equiv \frac{1}{N} \operatorname{Tr} \prod_{\tau=1}^{N} U_{\mathbf{x};\tau,\tau+1}, \qquad (14.41)$$

can serve as an order parameter. Equation (14.41) defines a *Wilson–Polyakov loop*, which is a closed loop because of the periodic boundary conditions (14.39) on the temporal axis; therefore, it is gauge invariant. It is related to the free energy F_q of an isolated static quark through

$$\langle L(\mathbf{x}) \rangle \simeq e^{-\beta F_q}. \qquad (14.42)$$

Thus $\langle L \rangle = 0$ implies confinement ($F_q \to \infty$) and $\langle L \rangle \neq 0$ implies deconfinement (McLerran and Svetitsky, 1981). This quantity is shown in Fig. 14.8 for the $SU(2)$ Yang–Mills system, and we conclude that the result in Fig. 14.7 is indicative of a phase transition; further analysis suggests that it is a second order transition.

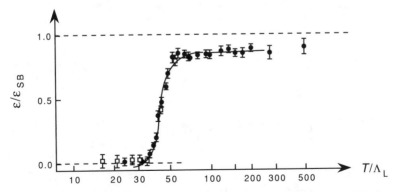

Fig. 14.7 $SU(2)$ gauge field energy density normalized to the (ideal gas) Stefan–Boltzmann form ε_{SB} [eq. (14.9) with $g(T) = 6$ gluon states for $SU(2)$], as derived from finite temperature calculations on a $10^3 \times 3$ lattice (Satz, 1985a). The temperature is given in terms of the scaling constant Λ_L, which is a lattice analog of the scale setting parameter introduced for the continuum theory in (10.41). The phase transition corresponds to a physical temperature $T_c \simeq 150 - 200$ MeV.

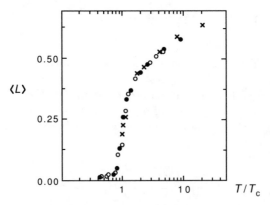

Fig. 14.8   Confinement order parameter (14.41) for $SU(2)$ Yang–Mills system evaluated on a $10^3 \times 3$ lattice with three different lattice actions (Satz, 1985a).

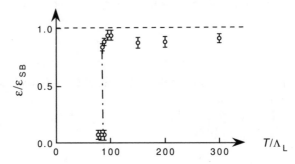

Fig. 14.9   Energy density ε of the $SU(3)$ Yang–Mills system evaluated on an $8^3 \times 3$ lattice and normalized to the ideal gas limit (Satz, 1985a).

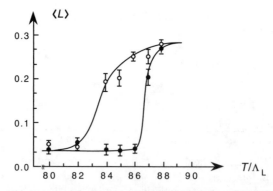

Fig. 14.10   Confinement order parameter (14.41) for $SU(3)$ Yang–Mills system evaluated on an $8^3 \times 3$ lattice with ordered (∘) and random (•) starting configurations (Satz, 1984). Compare Fig. 13.7 and the discussion in §13.1.9.

14.4.2 $SU(3)$ Yang–Mills Theory

In Figs. 14.9 and 14.10 the energy density and order parameters for pure $SU(3)$ glue are shown as a function of temperature. Fig. 14.10 exhibits a well-defined hysteresis loop in the thermal cycle, suggesting a phase transition (see §13.1.9). Figure 14.11 indicates that the phase transition for pure $SU(3)$ gauge fields is *first order*, as opposed to the higher order transition in $SU(2)$, and that the critical physical temperature is $T \simeq 86\Lambda_L \simeq 150 - 200$ MeV. The latent heat of deconfinement for the first order transition is found to be approximately 900 MeV/fm^3.

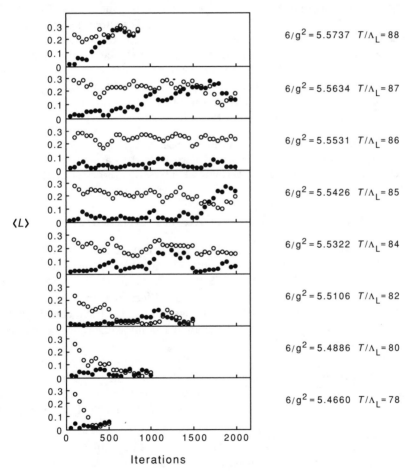

Fig. 14.11 Lattice average for the order parameter (14.41) of the $SU(3)$ Yang–Mills system as a function of the number of iterations, evaluated on an $8^3 \times 3$ lattice with ordered (∘) and random (•) starting configurations (Satz, 1984). Compare Fig. 13.8 and the discussion in §13.1.9.

14.5 Chiral Phase Transition

The question was posed earlier of whether deconfinement and chiral symmetry restoration should coincide. Kogut et al. (1982, 1983) have summarized two opinions on this matter. The first was introduced in the discussion of the bag model: confinement necessarily implies chiral symmetry breaking (see Fig. 12.7b and Casher, 1979). The second point of view is that chiral symmetry breaking originates in a short-range interaction that is independent of the long-range mechanism responsible for confinement.

The impetus for this second point of view was the Nambu, Jona-Lasinio model (1961), which is a non-renormalizable theory with spontaneous symmetry breaking in which massless nucleons constitute the fundamental fields. The cause of the spontaneous symmetry breaking is a direct nucleon–nucleon interaction, in analogy with the effective electron–electron interaction responsible for Cooper pairing in superconductors.[‡] If this attractive, zero-range interaction exceeds a critical strength a condensate violating chiral symmetry is formed, the fermions develop a non-perturbative dynamical mass, and a massless Goldstone pion triplet is produced as a consequence of the spontaneously broken chiral symmetry in the condensate.

> **EXERCISE 14.5** Show that a direct quark–quark interaction of the form used in the Nambu, Jona-Lasinio model would lead to a non-renormalizable theory. *Hint*: compare with the Fermi weak interaction theory of Ch. 6.

14.5.1 Origin of Chiral Symmetry Breaking

The chiral restoration problem has been studied in $SU(2)$ and $SU(3)$ lattice gauge theory by calculating the temperature at which changes in the chiral order parameter $\langle \bar{\psi}\psi \rangle$ and the Wilson–Polyakov line $\langle L \rangle$ signal a phase transition. The outcomes neglecting dynamical fermions are summarized for $SU(2)$ in Fig. 14.12 and for $SU(3)$ in Figs. 14.13a,b. (The calculations of Fig. 14.13c include the influence of dynamical fermions and will be discussed in the next section.)

For $SU(2)$ the chiral symmetry breaking [finite value of $\langle \bar{\psi}\psi \rangle$] may persist somewhat beyond the deconfinement transition [rise in $\langle L \rangle$], but for $SU(3)$ without dynamical fermions the transitions are first order and appear to coincide. Thus, although the $SU(2)$ calculations in quenched approximation

---

[‡]If we attempt to express this in QCD terms (see Huang, 1982, §12.4), a direct quark–quark interaction is precluded by the requirement of renormalizability (see Exercises 6.5b and 14.5 concerning the renormalizability of theories with 4-fermion vertices). However, just as for Cooper pairs where an effective $e^- e^-$ interaction is generated through electron–phonon lattice couplings, some effective quark–quark interaction might provide an acceptable symmetry-breaking mechanism.

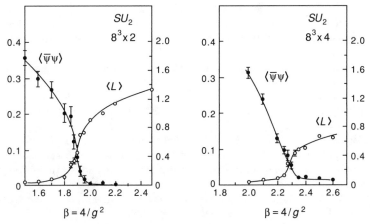

Fig. 14.12   Chiral order parameter $\langle \overline{\psi}\psi \rangle$ and Wilson–Polyakov line $\langle L \rangle$ for $SU(2)$ gauge theory; dynamical fermions have been neglected (Kogut, 1983).

show that it may be possible to break chiral symmetry on a scale different from that for deconfinement, the corresponding $SU(3)$ calculations suggest first order chiral and deconfinement transitions in the pure gauge theory that are coincident.

The $SU(2)$ calculations of Fig. 14.12 have also been done with quarks in the adjoint and higher-dimensional representations. For quarks in these representations a stronger short-range attraction is experienced than in

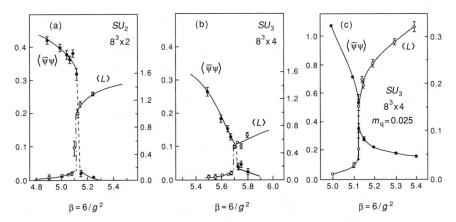

Fig. 14.13   (a)–(b) Same as Fig. 14.12, but for $SU(3)$ gauge theory (Kogut, 1983). (c) As for (a)–(b), but including dynamical fermions of mass $ma = 0.025$ on a $10^3 \times 6$ lattice (Kovacs, Sinclair, and Kogut, 1987).

the fundamental representation (larger color charge), but the color charge is screened at large distances so there is no confinement (and no deconfining transition). Chiral transitions are found in these calculations at much larger energy scales than for quarks in the fundamental representation, providing more evidence that strong, short-range forces can generate chiral phase transitions independent of confinement (Kogut et al., 1982).

By integrating over all degrees of freedom except the Wilson–Polyakov loop (14.41) in the partition function (14.37), we obtain from an $SU(N)$ theory one that is equivalent to a Z_N spin system of the same spatial dimensionality (Svetitsky and Yaffe, 1982; Satz, 1985a). The deconfining phase transition in $SU(N)$ gauge theory is associated with spontaneous breaking of the global center Z_N of the group (Yaffe and Svetitsky, 1982; Polonyi and Szlachanyi, 1982). The three-dimensional Ising model has symmetry Z_2 [center of $SU(2)$], and is known to exhibit second order transitions. The 3-state Potts models have symmetry Z_3 [center of $SU(3)$], and are known to have first order phase transitions. Therefore, the second order character of the $SU(2)$ deconfining transition and first order character of the $SU(3)$ transition were not a complete surprise (Kogut, 1983). More generally, it is hypothesized that for equivalent spatial dimensionalities an $SU(N)$ gauge theory, a Z_N gauge theory, and a Z_N spin system all belong to the same universality class (see §13.1.8).

14.5.2 Lattice Calculations with Dynamical Fermions

The deconfining phase transition of pure gauge fields may be ascribed to a color screening that converts the gluonic matter into a chromoplasma (§14.3). At a qualitative level this picture should survive the introduction of dynamical fermions (§13.1.10), but questions arise as to the sharpness of the transition and to the choice of an appropriate order parameter to distinguish phases in the presence of dynamical quarks. This is not surprising, given the previous analogies with condensed matter systems: dynamical fermions contribute terms to the functional integral that are mathematically similar to external fields for spin systems, and it is known that external fields smooth phase transitions in the latter case (Svetitsky, 1986).

As we have just noted, in pure glue $SU(N)$ theories the deconfining transition is characterized by a spontaneously broken global center Z_N of the group $SU(N)$. When Z_N is unbroken the expectation value $\langle L \rangle$ of the Wilson–Polyakov loop (14.41) is zero, and the spontaneous breaking of this symmetry yields a finite value of $\langle L \rangle$ in the deconfined region (see Fig. 14.8). However, the global Z_N symmetry is broken *explicitly* by fermion mass terms. As a result, in the presence of dynamical fermions the expectation value of the Wilson–Polyakov loop fails to vanish for the confinement region, so it no longer constitutes an order parameter for the deconfinement transition. The physical reason for this was discussed in §14.3: dynamical quarks allow the creation of colorless hadrons through $q\bar{q}$ excitations.

Only realistic calculations with dynamical fermions can show how badly Z_N is broken in the confinement regime. If $\langle L \rangle \simeq 0$ in the confined phase with dynamical quarks, it can still serve as a useful approximate order parameter. Satz (1985b) examines $SU(2)$ lattice gauge theory with dynamical fermions. The global center Z_2 corresponds to the symmetry of the 3-dimensional Ising model, for which the magnetization near the critical point is known to obey

$$M \simeq (\beta - \beta_c)^b \qquad (\beta > \beta_c),$$

with a critical exponent $b = 0.33$. It is found that the corresponding $SU(2)$ deconfinement order parameter $\langle L \rangle$ satisfies

$$\langle L \rangle \simeq (\beta - \beta_c)^{0.33},$$

implying that the $SU(2)$ theory with dynamical fermions still exhibits universal critical behavior, and that the global center symmetry is only weakly broken by the dynamical fermions. Thus Satz concludes that dynamical quarks have only a small effect on the deconfinement transition for $SU(2)$ lattice gauge theories.

In Fig. 14.13c we show a calculation of the $SU(3)$ phase transition that incorporates four species of light dynamical quarks (Kovacs, Sinclair, and Kogut, 1987; see also Karsch et al., 1987); it should be compared with Figs. 14.13a and 14.13b, which show the $SU(3)$ transition in quenched approximation. The chiral order parameter $\langle \bar{\psi}\psi \rangle$ and the Wilson–Polyakov loop $\langle L \rangle$ undergo rapid crossover near $\beta = 5.125$, indicating coincident (possibly first-order) transitions in the chiral and confinement properties of the system for nearly massless dynamical quarks. Evolution of the lattice near the transition β suggests the presence of coexisting states in the transition region, further supporting the interpretation of this transition as first order (compare §13.1.9 and Fig. 14.11).

More recent calculations indicate that there is no first-order transition for $SU(3)$ lattice gauge theory with four flavors and intermediate quark masses (Gottlieb et al., 1989; Kogut and Sinclair, 1989). This would suggest that the first-order deconfining transition of the pure gauge theory (corresponding to infinitely massive quarks) and the first-order chiral transition observed for four flavors of very light quarks are distinct. However, some confusion remains because different calculations do not always reach the same conclusions about similar systems. It is speculated that such discrepancies may be associated with finite-size effects in current lattice calculations (Gottlieb et al., 1989).

Since the exact path integral contains a fermion determinant raised to the power N_f (number of quark flavors), the quenched approximation with its neglect of internal quark loops corresponds to the limit $N_f \to 0$ and may represent the first term in a systematic expansion (Kogut et al., 1982). Alternatively, the quenched approximation may be viewed as the large quark-mass limit of QCD. The proper inclusion of dynamical fermions in lattice QCD is a developing field, and the details of how realistic fermions will modify pure glue calculations remain to be seen. The results presented here offer some hope that the revision will not be too radical. Duke and Owens (1985); Cleymans, Gavai, and Suhonen (1986); Hasenfratz (1987); Karsch (1988); Gottlieb et al. (1989); Kogut and Sinclair (1989); Fukugita (1988, 1989); and the recent literature may be consulted for the status of algorithms used to incorporate dynamical fermions in lattice calculations.

14.6 Hydrodynamics at 1 GeV/Nucleon

From Fig. 14.2 and the discussion in §14.2, we may expect that a quark–gluon plasma could be formed by compressing baryonic matter or by increasing the energy density sufficiently in baryon-free regions. Heavy-ion collisions are thought to approximate each of these limits: at a few GeV per nucleon the colliding ions stop each other and high baryon densities are achieved; at much larger energies the heavy ions pass through each other and the region between the receding ions is high in energy density, but low in baryons. In this section and the next we discuss these expected regimes of heavy-ion collisions.

If L is a typical linear dimension of the colliding nuclei and ℓ is a characteristic nucleon mean free path under specified conditions, we may expect that in a heavy-ion collision:

- For $\ell \gg L$, a single-nucleon knockout model is applicable.
- If $\ell \simeq L$, several binary collisions are likely and intranuclear cascade models are appropriate.
- If $\ell \ll L$, the intranuclear cascade is complex and hydrodynamics with a realistic equation of state may give a more economical description.

This last domain ($\ell \ll L$) is thought to be relevant for 1 GeV/nucleon (lab) heavy ion collisions, and hydrodynamics with the traditional Landau boundary conditions may be applicable for these reactions.

For the baryon-rich environment expected in a 1 GeV per nucleon heavy ion collision the hydrodynamical energy–momentum conservation equation

$$\partial_\mu T^{\mu\nu} = 0 \qquad (14.43)$$

[the energy–momentum tensor $T^{\mu\nu}$ is defined in eq. (14.47)] must be supplemented by an equation of state and the baryon-number conservation condition

$$\partial_\mu N^\mu = 0, \qquad (14.44)$$

where N^μ is a baryon current: $N^\mu = nu^\mu$, where n is a local baryon density in the fluid rest frame and u^μ is the four-velocity. These equations have been solved numerically for the collision of equal-mass nuclei (Kapusta and

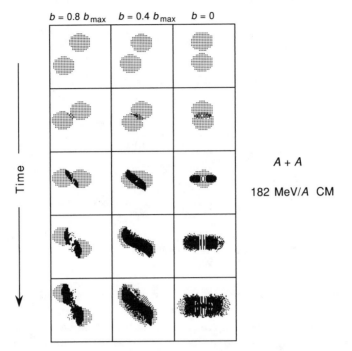

Fig. 14.14 Development in equal time steps of the projected center-of-mass baryon density at three different impact parameters b, obtained from a three-dimensional relativistic hydrodynamic model. The equivalent laboratory beam energy is 800 MeV per nucleon (Kapusta and Strottman, 1981).

Strottman, 1981); some examples are shown in Fig. 14.14. (For those who prefer action, "Super-Ion, the Movie" is available from the Los Alamos National Laboratory film library.) Notice the pronounced sideward splashing for the more central collisions as the nuclei largely come to rest in the center-of-mass of the colliding systems.

In collisions of this type it is expected that a quark–gluon plasma could be formed by following a trajectory near the horizontal axis in Fig. 14.2. It is estimated that baryon densities as large as 10 times that of ordinary nuclear matter could be achieved in this way (Gyulassy, 1988).

14.7 Ultrarelativistic Hydrodynamics

The situation is different for collisions at much higher energies. Figure 14.15 gives a view from the center-of-mass of a very high energy heavy-ion collision. At these energies the nuclei are Lorentz contracted and highly transparent. After they pass through each other there are two regions in which we might ex-

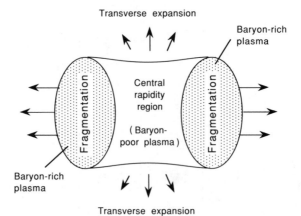

Fig. 14.15   Ultrarelativistic heavy ion collision in the center-of-mass (after Baym, 1982b). Rapidity is defined in Appendix C.

pect to find quark–gluon plasma: (1) The target and projectile fragmentation regions may experience a phase transition through compression and heating effects; any plasma formed there is likely to be rich in baryon contaminants. (2) The central rapidity region may undergo a phase transition through intense heating (production of a "hot vacuum" that materializes quark pairs and gluons), and the resulting plasma is expected to be baryon poor. Thus, the baryon-poor plasma in the central rapidity region is expected to resemble more closely the quark–gluon phase of the early universe, while the plasma in the fragmentation regions may be closer to that which might be produced in superdense objects like collapsed stars.

Hot Central-Rapidity Vacuum.   The origin of the hot vacuum in the central rapidity region may be understood qualitatively as follows (McLerran, 1982, 1984; Shuryak, 1980; Van Hove and Pokorski, 1975). The colliding ultrarelativistic nuclei are Lorentz contracted to a width no less than $\Delta x \simeq 1$ fm, with the limiting width a consequence of uncertainty principle arguments. The low momentum components of the nucleon wavefunction are composed of $q\bar{q}$ pairs (virtual pions) and gluons (see Fig. 10.15), and have $\Delta p \simeq p \simeq 200$ MeV; therefore,

$$\Delta x \geqslant \frac{1}{\Delta p} \simeq 1 \text{ fm.} \tag{14.45}$$

Higher momentum components would have smaller extent; for example, the valence quark (nucleon) component of the wavefunction would have $\Delta x \simeq R/\gamma$, where R is the rest frame nuclear radius and γ is the energy per nucleon in the center-of-mass system (McLerran, 1982). When the nuclei pass through each other, the low momentum components of the nuclear wavefunction interact more strongly than the high momentum ones. The low momentum com-

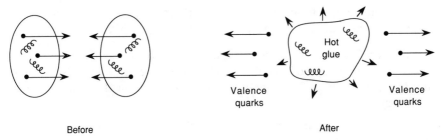

Fig. 14.16 Ultrarelativistic heavy ion collision.

ponents come to rest in the center-of-mass frame, but the high momentum components pass through—the ultrarelativistic heavy nucleus is a momentum filter for the nuclear wavefunction. The "scraped off" low momentum component of gluons and $q\bar{q}$ pairs is responsible for the hot vacuum between the receding Lorentz-contracted nuclei, as illustrated in Fig. 14.16.[‡]

Proper Time Scale. Once the plasma in both the central rapidity and fragmentation regions comes to an equilibrated, short mean-free-path regime, we may expect hydrodynamics to characterize the subsequent spacetime evolution. However, the hydrodynamical boundary conditions in the ultrarelativistic case are quite different from those that are appropriate for 1 GeV/nucleon collisions. The essential difference lies in the nuclear transparency of the ultrarelativistic collision, which results from Lorentz contraction and time dilation effects. This follows from the observation that a characteristic *proper time scale* should exist for relativistic strong interactions:

$$\tau_0 \simeq \frac{1}{\Lambda_{\text{QCD}}} \simeq 1 \text{ fm}/c, \tag{14.46}$$

where Λ_{QCD} is the QCD scale parameter [eq. (10.42)]. In any hadronic collision the fragments produced can interact only after a time of order τ_0 has elapsed. As the collision becomes ultrarelativistic the nuclei become so Lorentz contracted (large γ factor) that the pancake-like nuclei pass through each other in a time comparable to τ_0. Indeed, this provides a convenient working definition of an ultrarelativistic heavy ion collision (Kajantie, 1984): the nuclei in such collisions are Lorentz contracted to a longitudinal dimension of about one fermi. Conversely, in the 1 GeV/nucleon regime the total

---

[‡]There are two reasons to expect enhanced stopping of the gluons relative to the valence quarks: (1) the gluons have lower average momentum than the valence quarks so they become penetrable at higher energy than the quarks; (2) gluon–gluon scattering has larger cross section than quark–quark scattering because of color degeneracies (Shuryak, 1980).

collision time is much larger than τ_0 and the nuclei gradually overlap, with thermalization taking place in the overlap region (see Fig. 14.14).

Spacetime Picture. Bjorken (1983) discusses ultrarelativistic heavy ion collisions in a simple hydrodynamical model (see also McLerran, 1982; Kajantie and McLerran, 1987). The energy–momentum tensor in the absence of dissipation is

$$T_{\mu\nu} = (\varepsilon + P)u_\mu u_\nu - g_{\mu\nu}P, \tag{14.47}$$

where

$$\begin{cases} \varepsilon(x) = \text{local energy density} \\ P(x) = \text{local pressure} \\ u_\mu(x) = \text{local 4-velocity } (u_\mu u^\mu = 1), \end{cases}$$

and it is conserved,

$$\partial^\mu T_{\mu\nu} = 0. \tag{14.48}$$

The resulting first-order differential equations, supplemented by boundary conditions and an equation of state

$$\varepsilon = \varepsilon(P), \tag{14.49}$$

determine the behavior of the state variables for all time.

The normal Landau boundary conditions are total arrest and rapid equilibration of the fluid in the initial collision; Fig. 14.17a displays these boundary conditions in a spacetime diagram. The Landau conditions may be appropriate for lower energy collisions (§14.6), but the nuclear transparency in ultrarelativistic collisions implies different constraints on the hydrodynamical expansion of the baryon-poor plasma. Bjorken (1983) assumes boundary con-

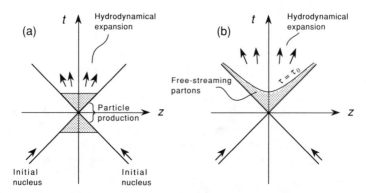

Fig. 14.17   Spacetime diagrams illustrating different initial conditions on the hydrodynamical expansion for simple Landau and ultrarelativistic models (see McLerran, 1982). Matter is produced in the hatched regions: in the Landau model matter forms near the origin; in the ultrarelativistic collision matter instead forms along the edge of the forward light cone.

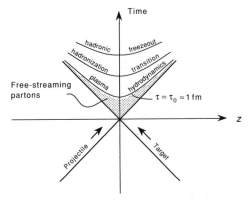

Fig. 14.18 Spacetime view of the stages in an ultrarelativistic central collision (after Baym, 1984).

ditions corresponding to the onset of hydrodynamical behavior at a proper time (Fig. 14.17b)

$$\tau_0 = \sqrt{t^2 - z^2} \simeq 1 \text{ fm,} \tag{14.50a}$$

with an initial longitudinal flow velocity

$$u_z = \frac{z}{t} \tag{14.50b}$$

and energy density

$$\varepsilon_0 \simeq 1 - 10 \text{ GeV/fm}^3. \tag{14.50c}$$

Until this proper time τ_0 has elapsed the system is a free-streaming parton gas; as this system evolves, collisions among the partons bring it to local thermal equilibrium and the hydrodynamical regime is entered at $\tau \simeq \tau_0$. Figure 14.18 illustrates in a spacetime diagram: the nuclei collide along the z-axis at $t = 0$, producing a free parton gas; by $\tau \simeq \tau_0$ local thermodynamical equilibrium has been established and the hydrodynamical expansion commences with initial conditions given by eqns. (14.50). With these assumptions the initial expansion is *one-dimensional* (longitudinal). This decoupling of the longitudinal and transverse expansions is quite different from the situation in Fig. 14.17a, which generally must be treated as $(3 + 1)$-dimensional hydrodynamics.

Hadronic Freezeout. The one-dimensional expansion continues for perhaps $\tau = 5 - 10$ fm, at which time the separation of the outgoing pancakes exceeds their diameter and the system begins to experience radial three-dimensional expansion with rapid cooling. At this stage the energy density is some $100 - 300$ MeV/fm^3, the temperature is $T \simeq 200$ MeV, and the mass density is 2–3 times that of normal nuclear matter. These conditions are near those expected

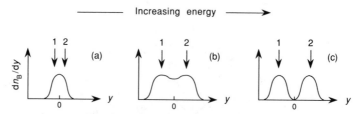

Fig. 14.19 Schematic baryon rapidity distributions in heavy ion collisions. At sufficient energy the ultrarelativistic nuclei pass through each other, leaving the baryon density concentrated at the incident rapidities of ions 1 and 2.

for the confining phase transition, and we may assume that the system is back in the hadronic gas phase soon after the three-dimensional expansion begins. This would imply that in our highly idealized picture the three-dimensional stage of the expansion entails rather conventional and uneventful physics, with the hadrons quickly moving from the collision-dominated to free-streaming regime (freezeout).

14.8   Production of Baryon-Poor Plasmas

Studies of reactions on heavy nuclei precipitated by 100 GeV protons indicate that the longitudinal momentum transfer to the leading proton is considerably higher than that found in proton–proton scattering (Busza, 1984; Busza and Ledoux, 1988). Specifically, a rapidity loss $\Delta y \simeq -1$ is observed in proton–proton reactions, but it is found that $\Delta y \simeq -(2-3)$ in proton plus lead collisions (rapidity is a relativistic analog of velocity; see Appendix C). This suggests that the formation of a clean central rapidity region uncontaminated by baryons may require higher energies than had been estimated originally from proton–proton scattering. Figure 14.19 illustrates schematically; only the situation depicted in the rightmost example allows a baryon-free region of high energy density near $y = 0$ where a baryon-poor plasma could be formed. Present estimates are that adequate baryon-free width in the central rapidity region for heavy ion collisions may require $E_{CM} \geqslant 50$ GeV/nucleon. Recent experimental results that are discussed in §14.9 may illuminate this issue.

14.9   Experimental Evidence

There is no unequivocal evidence that a quark–gluon plasma has been observed. However, the experimental situation is changing rapidly and new accelerator facilities may be on the verge of creating observable effects from a quark–gluon plasma. The status of these experiments is reviewed in McLerran (1987) and Gyulassy (1988). Despite the instability of the current situation, it is useful to summarize evidence that may have a bearing on whether it is possible to produce explicit and unmistakable QCD effects in extended systems. This brief survey is included to call attention to some issues that are relevant to the material in this chapter; it is neither complete nor likely to remain current, and the reader is urged to consult the recent literature on these and related topics.

The EMC Effect. A result of particular relevance for any discussion of quark effects in nuclei is the EMC effect (discovered by the European Muon Collaboration—see Berger and Coester, 1987, for a review). Basically, the EMC effect indicates that the $F_2(x)$ structure function (§10.5) for composite nuclei differs from the corresponding structure function for deuterium. Figure 14.20 summarizes: compared with deuterium the structure function of the composite nucleus is depleted at intermediate x and slightly enhanced near $x = 0.15$ (the rise at large x is a consequence of Fermi motion; the decrease at very small x is not well understood).

This suggests that a nucleus probed at the quark level is more than an assembly of independent nucleons, but there is not general agreement on the

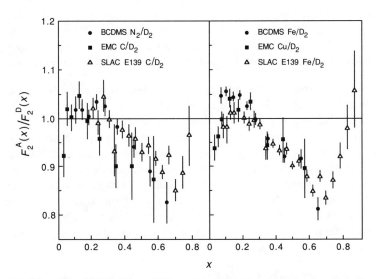

Fig. 14.20 The EMC effect. The ratio of the F_2 structure function for composite nuclei (A) relative to that for deuterium (D) is plotted. The SLAC data are from deep inelastic electron scattering, assuming the ratio (10.16) to be independent of the target; the remainder are from deep inelastic μ scattering. The left figure is for carbon and nitrogen targets; the right figure is for iron and copper targets. Data and references are compiled in Aguilar-Benitez et al. (1988).

detailed cause of the effect. Indeed, explanations have been proposed involving such disparate agents as partially free quarks, pion clouds, and α-particle clusters. Which of these are correct, and how many of the proposed explanations are actually independent is not completely obvious. A summary of some of these explanations may be found in Berger and Coester (1987) and Rith (1988).

If the EMC effect turns out to signal a density dependent quark degree of freedom in nuclear matter (or if it does not), it may be an important piece of information about the likelihood and nature of a deconfining transition in ultrarelativistic heavy ion collisions.

Quark Effects in Mass-3 Systems? Another set of nuclear phenomena for which it has been argued that explicit quark effects may play a role is in the properties of the nuclei ^{3}He and ^{3}H. These nuclei are rather simple, and precise calculations employing conventional nuclear physics are available. There are three areas in which discrepancies between conventional theories and the data have encouraged calculations that employ quark degrees of freedom: (1) the central charge density of ^{3}He, as probed by electron scattering; (2) ground state magnetic moments; and (3) the binding energy difference between ^{3}He and ^{3}H.

Miller (1984) discusses these discrepancies and possible improvements. The appearance of explicit quark effects in mass-3 nuclei could be relevant to the question of deconfinement, but the effects are not large and there are competing explanations using more traditional degrees of freedom, such as pions. As in the case of the EMC effect, measurements in mass-3 systems may indicate explicit QCD effects, but the interpretation is not conclusive.

Neutron Stars and Quark Matter. The presence of quark matter in the cores of neutron stars could have a significant influence on their cooling rates (Baym, 1981, 1982c). The initial cooling of neutron stars is thought to occur through neutrino emission from their interior. By phase-space considerations, degenerate quark matter (and pion condensed matter) would be more efficient at emitting neutrinos than normal nuclear matter, and neutron stars containing such exotic material should be cooler than otherwise expected. Present X-ray measurements indicate rapid cooling of neutron stars, but the experimental bounds on neutron star surface temperatures must be improved by an order of magnitude before anomalous cooling can be distinguished from cooling by more normal methods.

Evidence from Cosmic Ray Studies. It can be argued that nature has provided two "accelerator" experiments in which we might search for evidence of a quark–gluon plasma while we wait for man-made accelerators large enough to produce such a phase of matter in the laboratory. One is the *big bang*, which will be discussed in Ch. 15; the other is the flux of high-energy *cosmic rays* that continuously bombard the upper atmosphere. A few dozen interactions of cosmic ray heavy-ions with energies in excess of 100 GeV/nucleon have been reported (see Jones, 1984; and Jones et al., 1987, for a summary). For some of those events large charged-particle multiplicities and transverse momenta have been recorded, with estimated energy densities of $2-4.5$ GeV/fm^3 and charged-particle multiplicities of order 10^3. Such energy densities would exceed the critical value of approximately 2 GeV/fm^3 thought necessary to produce a chromoplasma. Thus it is possible that this new state of matter is being formed in cosmic ray heavy-ion events.

Accelerator Results. The SPS at CERN has produced 200 GeV/nucleon oxygen and sulfur beams, while the Brookhaven National Laboratory AGS has produced beams of 15 GeV/nucleon oxygen and silicon. The results from the first (fixed target) experiments with these beams are discussed extensively in a sequence of papers that can be found in *Zeit. Physik* **C38**, 1988.

Conservative estimates put the energy density at 1 GeV/fm^3 for central collisions in these experiments (Gyulassy, 1988)—seven times normal nuclear matter density and within a factor of about two of the density thought necessary to ensure the transition to a quark–gluon plasma. In addition, data analysis suggests that the production of J/ψ may be suppressed (CERN; Bussiere et al., 1988) and the ratio K^+/π^+ enhanced (AGS; Miake et al., 1988) in heavy-ion central collisions. These have been proposed as signals of quark–gluon plasma formation (see Exercise 14.6), or at least an indication that high mass or energy densities are being produced, but considerable work will be required to rule out mundane explanations [see Gyulassy (1988) and references therein].

EXERCISE 14.6 Use the Debye screening model of §14.3 to argue that J/ψ production may be suppressed by the formation of a quark–gluon plasma in a heavy-ion collision.

The rapidity interval in the CERN experiments is about six, with the fragmentation regions approximately two units wide. Therefore, 200 GeV/nucleon light heavy ions on fixed targets may approximate the intermediate transparency situation in Fig. 14.19b. The rapidity interval for the AGS experiments is only about three units, so the 15 GeV/nucleon fixed target experiments probably resemble the complete-stopping regime of Fig. 14.19a. Therefore, if any quark–gluon plasma is being produced in these fixed-target experiments it is expected to be baryon-rich, and probably transient since the spacetime extent of the collision region is limited for light heavy ions. The transparency regime of Fig. 14.19c, with the possibility of manufacturing both baryon-rich and baryon-poor plasma, will be reached only in the proposed Brookhaven RHIC collider. There the rapidity interval is expected to exceed 10 units for 100 GeV/nucleon colliding beams as heavy as uranium, with central energy densities as large as 10 GeV/fm^3. For the heaviest ions the spatial extent of the deconfined region should also be much larger than for oxygen or sulfur, which may be expected to enhance the stability of the plasma.

14.10 Background and Further Reading

A good introduction to the material in this chapter may be found in a series of conference proceedings related to the formation of quark–gluon plasma. The articles by Baym appearing in these proceedings are especially pedagogical: Baym (1981, 1982b, 1984). Also recommended are Satz (1983, 1984, 1985a), Baym (1982a,c), and Shuryak (1988). Recent summaries of the field are given in Cleymans, Gavai, and Suhonen (1986), McLerran (1986, 1987), Kajantie and McLerran (1987), and Gyulassy (1988). For discussions of field theory at finite temperature see Kogut (1983), Shuryak (1980, 1988), Satz (1984, 1985a), Brandenberger (1985), Kapusta (1989), and Bernard (1974). Short discussions and references to the literature on cosmological implications of a quark–gluon phase transition may be found in Cleymans, Gavai, and Suhonen (1986), Olinto (1988), Schramm and Olive (1984), and Shuryak (1988), §13.4.

CHAPTER 15

Cosmology and Gauge Theories

One consequence of the gauge theory approach to fundamental interactions has been a renewed dialog between the disciplines of elementary particle physics and astrophysics. On the one hand, gauge theories allow processes to be studied that are of astrophysical interest. These fall mostly into two categories: (1) processes of importance in cosmology such as baryogenesis, and (2) processes relevant to the physics of superdense matter (collapsed objects such as white dwarfs or neutron stars).

On the other hand, phenomena such as the big bang, supernovae, or neutron stars may provide the only laboratories for elementary particle physicists to test certain consequences of gauge theories. Accordingly, in this chapter we discuss some important interfaces between elementary particle physics and astrophysics. Our starting point is the standard cosmological model and the influence of grand unified theories (GUTs) on the evolution of the universe.

15.1 The Hot Big Bang

Just as there is a Standard Model for elementary particle physics, there is a *standard cosmological model* based on the idea that the universe we see about us was created in a primordial *big bang*. In this section we summarize the basic features of this model.

15.1.1 Evidence

Our standard cosmological view is rather widely accepted, but it is shaped by only a limited set of observations (Tayler, 1981; Kolb, 1987):

1. The red shift of distant galaxies implies an expansion of the universe (Hubble Law).
2. A highly isotropic large-scale distribution of galaxies and radio sources is found when account is taken of obscuration by interstellar dust.

3. A homogeneous distribution is observed for nearby galaxies within sufficiently large volumes of space (diameters $\geqslant 10^8$ light years).

4. The properties of distant radio sources, carrying information about an earlier epoch of the universe because of finite light velocity, suggest that these sources were once more closely spaced and more powerful.

5. The cosmic microwave background radiation seems to be all-pervading, highly isotropic, and well approximated by a 2.75 ± 0.05 K blackbody spectrum.

6. The elemental composition of the universe indicates (by mass) three parts hydrogen to one part helium, with only a trace of heavier elements.

7. Charge symmetry is implied by the observation that gravitation, not electromagnetism, governs the large-scale structure of the universe. In contrast, there appears to be a strong matter–antimatter asymmetry. In addition to the global predominance of matter over antimatter, there seem to be many fewer baryons than microwave photons.

8. More recent information suggests ordering of matter on scales much larger than galaxies or clusters of galaxies. For example, it has been found that distributions of galactic densities may resemble a collection of giant soap bubbles, with galaxies concentrated on the edges of bubbles surrounding enormous voids that are typically 10^8 light years in diameter and contain negligible luminous matter.

The conventional wisdom is that observations (1)–(3) imply an expanding, isotropic, homogeneous universe; observations (4)–(5) argue strongly against the *steady state theory* of the universe but are consistent with the *big bang;* observations (5)–(6) favor a particular class of homogeneous, isotropic universes called *hot big bang models;* and observation (7) restricts consideration to a particular subset of hot big bang models involving GUTs and *CP* violation that will be discussed further below. Observation (8) may require an exotic explanation, with the most attention focused on the detailed consequences of GUT phase transitions within an expanding universe, or on the existence of cold, dark matter consisting of unobserved elementary particles.

15.1.2 The Standard Cosmological Model

For orientation we present a quick review of the standard hot big bang model (Kolb and Turner, 1983; Kolb, 1987; Collins, Martin, and Squires, 1989).

Metric and Equations of Motion. Assuming a homogeneous, isotropic cosmology (*the Cosmological Principle*), the metric of the universe takes the Friedmann–Robertson–Walker (FRW) form

$$ds^2 = dt^2 - R(t)^2 \left[\frac{dr^2}{1 - kr^2} + r^2 \left(d\vartheta^2 + \sin^2 \vartheta d\varphi^2 \right) \right]. \qquad (15.1)$$

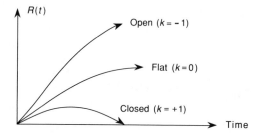

Fig. 15.1 Evolution of the universe for different values of k in eq. (15.1).

The constant k determines the global structure of the universe. It is conventional to rescale R and r so that k takes only three discrete values. From (15.1) and the equations of motion discussed below:

1. If $k = +1$ the universe is closed and finite, with positive curvature. The scale of length is set by the *cosmic scale factor* $R(t)$, which increases to a maximum and then decreases to zero again in a finite time.
2. If $k = 0$ the universe is flat and infinite; $R(t)$ increases without limit but $dR/dt \to 0$ as t goes to infinity.
3. If $k = -1$ the universe is open, negatively curved, and infinite, and $R(t)$ increases without limit.

Fig. 15.1 illustrates these three general scenarios for the universe.

The coordinates (r, ϑ, φ) in (15.1) are called *comoving coordinates*. Distances measured by these coordinates are termed *coordinate distances;* the corresponding physical distances are obtained by multiplying with the scale factor $R(t)$. The behavior of $R(t)$, which is like a radius for a closed universe, depends on the field equations of general relativity and the equation of state for the universe. From the Einstein field equations

$$H^2 \equiv \left(\frac{\dot{R}}{R}\right)^2 = \left(\frac{8\pi G}{3}\right)\rho - \frac{k}{R^2} + \frac{\Lambda}{3}, \qquad (15.2)$$

where G is the gravitational constant, ρ is the energy density of the universe, and Λ is the *cosmological constant*, which we will neglect for the remainder of this part of the discussion. The inverse expansion time H is the *Hubble constant*, which is not constant. Presently, $H = 50$–100 km/sec/Mpc, or $H^{-1} = 10$–20×10^9 years, where Mpc denotes a megaparsec [1 parsec (pc) = 3.26 light years = 3.09×10^{16} m; 1 Mpc = 10^6 pc; 1 yr = 3.16×10^7 sec].

The evolution of the universe is determined by (15.2), the relation

$$\frac{\dot{\rho}}{\rho} + 3\left(1 + \frac{p}{\rho}\right)\frac{\dot{R}}{R} = 0 \qquad (15.3)$$

(p is the isotropic pressure) which follows from energy conservation, and an *equation of state* for the universe that provides a relation between ρ and p. If the universe is dominated by relativistic matter (*radiation dominated*), the equation of state is[‡]

$$p = \tfrac{1}{3}\rho, \tag{15.4}$$

and (15.3) has a solution

$$\rho(t) \simeq \frac{1}{R^4(t)} \qquad \text{(radiation dominated)}. \tag{15.5a}$$

On the other hand, if the universe is dominated by nonrelativistic matter (*matter dominated*), $p \simeq 0$ and (15.3) has a solution

$$\rho(t) \simeq \frac{1}{R^3(t)} \qquad \text{(matter dominated)}. \tag{15.5b}$$

In either extreme these results may be used to solve the equation of motion (15.2).

Because of (15.5), the last two terms of (15.2) become relatively more important with time. Observations indicate that they presently do not dominate the first term, so they did not dominate it in the past. Therefore, in most practical applications we can neglect the last two terms of (15.2). This leads immediately to the behavior (Exercise 15.2a)

$$R(t) \simeq \begin{cases} t^{1/2} & \text{(radiation dominated)} \\ t^{2/3} & \text{(matter dominated)}. \end{cases} \tag{15.6}$$

These equations, in conjunction with the known properties of atoms, nuclei, and elementary particles, allow a history of the universe to be constructed (see Fig. 14.3).

The observed mass of the present universe is contained mostly in nonrelativistic matter, but eqns. (15.5) imply that at much earlier times the universe was radiation dominated. Specifically, the universe was pervaded by relativistic matter when the scale factor $R(t)$ was less than 10^{-4} times its present value, corresponding to temperatures greater than 10 eV and times earlier than about 10^{10} sec after the big bang (see Fig. 14.3). This early radiation dominated epoch is of primary interest for baryogenesis.

[‡]By relativistic matter in the present universe we mean photons, very light or massless neutrinos, or other such particles; by nonrelativistic matter we mean baryons and other massive particles. Although the observed baryon to photon ratio is about 10^{-9}, most photons are in the $T \simeq 2.3 \times 10^{-4}$ eV microwave background, while the rest mass energy of a typical baryon is about 10^9 eV. Thus the energy density contributed by baryons to the universe is approximately $10^3 - 10^4$ times that contributed by relativistic matter and the current universe is *matter dominated;* at much earlier times the universe was *radiation dominated*, with the energy density concentrated in relativistic particles.

Equilibrium in the Universe. We may distinguish two kinds of equilibrium in the universe: (1) thermal equilibrium and (2) chemical equilibrium. If a system is in thermal equilibrium the phase-space number density for particles is

$$f(p) = \frac{g}{e^{(\mu+E)/T} + \Theta} \tag{15.7}$$

where $\Theta = 0$ for Maxwell–Boltzmann statistics, $\Theta = +1$ for Fermi–Dirac statistics, $\Theta = -1$ for Bose–Einstein statistics, g is the total number of degrees of freedom for the particles, μ is the chemical potential, and $E^2 = \mathbf{p}^2 + m^2$. Generally, for a two-body reaction the rate is

$$\Gamma \simeq nv\sigma,$$

where n is the density, v is the relative speed, and σ is the interaction cross section. Thus the reaction rates for interacting particles decrease as the universe expands, finally decoupling the particles from thermal equilibrium; this decoupling of a species from equilibrium is called *freezeout* (see Exercise 15.5; a corresponding phenomenon was discussed for heavy-ion collisions in §14.7). In addition, if for the reaction $a + b \leftrightarrow c + d$ the chemical potentials satisfy

$$\mu_a + \mu_b = \mu_c + \mu_d, \tag{15.8}$$

we say that the species a, b, c, and d are in *chemical equilibrium*. These two kinds of equilibrium may not coincide. For example, if reactions that equilibrate energy (e.g., inelastic scattering) are occurring faster than the expansion, but reactions changing the particle number (e.g., annihilation) are slower than the expansion, the universe is expected to be in thermal equilibrium but not chemical equilibrium.

Strictly speaking, neither thermal nor chemical equilibrium occurs in an expanding universe. However, a practical equilibrium can exist if

$$\Gamma >> H, \tag{15.9}$$

where Γ denotes the rate of reactions leading to equilibrium and the Hubble parameter $H = \dot{R}/R$ characterizes the expansion rate for the universe. If (15.9) is satisfied, the universe may be expected to evolve through a sequence of nearly equilibrated states.

Assuming weakly interacting relativistic particles ($m << T$) and thermal equilibrium, and neglecting chemical potentials, the number density n_i and energy density ρ_i for a species i are given by (Kolb and Turner, 1983)

$$n_i = g_i \frac{\zeta(3)}{\pi^2} T^3 \times \begin{cases} 1 & \text{(Bose–Einstein)} \\ 3/4 & \text{(Fermi–Dirac)} \\ \zeta(3)^{-1} & \text{(Maxwell–Boltzmann)} \end{cases} \tag{15.10}$$

$$\rho_i = g_i \frac{\pi^2}{30} T^4 \times \begin{cases} 1 & \text{(Bose–Einstein)} \\ 7/8 & \text{(Fermi–Dirac)} \\ 90/\pi^4 & \text{(Maxwell–Boltzmann)} \end{cases} \tag{15.11}$$

where $\zeta(3) = 1.202$ (Riemann zeta-function) and g_i counts degrees of freedom (see §14.2). The energy density contributed by all relativistic particles is [see eq. (14.9)]

$$\rho = g_* \frac{\pi^2}{30} T^4, \tag{15.12}$$

with

$$g_* = \sum_{\text{bosons}} g_b + \tfrac{7}{8} \sum_{\text{fermions}} g_f, \tag{15.13}$$

where g_b and g_f denote boson and fermion degrees of freedom. If all species are in equilibrium, the entropy density s is

$$s = \frac{\rho + p}{T} = \frac{4}{3} \left(\frac{\rho}{T} \right) = \frac{2\pi^2}{45} g_* T^3, \tag{15.14}$$

and the entropy per comoving volume (total entropy for a closed universe)

$$S \simeq sR^3 \simeq \text{ constant} \tag{15.15}$$

does not change [adiabatic expansion: $d(sR^3)/dt = 0$]. Utilizing (15.14) and (15.10), the entropy density s is proportional to the number density of relativistic particles

$$s \simeq \sum n_i \simeq \sum g_i T^3. \tag{15.16}$$

In the present universe $s \simeq 2.8 \times 10^3/\text{cm}^3$.

From (15.14) and (15.15), as long as the number of degrees of freedom g_* is constant (whenever kT is comparable with the mass for a species, g_* will change), the temperature of a radiation dominated universe varies inversely with the scale factor R:

$$T(t) \simeq \frac{1}{R(t)}, \tag{15.17}$$

and from (15.2), neglecting the last two terms,

$$T \simeq t^{-1/2}, \tag{15.18}$$

with t the time since the bang. In particular, under these conditions the universe evolves according to (Kolb and Turner, 1983)

$$\frac{\dot{R}}{R} = -\frac{\dot{T}}{T} = \alpha T^2$$

$$t = \frac{1}{2\alpha T^2} = \frac{2.4 \times 10^{-6}}{\sqrt{g_*}\, T^2} \text{ GeV}^2 \cdot \text{s} \tag{15.19}$$

$$\alpha = \sqrt{\frac{4\pi^3 g_*}{45 M_P^2}},$$

where the Planck mass is

$$M_{\mathrm{P}} \equiv \sqrt{\frac{\hbar c}{G}} = G^{-1/2} = 1.2 \times 10^{19} \text{ GeV}. \tag{15.20}$$

Equations (15.19) are expected to be valid for temperatures 10^{19} GeV $\gtrsim T \gtrsim$ 10 eV; that is, for the radiation-dominated period from the epoch of quantum gravitation to the decoupling of matter and radiation (see Fig. 14.3).

15.2 Baryogenesis

Pre-GUT big bang cosmology was hard pressed to account for the observed preponderance of matter over antimatter, or of photons over baryons. These had to be introduced as *initial conditions*. Sakharov (1967) listed the three ingredients essential for generation of these baryon asymmetries within the big bang model:

1. There must exist baryon nonconserving interactions in nature.
2. Both C and CP symmetries must be violated.
3. There must be a departure from thermal equilibrium.

Departures from thermal equilibrium are likely to have occurred in the evolution of the universe, and at the time of Sakharov's original proposal it was known that C and CP were violated in weak interactions. However, there was little reason other than baryogenesis to postulate baryon nonconserving reactions. Today, GUTs provide a theoretical framework that can accommodate C and CP violation and has baryon non-conservation as a centerpiece (Ch. 11). Thus grand unified theories, in conjunction with nonequilibrium evolution of the universe, have finally supplied us with a plausible mechanism for baryogenesis. For a review of attempts to account for the baryon excess see Kolb and Turner (1983) and references cited there. We will restrict discussion to a schematic model that illustrates the basic idea (Weinberg, 1979; Toussaint et al., 1979; Kolb and Turner, 1983; Turner and Schramm, 1979).

In this model baryon decay at a temperature $T \simeq 10^{28}$ K (corresponding to a time 10^{-35} sec after the big bang; see Fig. 14.3) is assumed to be mediated by a very heavy boson X that has baryon nonconserving interactions. It could either be a GUT gauge particle or a Higgs particle associated with spontaneous symmetry breaking in a GUT. Its mass is denoted by M, with $M \geqslant 10^{14}$ GeV; its coupling strength to fermions is $\alpha^{1/2}$, and on dimensional grounds the decay rate of the X is

$$\Gamma_X \simeq \alpha M. \tag{15.21}$$

The universe is taken to be baryon symmetric at the Planck time $t_{\mathrm{P}} \simeq 10^{-43}$

sec, and very soon thereafter the baryon-symmetric soup of fundamental particles is assumed to be in thermal equilibrium. As the temperature of the universe plummets the X and $\overline{X}$ bosons are relativistic, in thermal equilibrium, and approximately as abundant as photons: $n_X = n_{\bar{x}} \simeq n_\gamma$. Eventually the temperature falls to the point where $T < M$; below that temperature the *equilibrium* abundance of X particles is

$$n_X \simeq e^{-M/T} n_\gamma, \tag{15.22}$$

but this equilibrium abundance can be maintained only if there are reactions decreasing the population of X particles faster than the expansion rate H. The most important reaction decreasing X is its decay, with the rate given by (15.21): reactions such as annihilation are slower by additional factors of α when $T < M$. Thus, the maintenance of thermal equilibrium requires that $\Gamma_X > H$, and the evolution of the universe for $T \leqslant M$ depends critically on the values of α and M.

Figures 15.2 and 15.3 depict two qualitatively different possibilities. If $\Gamma_X > H$ for $T < M$, the X bosons can maintain thermal equilibrium until they have decayed away: there is no departure from equilibrium and hence no generation of a baryon excess (Fig. 15.2). On the other hand, if $\Gamma_X < H$ for a period after $T < M$ the situation applies that is depicted schematically in Fig. 15.3a. When T drops below M, neither decay nor annihilation can maintain X and $\overline{X}$ in equilibrium; they will remain approximately as abundant as photons and out of equilibrium until $\Gamma_X > H$ (until the age of the universe $\sim 1/H$ is greater than the mean life of X). At that point X and $\overline{X}$ decay freely because

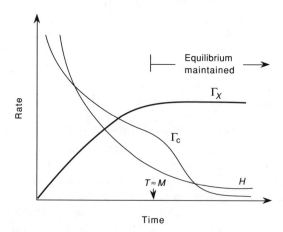

Fig. 15.2   Important rates in baryon evolution for a case where $\Gamma_X > H$ after $T = M$; H is the expansion rate, Γ_X is the decay rate for the X boson, and Γ_c is the rate for baryon nonconserving collisions. Equilibrium is maintained so no baryon excess can be generated. Γ_X is small at early times because X is highly relativistic then and its lifetime is dilated (after Turner and Schramm, 1979).

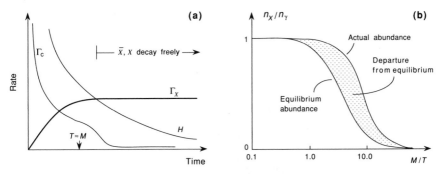

Fig. 15.3 (a) As in Fig. 15.2, but for the case $\Gamma_X < H$ for a time after $T = M$. (b) The corresponding departure from equilibrium: X is overabundant relative to its equilibrium population by the amount in the shaded region.

their decay products lack sufficient energy to regenerate them $(T < M)$. The corresponding departure from equilibrium is shown in Fig. 15.3b.

In this nonequilibrium case a baryon excess might arise; a plausible mechanism is to allow X and $\overline{X}$ each to decay by two different paths leading to states with different baryon number:

$$X \begin{array}{c} \xrightarrow{r} B_1 \\ \xrightarrow[1-r]{} B_2 \end{array} \qquad \overline{X} \begin{array}{c} \xrightarrow{\bar{r}} -B_1 \\ \xrightarrow[1-\bar{r}]{} -B_2 \end{array} \qquad (15.23)$$

The net baryon numbers for the decay products of X and $\overline{X}$ are

$$B_X = rB_1 + (1-r)B_2 \qquad B_{\overline{X}} = -\bar{r}B_1 - (1-\bar{r})B_2,$$

so for the decay of an X and $\overline{X}$ pair

$$\Delta B = B_X + B_{\overline{X}} = (r - \bar{r})(B_1 - B_2). \qquad (15.24)$$

If C and CP are not conserved, the branching ratios r and $\bar{r}$ can be different and a net baryon number results from X and $\overline{X}$ decays. This baryon excess survives under the nonequilibrium scenario of Fig. 15.3 because the rate Γ_c of baryon nonconserving collisions is smaller than the expansion rate H. The results of a model calculation are shown in Fig. 15.4, and we see that it is possible to generate a baryon excess by this method (Kolb and Turner, 1983).

Thus baryon nonconservation appears to have played a leading role in determining the structure of our universe: it has been said that *the proton would not exist if it were stable*, since there is no plausible way to generate a baryon excess in the early universe without baryon nonconservation ("I decay, therefore I am!"). Indeed, one might argue that the most convincing evidence for

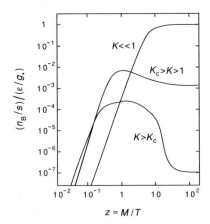

$z = M/T$

Fig. 15.4   The evolution of n_B/s, where n_B is the baryon number density and s is the entropy density, as a function of $z \equiv M/T \simeq t^{1/2}$. The mass of the X boson is M, T is the temperature, t is the time, ε measures the CP violation, K is proportional to the ratio of the decay rate and the expansion rate when $T = M$, and K_c is a critical value of K beyond which n_B/s falls rapidly if K is increased (Kolb and Turner, 1983). The entropy of the present universe is proportional to the number of relativistic particles (mostly photons in the 2.7 K background, and light neutrinos), so n_B/s is approximately the ratio of baryons to photons.

the instability of the proton is that there are physicists, made of baryons, searching for its decay.

The conjectured existence of black holes has sometimes been used as an argument against baryon conservation (De Rújula, Georgi, and Glashow, 1975; Turner and Schramm, 1979). A black hole destroys all detailed memory of the matter that went to make it, exhibiting only charge, mass, and angular momentum as distinguishing features to an external observer (sometimes summarized by "the black hole has no hair"). Thus black holes do not conserve baryon number. The ultimate reason is that baryon conservation is believed to be a global symmetry that has no long-range force associated with it: black holes made from equally charged baryons or antibaryons would be indistinguishable to an outside observer if they had the same mass and angular momentum. Thus baryon number falling into a classical black hole disappears from the universe.

The argument is more striking if we conjecture microscopic Hawking black holes that evaporate by radiating a thermal spectrum of particles. Such a black hole created from charged baryons would radiate a net charge equal to that which went to make it; however, it would be expected to radiate approximately equal numbers of baryons and antibaryons. Thus in the creation and ultimate evaporation of a Hawking black hole the charge, angular momentum, and energy are all conserved, but baryon number could be substantially modified.

15.3 Problems with the Hot Big Bang

The hot big bang has been quite successful in accounting for the nature of our universe. However, amid the triumphs of the standard cosmological model there remain significant conceptual problems. The fundamental ones are:

1. *The Horizon Problem:* there is an apparent conflict between causality and both the large-scale isotropy and homogeneity and smaller scale density fluctuations of the universe.

2. *The Flatness Problem:* why does the universe appear to be so flat? That is, why is the amount of matter in the universe apparently so close to the critical density required to close the universe?

3. *The Baryon Number/Monopole Problem:* where is the horde of magnetic monopoles that GUTs, which seem to be necessary for a plausible theory of baryogenesis, predict for a hot big bang?

Let us consider each of these problems in turn, and speculate on what would be required to provide a solution for each (Sasaki, 1983; Steinhardt, 1986; Blau and Guth, 1987).

15.3.1 The Horizon Problem

The ratio of the horizon $l(t)$ to the physical size $L(t)$ of the universe at a time t is (Brandenberger, 1985)

$$\frac{l(t)}{L(t)} \simeq \frac{1}{\alpha(t)T(t)T(t_0)L(t_0)}, \tag{15.25}$$

where α is defined in (15.19) and t_0 is the present time. Taking $L(t_0) = 10^{41}$ fm and $T(t_0) = 2.7$ K, we estimate that for the Planck time $t_{\mathrm{P}} = 10^{-43}$ sec with $T(t_{\mathrm{P}}) = 10^{19}$ GeV and $g_* \simeq 100$,

$$\ell(t) \simeq 10^{-30}L(t). \tag{15.26}$$

This implies that at the Planck time the radius of the universe was 10^{30} times that of the event horizon. But the anisotropy of the 2.7 K microwave background is not more than one part in 10^4 once correction is made for the earth's motion, which argues that our universe is highly isotropic to the present horizon. If our universe expanded from one just after the Planck time, why is it isotropic over large domains that should not have been causally connected since the Planck time (see Exercise 15.3)?

This is one aspect of the *horizon problem*, but there is another: what is the origin of inhomogeneities such as galaxies and clusters of galaxies? The typical size of a cluster of galaxies is about 10^{-3} times the present horizon.

Therefore, coherent perturbations finely tuned over a scale many orders of magnitude larger than the horizon at the Planck time seem to be required, and the horizon problem makes it difficult to account for the energy-density fluctuations responsible for galaxy formation. As a consequence, seed fluctuations for large scale structure cannot be generated dynamically in the standard cosmology—they must be specified as initial data. The two-fold horizon problem is solved only when we understand how both global homogeneity and inhomogeneities of galactic size and larger arise over domains that were not causally connected since the Planck time, according to the standard model.

A schematic solution of the horizon problem would result from a period early in the history of the universe when the expansion rate was greater than the rate of horizon growth, which would allow points currently outside the horizon to have been inside at some past time. Specifically, it may be shown that the horizon problem appears in any model that satisfies

$$R(t) \propto t^n \qquad (n < 1), \qquad (15.27)$$

and that the problem could be alleviated if the universe experienced a period in which the expansion rate was more rapid:

$$R(t) \propto t^n \qquad (n > 1). \qquad (15.28)$$

It is easily demonstrated that such behavior requires the pressure to become *negative*, which makes it inadmissible in a standard model with positive pressure (see Exercise 15.2c).

15.3.2 The Flatness Problem

The importance of the flatness problem can be appreciated by recognizing that the low geometrical curvature of the present universe is an unstable situation that can be guaranteed only if the initial curvature radius is fine-tuned to order 10^{-30} (Sasaki, 1983). If this is not done, the universe either collapses to black holes or flies apart too rapidly for stars and galaxies and cosmologists to form. The flatness problem may also be viewed as an entropy problem— the present universe has too much relative to that expected for the early universe (Exercise 15.1b). This difficulty might be resolved if the universe went through a period of rapid heating early in its history. A phase transition could accomplish this, but in the standard cosmology it is assumed that phase transitions occur quickly, without significant entropy generation.

15.3.3 The Monopole Problem

The monopole problem is intimately connected with the baryon asymmetry problem. We have seen that the only plausible baryogenesis scheme yet devised requires a GUT. However, the same GUTs that provide the universe with baryons (at least for a time) are also expected to populate it in unacceptable numbers with massive topological excitations such as magnetic monopoles. Specifically, in any grand unified theory where the symmetry is broken from a simple group G to a smaller group $H = h \times U(1)$ that contains an explicit $U(1)$ factor, there will occur topologically stable monopole solutions of the 't Hooft–Polyakov type (Preskill, 1984). Their masses will be approximately M/e^2, where M is the GUT symmetry-breaking scale. For $SU(5)$ this gives a monopole mass of about 10^{16} GeV. From calculated production and depletion rates, the mass of so many monopoles would have produced an early universe that was matter dominated rather than radiation dominated, nullifying the successes of the hot big bang. Something must dilute the monopoles.

The triumphs of the standard cosmological model amid these difficulties are less puzzling once we realize that *the correct predictions of the standard model rest on a proper description of the universe from approximately one second after the big bang.* This suggests that modifications of the standard cosmology operating in the first second of time might retain the positive aspects of the hot big bang, while simultaneously alleviating the difficulties. The inflationary models introduced in the next section alter the standard model by treating matter in terms of quantum fields, rather than as an ideal gas. As we shall see, this modification can have a profound influence on the evolution of the very early universe.

EXERCISE 15.1 (a) Use the 2.7 K microwave background to place a lower limit on the entropy of the observable universe. (b) Show that the total entropy (15.15) can be written as

$$S = \left(\frac{k}{H^2(\Omega - 1)} \right)^{3/2} s,$$

where s is the entropy density, H is the Hubble constant, k is the curvature constant appearing in (15.1), and Ω is the ratio of the density ρ to the critical density ρ_c required to close the universe. Use this to place a lower limit on the entropy of the universe. Convince yourself that the flatness problem may also be viewed as an entropy problem. Show that the critical density ratio varies with time: $(\rho - \rho_c)/\rho \simeq t^{2/3}$ for a matter dominated universe and $(\rho - \rho_c)/\rho \simeq t$ for a radiation-dominated universe. *Hint:* First show that the critical density is $\rho_c = 3H^2/8\pi G$, and that

$$\frac{(\Omega - 1)}{\Omega} = \frac{3ks^{2/3}}{8\pi G \rho S^{2/3}},$$

where $\Omega \equiv \rho/\rho_c$. Assume $H \simeq (10^{10} \text{ yr})^{-1}$, that $|\Omega - 1| < 1$, and that the present entropy density is $s \simeq 2.8 \times 10^3 \text{ cm}^{-3}$.

EXERCISE 15.2 (a) Demonstrate that in a radiation-dominated universe $R(t) \simeq t^{1/2}$, but in a universe dominated by nonrelativistic matter $R(t) \simeq t^{2/3}$.

(b) Prove that in standard cosmologies the horizon distance

$$l_{\mathrm{H}}(t) \equiv R(t) \int_0^t \frac{cdt'}{R(t')}$$

is $l_{\mathrm{H}}(t) \simeq 2t$ for a radiation dominated universe and $l_{\mathrm{H}}(t) \simeq 3t$ for a matter dominated one. (c) Show that $\rho + 3p < 0$ is a necessary condition for $\dot{R}(t)$ to increase with time. Hence show that if $\rho + 3p > 0$, objects currently outside the horizon have always been outside. *Hint*: approximate the horizon as $1/H$ and compare this with $R(t)$.

EXERCISE 15.3 Assume that two microwave antennae point in opposite directions and receive radiation last emitted or scattered at the time of hydrogen atom recombination (about 10^5 years after the big bang; see Fig. 14.3). Show that the number of horizon distances separating the two sources at the time of emission was about 90, so the observed homogeneity of the 2.7 K microwave background coupled with this result produces the horizon problem. *Hint*: assume the universe to have been matter dominated for most of the history relevant to this problem.

15.4 Inflationary Cosmologies

Remarkably, there are new improved cosmologies on the market that modify the very early universe and that may be able to cure all of these diseases simultaneously. These go under the generic name *inflationary universe*. The central idea is quite simple: at the GUTs phase transition the expansion of the universe was driven for a short time at an exponential rate. This is precisely what is required to cure the difficulties of the hot big bang.

15.4.1 Old Inflation

In the original inflationary GUT scenario the spontaneous symmetry breakdown to $SU(3) \times SU(2) \times U(1)$ is associated with an effective potential $U(\phi, T)$ of the Higgs field ϕ that is of the form

$$U(\phi, T) = U(\phi) + \delta U(\phi, T), \tag{15.29}$$

where $U(\phi)$ has an absolute minimum at $\phi \neq 0$ and $\delta U(\phi, T)$ is a temperature-dependent correction

$$\delta U(\phi, T) \simeq T^2 \phi^2 \tag{15.30}$$

that stabilizes the $\phi = 0$ state. Such a potential is shown in Fig. 15.5 for different values of the temperature (for a discussion of calculating such potentials see Coleman and Weinberg, 1973; Dolan and Jackiw, 1974; the review by Brandenberger, 1985; and the Appendix of Abbott and Pi, 1986).

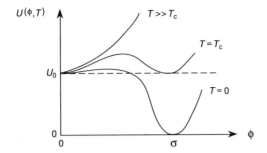

Fig. 15.5   An effective potential responsible for spontaneous symmetry break-ing in a GUT $\rightarrow SU(3) \times SU(2) \times U(1)$ phase transition. The field ϕ is a multicomponent field, so the figure is schematic. A potential of this form leads to the first-order phase transition of the original inflationary model.

At high temperature the expectation value of ϕ is zero, so there is no sym-metry breaking. At $T = 0$ the classical minimum of U occurs at $\phi = \sigma \neq 0$ and the symmetry is broken spontaneously. At $T = T_c$ the minimum at $\phi \neq 0$ is degenerate with that at $\phi = 0$, but a barrier separates the broken symme-try minimum from the symmetric minimum. Therefore, as the temperature decreases to T_c the universe becomes unstable against a phase transition to the broken symmetry phase. However, because of the barrier the two phases are connected only by tunneling processes and the transition is delayed.

This sequence is characteristic of a *first-order phase transition;* an analo-gous situation occurs in the cooling of water through the freezing point. The metastable state at $\phi = 0$ is called a *supercooling state* and the first-order phase transition proceeds by nucleation of bubbles of the stable phase (ice crystals in the case of water). Thus the GUTs phase transition may be pic-tured metaphorically as a "crystallization" of the Higgs field that leads to spontaneous symmetry breaking, just as crystallization implies the choice of preferred directions in condensed matter. As in the case of water freezing, we may expect the release in such a first-order phase transition of a *latent heat* associated with the difference between the metastable and stable potential minima. In addition, there may be supercooling effects, as in the case of wa-ter, where it is possible to cool the liquid in a metastable state well below the freezing point without the formation of ice. If the universe is in a supercooling state, it is said to be in a *false vacuum* at $\phi = 0$ instead of the true minimum at $\phi = \sigma$.

The inflationary hypothesis is now based on the following observation (Guth, 1981). The energy density of the universe is [see eq. (15.12) and Fig. 15.5]

$$\rho \simeq U_0 + \frac{\pi^2}{30} g_*(T) T^4 \equiv \rho_0 + \rho', \tag{15.31}$$

where $g_*(T)$ is the number of effectively massless spin degrees of freedom at temperature T. Equation (15.2) can be written as

$$\left(\frac{\dot{R}}{R}\right)^2 = \frac{8\pi G}{3}\rho \tag{15.32}$$

when the entropy

$$S \simeq (RT)^3 \simeq \text{number of relativistic particles} \tag{15.33}$$

is large (so k/R^2 is negligible) and the cosmological constant is ignored. If the universe supercools for $T < T_c$ the solutions of (15.32) are (Exercise 15.4)

$$R(t) \simeq \begin{cases} t^{1/2} & (T \gg T_c) \\ e^{t/\ell} & (T \ll T_c) \end{cases} \tag{15.34}$$

$$\ell \equiv \left(\frac{8\pi G}{3}U_0\right)^{-\frac{1}{2}} = \frac{1}{H}, \tag{15.35}$$

where U_0 is the energy density of the false vacuum (Fig. 15.5) and typically (Blau and Guth, 1987)

$$H \simeq 10^{10} \text{ GeV} \simeq 10^{34} \text{ sec}^{-1}.$$

Thus, in the time required for the phase transition the supercooled universe expands at an *exponential rate*, driven by the energy of the false vacuum. The solution e^{Ht} is called a *de Sitter solution* of the Einstein equations; such solutions entail a *negative pressure* for the universe.[‡] Therefore, the exponential inflationary expansion is the response of gravity to an unusual state of matter with constant energy density and negative pressure. Under these conditions the expansion of the universe *accelerates*, rather than decelerates, and the expansion is exponential because ρ and p are constant.

To summarize: in the most general terms, inflation corresponds to a period in which $\rho + 3p$ is *negative*; under such conditions the gravitational force becomes *repulsive*, resulting in an exponential expansion of the universe (Steinhardt, 1986; Abbott and Pi, 1986).

[‡]Notice from (15.2) and (15.31) that if the universe is in a supercooling state so that $U(\phi, T) \simeq U_0$ when $T \ll T_c$ (see Fig. 15.5), the vacuum energy U_0 plays the role of a cosmological constant: $\Lambda \simeq U_0$. Thus in the supercooling state the universe evolves in de Sitter space with large cosmological constant, before making the transition to the true vacuum with small (zero?) cosmological constant. A phase with constant positive energy density has a *negative pressure* with magnitude equal to the energy density (the equation of state is $p = -\rho$). *Proof*: if a bubble of this phase is placed in a background of the zero energy-density phase it will collapse to lower the energy. Since there is no energy density outside the bubble, the force to do this must come from inside the bubble. Therefore, the phase inside the bubble has a negative pressure of magnitude equal to the energy density, by energy conservation (Steinhardt, 1986).

EXERCISE 15.4 Demonstrate that eqns. (15.34)–(15.35) represent approximate solutions of eq. (15.32) whenever the energy density is specified by eq. (15.31).

EXERCISE 15.5 Estimate the temperature at which massless neutrinos decouple from thermal equilibrium in the early (radiation dominated) universe. *Hint*: determine the temperature at which the two-body weak interaction rate is comparable with the expansion rate.

In the original inflationary model the universe expands exponentially in the symmetric GUT phase for a period of time before making a transition to the $\phi = \sigma$ true vacuum state where the symmetry is broken to $SU(3) \times SU(2) \times U(1)$. Schematically at least, this will solve the basic difficulties of the standard model: the horizon problem is nullified by the rapid expansion [eq. (15.28)] because the universe that we see *was* in causal contact before inflation; the flatness problem is alleviated because in the rapid expansion local curvature becomes negligible; the monopole problem is solved because the uniformity of the Higgs field in the original inflating region suppresses their production, and because the expansion quickly dilutes any monopole density that does form.

The temperature of the original thermal state will decrease exponentially in the inflationary period (Fig. 15.6). At the end of inflation, provided that an efficient coupling mechanism exists, there is a short transition period $\delta\tau << \tau$ in which vacuum energy is converted rapidly to thermal energy and the entropy increases by a factor $\sim Z^3$, where

$$Z = e^{\tau/\ell}, \tag{15.36}$$

with τ the period of inflationary growth and $\delta\tau$ the time for reheating after inflation. If reheating increases the temperature to the same order of magnitude as before inflation, the universe can then evolve in standard model fashion, as illustrated in Fig. 15.6. In particular, a baryon excess might then be produced by the method discussed in §15.2. This reheating process would explain the entropy and baryon asymmetry of the present universe.

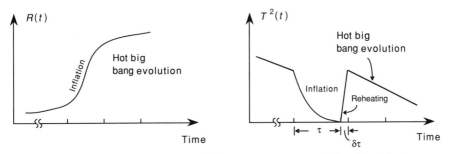

Fig. 15.6   Cosmic scale factor $R(t)$ and temperature T for the inflationary universe (compare Fig. 15.1).

The beauty of the inflationary scenario is that the original state of the universe is relatively immaterial. The exponential expansion so dilutes curvature and particle density that almost any set of initial conditions capable of triggering inflation can lead to a universe that solves the flatness, monopole, and horizon problems. In essence, the universe looks the way that it does because we see only a small part of it. As a bonus, the inflationary hypothesis provides an answer to the question of origin for matter and energy: our universe is composed almost entirely of thermalized vacuum energy left over from the inflationary epoch.

These ideas are attractive but the original inflationary universe had a serious problem, as Guth realized in his initial paper. In the original model it was assumed that

$$m^2 \equiv \left(\frac{\partial^2 U}{\partial \phi^2} \right)_{\phi=0} \gg \ell^{-2}, \tag{15.37}$$

in which case the barrier separating the stable and metastable vacuum is comparable with U_0. The difficulty with the resulting picture concerns bubbles of true vacuum expanding in the false vacuum. It was shown that two bubbles formed at the same time and separated by a distance greater than 2ℓ would never touch, and the phase transition could not be completed because the exponentially expanding universe would never be covered by bubbles of the new phase. Furthermore, the bubble collisions that did occur would generate a universe that is too inhomogeneous. Thus the original inflationary theory found no graceful exit from the phase transition into a universe that resembles our universe.

15.4.2 New Inflation

This "graceful exit" problem is at least schematically cured in the *new inflationary universe* (Linde, 1982; Albrecht and Steinhardt, 1982), where generally $m^2 \ll \ell^{-2}$. An approximate sketch of the effective potential at the phase transition in this case is shown in Fig. 15.7. The transition may now

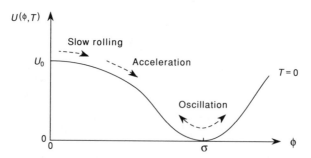

Fig. 15.7 Schematic effective potential for new inflationary cosmologies.

be viewed as a "rolling" of the system from $\phi = 0$ to $\phi = \sigma$, rather than tunneling as in Fig. 15.5. In the new inflationary universe the bubbles expand at an exponential rate for a time after they are formed, as the system slowly rolls down the almost flat part of the potential.[‡] Our present universe is part of a *single bubble,* which expanded from a size of about 10^{-25} cm to greater than 10^{20} cm in a period $\tau \simeq 10^{-31}$ sec (Steinhardt, 1986). The currently observable universe was about 10 cm in diameter then, so it is part of a single bubble and there are no inhomogeneities resulting from wall collisions.

Unlike in the original inflationary model, the phase transition in new inflation can be completed when the field ϕ finally rolls to the true minimum σ in Fig. 15.7 and begins to oscillate about it. Then the transformation of vacuum energy to thermal energy occurs rapidly through the formation of massive ($\sim 10^{14}$ GeV) Higgs particles, which correspond to quantized oscillations of the ϕ field about the minimum at σ. These subsequently decay to more familiar particles. The reheated and repopulated universe then evolves according to the standard hot big bang (Fig. 15.6). It is necessary that this mechanism reheat the universe to a temperature $T \geqslant 10^{11}$ GeV (perhaps higher) to ensure that the subsequent evolution recovers the successes of the hot big bang; in particular, reheat temperatures that are too low could jeopardize the production of a baryon asymmetry by the mechanism of §15.2.

The new inflationary universe provides a schematic resolution of the difficulties previously discussed for the standard cosmology. All points in the initial bubble lie within the event horizon and are causally connected; with inflation this causally related domain expands beyond the known universe, which solves the horizon problem. The density fluctuations inside a bubble are small immediately after its formation, and the exponential expansion of this uniformity explains the large-scale homogeneity and isotropy of the universe. After the phase transition and subsequent reheating the size of the universe is $R > 10^{20}$ cm and the temperature is $T \simeq 10^{14}$ GeV, generating an

---

[‡]This special kind of phase transition is called a *slow rollover transition.* The original inflationary models attempted to produce inflation with the same Higgs field used to spontaneously break the GUT (see Fig. 15.5). However, the generation of slow rollover transitions with mass density perturbations small enough to agree with the present understanding of galaxy formation requires an exquisitely flat potential energy function for the scalar field; such behavior does not come easily in the standard particle physics models (see §15.4.3). Accordingly, in many newer models inflation is driven by special scalar fields that are weakly coupled singlets under the gauge group. [In *chaotic inflation* the scalar field need not even be associated with high temperature phase transitions (Linde, 1983).] The drawback for such approaches is the need to justify these fields from the particle physics perspective. This has motivated various attempts to link inflation-driving scalar fields to supersymmetries, axions, higher-dimensional geometries, and other phenomena having an independent theoretical justification in elementary particle physics. An introduction to many of these ideas may be found in Abbott and Pi (1986), Chs. 4–8.

entropy

$$S \simeq (RT)^3$$

that accounts for the flatness of the present universe (Exercise 15.1b). Finally, primordial monopoles or other topological beasts such as domain walls are created at the interface of regions evolving toward different spontaneously broken Higgs minima [see Ch. 17 of Collins, Martin, and Squires (1989) for an introduction to such objects and their significance in cosmology]. Since our universe is deep within a single bubble, there is little chance for the occurrence of primordial monopoles or domain walls; even if some were formed the exponential expansion would lower their densities to acceptable levels.

15.4.3 Fluctuations and the Growth of Structure

Finally, we may note that the inflationary paradigm supplies a plausible first-principles explanation of the energy-density fluctuations responsible for galaxy formation (Brandenberger, 1985; Abbott and Pi, 1986). The required density perturbations originate in the zero-point fluctuations of the quantized fields: the wavelengths of these microscopic fluctuations are stretched to (literally) astronomical scales by the rapid inflation. As we have seen, if this inflationary period were sufficiently long a region initially in a single horizon volume could expand to a size much larger than the region corresponding to the present universe, which would solve the horizon problem. In addition, a fluctuation could be expanded *outside* the horizon during inflation and then *reenter* the horizon at a much later stage in standard-model cosmic evolution to provide a seed for galaxy formation. (In Exercise 15.2c it is demonstrated that this would be impossible in the usual cosmology.)

Unlike in the standard big bang, these seed fluctuations originate inside the causal horizon. In addition, they have a *scale-invariant spectrum of density fluctuations,* which is the spectrum in vogue for galaxy formation (see Abbott and Pi, 1989, Ch. 4). However, obtaining even the proper order of magnitude for the density perturbations is difficult, and generally requires fine-tuning of the particle physics models. Thus it may be argued that to a certain extent the present versions of inflation merely transfer unphysical assumptions and fine-tuning from the cosmological to elementary particle physics sector.

15.4.4 Prospects for Inflationary Theories

Difficulties such as the failure to reproduce the correct size for density fluctuations without unnatural parameter adjustment, or the introduction of special fields, indicate that the specific models of inflation that have been proposed are probably not acceptable. Nevertheless, the inflationary paradigm is economical and powerful in its treatment of long-standing cosmological difficulties, which suggests that the essential idea may be correct but the details remain to be understood.

There are many important aspects of inflation that we have yet to touch on: supersymmetric models, chaotic inflation, and possible connections with higher-dimensional geometry. In addition, we have not discussed such important topics as neutrino masses in GUTs and cosmology, the solar neutrino problem, or dark matter in the universe. However books—no less than universes—are prone to inflation, so we reluctantly leave these matters for the reader to pursue independently. Those so inclined will find excellent discussions of these and other interesting topics in the references listed below.

15.5 Background and Further Reading

Extensive and pedagogical discussions of the interface between particle physics and cosmology have been given recently by Kolb (1987); Collins, Martin, and Squires (1989); and Kolb and Turner (1990). The introductory material on the standard cosmology is from Kolb and Turner (1983), Kolb (1987), and Turner and Schramm (1979). The baryogenesis discussion is based on Kolb and Turner (1983) and Turner and Schramm (1979). Inflationary cosmologies are described in Sasaki (1983), Brandenberger (1985), Steinhardt (1984, 1986), Pi (1985), Olive and Schramm (1985), Blau and Guth (1987), and Olive (1990). Finally, Abbott and Pi (1986) reprint many of the original papers relevant to inflation and provide clear introductions to their significance.

APPENDICES

APPENDIX A

Natural Units

The use of $\hbar=c=1$ units (*natural units*) can simplify particle physics notation considerably. Since one typically deals with particles that are both relativistic and quantum mechanical, a multitude of $\hbar$'s and c's will encumber the equations if natural units are not adopted.

Let us consider a few examples of how this works. Set $\hbar=c=1$; since

$$[c] = [L][T]^{-1}$$

(where the symbol $[\ ]$ means "the dimension of"), we have $[L] = [T]$, and since $E^2 = \mathbf{p}^2 + M^2 c^4$ we find

$$[E] = [M] = [\mathbf{p}] = [\mathbf{k}], \tag{A.1}$$

where $\mathbf{p} = \hbar\mathbf{k}$. Because $[\hbar] = [M][L]^2[T]^{-1}$, setting $\hbar=c=1$ yields

$$[M] = [L]^{-1} = [T]^{-1}. \tag{A.2}$$

Hence $[M]$ can be chosen as the single independent dimension of our set of natural units.

For electromagnetic interactions it is convenient to introduce a set of units in which the MKS constants ϵ_0 and μ_0 are set to unity, and factors of 4π are expunged from the fields, appearing in the forces instead. This is done in the rationalized Heaviside–Lorentz system of units, which is common in high energy physics. Further, with natural units $c = 1$, so the c's would not appear in the equations of electrodynamics. The Maxwell equations in these units are given by eqns. (2.112).

The fine structure constant is a dimensionless ratio of the electrostatic repulsion between two electrons separated by one Compton wavelength and the electron rest mass:

$$\alpha = \frac{1}{4\pi}\left(\frac{e^2}{\hbar/mc}\right)\left(\frac{1}{mc^2}\right) = \frac{e^2}{4\pi\hbar c} \simeq \frac{1}{137}$$

($1/137$ corresponds to the asymptotic value of the running coupling constant; see §6.4.4). Therefore, $e = (4\pi\alpha)^{1/2}$ in natural units.

511

Example: The pion Compton wavelength in natural units is

$$\lambda_\pi = \frac{\hbar}{M_\pi c} \rightarrow \frac{1}{M_\pi} \simeq (140 \text{ MeV})^{-1}. \qquad (A.3)$$

We may convert this to conventional units by multiplying with a combination of $\hbar$ and c to give a distance unit. Since $\hbar c = 197.3$ MeV $\cdot$ fm (with 1 fm $\equiv 1 \times 10^{-13}$ cm), in natural units we have

$$1 \text{ fm} = \frac{1}{197.3} \text{ MeV}^{-1} = 5.068 \text{ GeV}^{-1}$$

$$\qquad (A.4)$$

$$1 \text{ fm}^{-1} = 197.3 \text{ MeV} \qquad 1 \text{ GeV} = 5.068 \text{ fm}^{-1}.$$

Hence

$$\lambda_\pi = \left(\frac{1}{140} \text{ MeV}^{-1}\right)(197.3 \text{ MeV} \cdot \text{fm}) = 1.41 \text{ fm}.$$

For cross sections we have the dimensions

$$[\sigma] = [L]^2 = [M]^{-2},$$

and from (A.4), 1 fm$^2 = 25.7$ GeV^{-2} in natural units. But 1 b $= 10^{-24}$ cm$^2 = 100$ fm^2 and 1 mb $= 0.1$ fm^2, so

$$1 \text{ GeV}^{-2} = 0.3894 \text{ mb} \qquad 1 \text{ mb} = 2.568 \text{ GeV}^{-2} \qquad 1 \text{ fm}^2 = 10 \text{ mb}.$$

Example: A typical hadronic cross section is of order

$$\sigma \simeq \lambda_\pi^2 \simeq \frac{1}{M_\pi^2} \simeq \frac{1}{(140)^2} \text{ MeV}^{-2}.$$

If σ is to be in fm^2, we must multiply by a combination of $\hbar$ and c with units MeV$^2 \cdot$ fm^2. This is just $\hbar^2 c^2 = (197.3)^2$ MeV$^2 \cdot$ fm^2 and

$$\sigma = \frac{\hbar^2 c^2}{(140)^2 \text{ MeV}^2} \simeq 2 \text{ fm}^2 \simeq 20 \text{ mb}.$$

Example: In natural units the mean lifetime for the decay $\Sigma^0 \rightarrow \Lambda + \gamma$ is (Sakurai, 1967, §4.1)

$$\tau \simeq \frac{\pi (M_\Lambda + M_\Sigma)^2}{e^2 E_\gamma^3}.$$

Since $[M] = [E]$, this has dimension $[M]^{-1}$ and we must multiply the right side by $\hbar = 6.58 \times 10^{-22}$ MeV $\cdot$ sec to make it dimensionally correct. Using experimental values $(M_\Lambda + M_\Sigma) = 2307$ MeV, $E_\gamma = 74.5$ MeV, and $\pi/e^2 = 1/4\alpha = 137/4$,

$$\tau = \frac{137}{4} \frac{(2307 \text{ MeV})^2}{(74.5 \text{ MeV})^3} \left(6.58 \times 10^{-22} \text{ MeV} \cdot \text{sec}\right) \simeq 2.9 \times 10^{-19} \text{ sec}.$$

It is important to know the dimensions of various field operators in these natural units. Assume $d = 4$ spacetime dimensions; the Lagrangian has the dimension of mass

$$[\text{Lagrangian}] = [M], \tag{A.5}$$

and the Lagrangian density then has units

$$[\mathcal{L}] = [M][L]^{-3} = [M]^4. \tag{A.6}$$

The Hamiltonian and Hamiltonian density have the same dimensions as the Lagrangian and Lagrangian density, respectively, and the action is dimensionless in natural units. From the free field equations (see §2.3), we can then infer the dimensions of various fields. For a spinor field

$$[\psi] = [M]^{3/2}, \tag{A.7}$$

for scalar fields

$$[\phi] = [M], \tag{A.8}$$

and for the photon and massive vector fields

$$[A_\mu] = [M]. \tag{A.9}$$

Likewise, for derivative or covariant derivative operators

$$[\partial^\mu] = [D^\mu] = [M]. \tag{A.10}$$

Using these dimensions and (A.6), we can find the dimension of any coupling constant appearing in the interaction Lagrangian density (see Exercise 6.4).

Example: Consider the Skyrme Lagrangian density (13.94), for which ε^2 is dimensionless and $[f_\pi] = [M]$. From the above considerations,

$$[\sigma] = [M] \quad \text{(scalar field)} \qquad [\partial_\mu] = [M].$$

Therefore U and $U^\dagger$ are dimensionless, $[L_\mu] = [M]$, and $[\tilde{\mathcal{L}}] = [M]^4$, as befits a Lagrangian density.

The preceding equations assume spacetime has $d = 4$ dimensions. Examples where $d \neq 4$ (see §6.4.5) are considered in Exercises 6.3–6.4.

In cosmology one often sets $\hbar = c = k_B = 1$, where k_B is the Boltzmann constant. Then,

$$1 \text{ GeV} = 1.2 \times 10^{13} \text{ K}, \tag{A.11}$$

and the gravitational constant G is

$$G = \frac{1}{M_P^2}, \tag{A.12}$$

where the Planck mass is

$$M_P = 1.2 \times 10^{19} \text{ GeV}. \tag{A.13}$$

From (A.4) the corresponding Planck length is

$$\ell_P = \frac{1}{M_P} = 1.6 \times 10^{-33} \text{ cm}. \tag{A.14}$$

Multiplying by $1/c$ gives the corresponding Planck time

$$t_P = 5.4 \times 10^{-44} \text{ sec}, \tag{A.15}$$

and using (A.11) in (A.13) gives the Planck temperature

$$T_P = 1.4 \times 10^{32} \text{ K}. \tag{A.16}$$

APPENDIX B

Hadronic Properties

In this appendix we list some basic properties of selected baryons and mesons. A more extensive listing may be found in the data compilation of Hernández et al. (1990).

Table B.1
Properties of Selected Baryons

Particle	Mass (MeV)	J^P	Valence Quarks	Flavor $SU(3)$
p	938.3	$\frac{1}{2}^+$	uud	8
n	939.6	$\frac{1}{2}^+$	udd	8
Σ^+	1189.4	$\frac{1}{2}^+$	uus	8
Σ^0	1192.5	$\frac{1}{2}^+$	uds	8
Λ	1115.6	$\frac{1}{2}^+$	uds	8
Σ^-	1197.4	$\frac{1}{2}^+$	dds	8
Ξ^0	1314.9	$\frac{1}{2}^+$	uss	8
Ξ^-	1321.3	$\frac{1}{2}^+$	dss	8
Δ^{++}	1230.6	$\frac{3}{2}^+$	uuu	10
Δ^+	1234.9	$\frac{3}{2}^+$	uud	10
Δ^0	1232.5	$\frac{3}{2}^+$	udd	10
Δ^-	1232	$\frac{3}{2}^+$	ddd	10
$\Sigma^+(1385)$	1382.8	$\frac{3}{2}^+$	uus	10
$\Sigma^0(1385)$	1383.7	$\frac{3}{2}^+$	uds	10
$\Sigma^-(1385)$	1387.2	$\frac{3}{2}^+$	dds	10
$\Xi^0(1530)$	1531.8	$\frac{3}{2}^+$	uss	10
$\Xi^-(1530)$	1535.0	$\frac{3}{2}^+$	dss	10
Ω^-	1672.4	$\frac{3}{2}^+$	sss	10

Table B.2
Properties of Selected Mesons

Particle	Mass (MeV)	J^P	Valence Quarks	Flavor $SU(3)$
π^+, π^-	139.6	0^-	$u\bar{d}\,(d\bar{u})$	**8**
π^0	135.0	0^-	$(u\bar{u} - d\bar{d})/\sqrt{2}$	**8**
K^+, K^-	493.7	0^-	$u\bar{s}\,(s\bar{u})$	**8**
$K^0, \overline{K^0}$	497.7	0^-	$d\bar{s}\,(s\bar{d})$	**8**
η	548.8	0^-	$(u\bar{u} + d\bar{d})/\sqrt{2}$	**8**
η'	957.5	0^-	$s\bar{s}$	**1**
ρ^+, ρ^-	770.0	1^-	$u\bar{d}\,(d\bar{u})$	**8**
ρ^0	770.0	1^-	$(u\bar{u} - d\bar{d})/\sqrt{2}$	**8**
K^{*+}, K^{*-}	892.1	1^-	$u\bar{s}\,(s\bar{u})$	**8**
$K^{*0}, \overline{K}^{*0}$	896.2	1^-	$d\bar{s}\,(s\bar{d})$	**8**
ω	782.0	1^-	$(u\bar{u} + d\bar{d})/\sqrt{2}$	**8**
ϕ	1019.4	1^-	$s\bar{s}$	**1**
D^+, D^-	1869.3	0^-	$c\bar{d}\,(d\bar{c})$	$\bar{\mathbf{3}}\,(\mathbf{3})$
$D^0, \overline{D}^0$	1864.6	0^-	$c\bar{u}\,(u\bar{c})$	$\bar{\mathbf{3}}\,(\mathbf{3})$
D_s^+, D_s^-	1969.4	0^-	$c\bar{s}\,(s\bar{c})$	$\bar{\mathbf{3}}\,(\mathbf{3})$
B^+, B^-	5271.2	0^-	$u\bar{b}\,(b\bar{u})$	$\mathbf{3}\,(\bar{\mathbf{3}})$
$B^0, \overline{B}^0$	5275.2	0^-	$d\bar{b}\,(b\bar{d})$	$\mathbf{3}\,(\bar{\mathbf{3}})$
η_c	2980	0^-	$c\bar{c}$	**1**
J/ψ	3096.9	1^-	$c\bar{c}$	**1**
Υ	9460.3	1^-	$b\bar{b}$	**1**

APPENDIX C

Relativistic Kinematic Variables

The momentum of a system may be decomposed into longitudinal and transverse components with respect to an arbitrary axis:

$$p_L = p\cos\theta \qquad p_T = p\sin\theta. \tag{C.1}$$

It is common to use a set of kinematic variables s, p_T, and y where $\sqrt{s}$ is the center-of-mass energy, p_T is the transverse momentum, and the *rapidity* y is defined by

$$y = \frac{1}{2}\log\left(\frac{E+p_L}{E-p_L}\right) = \tanh^{-1}\beta_L, \tag{C.2}$$

where $E^2 = \mathbf{p}^2 + m^2$ and $\beta_L \equiv v_L/c$ (longitudinal velocity). For small β_L we have $y \simeq v_L$, so rapidity is a relativistic analog of velocity. A Lorentz boost along the longitudinal axis from a frame K to a frame K' changes the rapidity in a simple way:

$$y' = y + \tanh^{-1}\beta, \tag{C.3}$$

where β is the velocity of K' with respect to K. Thus distributions expressed in terms of y have shapes that are invariant under such boosts.

The *pseudorapidity* η is defined by

$$\eta = -\log\left(\tan\frac{\theta}{2}\right). \tag{C.4}$$

From (C.2), if the masses are negligible,

$$y \simeq \frac{1}{2}\log\left(\frac{p+p_L}{p-p_L}\right) = \frac{1}{2}\log\left(\frac{1+\cos\theta}{1-\cos\theta}\right)$$

$$= -\log\left(\tan\frac{\theta}{2}\right) = \eta, \tag{C.5}$$

so at high energy the rapidity and pseudorapidity are equivalent. Pseudorapidity is useful because it can be determined directly from the production angle θ measured with respect to the beam axis in an experiment.

Other kinematic variables in common use for relativistic problems include the Mandelstam variables discussed in Exercise 3.5d, and the deep inelastic scattering variables defined in conjunction with eqns. (10.10) and (10.15).

APPENDIX D

Exercise Solutions

CHAPTER 1

1.1 Using (1.38) and (1.39),

$$x'_\mu x'^\mu = \alpha^\lambda_\mu \alpha^\mu_\epsilon x_\lambda x^\epsilon = x_\mu x^\mu,$$

which implies that $\alpha^\lambda_\mu \alpha^\mu_\epsilon = \delta^\lambda_\epsilon$. In matrix notation the scalar product (1.28) is $a \cdot b = a^{\mathrm{T}} g b$, where a^{T} denotes the transpose of the matrix a and g is the matrix (1.21). Thus, (1.38) is $x^{\mathrm{T}} g x = (x')^{\mathrm{T}} g x'$, or on employing (1.42),

$$x^{\mathrm{T}} g x = x^{\mathrm{T}} A^{\mathrm{T}} g A x,$$

which means that $A^{\mathrm{T}} g A = g$. Take the determinant of both sides and use

$$\mathrm{Det}\,(ab) = (\mathrm{Det}\,a)(\mathrm{Det}\,b) \qquad \mathrm{Det}\,g = -1 \neq 0 \qquad \mathrm{Det}\,a = \mathrm{Det}\,a^{\mathrm{T}},$$

to show that $\mathrm{Det}\,A = \pm 1$.

1.2 From eqns. (1.12), (1.13), (1.17), and (1.43)

$$A \cdot B \equiv A^\mu B_\mu \qquad A'^\mu = \alpha^\mu_\nu A^\nu \qquad B'_\mu = \alpha^\lambda_\mu B_\lambda$$
$$A'^\mu B'_\mu = \alpha^\mu_\nu \alpha^\lambda_\mu A^\nu B_\lambda = \delta^\lambda_\nu A^\nu B_\lambda = A^\mu B_\mu = A \cdot B,$$

so the scalar product is Lorentz invariant. The transformation to a new differential volume element is given by $d^4 x' = J d^4 x$, where J is the Jacobian determinant of the transformation matrix. But for proper Lorentz transformations $J = \mathrm{Det}\,A = 1$ [see (1.42) and Exercise 1.1], and $d^4 x' = d^4 x$.

1.3 (a) $\Box = \partial_\mu \partial^\mu \qquad \partial'_\mu = \alpha^\nu_\mu \partial_\nu \qquad \Box' = \partial'_\mu \partial'^\mu = \alpha^\nu_\mu \alpha^\mu_\lambda \partial_\nu \partial^\lambda = \partial_\mu \partial^\mu = \Box.$
(b) From (1.13),

$$\partial'_\mu A'_\nu = \alpha^\lambda_\mu \alpha^\epsilon_\nu \partial_\lambda A_\epsilon,$$

which is the transformation law of a covariant rank-2 tensor [eq. (1.15)]. Thus, $F_{\mu\nu}$ is a rank-2 tensor, and it is clearly antisymmetric: $F_{\mu\nu} = -F_{\nu\mu}$.

1.4 (a) The KG equation and its complex conjugate are (1.58)

$$(\Box + m^2)\phi = 0 \qquad (\Box + m^2)\phi^* = 0.$$

Multiply the first from left by ϕ^*, the second from left by ϕ, and subtract

$$\phi^* \frac{\partial^2}{\partial x_0 \partial x^0}\phi - \phi^* \boldsymbol{\nabla}^2\phi - \phi\frac{\partial^2}{\partial x_0 \partial x^0}\phi^* + \phi\boldsymbol{\nabla}^2\phi^* = 0.$$

But $\partial/\partial x_0 = \partial/\partial x^0 = \partial/\partial t$, and using the definitions (1.63)–(1.64) for ρ and **j** we arrive at (1.62). From (1.31) and (1.32), (1.65)–(1.66) are equivalent to (1.62)–(1.64). (b) From (1.63) for plane waves

$$\phi \simeq e^{-ip\cdot x} = e^{i\mathbf{p}\cdot\mathbf{x}}e^{-iEt} \qquad \rho = \pm 2\sqrt{\mathbf{p}^2 + m^2}\,\phi^*\phi,$$

so ρ can be positive or negative.

1.5 From $E^2 = \mathbf{p}^2 + m^2$ and (1.72)

$$H^2\psi = \left(\mathbf{p}^2 + m^2\right)\psi$$
$$(\boldsymbol{\alpha}\cdot\mathbf{p} + \beta m)\,(\boldsymbol{\alpha}\cdot\mathbf{p} + \beta m)\,\psi = \left(\mathbf{p}^2 + m^2\right)\psi$$
$$\left(\alpha^i\alpha^j p^i p^j + m\left(\alpha^i\beta + \beta\alpha^i\right)p^i + m^2\beta^2\right)\psi = \left(\mathbf{p}^2 + m^2\right)\psi.$$

Equating terms on the two sides of the equation requires

$$\{\alpha^i, \alpha^j\} = 2\delta_{ij} \qquad \{\alpha^i, \beta\} = 0 \qquad \beta^2 = 1,$$

which are (1.73)–(1.75). These equations and the definitions (1.85) are equivalent to (1.86), as may be shown by judicious use of $\gamma^0\gamma^0 = 1$. From (1.88)–(1.89)

$$\sigma^{ij} = \frac{-i}{2}\begin{pmatrix} [\sigma_i, \sigma_j] & 0 \\ 0 & [\sigma_i, \sigma_j] \end{pmatrix} = \begin{pmatrix} \epsilon_{ijk}\sigma_k & 0 \\ 0 & \epsilon_{ijk}\sigma_k \end{pmatrix},$$

where $i, j = 1, 2, 3$, and where the Pauli commutator $[\sigma_i, \sigma_j] = 2i\epsilon_{ijk}\sigma_k$ has been used. From (1.89), (1.85), and (1.86), $\sigma^{0i} = i\alpha^i$, which is (1.92). By inspection the Pauli matrices (1.78) are Hermitian: $\sigma_i = \sigma_i^\dagger$. For the energy eigenvalues of (1.72) to be real, α_i and β must be Hermitian. The properties of the other Dirac matrices under Hermitian conjugation follow from this, (1.85), and (1.73)–(1.75) or (1.86). For example (see Exercise 1.6a),

$$\left(\gamma^i\right)^\dagger = \left(\beta\alpha^i\right)^\dagger = \left(\alpha^i\right)^\dagger\beta^\dagger$$
$$= \alpha^i\beta = -\beta\alpha^i = -\gamma^i.$$

Alternatively, one may Hermitian conjugate the explicit matrices in the representation (1.79) and (1.88)–(1.93), recalling that the Hermitian conjugate of a matrix is obtained by transposition and complex conjugation of the elements.

1.6 (a) These may be proved by writing the matrix product for two matrices out explicitly, and then generalizing the result to three matrices, four matrices, ... (b) From Table 1.2 we have $\gamma_5\gamma_\mu = -\gamma_\mu\gamma_5$ and $\gamma_5^2 = 1$. Consider a product of an odd number of γ-matrices: $\Gamma \equiv \gamma_\alpha\gamma_\beta\gamma_\gamma\ldots\gamma_\epsilon$. Then $\mathrm{Tr}\,(\gamma_5\Gamma\gamma_5) = -\,\mathrm{Tr}\,\Gamma$. But by the cyclic property of the trace

$$\mathrm{Tr}\,(\gamma_5\Gamma\gamma_5) = \mathrm{Tr}\,(\gamma_5\gamma_5\Gamma) = \mathrm{Tr}\,\Gamma.$$

Thus $\mathrm{Tr}\,\Gamma = -\,\mathrm{Tr}\,\Gamma$, which can be satisfied only if $\mathrm{Tr}\,\Gamma = 0$. (c) From the cyclic property of the trace (Exercise 1.6a)

$$
\begin{aligned}
\mathrm{Tr}\,(A\!\!\!/\,B\!\!\!/) &= \tfrac{1}{2}\,\mathrm{Tr}\,(A\!\!\!/\,B\!\!\!/ + B\!\!\!/\,A\!\!\!/) \\
&= \tfrac{1}{2}\,\mathrm{Tr}\,(\gamma_\mu\gamma^\nu A^\mu B_\nu + \gamma^\nu\gamma_\mu B_\nu A^\mu) \\
&= A\cdot B\,\mathrm{Tr}\,1 \\
&= 4A\cdot B,
\end{aligned}
$$

where (1.87) has been used.

1.7 From (1.105) and (1.114), under Lorentz transformations

$$\overline{\psi}'\psi' = \overline{\psi}S^{-1}S\psi = \overline{\psi}\psi.$$

Therefore, $\overline{\psi}\psi$ is a scalar [note that $\overline{\psi}\psi$, not $\psi^\dagger\psi$, is the scalar; $\psi^\dagger\psi$ is the timelike component of a 4-vector—see (1.115)–(1.118)]. Under proper Lorentz transformations

$$\overline{\psi}'\gamma_5\psi' = \overline{\psi}\,S^{-1}\gamma_5 S\psi = \overline{\psi}\gamma_5\psi,$$

where we have used $[S,\gamma_5] = 0$ for proper transformations (see Bjorken and Drell, 1964, §2.4). Under parity P we have

$$\overline{\psi}P^{-1}\gamma_5 P\psi = -\overline{\psi}\gamma_5\psi,$$

where $\gamma^\mu\gamma_5 + \gamma_5\gamma^\mu = 0$ and $P \sim \gamma^0$ (Table 1.2 and Exercise 1.8) have been used. Hence $\overline{\psi}\gamma_5\psi$ is a pseudoscalar. Finally,

$$\overline{\psi}'\sigma^{\mu\nu}\psi' = \overline{\psi}S^{-1}\sigma^{\mu\nu}S\psi.$$

Using (1.89), inserting factors $SS^{-1} = 1$, and employing (1.110) we find

$$\overline{\psi}'\sigma^{\mu\nu}\psi' = \alpha^\mu_\lambda\alpha^\nu_\beta\overline{\psi}\sigma^{\lambda\beta}\psi,$$

which is the transformation law (1.14) for a second-rank tensor.

1.8 (a) With α^ν_μ given by (1.45) and using the γ-matrix properties in Table 1.2, we find $S = \text{phase} \times \gamma^0$ satisfies (1.110). (b) Apply the intrinsic parity operator $\gamma_0 = \beta$ in the representation (1.79) to the rest spinors (1.121)–(1.122), demonstrating immediately that the positive-energy solutions have positive parity and the negative-energy solutions have negative parity. (c) See Bjorken and Drell (1964), §5.2 and work in the Pauli–Dirac representation. From (1.101) the Dirac equation for a particle of charge q is $(i\partial\!\!\!/-qA\!\!\!/-m)\psi = 0$, and for the $(q \to -q)$ antiparticle, $(i\partial\!\!\!/ + qA\!\!\!/ - m)\psi_c = 0$. Complex conjugate the first equation and multiply it by $C\gamma_0$, where the matrix C satisfies

$$-(C\gamma_0)\gamma^{\mu^*} = \gamma^\mu(C\gamma_0)$$

[which implies $\gamma^\mu = -C(\gamma^\mu)^{\mathrm{T}}C^{-1}$] to obtain

$$(i\partial\!\!\!/ + qA\!\!\!/ - m)\psi_c = 0 \qquad \psi_c \equiv C\gamma_0\psi^* = C\overline{\psi}^{\mathrm{T}}.$$

Thus the particle wavefunction ψ is converted to an antiparticle wavefunction by complex conjugating and multiplying by $C\gamma_0$. As may be verified by explicit multiplication, in the Pauli–Dirac representation a suitable choice (up to an arbitrary phase) is $C\gamma_0 = i\gamma^2$. This allows definition of an invariance (*charge conjugation*) for the Dirac theory: the sequence of operations (1) take the complex conjugate, (2) multiply by $C\gamma_0 = i\gamma^2$, and (3) replace $A_\mu \to -A_\mu$ interconverts particle and antiparticle Dirac equations. The action on the spinors (1.123) is easily found by direct matrix multiplication in the Pauli–Dirac representation. For example,

$$\left(\psi_+^{(1)}\right)_c = C\gamma^0 \left(u^{(1)}e^{-ip\cdot x}\right)^*$$
$$= i\gamma^2 u^{(1)^*} e^{ip\cdot x} = v^{(1)}e^{ip\cdot x} = \psi_-^{(1)}.$$

(d) Apply the matrices (Pauli–Dirac representation)

$$C\gamma^0 = i\gamma^2 = \begin{pmatrix} 0 & 0 & 0 & 1 \\ 0 & 0 & -1 & 0 \\ 0 & -1 & 0 & 0 \\ 1 & 0 & 0 & 0 \end{pmatrix} \qquad \frac{\hbar}{2}\Sigma_3 = \begin{pmatrix} 1 & 0 & 0 & 0 \\ 0 & -1 & 0 & 0 \\ 0 & 0 & 1 & 0 \\ 0 & 0 & 0 & -1 \end{pmatrix}$$

to the rest spinors (1.121)–(1.122). Thus show that

$$\psi_\pm^{(1)} \leftrightarrow \text{spin up} \qquad \psi_\pm^{(2)} \leftrightarrow \text{spin down},$$

but after charge conjugation

$$\left(\psi_\pm^{(1)}\right)_c \leftrightarrow \text{spin down} \qquad \left(\psi_\pm^{(2)}\right)_c \leftrightarrow \text{spin up}.$$

This reversal of sign is the reason for the "backward" label correlation $(1 \leftrightarrow 4$ and $2 \leftrightarrow 3)$ in the definition (1.123e). Physically, this can be understood in terms of Dirac's hole theory for antiparticles (§1.6.1): the absence of a spin-up particle is equivalent to the presence of a spin-down particle. The rather confusing subject of the relationship between negative-energy spinors, positive-energy antiparticles, and Dirac hole theory probably is discussed best in Sakurai (1967), §3–9 and §3–10.

1.9 (a) This is most easily seen by multiplication of the spinors (1.123). For example, omitting the normalization,

$$\bar{u}^{(1)}(\mathbf{p})v^{(1)}(\mathbf{p}) = u^{(1)\dagger}(\mathbf{p})\gamma^0 v^{(1)}(\mathbf{p})$$

$$= \left(1 \; 0 \; \frac{p_3}{\lambda} \; \frac{p_-}{\lambda}\right) \begin{pmatrix} 1 & 0 & 0 & 0 \\ 0 & 1 & 0 & 0 \\ 0 & 0 & -1 & 0 \\ 0 & 0 & 0 & -1 \end{pmatrix} \begin{pmatrix} p_-/\lambda \\ -p_3/\lambda \\ 0 \\ 1 \end{pmatrix} = 0,$$

where $\lambda \equiv E + m$. A more general proof may be found in Itzykson and Zuber (1980), §2–2–1. (b) The probability of finding a free fermion of mass m and energy E in a box of volume V is $\int \rho dV = \psi^\dagger \psi V = u^\dagger u V$, where (1.123) was used. The Lorentz contracted volume for a frame in which the particle has energy E is $V = V_0 m/E$, where V_0 is the box volume in the particle restframe. Hence, using (1.125c)

$$\int \rho dV = u^\dagger u V_0 \frac{m}{E} = \left(\frac{E}{m}\right)\left(\frac{m}{E}\right)V_0 = V_0,$$

and the normalization is independent of the Lorentz frame. Notice that a normalization such as $u^\dagger u = 1$ is not covariant: the right side is a scalar but the left side is the timelike component of a 4-vector.

(c)
$$\boldsymbol{\sigma} \cdot \mathbf{p} = \sigma_1 \cdot p_1 + \sigma_2 \cdot p_2 + \sigma_3 \cdot p_3 = \begin{pmatrix} p_3 & p_1 - ip_2 \\ p_1 + ip_2 & -p_3 \end{pmatrix},$$

where (1.78) was used. Then by matrix multiplication,

$$(\boldsymbol{\sigma} \cdot \mathbf{p})^2 = \mathbf{p}^2 \begin{pmatrix} 1 & 0 \\ 0 & 1 \end{pmatrix}.$$

(d) Use the Dirac equation to show that

$$u_B = \boldsymbol{\sigma} \cdot \mathbf{p} \, \frac{u_A}{(E + m)}.$$

Then using $(\boldsymbol{\sigma} \cdot \mathbf{p})^2 = \mathbf{p}^2$ and $E \simeq m$, show that $|u_B|/|u_A| \simeq v/c$. For negative energy solutions the situation is reversed: the upper components are small and the lower components are large.

1.10 From the Feynman recipe $p_i = \left(-E^{(-)}, -\mathbf{p}_i\right)$ and $p_f = \left(E^{(+)}, \mathbf{p}_f\right)$. Considering only the time-dependent part of the plane waves, the amplitude in first-order perturbation theory is

$$a_{fi} \simeq \int dt\, e^{iE_f t} V(t) e^{-iE_i t} = \int dt\, e^{i\left(E^{(+)} + E^{(-)} - \omega\right)t}$$

$$= 2\pi\delta\left(E^{(+)} + E^{(-)} - \omega\right),$$

which implies that $\hbar\omega = E^{(+)} + E^{(-)}$.

1.11 (a) The KG equation is $(\partial^\mu \partial_\mu + m^2)\phi = 0$. Making the gauge invariant replacement $\partial^\mu \rightarrow \partial^\mu + ieA^\mu$ gives (1.139)–(1.140). (b) Integrating by parts

$$\int dt\, e^{ip_f \cdot x} \partial_0 (A^0 e^{-ip_i \cdot x}) = e^{ip_f \cdot x} A^0 e^{-ip_i \cdot x}\Big|_{-\infty}^{+\infty} - i(p_f)_0 \int dt\, e^{ip_f \cdot x} A^0 e^{-ip_i \cdot x}$$

$$= -i(p_f)_0 \int dt\, e^{ip_f \cdot x} A^0 e^{-ip_i \cdot x},$$

where $A_0 \rightarrow 0$ as $t \rightarrow \pm\infty$ was used. By an analogous procedure,

$$\int d^3x\, e^{ip_f \cdot x} \mathbf{\nabla} \cdot (\mathbf{A} e^{-ip_i \cdot x}) = i\mathbf{p}_f \cdot \int d^3x\, e^{ip_f \cdot x} \mathbf{A} e^{-ip_i \cdot x},$$

while by explicit differentiation

$$\int d^3x\, dt\, e^{ip_f \cdot x} A^\mu \partial_\mu e^{-ip_i \cdot x} = -i(p_i)_\mu \int d^3x\, dt\, e^{ip_f \cdot x} A^\mu e^{-ip_i \cdot x}.$$

These equations and (1.32) give the final result. (c) Inserting plane waves in (1.143) gives $j_\mu = e(p_f + p_i)_\mu e^{ip_f \cdot x} e^{-ip_i \cdot x}$ (with $p_1 = p_i$ and $p_3 = p_f$). Comparing with part (b), we see that (1.42) is equivalent to (1.41) with the definition (1.43).

CHAPTER 2

2.1 From eq. (2.1)

$$\delta S = \int_{t_1}^{t_2} dt \left(\frac{\partial L}{\partial q} \delta q + \frac{\partial L}{\partial \dot{q}} \delta \dot{q} \right).$$

Integrate the second term by parts:

$$\int_{t_1}^{t_2} dt \frac{\partial L}{\partial \dot{q}} \delta \dot{q} = \frac{\partial L}{\partial \dot{q}} \delta q \Big|_{t_1}^{t_2} - \int_{t_1}^{t_2} dt \delta q \frac{d}{dt} \frac{\partial L}{\partial \dot{q}}.$$

But the first term vanishes because the boundary conditions require $\delta q(t_1) = \delta q(t_2) = 0$. Thus we obtain (2.4), which can be satisfied for general variations δq only if the integrand within the parentheses vanishes; this gives (2.5).

2.2 Hamilton's principle requires (2.36), but

$$\frac{\partial \mathcal{L}}{\partial(\partial_\mu \phi)} \delta(\partial_\mu \phi) = -\partial_\mu \left(\frac{\partial \mathcal{L}}{\partial(\partial_\mu \phi)} \right) \delta\phi + \partial_\mu \left(\frac{\partial \mathcal{L}}{\partial(\partial_\mu \phi)} \delta\phi \right),$$

where $\partial_\mu(\delta\phi) = \delta(\partial_\mu \phi)$, and (2.36) can be written

$$\delta S = \int_\Omega d^4x \left[\frac{\partial \mathcal{L}}{\partial \phi} - \partial_\mu \left(\frac{\partial \mathcal{L}}{\partial(\partial_\mu \phi)} \right) \right] \delta\phi + \int_\Omega d^4x\, \partial_\mu \left(\frac{\partial \mathcal{L}}{\partial(\partial_\mu \phi)} \delta\phi \right).$$

But the last term vanishes because by (2.40) it can be written as a surface integral $\oint_S d\sigma_\mu \, (\partial \mathcal{L}/\partial(\partial_\mu \phi)) \, \delta\phi$, which makes no contribution because the variation $\delta\phi$ is assumed to vanish on the boundary. Thus

$$\delta S = \int_\Omega d^4x \left[\frac{\partial \mathcal{L}}{\partial \phi} - \partial_\mu \left(\frac{\partial \mathcal{L}}{\partial(\partial_\mu \phi)} \right) \right] \delta\phi = 0.$$

For general variations $\delta\phi$, this can be assured only if $\mathcal{L}$ obeys (2.39).

2.3 The Euler–Lagrange field equations are given by (2.39). For a complex scalar field (2.44),

$$\frac{\partial \mathcal{L}}{\partial(\partial_\mu \phi)} = \partial^\mu \phi^\dagger \qquad \frac{\partial \mathcal{L}}{\partial \phi} = -m^2 \phi^\dagger$$

$$\frac{\partial \mathcal{L}}{\partial(\partial^\mu \phi^\dagger)} = \partial_\mu \phi \qquad \frac{\partial \mathcal{L}}{\partial \phi^\dagger} = -m^2 \phi.$$

Insertion in (2.39) gives the two independent equations (2.45). For the Dirac field (2.46)

$$\frac{\partial \mathcal{L}}{\partial(\partial_\mu \bar{\psi})} = 0 \qquad \frac{\partial \mathcal{L}}{\partial \bar{\psi}} = (i\slashed{\partial} - m)\psi,$$

and from (2.39), $(i\slashed{\partial} - m)\psi = 0$, which is (1.99). For the free Maxwell field

$$\mathcal{L} = -\tfrac{1}{4} F_{\mu\nu} F^{\mu\nu} = -\tfrac{1}{4}(\partial_\mu A_\nu - \partial_\nu A_\mu) g^{\mu\alpha} g^{\beta\nu} (\partial_\alpha A_\beta - \partial_\beta A_\alpha),$$

where (2.47) and (1.29) have been used. Then

$$\frac{\partial \mathcal{L}}{\partial(\partial_\rho A_\sigma)} = -F^{\rho\sigma} \qquad \frac{\partial \mathcal{L}}{\partial A_\sigma} = 0.$$

Thus from (2.39) with ϕ replaced by A_μ we have $\partial_\mu F^{\mu\nu} = 0$, which is (2.132) for a free field. For a massive vector field the Lagrangian density is (2.49). The first term of (2.49) is the same as for a Maxwell field so $\partial \mathcal{L}/\partial(\partial_\mu A_\nu) = -F^{\mu\nu}$. From the second term,

$$\frac{\partial \mathcal{L}}{\partial A_\rho} = \tfrac{1}{2} m^2 g^{\lambda\mu} \frac{\partial}{\partial A_\rho} (A_\lambda A_\mu)$$

$$= \tfrac{1}{2} m^2 g^{\lambda\mu} (A_\lambda \delta^\rho_\mu + A_\mu \delta^\rho_\lambda) = m^2 A^\rho.$$

Inserting in (2.39) gives the Proca equation $\partial_\mu F^{\mu\nu} + m^2 A^\nu = 0$. For a massive vector field $\partial_\mu A^\mu = 0$; this can also be written $(\Box + m^2) A^\mu = 0$, which is (2.50).

2.4 (a) See Bjorken and Drell (1965), §16.12; or Mandl (1959), Ch. 5 and Ch. 8; see also Exercise 2.9b. (b) Use $[H, \Pi(\mathbf{x}, t)] = i\delta H / \delta\phi(\mathbf{x}, t)$, the definition of the functional derivative (4.42), eqns. (2.59)–(2.60), and integration by parts (see Exercise 2.5b) to obtain

$$\frac{\delta H}{\delta \phi} = -i[H, \Pi] = -\dot{\Pi} = -\frac{\partial^2 \phi}{\partial t^2} = -\boldsymbol{\nabla}^2 \phi + \frac{dV}{d\phi}.$$

The Klein–Gordon equation (1.57) follows from this and (2.52) with $\mathcal{H}_{int} = 0$.
(c) The Lagrangian densities for free complex scalar $[\mathcal{L}_0(\phi)]$ and electromagnetic $[\mathcal{L}_0^\gamma(A)]$ fields are given by (2.44) and (2.48). Making the minimal substitution (1.83) gives

$$\mathcal{L} = \mathcal{L}_0(\phi) + \mathcal{L}_0^\gamma(A) + \mathcal{L}_{int}(A, \phi),$$

where

$$\mathcal{L}_{int}(A, \phi) = -iqA^\mu(\phi^\dagger \overset{\leftrightarrow}{\partial_\mu} \phi) + q^2 A^2 \phi^\dagger \phi.$$

Using (2.57), the canonical momenta are

$$\Pi^\dagger = \frac{\partial \mathcal{L}}{\partial(\partial_0 \phi^\dagger)} = \partial^0 \phi + iqA^0 \phi \qquad \Pi = \frac{\partial \mathcal{L}}{\partial(\partial_0 \phi)} = \partial^0 \phi^\dagger - iqA^0 \phi^\dagger.$$

From this and $\mathcal{H} = \mathcal{H}_0 + \mathcal{H}_{int} = \Pi \dot{\phi} + \Pi^\dagger \dot{\phi}^\dagger - \mathcal{L}$ we find

$$\mathcal{H}_{int} = -\mathcal{L}_{int}(A, \phi) - q^2(A^0)^2 \phi^\dagger \phi$$

(see Itzykson and Zuber, 1981, §6–1–4).

2.5 (a) See the solution of Exercise 2.9c. (b) Set $\mathcal{H}_{int} = 0$ in (2.60) and use integration by parts to write

$$H_0 = \tfrac{1}{2} \int d^3 x \left[\Pi^2 + \phi(-\boldsymbol{\nabla}^2 + m^2)\phi \right].$$

Substitute (2.68) and use (2.70) and some algebra to arrive at (2.74).

2.6 The commutator with $a_k^\dagger$ is $[P^\mu, a_k^\dagger] = k^\mu a_k^\dagger$. Multiply from the right with a state $|q^\mu\rangle$ of 4-momentum q^μ and use $P^\mu |q^\mu\rangle = q^\mu |q^\mu\rangle$ to obtain

$$P^\mu \left(a_k^\dagger |q^\mu\rangle \right) = \left(k^\mu + q^\mu \right) \left(a_k^\dagger |q^\mu\rangle \right),$$

so $a_k^\dagger$ adds a 4-momentum k^μ to a state.

2.7 Invoking (2.85) and (1.99), we find

$$\mathcal{L} + \mathcal{H}_{int} = \overline{\psi}(i\slashed{\partial} - m)\psi = 0.$$

Employing this and (1.72) in (2.87) gives (2.88).

2.8 (a) Use (1.86) to show that $\not{p}\gamma_0 + \gamma_0\not{p} = 2p_\mu\delta_0^\mu = 2p_0 = 2E$. (b) Use (1.86) to show that

$$(\not{p} + m)\gamma^0(\not{p} + m) = 2p_0(\not{p} + m) + \gamma_0(-\not{p}\not{p} + m^2) = 2E(\not{p} + m),$$

if $p^2 = m^2$, where $\not{p}\not{p} = p^2$ and $p_0 = E$ have been used. (c) From the equations given in the hint,

$$\Lambda_+(p) = \frac{1}{2m(m + E)} \sum_{\alpha=1,2} (\not{p} + m)u_{0\alpha}(m)\bar{u}_{0\alpha}(m)(\not{p} + m).$$

From the explicit forms (1.121) for the rest spinors and (1.88) for γ_0,

$$\sum_\alpha u_{0\alpha}\bar{u}_{0\alpha} = \begin{pmatrix} 1 & 0 & 0 & 0 \\ 0 & 1 & 0 & 0 \\ 0 & 0 & 0 & 0 \\ 0 & 0 & 0 & 0 \end{pmatrix} = \frac{1 + \gamma^0}{2}$$

[$u\bar{u}$ means the *matrix multiplication* of the row matrix $\bar{u}$ and the column matrix u: $(u\bar{u})_{\lambda\mu} = u_\lambda\bar{u}_\mu$]. Therefore,

$$\Lambda_+ = \frac{1}{2m(m + E)}(\not{p} + m)\frac{1 + \gamma^0}{2}(\not{p} + m)$$

$$= \frac{1}{2m(m + E)}\left(\frac{(\not{p} + m)^2}{2} + E(\not{p} + m)\right),$$

where the results of Exercise 2.8b were employed. Utilizing $\not{p}^2 = p^2 = m^2$, we have $(\not{p} + m)^2 = 2m(\not{p} + m)$ and finally $\Lambda_+ = (\not{p} + m)/2m$. Furthermore,

$$\Lambda_+^2 = \sum_{\alpha\beta} u_{p\alpha}\bar{u}_{p\alpha}u_{p\beta}\bar{u}_{p\beta} = \sum_\alpha u_{p\alpha}\bar{u}_{p\alpha} = \Lambda_+,$$

so Λ_+ is a projection operator; from (1.125a,b), it projects positive-energy states. By a similar proof

$$\Lambda_- = \frac{-\not{p} + m}{2m} \qquad \Lambda_-^2 = \Lambda_- \qquad \Lambda_+ + \Lambda_- = 1,$$

and Λ_- is a projector for negative-energy states.

2.9 (a) Eq. (2.88) gives the Dirac Hamiltonian density. Inserting the expansions (2.94)–(2.95) and using (2.92) and the orthonormality of the bispinors we obtain

$$\mathcal{H} = \frac{1}{\Omega}\sum_{\mathbf{p}s}(E_p a_{\mathbf{p}s}^\dagger a_{\mathbf{p}s} - E_p b_{\mathbf{p}s} b_{\mathbf{p}s}^\dagger)$$

$$H = \int d^3x\,\mathcal{H} = \Omega\mathcal{H} = \sum_{\mathbf{p}s} E_p\left(a_{\mathbf{p}s}^\dagger a_{\mathbf{p}s} - b_{\mathbf{p}s} b_{\mathbf{p}s}^\dagger\right).$$

Use the anticommutator on the second term and drop the infinite constant $\sum E_p$ to give the second form of (2.97). By analogous steps we may prove (2.98)–(2.99). (b) If the operators commute instead of anticommute the Hamiltonian (2.97) will not be positive definite, even after normal ordering: the second term makes a negative contribution and there is no stable ground state. (c) From the Heisenberg equations (2.12) we have $-i\dot{a}_{\mathbf{ps}}(t) = [H, a_{\mathbf{ps}}(t)]$. Insert the expression (2.97) for H and evaluate the resulting commutators by using the anticommutator (2.96).

2.10 Write out (2.113) explicitly in components and use (1.31). For example,

$$B^2 = \partial^1 A^3 - \partial^3 A^1 = F^{13} = -F^{31}.$$

The equivalence of (2.132) and (2.112) may be established by multiplying the terms of (2.132) out explicitly using (2.128a) and (2.129), and employing (1.31). For example, setting $\nu = 0$ in (2.132a) and using (2.128a) gives

$$\partial_1 E^1 + \partial_2 E^2 + \partial_3 E^3 = j^0,$$

which is (2.112a).

2.11 Just some algebra; see §8.2 of Aitchison and Hey (1982) for help.

2.12 Use

$$\mathcal{L} = -\tfrac{1}{4} F_{\mu\nu} F^{\mu\nu} \qquad \pi^\lambda = \frac{\partial \mathcal{L}}{\partial(\partial_0 A_\lambda)},$$

eq. (2.127), and properties of the metric tensor to show that $\pi^\lambda = -g^{\lambda\beta} F_{0\beta}$, which gives (2.153). Equation (2.154) comes from explicit calculation of $F^{\mu\nu} F_{\mu\nu}$, using (2.128). From (2.153b),

$$\pi^k \dot{A}_k = E^2 + \mathbf{E} \cdot \boldsymbol{\nabla} A_0,$$

which leads to (2.155). The term $\mathbf{E} \cdot \boldsymbol{\nabla} A_0$ vanishes on integration over all space [integrate by parts, assume that the fields vanish at infinity, and use eq. (2.112a)]; this gives the Hamiltonian (2.156).

2.13 See Ryder (1985), §4.4; Sakurai (1967), §4–6; Mandl (1959), Chs. 9–10; Itzykson and Zuber (1981), §3–2–1.

2.14 The Lagrangian density (2.46) is unchanged by the global (α independent of coordinates) rotation $\psi \to e^{i\alpha}\psi$ and $\overline{\psi} \to \overline{\psi}e^{-i\alpha}$. The Noether current is given by (2.192):

$$J_\mu = -i\frac{\partial \mathcal{L}}{\partial(\partial^\mu \psi)}(1)\psi = \overline{\psi}\gamma_\mu\psi,$$

where comparison of $U = e^{i\alpha}$ with (5.8) or (5.16) indicates that the single $U(1)$ generator is just $t = 1$. But $\partial^\mu J_\mu = 0$ (see §1.5.2), which is easily proved using the Dirac equation (1.99) and its adjoint,

$$\overline{\psi}(i\gamma^\mu \overleftarrow{\partial}_\mu + m) = 0$$

[the notation means that $\overline{\psi}$ is a row vector and $\overleftarrow{\partial}$ operates on it to the left]. Therefore, the charge Q defined in (2.193) is conserved [see §2.8.3].

2.15 (a) The $SO(2)$ rotations on the fields are

$$\phi' = R\phi = \begin{pmatrix} \cos\alpha & -\sin\alpha \\ \sin\alpha & \cos\alpha \end{pmatrix} \begin{pmatrix} \phi_1 \\ \phi_2 \end{pmatrix} \simeq \begin{pmatrix} 1 & -\alpha \\ \alpha & 1 \end{pmatrix} \begin{pmatrix} \phi_1 \\ \phi_2 \end{pmatrix}.$$

Thus, to order α we have $\phi_1 \to \phi_1 - \alpha\phi_2$ and $\phi_2 \to \phi_2 + \alpha\phi_1$. Substituting in $\mathcal{L}$ we find $\mathcal{L} \to \mathcal{L} + \mathcal{O}(\alpha^2)$ if $m_1 = m_2$, so $\mathcal{L}$ is invariant under infinitesimal rotations by

$$R \simeq \begin{pmatrix} 1 & -\alpha \\ \alpha & 1 \end{pmatrix} = 1 + \begin{pmatrix} 0 & -\alpha \\ \alpha & 0 \end{pmatrix}.$$

Comparing with the infinitesimal expansion of (5.8), $R \simeq 1 + i\alpha J$, we identify the $SO(2)$ infinitesimal generator

$$J = \begin{pmatrix} 0 & i \\ -i & 0 \end{pmatrix}.$$

Inserting in (2.192), the Noether current is

$$J_\mu = -i\frac{\partial\mathcal{L}}{\partial(\partial^\mu\phi_i)}J_{ij}\phi_j = (\partial_\mu\phi_1)\phi_2 - (\partial_\mu\phi_2)\phi_1.$$

(b) Letting $\phi = \frac{1}{\sqrt{2}}(\phi_1 + i\phi_2)$ and $\phi^\dagger = \frac{1}{\sqrt{2}}(\phi_1 - i\phi_2)$, the Lagrangian density is

$$\mathcal{L} = (\partial_\mu\phi)^\dagger(\partial^\mu\phi) - m^2(\phi^\dagger\phi) - \lambda(\phi^\dagger\phi)^2,$$

which is clearly invariant under global $U(1)$ rotations (see Exercise 2.14). Proceeding as in Exercise 2.14 with $\phi \to \phi + i\alpha$ and $\phi^\dagger \to \phi^\dagger - i\alpha$, the conserved Noether current is given by (2.192):

$$J_\mu = -i\frac{\partial\mathcal{L}}{\partial(\partial^\mu\phi)}(1)\phi - i\frac{\partial\mathcal{L}}{\partial(\partial_\mu\phi^\dagger)}(-1)\phi^\dagger$$
$$= i[(\partial_\mu\phi)\phi^\dagger - (\partial_\mu\phi^\dagger)\phi],$$

which is the same result as obtained in part (a).

2.16 The Lagrangian density is given by (2.202). The variations with respect to the fields give

$$\frac{\partial\mathcal{L}}{\partial(\partial_\mu\overline{\psi})} = 0 \qquad\qquad \frac{\partial\mathcal{L}}{\partial\overline{\psi}} = (i\gamma^\mu\partial_\mu - m_0)\psi - e_0\gamma_\mu A^\mu\psi$$

$$\frac{\partial \mathcal{L}}{\partial A_\nu} = -e_0 \overline{\psi} \gamma^\nu \psi \qquad \frac{\partial \mathcal{L}}{\partial(\partial_\mu A_\nu)} = -F^{\mu\nu} \quad \text{(see Exercise 2.3).}$$

Insert in the Euler–Lagrange equations (2.39) to obtain the coupled equations (2.204).

CHAPTER 3

3.1 (a) Equation (3.44) follows from the definitions (3.43), (3.39), (3.40), and (3.28). Equations (3.45) and (3.46) follow from (3.41)–(3.43) and (3.30)–(3.31). Equations (3.47) result from (3.28), (3.44), and (2.72). (b) Using the anticommutator repeatedly we have

$$\langle 0| \, a_i a_j a_k^\dagger a_l^\dagger \, |0\rangle = \langle 0| \, a_i \left(\delta_{jk} - a_k^\dagger a_j \right) a_l^\dagger \, |0\rangle$$
$$= \delta_{il}\delta_{jk} - \langle 0| \, a_i a_k^\dagger \left(\delta_{jl} - a_l^\dagger a_j \right) |0\rangle = \delta_{il}\delta_{jk} - \delta_{ik}\delta_{jl}.$$

Alternatively, by Wick's theorem only the complete contractions survive:

$$\langle 0| \, a_i a_j a_k^\dagger a_l^\dagger \, |0\rangle = \overline{a_i a_j a_k^\dagger a_l^\dagger} - \overline{a_i a_k^\dagger} \, \overline{a_j a_l^\dagger} + \overline{a_i a_l^\dagger} \, \overline{a_j a_k^\dagger}$$
$$= \delta_{il}\delta_{jk} - \delta_{ik}\delta_{jl},$$

where (3.48) for an ordinary product was used to write $\overline{a_i a_k^\dagger} = \langle 0| \, a_i a_k^\dagger \, |0\rangle = \delta_{ik}$, and so on.

3.2 (a) The Dirac equation is given by (2.146) and (2.138):

$$(i\not{\partial} - m)\, \psi = q\not{A}\psi.$$

Clearly, a formal solution that can be approximated by iteration is

$$\psi(x) = q \int d^4x' \, G(x - x') \not{A} \psi(x'),$$

provided $G(x - x')$ satisfies (3.60). Let

$$G(x - x') = \frac{1}{(2\pi)^4} \int G(k) e^{-ik\cdot(x-x')} d^4k \,,$$

with $G(k)$ to be determined. Show by substituting in (3.60) and using the integral representation of $\delta^{(4)}(x' - x)$ that $G(k)$ is given by $S_F(k)$ in (3.63), so $G(x - x')$ is given by $S_F(x - x')$ in (3.54). (b) For the 4-momentum integration variable k in (3.54) we have

$$k^2 - m^2 = k_0^2 - E^2 = (k_0 + E)(k_0 - E),$$

where k_0 and $E = \sqrt{\mathbf{k}^2 + m^2}$ are independent variables since we are dealing with particles that may be off mass shell. Then from (3.54) (without the $i\epsilon$ term)

$$G(x - x') = \frac{1}{(2\pi)^4} \int d^3k \, e^{i\mathbf{k}\cdot(\mathbf{x}-\mathbf{x}')} \int_{-\infty}^{+\infty} dk_0 \frac{\not{k} + m}{(k_0 + E)(k_0 - E)} e^{-ik_0(t-t')},$$

which has poles on the real k_0 axis at $\pm E$. If we choose contour I, for $t > t'$ the semicircle at infinity gives no contribution and the dk_0 integration can be replaced by the contour integral I, which by the Cauchy residue theorem gives

$$G(x - x')_{t>t'} = \frac{-i}{(2\pi)^3} \int \frac{d^3k}{2E} e^{-ik\cdot(x-x')}(\not{k} + m).$$

But $\not{k} + m$ (more precisely: $(\not{k} + m)/2m$) projects positive-energy states (Exercise 2.8c), so only such states propagate forward in time $(t > t')$. Likewise, for the case $t < t'$ the contour II may be used, leading to

$$G(x - x')_{t<t'} = \frac{-i}{(2\pi)^3} \int \frac{d^3k}{2E} e^{ik\cdot(x-x')}(-\not{k} + m),$$

where the invariance of the integral under $\mathbf{k} \to -\mathbf{k}$ has been invoked. But $(-\not{k} + m)/2m$ projects negative-energy states (Exercise 2.8c), so only those propagate backward in time $(t < t')$. Use of the contours I and II is equivalent to integration on the real k_0 axis with the poles displaced by adding to the denominator an infinitesimal $i\epsilon$ with $\epsilon > 0$:

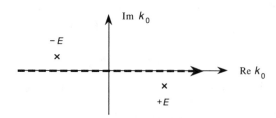

and allowing $\epsilon \to 0$ after the integration; thus we arrive at the denominator prescription in (3.54). (c) The particles propagating backward in time (contour integral II) are a consequence of the pole at Re $k_0 = -E$, which is not present for a nonrelativistic theory. That the correct helicity states are propagated follows from the completeness relations (1.126). (d) Evaluate (3.56) by contour integration using the same contours and analogous steps as in part (b). Then use

$$\frac{1}{\Omega} \sum_{\mathbf{k}} \longrightarrow \frac{1}{(2\pi)^3} \int d^3k$$

(see §2.4.2) and the invariance of the sum under $\mathbf{k} \to -\mathbf{k}$ to establish the equivalence to the right sides of (3.47a) (contour I) and (3.47b) (contour II).

3.3 Only complete contractions such as

$$\langle 34| \, S \, |12\rangle = \frac{-i}{4!} f_0 \int d^4x \, \langle 0| \, a_3 a_4 \overbrace{:\phi(x)\phi(x)}\overbrace{\phi(x)\phi(x):a_1^\dagger a_2^\dagger} |0\rangle$$

can contribute. We may construct 4!=24 equivalent such pairings. There are four equivalent ways of choosing the pairing with a_3,

$$\langle 0| \, a_3 a_4 \overbrace{:\phi(x)}\phi(x)\phi(x)\underbrace{\phi(x):a_1^\dagger a_2^\dagger} |0\rangle \, .$$

Once the pairing with a_3 has been chosen, there are three ways a_4 can be paired with the remaining ϕ's. Once the pairing with both a_3 and a_4 is chosen, there remain two ways $a_1^\dagger$ can be paired with the remaining ϕ's. Finally, once the pairing of a_3, a_4, and $a_1^\dagger$ have been chosen there remains only one choice for the pairing with $a_2^\dagger$. Therefore, the total number of equivalent combinations is $4 \cdot 3 \cdot 2 \cdot 1 = 4! = 24$. From (3.48) and (3.28)

$$\overbrace{\phi a_{\mathbf{k}}^\dagger} = (2\omega_k \Omega)^{-1/2} e^{-ik\cdot x} \qquad \overbrace{a_{\mathbf{k}} \phi} = (2\omega_k \Omega)^{-1/2} e^{ik\cdot x},$$

and

$$\langle 34| \, S \, |12\rangle = \frac{-if_0}{\Omega^2} \frac{4!}{4!} \frac{1}{4\sqrt{\omega_1 \omega_2 \omega_3 \omega_4}} \int d^4x \, e^{i(k_3+k_4-k_1-k_2)\cdot x}.$$

The integral is $(2\pi)^4 \delta^{(4)}(k_3 + k_4 - k_1 - k_2)$, which gives (3.71)–(3.72) and the corresponding graph in Fig. 3.8.

3.4 (a) One $\phi(x)$ must be contracted with one $\phi(y)$. Using (3.55)–(3.56),

$$\langle 34| \, T \left(:\phi(x)\phi(x)\overbrace{\phi(x)::\phi(y)}\phi(y)\phi(y): \right) |12\rangle$$

$$= i \int \frac{d^4k}{(2\pi)^4} \frac{e^{-ik\cdot(x-y)}}{k^2 - m^2 + i\epsilon} \, \langle 34| :\phi(x)\phi(x)::\phi(y)\phi(y): |12\rangle \, .$$

There are nine equivalent ways to make this pairing. Thus, from the first equation in §3.5.4,

$$\langle 34| \, S_2 \, |12\rangle = \frac{-9ig_0^2}{2!3!3!} \int d^4x \, d^4y \int \frac{d^4k}{(2\pi)^4} \frac{e^{-ik\cdot(x-y)}}{k^2 - m^2 + i\epsilon}$$

$$\times \langle 0| \, a_3 a_4 :\phi(x)\phi(x)::\phi(y)\phi(y):a_1^\dagger a_2^\dagger |0\rangle \, .$$

Only complete contractions contribute to vacuum expectation values. There are three independent kinds (see Fig. 3.11):

(1) $\langle 0|\, a_3 a_4 {:}\phi(x)\phi(x){:}{:}\phi(y)\phi(y){:} a_1^\dagger a_2^\dagger\,|0\rangle$

(2) $\langle 0|\, a_3 a_4 {:}\phi(x)\phi(x){:}{:}\phi(y)\phi(y){:} a_1^\dagger a_2^\dagger\,|0\rangle$

(3) $\langle 0|\, a_3 a_4 {:}\phi(x)\phi(x){:}{:}\phi(y)\phi(y){:} a_1^\dagger a_2^\dagger\,|0\rangle\,.$

Each of these can be chosen in eight ways, so inserting the contractions (see Exercise 3.3)

$$\langle 34|\, S_2\,|12\rangle = \frac{-ig_0^2}{4\Omega^2\sqrt{\omega_1\omega_2\omega_3\omega_4}}\int d^4x\, d^4y$$

$$\int \frac{d^4k}{(2\pi)^4}\,\frac{1}{k^2-m^2+i\epsilon}\left(e^{i(-k+k_3+k_4)\cdot x}e^{i(k-k_1-k_2)\cdot y}\right.$$

$$\left.+e^{i(-k+k_3-k_1)\cdot x}e^{i(k+k_4-k_2)\cdot y}+e^{i(-k+k_4-k_1)\cdot x}e^{i(k+k_3-k_2)\cdot y}\right).$$

The d^4x and d^4y integrals give $\delta^{(4)}$ functions times $(2\pi)^4$; for the three terms the $\delta^{(4)}$ functions imply (1) $k = k_1+k_2 = k_3+k_4$, (2) $k = k_2-k_4 = k_3-k_1$, and (3) $k = k_2 - k_3 = k_4 - k_1$, respectively, which gives (3.82) when (3.71) is used. It is clear that the scalar field Feynman rules of §3.5.5 also give (3.82) when applied to the diagrams in Fig. 3.11. (b) This is a combinatorial problem in the number of ways that Wick's theorem contractions can be made. Some specific examples of obtaining this factor have been given in Exercises 3.3 and 3.4a. The general rules are obtained in Bogoliubov and Shirkov (1983), §20.3: (1) A factor $n!$ for n vertices [this factor is canceled by the $1/n!$ factor in the nth-order term of the S-matrix expansion (3.22)]. (2) A factor $k!$ for each vertex, where k is the number of identical field operator factors in the interaction Lagrangian [this factor is canceled by the coefficients we chose for the g_0 and f_0 terms in (3.65)]. (3) A factor $1/\nu!$ for ν topologically equivalent ways of making the internal line pairings once the external lines are specified. For example, the second-order ϕ^4 diagrams of Fig. 3.10 have $n = 2$, $k = 4$, $\nu = 2$, and the combinatorial factor is

$$C = n!\,k!\,k!/\nu! = 2!4!4!/2!$$

In like manner, $C = 2!3!3!/1$ for the second-order ϕ^3 diagrams in Fig. 3.11, and $C = 1!4!/1$ for first-order ϕ^4 diagrams. These agree with the results in Exercises 3.3 and 3.4a. (c) From Exercise 1.4b, $\rho = 2EN^2$. If we normalize to $2E$ particles in a volume Ω so that $\int_\Omega \rho dV = 2E$, then $N = 1/\sqrt{\Omega}$, so $\rho = 2E/\Omega$ for bosons [for fermions, utilizing (1.125c), $\rho = (E/m)/\Omega$]. For the reaction $1+2 \to 3+4$, the transition rate per unit volume is $R_{\mathrm{fi}} = |\mathcal{A}_{\mathrm{fi}}|^2/\Omega T$, where T is the time interval and Ω the volume, and

$$\mathcal{A}_{\mathrm{fi}} \equiv -iN_1N_2N_3N_4(2\pi)^4\delta^{(4)}(p_3+p_4-p_1-p_2)\mathcal{M},$$

where $\mathcal{M}$ is the invariant amplitude [see (1.158a)]. Therefore,

$$R_{\mathrm{fi}} = \frac{1}{\Omega^4}(2\pi)^4\delta^{(4)}(p_3+p_4-p_1-p_2)\frac{1}{\Omega T}|\mathcal{M}|^2\int d^4x\, e^{i(p_3+p_4-p_1-p_2)\cdot x},$$

where one of the δ-functions arising from squaring $\mathcal{A}_{\mathrm{fi}}$ has been written in integral representation using (1.152). But the integral is just ΩT, by virtue of the pre-exponential δ-function, and

$$R_{\mathrm{fi}} = \frac{(2\pi)^4}{\Omega^4}\delta^{(4)}(p_3+p_4-p_1-p_2)|\mathcal{M}|^2.$$

The cross section is

$$d\sigma = \frac{dN_{\mathrm{f}}}{F_{\mathrm{i}}}R_{\mathrm{fi}} = \left[\frac{\text{number of final states}}{\text{initial particle flux} \times (\text{target particles / volume})}\right]R_{\mathrm{fi}}.$$

By imposing periodic boundary conditions, for a single particle in a spatial volume Ω and momentum volume d^3p there are $\Omega d^3p/(2\pi)^3$ states available (see, e.g., Schiff, 1968, §3.11; Halzen and Martin, 1984, Exercise 4.1). With our normalization there are $2E$ particles per volume Ω and for a two-body final state

$$dN_{\mathrm{f}} = \frac{\Omega d^3p_3}{(2\pi)^3(2E_3)}\frac{\Omega d^3p_4}{(2\pi)^3(2E_4)}.$$

In the lab frame, the initial particle flux times the target particle density is

$$F_i = \left(|\mathbf{v}_1|\frac{2E_1}{\Omega}\right)\frac{2E_2}{\Omega},$$

where 1 is projectile and 2 is target, and we have again used $\rho = 2E/\Omega$ for bosons. For fermions with the normalization (1.125c) the factors $2E_i$ in the above two expressions would be replaced by factors E_i/m. Combining everything gives (3.83a) for two-body boson scattering (see, e.g., Halzen and Martin, 1984, §4.3; Aitchison and Hey, 1982, §2.5).

3.5 (a) From Exercise 3.4c and eq. (3.83)

$$d\sigma = \frac{\mathrm{dLips}}{F}|\mathcal{M}|^2 \qquad F = 4|\mathbf{v}_1 - \mathbf{v}_2|E_1E_2$$

$$\mathrm{dLips} = \frac{1}{4\pi^2}\frac{d^3p_3}{2E_3}\frac{d^3p_4}{2E_4}\delta^{(4)}(p_1+p_2-p_3-p_4).$$

But for the two-body collision, total momentum conservation allows us to eliminate one of the d^3p integrals occurring when (3.83) is integrated over the outgoing momenta:

$$\int \frac{d^3p_4}{2E_4} \delta^{(4)}(p_1 + p_2 - p_3 - p_4) = \frac{1}{2E_4}\delta(E_1 + E_2 - E_3 - E_4).$$

Convert to angular variables, $d^3p_3 \rightarrow d\Omega\, p_3^2\, dp_3$, where $d\Omega$ is the solid angle element around $\mathbf{p}_3$ and $p_3 \equiv |\mathbf{p}_3|$. But $E_3^2 = p_3^2 + m_3^2$ so that $E_3 dE_3 = p_3 dp_3$, and (Aitchison and Hey, 1982, §2.6.1)

$$d\mathrm{Lips} = \frac{1}{4\pi^2} \frac{d\Omega\, p_3^2\, dp_3}{2E_3\, 2E_4} \delta(E_1 + E_2 - E_3 - E_4)$$

$$= \frac{1}{(4\pi)^2} \frac{p_3\, dE_3}{E_4} \delta(E_1 + E_2 - E_3 - E_4)\, d\Omega.$$

This is valid for two-boson phase space in any frame. (b) Now, specializing to the center-of-mass frame:

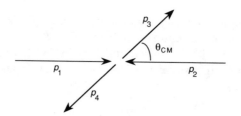

$$p_1 = (E_1, \mathbf{p}) \qquad p_2 = (E_2, -\mathbf{p}) \qquad p_3 = (E_1, \mathbf{p'}) \qquad p_4 = (E_2, -\mathbf{p'})$$

$$|\mathbf{p}| = |\mathbf{p'}| \equiv p \qquad \sqrt{s} = E_1 + E_2 \qquad W = E_3 + E_4.$$

But $E^2 = \mathbf{p}^2 + m^2$ and $E_3 dE_3 = p dp = E_4 dE_4$; therefore,

$$\frac{dW}{dp} = \frac{p}{E_3} + \frac{p}{E_4},$$

and $dp = E_3 E_4 dW/pW$. Substituting in the first form of the final result for part (a) with $p = p_3$,

$$d\mathrm{Lips} = \frac{1}{(4\pi)^2} \frac{pdW}{W} d\Omega\, \delta\left(\sqrt{s} - W\right) = \frac{1}{(4\pi)^2} \frac{p}{\sqrt{s}} d\Omega.$$

Also, from part (a)

$$F = 4\,|\mathbf{v}_1 - \mathbf{v}_2|\, E_1 E_2 = 4p\sqrt{s}.$$

Combining these results,

$$\left(\frac{d\sigma}{d\Omega}\right)_{\text{CM}} = \frac{1}{64\pi^2 s}|\mathcal{M}|^2,$$

which is valid for bosons. For two-fermion scattering with the normalization (1.125c) the right side is multiplied by a factor $(2m_1)^2(2m_2)^2 = 16m_1^2 m_2^2$ [see the comment after eq. (3.84)].   (c) The typical one-photon exchange QED amplitude is of the form [see (3.97)]

$$\mathcal{M} \simeq -e^2 \bar{u}_{k's'}\gamma^\mu u_{ks}\left(\frac{1}{q^2}\right)\bar{u}_{p'\sigma'}\gamma_\mu u_{p\sigma},$$

and the spin-averaged square modulus is

$$\overline{|\mathcal{M}|^2} \simeq \left(\frac{e^4}{q^4}\right) L^{\mu\nu} L'_{\mu\nu},$$

with $L^{\mu\nu}$ and $L'_{\mu\nu}$ given by expressions of the form

$$L^{\mu\nu} = \tfrac{1}{2}\sum_{ss'}(\bar{u}_{k's'}\gamma^\mu u_{ks})(\bar{u}_{k's'}\gamma^\nu u_{ks})^*.$$

But the second factor can be viewed as a 1×1 matrix, for which complex conjugation and Hermitian conjugation are equivalent:

$$(\bar{u}_{k's'}\gamma^\nu u_{ks})^* = (\bar{u}_{k's'}\gamma^\nu u_{ks})^\dagger = u_{ks}^\dagger \gamma^{\nu\dagger}\gamma^{0\dagger} u_{k's'}^\dagger = \bar{u}_{ks}\gamma^\nu u_{k's'}$$

(see Exercise 1.6 and Table 1.2), so

$$L^{\mu\nu} = \tfrac{1}{2}\sum_{ss'}\sum_{\alpha\beta\gamma\delta}(\bar{u}_{k's'})_\alpha(\gamma^\mu)_{\alpha\beta}(u_{ks})_\beta(\bar{u}_{ks})_\gamma(\gamma^\nu)_{\gamma\delta}(u_{k's'})_\delta,$$

where α, β, γ, and δ are matrix indices. These are sums over *matrix elements* (numbers), so we may reorder the factors at will. Moving the last factor to the front and using the results of Exercise 2.8c for the matrix products $u\bar{u}$ yields

$$L^{\mu\nu} = \tfrac{1}{2}\sum_{\alpha\beta\gamma\delta}\frac{(\not{k}' + m)_{\delta\alpha}}{2m}(\gamma^\mu)_{\alpha\beta}\frac{(\not{k} + m)_{\beta\gamma}}{2m}(\gamma^\nu)_{\gamma\delta}$$

$$= \frac{1}{8m^2}\text{Tr}\left[(\not{k}' + m)\gamma^\mu(\not{k} + m)\gamma^\nu\right],$$

which is (3.101). Now exploit the trace theorems of Table 1.2: from theorem (B),

$$L^{\mu\nu} = \frac{1}{8m^2}\text{Tr}\left(\not{k}'\gamma^\mu\not{k}\gamma^\nu + m^2\gamma^\mu\gamma^\nu\right).$$

From theorem (E) we obtain $\mathrm{Tr}\,(\gamma^{\mu}\gamma^{\nu}) = 4g^{\mu\nu}$, and from theorem (F)

$$\mathrm{Tr}\,(\not{k}'\gamma^{\mu}\not{k}\gamma^{\nu}) = 4(k'^{\mu}k^{\nu} + k'^{\nu}k^{\mu} - k \cdot k'g^{\mu\nu}).$$

Equation (3.102) follows from these relations, $q^2 = (k - k')^2$, and $k^2 = k'^2 = m^2$. (d) See Halzen and Martin (1984), §6.5. In the ultrarelativistic limit we may ignore masses and the leptonic tensor (3.102) can be written [see part (c)]

$$L_{\mu\nu} = \frac{1}{2m^2}\left(k'_{\mu}k_{\nu} + k'_{\nu}k_{\mu} - (k \cdot k')g_{\mu\nu}\right).$$

From the Feynman rules for $e^-\mu^- \to e^-\mu^-$ we find that the spin-averaged, squared matrix element is proportional to the complete contraction of two leptonic tensors:

$$\overline{|\mathcal{M}|^2} = \left(\frac{e^4}{q^4}\right)L^{\mu\nu}L'_{\mu\nu}.$$

Multiply out the contraction explicitly and use results such as

$$k'^{\mu}k^{\nu}g_{\mu\nu} = k'^{\mu}k_{\mu} = k \cdot k' \qquad g^{\mu\nu}g_{\mu\nu} = 4$$

to obtain

$$L^{\mu\nu}L'_{\mu\nu} = \frac{1}{4m_1^2 m_2^2}\left(2(k' \cdot p')(k \cdot p) + 2(k' \cdot p)(k \cdot p')\right).$$

If the masses are neglected the Mandelstam variables are

$$s \simeq 2k \cdot p = 2k' \cdot p' \qquad t \simeq -2k \cdot k' = -2p \cdot p' \qquad u \simeq -2k \cdot p' = -2k' \cdot p,$$

and we find

$$\overline{|\mathcal{M}(e^-\mu^- \to e^-\mu^-)|^2} = \frac{e^4}{q^4}L^{\mu\nu}L'_{\mu\nu} = \left(\frac{e^4}{8m_1^2 m_2^2}\right)\frac{s^2 + u^2}{t^2},$$

where $t = (k - k')^2 = q^2$ has been used. (e) By crossing symmetry, the amplitude for $e^-\mu^- \to e^-\mu^-$ is converted to the amplitude for $e^+e^- \to \mu^+\mu^-$

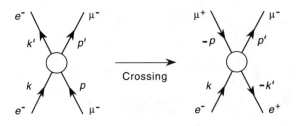

through the interchange $p \leftrightarrow -k'$. From the definitions of the Mandelstam variables, under this interchange $s \leftrightarrow t$ and $u \to u$. Thus, using the result from part (d) we obtain

$$\overline{|\mathcal{M}(e^+e^- \to \mu^+\mu^-)|^2} = \left(\frac{e^4}{8m_1^2 m_2^2}\right)\frac{t^2 + u^2}{s^2}.$$

(f) In the center-of-mass system,

$$k = (E, \mathbf{k_i}) \qquad p = (E, -\mathbf{k_i}) \qquad k' = (E, \mathbf{k_f}) \qquad p' = (E, -\mathbf{k_f}).$$

Thus, for example,

$$u = (k - p')^2 = (k_\mu - p'_\mu)(k^\mu - p'^\mu)$$
$$= -(\mathbf{k_i} + \mathbf{k_f})^2 = -2k^2(1 + \cos\theta),$$

where we have employed $\cos\theta = \mathbf{k_i}\mathbf{k_f}/k^2$, with $k \equiv |\mathbf{k_i}| = |\mathbf{k_f}|$. Therefore, we find that

$$\frac{t^2 + u^2}{s^2} = \tfrac{1}{2}\left(1 + \cos^2\theta\right),$$

and from the results of parts (b) and (e),

$$\left(\frac{d\sigma}{d\Omega}\right)_{\mathrm{CM}} = \frac{\alpha^2}{4s}(1 + \cos^2\theta) \qquad \sigma = \frac{4\pi\alpha^2}{3s},$$

where the total cross section σ results from integration over the solid angle. Up to constants, this result also follows immediately from dimensional analysis: the total cross section must have dimension $[M]^{-2}$ and be proportional to the square of the dimensionless QED coupling α; at high energies the masses are negligible and the CM energy $\sqrt{s}$ is the only quantity available with the dimension of mass, so $\sigma \propto \alpha^2/s$ [compare with the corresponding dimensional analysis of the weak interaction cross section (6.39)]. (g) The results of part (f) are in excellent agreement with the data for the total cross section. There are some small discrepancies in the differential cross section that can be removed by including a diagram corresponding to exchange of the neutral electroweak boson Z^0; see Fig. 2 in Bartel et al. (1982), the discussion in §6.6 of Perkins (1987), and Ch. 9.

3.6 Two independent contractions give the graphs (I) and (II) in Fig 3.15. Schematically they are of the form

$$\langle 0|\, a_4 b_3 T\left(:\overline{\psi}(x)\!\!\not{A}(x)\psi(x)::\overline{\psi}(y)\!\!\not{A}(y)\psi(y):\right) a_1^\dagger b_2^\dagger\,|0\rangle$$

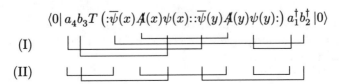

(I)

(II)

We may proceed essentially as for the example in §3.7.2. Evaluating contractions and using (3.57)–(3.58) leads to the usual momentum conserving delta functions; taking care to keep track of the permutation phases for the fermion operators, we arrive at (3.104).

3.7 Use the methods of Exercise 3.5c,d to show that (3.97) gives

$$\overline{|\mathcal{M}|^2} = \frac{e^4}{8m_1^2 m_2^2}\left(\frac{s^2 + u^2}{t^2} + \frac{2s^2}{tu} + \frac{s^2 + t^2}{u^2}\right).$$

(Itzykson and Zuber, 1980, §6–1–3 may be consulted for help with the required trace algebra.) Then express s, t, and u in terms of the scattering angles (see the hint in Exercise 3.5f), and use the two-body CM cross section formula derived in Exercise 3.5b (with the fermion normalization) to obtain (3.103).

3.8 By crossing symmetry the matrix element for $e^+ e^- \to e^+ e^-$ is obtained from the matrix element in Exercise 3.7 through the replacements $s \leftrightarrow u$, and $t \leftrightarrow t$ [see §1.7.3 and Exercise 3.5e]. Make this replacement and proceed as outlined in the solution of Exercise 3.7 to obtain (3.105).

3.9 (a) The scalar field is given by example in the exercise. For the massive vector field (2.49) leads to

$$\mathcal{L} = -\tfrac{1}{2}(\partial_\nu A_\mu \partial^\nu A^\mu - m^2 A_\mu A^\mu - \partial_\nu A_\mu \partial^\mu A^\nu),$$

where we have used (2.47) and that repeated indices are dummy indices. Remove the numerical factors, replace the fields with plane waves, operate with the derivatives, and lower indices with the metric tensor to leave

$$K_{\mu\nu} = (-k^2 + m^2)g_{\mu\nu} + k_\mu k_\nu.$$

The massive vector propagator is the inverse of this; write

$$W_{\mu\nu}(k) = K_{\mu\nu}^{-1} = A g_{\mu\nu} + B k_\mu k_\nu,$$

and require that $K^{-1}K = 1$:

$$(A g_{\nu\rho} + B k_\nu k_\rho)\left[(-k^2 + m^2)g^{\mu\nu} + k^\mu k^\nu\right] = \delta_\rho^\mu.$$

This is satisfied if

$$A = \frac{1}{-k^2 + m^2} \qquad\qquad B = \frac{-1/m^2}{-k^2 + m^2},$$

which yields the expression (6.49) for $W_{\mu\nu}(k)$. A similar exercise for the Dirac field gives (3.63). (b) Applying the procedure outlined in (a) to the massless vector Lagrangian (2.48) requires that

$$(A g_{\nu\rho} + B k_\nu k_\rho)(-k^2 g^{\mu\nu} + k^\mu k^\nu) = \delta_\rho^\mu,$$

which implies that $A(-k^2\delta^\mu_\rho + k^\mu k_\rho) = \delta^\mu_\rho$. This cannot be satisfied for any choice of A or B, so K has no inverse. (c) Proceeding as in (a), the propagator must be of the form $D_{\mu\nu}(k) = Ag_{\mu\nu} + Bk_\mu k_\nu$, where now

$$(Ag_{\nu\rho} + Bk_\nu k_\rho)\left[-k^2 g^{\mu\nu} + \left(1 - \frac{1}{\xi}\right)k^\mu k^\nu\right] = \delta^\mu_\rho.$$

This requires that $A = -1/k^2$ and $B = (1 - \xi)/k^4$, which gives the desired result. Parts (b) and (c) illustrate a fundamental difficulty in the analysis of gauge fields: such fields can be quantized only if the freedom associated with gauge transformations is suitably restricted (see §7.4).

3.10 (a)–(b) The propagators are found in Exercise 3.9. The interaction Lagrangian densities required for the vertices may be inferred from the equations cited in the hints. For example, from the second term of $\mathcal{L}_{\text{int}}$ given in Exercise 2.4c,

$$i\mathcal{L}_{\text{int}} = iq^2 A_\mu A_\nu g^{\mu\nu} \phi^* \phi$$

$$= iq^2 \underbrace{\epsilon^*_{k_1\mu} \epsilon^*_{k_2\nu} e^{ik_1\cdot x} e^{ik_2\cdot x}}_{} g^{\mu\nu} \underbrace{N_1 N_2 e^{-p_1\cdot x} e^{ip_2\cdot x}}_{}$$

$$+ \text{(same term but } k_1 \leftrightarrow k_2),$$

where external line factors are indicated by underbraces, and the two terms result from permutation on the $\phi^* \phi$ factor. Therefore, discarding the external line factors,

$$\text{vertex} = 2iq^2 g^{\mu\nu},$$

which is Feynman rule 5.B(iii) of §3.9.

CHAPTER 4

4.1 See Ryder (1985), §6.6, and the scalar field example in Exercise 4.3.

4.2 Apply the definition (4.42); for example, utilizing (4.36)

$$\frac{\delta Z[J]}{\delta J(y)} = \lim_{\epsilon \to 0} \frac{1}{\epsilon} \int [D\phi] \left(e^{iS} e^{i\epsilon \int d^4x\, \delta^{(4)}(x-y)\phi(x)} - e^{iS}\right),$$

where

$$S = \int d^4x \left(\mathcal{L} + \phi(x)J(x)\right).$$

Use the delta function to do the d^4x integral and expand the resulting factor $\exp(i\epsilon\phi(y))$ to give the desired result. Taking $\delta/\delta J(z)$ of this result by the same procedure leads to the expression for $\delta^2 Z[J]/\delta J(y)\delta J(z)$.

4.3 Expand the first exponential in (4.65):

$$e^{iW[J]} \simeq \left(1 - \frac{if_0}{4!} \int d^4z \left(\frac{1}{i}\frac{\delta}{\delta J(z)}\right)^4 + \cdots\right)$$

$$\times \exp\left(-\frac{i}{2}\int d^4x\, d^4y\, J(x)\Delta(x-y)J(y)\right).$$

The first term gives the free propagator [see (4.58) and (4.59)]. The second term may be evaluated by taking the functional derivatives of the exponential (see Exercise 4.2). For example, utilizing the result of Exercise 4.2 that $\delta J(y)/\delta J(x) = \delta^{(4)}(x-y)$, introducing a notation

$$\exp\left(-\frac{i}{2}\int d^4x\, d^4y\, J(x)\Delta(x-y)J(y)\right) \equiv \exp\left(-\frac{i}{2}\int J\Delta J\right),$$

and adopting a shorthand notation $d^4x \equiv dx$ allows us to write

$$\frac{1}{i}\frac{\delta}{\delta J(z)}e^{-\frac{i}{2}\int J\Delta J} = -\int dx\, \Delta(z-x)J(x)e^{-\frac{i}{2}\int J\Delta J}.$$

Taking the functional derivative again,

$$\left(\frac{1}{i}\frac{\delta}{\delta J(z)}\right)^2 e^{-\frac{i}{2}\int J\Delta J} = \left[i\Delta(0) + \left(\int\Delta(z-x)J(x)\,dx\right)^2\right]e^{-\frac{i}{2}\int J\Delta J}.$$

Continuing in this way, we find (see Ryder, 1985, §6.5)

$$\left(\frac{1}{i}\frac{\delta}{\delta J(z)}\right)^4 e^{-\frac{i}{2}\int J\Delta J} = \left[-3(\Delta(0))^2 + 6i\Delta(0)\left(\int\Delta(z-x)J(x)\,dx\right)^2\right.$$

$$\left. + \left(\int\Delta(z-x)J(x)\,dx\right)^4\right]e^{-\frac{i}{2}\int J\Delta J}$$

$$= \left(-3\,\infty + 6i\,\underline{\Omega} + \times\right)e^{-\frac{i}{2}\int J\Delta J},$$

where we have introduced a graphical notation $\Delta(x-y) \equiv x\!-\!\!-\!y$, so

$$\Delta(0) = \bigcirc \qquad [\Delta(0)]^2 = \infty \qquad \Delta(0)\left[\int\Delta(x-z)J(x)\,dx\right]^2 = \underline{\Omega}$$

and so on.

The vacuum graphs ∞ will cancel out if we compute a normalized generating functional $Z[J]/Z[J=0]$. To order f_0,

$$\frac{Z[J]}{Z[J=0]} = \frac{\left(1 - \frac{if_0}{4!}\int\left(-3\,\infty + 6i\,\ominus + \times\right)dz\right)e^{-\frac{i}{2}\int J\Delta J}}{1 + \frac{if_0}{4!}\int 3\,\infty\,dz}$$

$$\simeq \left(1 - \frac{if_0}{4!}\int\left(6i\,\ominus + \times\right)dz\right)e^{-\frac{i}{2}\int J\Delta J},$$

where in the last step the denominator was expanded by the binomial theorem and terms of order $f_0{}^2$ discarded. [Since the functional integral does not normal order the interaction, the graphs will include tadpoles (Fig. 3.5) unless counterterms are added to suppress them.] The Green's functions now follow by functional differentiation [eq. (4.43)]. The resulting expressions are straightforward, but voluminous; we do not reproduce them here but refer to Ryder (1985), §6.5, where it is shown that the standard rules for ϕ^4 fields result.

Notice that in the path integral method with external sources $J(x)$, the functional derivative rule $\delta J(y)/\delta J(x) = \delta^{(4)}(x - y)$ plays the same role as Wick's theorem (§3.4.4) does in canonical quantization (see Abers and Lee, 1973, §12).

4.4 The most important point is that $\theta_i^2 = 0$, because the Grassmann generators obey $\{\theta_i, \theta_j\} = 0$: a power series expansion on the elements of a Grassmann algebra with n generators has only 2^n terms. For example, if we have two generators $\theta = (\theta_1, \theta_2)$ there are $2^n = 4$ independent terms:

$$p(\theta) = a + b\theta_1 + c\theta_2 + d\theta_1\theta_2.$$

The exercise results follow from the finite power series expansion and anticommutation relations, and the hints given in the exercise. Details may be found in Cheng and Li (1984), §1.3; Ryder (1985), §6.7; and Berezin (1966).

4.5 For parts (a) and (b), diagonalize the matrix A. Each integral is of Gaussian form

$$\int dx \exp\left(-\tfrac{1}{2}ax^2\right) = \sqrt{\frac{2\pi}{a}},$$

where a is an eigenvalue of A, so the multiple integral gives a product of such factors. The desired result follows then from (4.49). Follow a similar procedure for part (c), but use the properties of Grassmann integration derived in Exercise 4.4. Further help may be found in Rajaraman (1982), §9.1; Huang (1982), §7.9; and Cheng and Li (1984), §1.3.

CHAPTER 5

5.1 Letting $a = c_4$, $a^2 = c_4 \cdot c_4$, and $a^3 = c_4 \cdot c_4 \cdot c_4$, the multiplication table for $A \cdot B$ is

B/A	1	a	a^2	a^3
1	1	a	a^2	a^3
a	a	a^2	a^3	1
a^2	a^2	a^3	1	a
a^3	a^3	1	a	a^2

which satisfies the requirements (1)–(4) of §5.1.1, so this is a group. The multiplication commutes so the group is Abelian.

5.2 Any group of order two must contain two elements, one of which is the identity, so $G = \{1, a\}$ and the multiplication table for $A \cdot B$ must be

B/A	1	a
1	1	a
a	a	1

where the result $a \cdot a = 1$ follows from the requirement that each group element have an inverse: by inspection, a must be its own inverse. This multiplication table was constructed without reference to a particular group; hence, all groups of order two have the same multiplication table and are isomorphic to S_2.

5.3 The multiplication table for $A \cdot B$ is

B/A	1	(12)	(23)	(13)	(123)	(321)
1	1	(12)	(23)	(13)	(123)	(321)
(12)	(12)	1	(123)	(321)	(23)	(13)
(23)	(23)	(321)	1	(123)	(13)	(12)
(13)	(13)	(123)	(321)	1	(12)	(23)
(123)	(123)	(13)	(12)	(23)	(321)	1
(321)	(321)	(23)	(13)	(12)	1	(123)

where 1 is the identity. This satisfies the group postulates of §5.1.1, and since the multiplication does not commute the group is non-Abelian. The identity and any of the single permutations form a two-object Abelian subgroup (S_2). The set of even permutations $A_3 = \{1, (123), (321)\}$ is also closed under multiplication, so it constitutes an Abelian subgroup. The multiplication table for $C_3 = \{1, c_3, c_3^2\}$, where c_3 rotates by $120°$, can be put into one-to-one correspondence with that of A_3, so C_3 and A_3 are isomorphic.

5.4 A two-dimensional rotation by an angle ϕ is defined by the one-parameter transformation

$$\begin{pmatrix} x' \\ y' \end{pmatrix} = R(\phi) \begin{pmatrix} x \\ y \end{pmatrix} = \begin{pmatrix} \cos\phi & -\sin\phi \\ \sin\phi & \cos\phi \end{pmatrix} \begin{pmatrix} x \\ y \end{pmatrix}.$$

By matrix multiplication

$$\begin{pmatrix} \cos\phi & -\sin\phi \\ \sin\phi & \cos\phi \end{pmatrix} \begin{pmatrix} \cos\phi' & -\sin\phi' \\ \sin\phi' & \cos\phi' \end{pmatrix} = \begin{pmatrix} \cos(\phi+\phi') & -\sin(\phi+\phi') \\ \sin(\phi+\phi') & \cos(\phi+\phi') \end{pmatrix},$$

where standard trigonometric identities have been used. Thus

$$R(\phi)R(\phi') = R(\phi+\phi')$$

and the matrices $R(\phi)$ form a group [see (5.4a)]. The multiplication depends smoothly on the continuous parameter ϕ and commutes, so this is an Abelian Lie group [the group of special (unit determinant) 2×2 orthogonal matrices, commonly denoted $SO(2)$].

5.5 (a) Equation (5.12) follows trivially from (5.10); eq. (5.13) follows from (5.11) when (5.9) and (5.12) are used. Write the left side of (5.15) in matrix components,

$$[T_a, T_b]_{ce} = (T_a)_{cd}(T_b)_{de} - (T_b)_{cd}(T_a)_{de}.$$

Then substitute (5.14) and use (5.13), (5.12), and (5.14) to obtain

$$[T_a, T_b]_{ce} = f_{abd}f_{dce} = if_{abd}(T_d)_{ce},$$

which is (5.15) in component form. (b) Substitute the exponential forms (5.8) for both sides of $U(\alpha)^{-1}U(\beta)^{-1}U(\alpha)U(\beta) = U(\gamma)$ and expand the exponentials around the origin. Term by term comparison then requires that the generators obey equations of the form $[X_k, X_\ell] = \sum_m C_{k\ell}^m X_m$, which is (5.9) with $C_{k\ell}^m = if_{k\ell m}$.

5.6 (a) The Lagrangian density is given by (2.202) and the Hamiltonian density by (2.87). Since L contains only fermion operators it commutes with the photon operators and we need examine only the effect of (5.17) on the fermion fields. First show that $[L, \psi(z)] = -\psi(z)$, by using (5.18) and (2.89). Then introduce

$$\psi_\theta(x) \equiv e^{iL\theta}\psi(x)e^{-iL\theta},$$

differentiate it with respect to θ, use the commutator $[L, \psi(z)] = -\psi(z)$, and integrate to prove that

$$e^{iL\theta}\psi(x)e^{-iL\theta} = e^{-i\theta}\psi(x).$$

In addition, L is Hermitian so $e^{iL\theta}$ is unitary and

$$e^{iL\theta}\psi^\dagger(x)e^{-iL\theta} = e^{i\theta}\psi^\dagger(x).$$

Now use these results, and that L commutes with photon operators and γ-matrices, to show that each term in the QED Hamiltonian is invariant under (5.17). Finally, differentiate (5.17) with respect to θ and set $\theta = 0$ to show that $[L, H] = 0$, so the lepton number is a constant of motion (see Lee, 1981, §10.1). (b) The form of σ follows immediately from (5.25) and (5.28). Let

$U = \begin{pmatrix} a & b \\ c & d \end{pmatrix}$. Unitarity requires

$$UU^\dagger = 1 = \begin{pmatrix} 1 & 0 \\ 0 & 1 \end{pmatrix} = \begin{pmatrix} aa^* + bb^* & ac^* + bd^* \\ ca^* + db^* & cc^* + dd^* \end{pmatrix}$$

from which $c = -b^*d/a^*$ and $aa^* + bb^* = 1$. Unit determinant requires $\text{Det}\, U = ad - bc = 1$. Combining these results gives $d = a^*$ and $c = -b^*$, so the most general two-dimensional $SU(2)$ matrix is

$$U = \begin{pmatrix} a & b \\ -b^* & a^* \end{pmatrix} \qquad aa^* + bb^* = 1.$$

Write $a = d_1 + id_2$ and $b = d_3 + id_4$; the constraint is then $\sum_{i=1}^4 d_i^2 = 1$, which is the equation of a 3-sphere, S^3.

5.7 The adjoint representation for $SU(2)$ consists of three 3×3 matrices T_a, since $SU(2)$ has three generators. From (5.14) and (5.31) the elements of these matrices are $(T_a)_{ij} = -i\epsilon_{ija}$, with $a, i, j = 1$–3. Therefore,

$$T_1 = \begin{pmatrix} 0 & 0 & 0 \\ 0 & 0 & -i \\ 0 & i & 0 \end{pmatrix} \qquad T_2 = \begin{pmatrix} 0 & 0 & i \\ 0 & 0 & 0 \\ -i & 0 & 0 \end{pmatrix} \qquad T_3 = \begin{pmatrix} 0 & -i & 0 \\ i & 0 & 0 \\ 0 & 0 & 0 \end{pmatrix}.$$

By explicit matrix multiplication, these satisfy (5.15) with (5.31). Since this is a rank-1 algebra we may diagonalize one matrix, say T_3:

$$\begin{vmatrix} 0 - \lambda & -i & 0 \\ i & 0 - \lambda & 0 \\ 0 & 0 & 0 - \lambda \end{vmatrix} = 0.$$

This implies that the characteristics λ satisfy $-\lambda^3 + \lambda = 0$, which has solutions $\lambda = 0, \pm 1$. The diagonalized matrix has these characteristics down the diagonal (in any order). Choose

$$J_3 = \begin{pmatrix} 1 & 0 & 0 \\ 0 & 0 & 0 \\ 0 & 0 & -1 \end{pmatrix}.$$

The matrix C generating this transformation, $J_3 = C^{-1}T_3C$, has columns that are the eigenvectors of T_3. Solving

$$(T_3 - \lambda \mathbf{1}) \begin{pmatrix} x \\ y \\ z \end{pmatrix} = 0$$

for $\lambda = 0, \pm 1$ gives the eigenvectors. After normalizing each eigenvector we may choose

$$C = \begin{pmatrix} -i/\sqrt{2} & 0 & i/\sqrt{2} \\ 1/\sqrt{2} & 0 & 1/\sqrt{2} \\ 0 & 1 & 0 \end{pmatrix} \qquad C^{-1} = \begin{pmatrix} i/\sqrt{2} & 1/\sqrt{2} & 0 \\ 0 & 0 & 1 \\ -i/\sqrt{2} & 1/\sqrt{2} & 0 \end{pmatrix}.$$

Using $J_a = C^{-1}T_a C$ to transform to the new basis gives

$$J_1 = \frac{1}{\sqrt{2}} \begin{pmatrix} 0 & -i & 0 \\ i & 0 & i \\ 0 & -i & 0 \end{pmatrix} \qquad J_2 = \frac{1}{\sqrt{2}} \begin{pmatrix} 0 & -1 & 0 \\ -1 & 0 & 1 \\ 0 & 1 & 0 \end{pmatrix} \qquad J_3 = \begin{pmatrix} 1 & 0 & 0 \\ 0 & 0 & 0 \\ 0 & 0 & -1 \end{pmatrix},$$

for an equivalent set of generators. If we use the same methods to diagonalize T_2 instead, we obtain eqns. (5.50).

5.8 (a) The commutator $[T_3, T_+]$ is evaluated by inserting (5.56) and using

$$[AB, CD] = [A, C]DB + C[A, D]B + A[B, C]D + AC[B, D],$$

eq. (2.96), $[A, B] = 2AB - \{A, B\}$, and that the neutron and proton operators commute. The commutator $[T_3, T_-]$ can be obtained in the same way, but it is simpler to Hermitian conjugate the preceding result to give

$$[T_3, T_-] = -(T_+)^\dagger = -T_-.$$

The commutator $[T_+, T_-]$ follows from insertion of (5.56), commuting neutron and proton operators, and the use of (2.96). **(b)** Under an $SU(2)$ transformation we have

$$\phi \to \phi' = U\phi \qquad \phi^\dagger \to \phi'^\dagger = \phi^\dagger U^\dagger.$$

The $SU(2)$ matrix U is unitary ($U^\dagger U = UU^\dagger = 1$), so

$$\phi'^\dagger \phi' = \phi^\dagger \phi \qquad (\partial_\mu \phi')^\dagger (\partial^\mu \phi') = (\partial_\mu \phi)^\dagger (\partial^\mu \phi),$$

and the Lagrangian density is invariant under $SU(2)$ transformations. The isospin current is given by (2.192); evaluating for this Lagrangian we obtain (see Exercises 2.14 and 2.15)

$$\mathbf{J}^\mu = \frac{-i}{2} \left[(\partial^\mu \phi^\dagger) \boldsymbol{\tau} \phi - \phi^\dagger \boldsymbol{\tau} (\partial^\mu \phi) \right].$$

The timelike components are

$$J_a^0 = \frac{-i}{2} \left[(\partial^0 \phi^\dagger) \tau_a \phi - \phi^\dagger \tau_a (\partial^0 \phi) \right] = \frac{-i}{2} \left(\Pi \tau_a \phi - \phi^\dagger \tau_a \Pi^\dagger \right),$$

where Π is the canonical momentum [see (2.57b)]; the corresponding charges are given by (2.193).

5.9 Complex conjugation of eq. (5.9) gives

$$[X_a^*, X_b^*] = [-X_a^*, -X_b^*] = i f_{abc} (-X_c^*).$$

Therefore, if the matrices X_a satisfy a Lie algebra the matrices $-X_a^*$ do also; the representation generated by $-X_a^*$ is said to be *conjugate* to that generated by X_a. For the $SU(2)$ fundamental representation the effect of replacing the generators X_a by $-X_a^*$ [see (5.29) and (5.40)–(5.43)] is

$$\sigma_3 \rightarrow -\sigma_3 \qquad \sigma_\pm \rightarrow -\sigma_\mp,$$

so the sign of the magnetic quantum number M is reversed and the actions of σ_+ and σ_- are interchanged. However, the $SU(2)$ weight diagrams are symmetric about $M = 0$ (Fig. 5.1) so $\overline{\mathbf{2}} = \mathbf{2}$ for $SU(2)$. For $SU(3)$ the same replacement in (5.68) has the effect

$$T_3 \rightarrow -T_3 \qquad Y \rightarrow -Y,$$

and interchanges raising and lowering roles for the operator pairs $V_\pm$, $U_\pm$, and $T_\pm$ (Fig. 5.3). This is equivalent to reflecting the weight diagram in the T_3 and Y axes. The $\mathbf{3}$ weight diagram is not symmetric under this reflection (see Fig. 5.5) and $\mathbf{3} \neq \overline{\mathbf{3}}$ for $SU(3)$. In $SU(3)$ the conjugate of a representation is obtained by interchanging p and q: $(p, q) \rightarrow (q, p)$. Therefore, $SU(3)$ representations such as $\mathbf{8}$ with $p = q$ are real, while those such as $\mathbf{3}$ with $p \neq q$ are complex. In contrast, all $SU(2)$ representations are real.

5.10 (a) These follow from the definitions (5.63) and (5.67)–(5.68), and the basic commutator (5.64a). (b) From (5.64a) and (5.65), prove that

$$\text{Tr} \left(\lambda_k [\lambda_i, \lambda_j] \right) = 4 i f_{ijk}.$$

Now use the cyclic property of the trace,

$$\text{Tr} \left(\lambda_i \lambda_j \lambda_k \right) = \text{Tr} \left(\lambda_k \lambda_i \lambda_j \right),$$

and (5.10) to show that $f_{ijk} = -f_{ikj}$. Combining this result with (5.12) gives

$$f_{ijk} = -f_{ikj} = -f_{jik} = -f_{kji}.$$

5.11 (a) The relations (5.42) and (5.43) lead to

$$J_3 \left(J_\pm |jm\rangle \right) = J_\pm (J_3 \pm 1) |jm\rangle = (m \pm 1) J_\pm |jm\rangle,$$

where $m \equiv j_3$. Therefore, $J_\pm |jm\rangle$ is an eigenvector of J_3 with eigenvalue $m \pm 1$ and

$$J_\pm |jm\rangle \propto |j, m \pm 1\rangle.$$

(b) From the commutators in Table 5.2 we see that $(V_\pm, V_3)$, $(U_\pm, U_3)$, and $(T_\pm, T_3)$ generate three independent $SU(2)$ subgroups [compare with (5.42)]. In a generalization of part (a), we may use the algebra of the operators $U_\pm, V_\pm$, and $T_\pm$ to define their action in the weight space. For example, from the commutators in Table 5.2

$$T_3\left(U_\pm |t_3 y\rangle\right) = \left(\mp\tfrac{1}{2}U_\pm + U_\pm T_3\right)|t_3 y\rangle$$
$$= U_\pm\left(t_3 \mp \tfrac{1}{2}\right)|t_3 y\rangle = \left(t_3 \mp \tfrac{1}{2}\right)U_\pm |t_3 y\rangle$$
$$Y\left(U_\pm |t_3 y\rangle\right) = \left(\pm U_\pm + U_\pm Y\right)|t_3 y\rangle$$
$$= U_\pm\left(y \pm 1\right)|t_3 y\rangle = \left(y \pm 1\right)U_\pm |t_3 y\rangle,$$

where t_3 and y are the eigenvalues of T_3 and Y, respectively. Therefore

$$U_\pm |t_3 y\rangle \propto \left|t_3 \mp \tfrac{1}{2}, y \pm 1\right\rangle,$$

which corresponds to the action of $U_\pm$ shown in Fig. 5.3. The properties of $V_\pm$ and $T_\pm$ follow by analogous proofs.

5.12 (a) The allowed diagrams are given in (5.88); applying the dimensionality rules (5.82)–(5.84) leads immediately to (5.89). (b) The allowed diagrams are given in (5.81); applying the rules (5.82)–(5.84), we find $\mathbf{\bar 3} \otimes \mathbf{3} = \mathbf{8} \oplus \mathbf{1}$. (c) Carry out the procedure in Fig. 5.6 for $\mathbf{3} \otimes \mathbf{3}$. (d) The $\mathbf{8}$ in $SU(3)$ is given by $(p,q) = (1,1)$, so upon applying the rules of §5.3.4

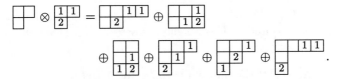

In determining representation dimensionality for $SU(N)$ the N-box columns may generally be omitted and

$$\square \square \otimes \square \square = \square \square \square \square \oplus \square \square \square \oplus \square \square \square \oplus \square \square \oplus \square \square \oplus \square.$$

This may also be written in terms of dimensionalities using (5.84):

$$\mathbf{8} \otimes \mathbf{8} = \mathbf{27} \oplus \overline{\mathbf{10}} \oplus \mathbf{10} \oplus \mathbf{8} \oplus \mathbf{8} \oplus \mathbf{1}.$$

As a simple check against omitting a necessary diagram or including a forbidden one, we note that the product of dimensionalities on the left side is equal to the sum of dimensionalities on the right side.

5.13 The u, d, s quantum numbers are given in Table 5.3, and $Y = B + S$ and T_3 are additive. Each baryon in the decuplet of Fig. 5.7 must consist of three quarks $(B = \tfrac{1}{3} + \tfrac{1}{3} + \tfrac{1}{3} = 1)$, such that the sum of contributions to Y and T_3 from the quarks gives the appropriate value of Y and T_3 for the baryon. For example, the baryon in the upper right corner $(\Delta^{++}$, see Fig. 5.7) has $T_3 = \tfrac{3}{2}$ and $Y = 1$. Consulting Table 5.3 and eq. (5.71), this can come only from a uuu structure (see Fig. 5.9), since then

$$T_3 = \tfrac{1}{2} + \tfrac{1}{2} + \tfrac{1}{2} = \tfrac{3}{2} \qquad Y = \tfrac{1}{3} + \tfrac{1}{3} + \tfrac{1}{3} = 1.$$

The baryon to its left (Δ^+) has $T_3 = \frac{1}{2}$ and $Y = 1$, corresponding to uud, since then

$$T_3 = \tfrac{1}{2} + \tfrac{1}{2} - \tfrac{1}{2} = \tfrac{1}{2} \qquad Y = \tfrac{1}{3} + \tfrac{1}{3} + \tfrac{1}{3} = 1.$$

Successive rows may be reached by replacing u or d quarks with s quarks. Considerations such as these give the quark content of the **10**; the actual flavor wavefunctions must incorporate the appropriate permutational symmetry. A technique for doing this is described in Exercise 12.14. Applying this method, the standard arrangements of Young tableaux for the **10** are

$$\boxed{1|1|1} \qquad \boxed{1|1|2} \qquad \boxed{1|1|3} \qquad \boxed{1|2|2} \qquad \boxed{1|2|3}$$

$$\boxed{1|3|3} \qquad \boxed{2|3|3} \qquad \boxed{3|3|3} \qquad \boxed{2|2|2} \qquad \boxed{2|2|3}$$

where the numbers are state labels: $(u, d, s) \equiv (1, 2, 3)$. These are one-row diagrams so all wavefunctions are completely symmetric. For example (see Exercise 12.14)

$$\psi\left(\boxed{1|1|2}\right) \equiv \psi\left(\boxed{u|u|d}\right) = \sqrt{\tfrac{1}{3}}\,(uud + duu + udu)$$

where $uud \equiv \psi_u(a)\psi_u(b)\psi_d(c)$ for particles a, b, and c, and we have normalized the wavefunction to unity. The weight vector is a sum of weight vectors for the constituent quarks:

$$\mathbf{m}\left(\boxed{1|1|2}\right) = (T_3, Y) = \mathbf{m}_u + \mathbf{m}_u + \mathbf{m}_d$$

$$= \left(\tfrac{1}{2}, \tfrac{1}{3}\right) + \left(\tfrac{1}{2}, \tfrac{1}{3}\right) + \left(-\tfrac{1}{2}, \tfrac{1}{3}\right) = \left(\tfrac{1}{2}, 1\right).$$

Comparing this weight with Fig. 5.7, we identify

$$\psi_{\Delta^+} = \sqrt{\tfrac{1}{3}}(uud + duu + udu).$$

The other nine states of the **10** follow in the same way; for example

$$\psi\left(\boxed{1|2|3}\right) = \sqrt{\tfrac{1}{6}}\,(uds + sud + sdu + usd + dsu + dus)$$

and $\mathbf{m}\left(\boxed{1|2|3}\right) = (0, 0)$, so $\boxed{1|2|3}$ is associated with Σ^{*0} in Fig. 5.7. More formally, the states of a multiplet may all be constructed from one of the states using the stepping operators (5.68b) [see Exercise 5.11b]. The same method may be applied to determine the quark content of other multiplets, except that when there are degeneracies (e.g., the center states of the **8**) we must orthogonalize the degenerate solutions.

5.14 (a) The **10** is a row of boxes, $\boxed{||}$, so it is completely symmetric. We assume the ground state orbital wavefunction is also symmetric ($L = 0$), and that the spin-$\frac{3}{2}$ baryons of the **10** must correspond to a symmetric spin wavefunction $|\!\uparrow\uparrow\uparrow\rangle$ for the three quarks. Thus

$$\psi(\text{space}) \cdot \psi(\text{spin}) \cdot \psi(\text{flavor}) = \text{symmetric}.$$

But the baryons must be totally *antisymmetric* (Pauli). If there is an additional $SU(3)$ degree of freedom (color—distinct from flavor) and the $SU(3)$ color state is

$$(0,0) = \mathbf{1} = \boxed{} = \text{antisymmetric},$$

the total wavefunction is antisymmetric. For a three-quark baryon a color singlet wavefunction is given by [apply the method of Exercise 12.14 to the above $(0,0)$ Young diagram]

$$(qqq)_\mathbf{1} = \sqrt{\tfrac{1}{6}}(RBG - BRG + BGR - GBR + GRB - RGB)$$

where R, B, and G are quark colors [eq. (10.35)], and RBG means quark 1 is red (R), quark 2 is blue (B), and quark 3 is green (G). This color wavefunction is manifestly antisymmetric under the exchange of any two particles. (b) From part (a) the ground state baryon wavefunction must have

$$\psi(\text{spin}) \cdot \psi(\text{flavor}) = \text{symmetric}.$$

The proton is uud; since $J = \tfrac{1}{2}$, two of the quarks must be in a spin singlet: $(\uparrow\downarrow - \downarrow\uparrow)$, which is antisymmetric under exchange of particles. Thus the corresponding two-quark flavor function must also be antisymmetric: $(ud - du)$. Combining and adding a third quark $u\uparrow$ gives $(u\uparrow d\downarrow - u\downarrow d\uparrow - d\uparrow u\downarrow + d\downarrow u\uparrow)u\uparrow$. This is symmetric in the first two quarks. Finally, take all cyclic permutations (three for each term) and normalize to obtain

$$\psi\,(p\uparrow) = \sqrt{\tfrac{1}{18}}\,(2u\uparrow u\uparrow d\downarrow + 2d\downarrow u\uparrow u\uparrow + 2u\uparrow d\downarrow u\uparrow - u\downarrow d\uparrow u\uparrow$$
$$- u\uparrow u\downarrow d\uparrow - d\uparrow u\uparrow u\downarrow - d\uparrow u\downarrow u\uparrow - u\uparrow d\uparrow u\downarrow - u\downarrow u\uparrow d\uparrow),$$

which is symmetric in the simultaneous exchange of spin and flavor for any two quarks. The corresponding neutron quark wavefunction may be obtained by the isospin rotation $u \leftrightarrow d$.

CHAPTER 6

6.1 The lepton numbers for the relevant particles are

	μ^+	$\bar\nu_\mu$	e^-	ν_e	e^+	$\bar\nu_e$	γ	p	n
L_μ	-1	-1	0	0	0	0	0	0	0
L_e	0	0	1	1	-1	-1	0	0	0

(a) Lepton family number L_i is conserved in the first reaction but not in the second. (b) A two-step mechanism that could occur in a composite nucleus is $n \to p + e^- + \bar\nu_e$, followed by $\bar\nu_e + n \to p + e^-$. However, the second step changes L_e.

6.2 (a) By explicit calculation, $\boldsymbol{\Sigma} \cdot \hat{\mathbf{p}}$ commutes with the Dirac Hamiltonian $H = \boldsymbol{\alpha} \cdot \mathbf{p} + \beta m$, but γ^5 commutes with H only if $m = 0$. However, helicity is frame dependent for massive particles; see §6.2.2. Set $m = 0$ in (1.100), multiply from the left by $\gamma^5 \gamma^0 = -i\gamma^1 \gamma^2 \gamma^3$, and use (1.28), (1.89), (1.91), and properties of the γ-matrices displayed in Table 1.2 to write

$$(\boldsymbol{\Sigma} \cdot \mathbf{p})\psi = \gamma^5 p_0 \psi \qquad \boldsymbol{\Sigma} \equiv \begin{pmatrix} \boldsymbol{\sigma} & 0 \\ 0 & \boldsymbol{\sigma} \end{pmatrix},$$

where we have employed steps such as

$$\gamma^5 \gamma^0 \gamma^1 = -i\gamma^1 \gamma^2 \gamma^3 \gamma^1 = i\gamma^2 \gamma^3 = \sigma_{23} = \begin{pmatrix} \sigma_1 & 0 \\ 0 & \sigma_1 \end{pmatrix}.$$

For $m = 0$ we have $E = p_0 = \pm|\mathbf{p}|$. Therefore,

$$(\boldsymbol{\Sigma} \cdot \hat{\mathbf{p}})\psi = \gamma^5 \psi \qquad \text{(positive energy)}$$
$$-(\boldsymbol{\Sigma} \cdot \hat{\mathbf{p}})\psi = \gamma^5 \psi \qquad \text{(negative energy)}$$

where $\hat{\mathbf{p}} = \mathbf{p}/|\mathbf{p}|$: the eigenvalues of $\boldsymbol{\Sigma} \cdot \hat{\mathbf{p}}$ and γ^5 are equal for positive-energy massless solutions and opposite in sign for negative-energy massless solutions.
(b) The *chiral representation* of the γ-matrices is obtained from the Pauli–Dirac representation (1.88)–(1.93) by the unitary transformation given in the footnote following (1.88):

$$\alpha_i = \begin{pmatrix} \sigma_i & 0 \\ 0 & -\sigma_i \end{pmatrix} \qquad \gamma^0 = \begin{pmatrix} 0 & -1 \\ -1 & 0 \end{pmatrix}$$
$$\gamma^i = \begin{pmatrix} 0 & \sigma_i \\ -\sigma_i & 0 \end{pmatrix} \qquad \gamma^5 = \begin{pmatrix} 1 & 0 \\ 0 & -1 \end{pmatrix}.$$

The Dirac equation is $(\boldsymbol{\alpha} \cdot \mathbf{p} + \beta m)\psi = E\psi$. Write ψ in terms of two (two-component) spinors:

$$\psi = \begin{pmatrix} \phi \\ \chi \end{pmatrix};$$

in the chiral representation the Dirac equation then is equivalent to

$$(E - \boldsymbol{\sigma} \cdot \mathbf{p})\phi = -m\chi \qquad (E + \boldsymbol{\sigma} \cdot \mathbf{p})\chi = -m\phi.$$

In the limit $m \to 0$ these equations decouple and we obtain the Weyl equations [see also part (a)]

$$(\boldsymbol{\sigma} \cdot \hat{\mathbf{p}})\phi = \frac{E}{|\mathbf{p}|}\phi \qquad (\boldsymbol{\sigma} \cdot \hat{\mathbf{p}})\chi = -\frac{E}{|\mathbf{p}|}\chi,$$

where $\hat{\mathbf{p}} = \mathbf{p}/|\mathbf{p}|$ and $E = p_0 = \pm|\mathbf{p}|$ for $m = 0$. From part (a), the massless spinors ϕ and χ are also eigenstates of chirality. (c) Using $(\gamma^5)^2 = 1$ we may easily show that $P_\pm \equiv \frac{1}{2}(1 \pm \gamma_5)$ satisfies

$$P_\pm^2 = P_\pm \qquad P_+ + P_- = 1 \qquad P_+ P_- = 0.$$

Thus $\frac{1}{2}(1 \pm \gamma_5)$ is a projection operator. From (6.16) it projects left- and right-handed chiral components. (d) A typical weak matrix element is $M \simeq \bar{u}_2 \gamma^\alpha (1 - \gamma_5) u_1$. But

$$\tfrac{1}{2}(1 - \gamma_5)\tfrac{1}{2}(1 - \gamma_5) = \tfrac{1}{2}(1 - \gamma_5);$$

using this, $\{\gamma^\mu, \gamma_5\} = 0$, and (6.17) we obtain

$$\bar{u}_2 \gamma^\alpha (1 - \gamma_5) u_1 = 2\bar{u}_2^{\mathrm{L}} \gamma^\alpha u_1^{\mathrm{L}},$$

and the $\frac{1}{2}\gamma^\alpha(1 - \gamma_5)$ structure restricts participation to left-handed particles. (e) The proofs follow from (6.17), $\{\gamma^\mu, \gamma^5\} = 0$, and $(\gamma^5)^2 = 1$. For example,

$$\begin{aligned}
\overline{R}\gamma_\mu L &= \tfrac{1}{4}\overline{\psi}(1 - \gamma_5)\gamma_\mu(1 - \gamma_5)\psi \\
&= \tfrac{1}{4}\overline{\psi}\gamma_\mu(1 + \gamma_5)(1 - \gamma_5)\psi \\
&= \tfrac{1}{4}\overline{\psi}\gamma_\mu\left(1 - \gamma_5^2\right)\psi \\
&= 0.
\end{aligned}$$

6.3 (a) Each loop gives $d^d k$ in the numerator and each internal line gives a k^2 from the propagator in the denominator, so

$$D = Ld - 2I,$$

where L is the number of internal loops (it is also the number of free loop momenta) and I is the number of internal lines. The $n - 1$ constraints from momentum conservation (overall and at each vertex) imply

$$L = I - (n - 1)$$

free loop momenta. Vertices have r legs so there are rn legs, both internal and external. Each internal line gives rise to two legs since it is connected to two vertices; hence $rn = E + 2I$. Combining these three equations gives the desired result. (b) Specializing to $d = 4$ gives

$$D\left(\phi^3\right) = 4 - E - n \qquad D\left(\phi^4\right) = 4 - E \qquad D\left(\phi^6\right) = 4 - E + 2n.$$

Thus, in four dimensions ϕ^3 divergences decrease with perturbative order (super-renormalizable: a finite number of divergent graphs for fixed E), ϕ^6 divergences get worse with increasing order (non-renormalizable), while ϕ^4 superficial degrees of divergence are independent of order (ϕ^4 turns out to be renormalizable). For $d = 2$ the degree of divergence is $D = 2 - 2n$, which is independent of r and E (see Ryder, 1985, §9.1).   (c) From the formula derived in part (b) we have $D = -2$, so the graph is superficially quite convergent. In actual fact it is *divergent*, because of the logarithmic divergence in the loop. This illustrates one way that a superficial and actual degree of divergence may differ: the divergence in the loop momentum integration is not altered by embellishing the diagram with additional lines and vertices that make the total diagram superficially convergent.

6.4   Generalizing (A.6) to d dimensions gives $[\mathcal{L}] = [M]^d$, and from the Lagrangian kinetic energy term and (A.10),

$$[\phi] = [M]^{d/2-1}.$$

Then from the interaction term $g\phi^r$ we must have

$$[g][\phi]^r = [M]^d \qquad [g] = [M]^\delta \qquad \delta = d + r - rd/2.$$

Solve for r and substitute in the equation for D derived in Exercise 6.3.

6.5   (a) Follow the procedure illustrated in Exercise 6.3, but count fermion and photon lines separately with $1/k$ and $1/k^2$ factors, respectively, from their propagators. (b) Analogous to Exercise 6.3, but count $1/k$ from each fermion propagator.

CHAPTER 7

7.1   Perform a gauge transformation on (7.6) using (7.3) to obtain an expression for $D'_\mu \Psi'(x)$ in terms of Ψ' and A'_μ. Insert (7.6) in (7.5) to obtain another expression for $D'_\mu \Psi'(x)$. Equating for arbitrary $\Psi(x)$ leads to (7.8). Use infinitesimal unitary transformations

$$U(\theta) \simeq 1 - i\boldsymbol{\tau} \cdot \boldsymbol{\theta} \qquad U^{-1}(\theta) \simeq 1 + i\boldsymbol{\tau} \cdot \boldsymbol{\theta}$$

in (7.8), keep terms only to first order in θ_j, and employ (5.9). Take the trace and use (5.65) and $f_{ijk} = -f_{kji}$ (Exercise 5.10b) to obtain (7.9).

7.2   (a) For a $U(1)$ field
$$U = e^{iq\chi(x)};$$
identify $g = q$ and $\boldsymbol{\tau} \cdot \mathbf{A}_\mu = A_\mu$, and substitute in (7.8).   (b) Use $A_\mu \equiv A^j_\mu \tau_j$

and (5.9) in (5.10a); take the trace of both sides and use (5.65b) and $f_{jkl} = f_{ljk}$ (Exercise 5.10b).   (c) Evaluate $[D_\mu, D_\nu]\psi$ for arbitrary ψ using (2.138) to show that

$$\frac{1}{iq}[D_\mu, D_\nu] = (\partial_\mu A_\nu - \partial_\nu A_\mu) + iq[A_\mu, A_\nu] = F_{\mu\nu},$$

where the last step follows from (2.127) and the vanishing Abelian commutator. For a non-Abelian theory $[A_\mu, A_\nu] \neq 0$ and we are left with the form (7.10).   (d) From (7.5), (7.3), and the preceding result, $F_{\mu\nu} = [D_\mu, D_\nu]/ig$, show that under a gauge transformation

$$\tau_l F^l_{\mu\nu} \to U \left(\tau_l F^l_{\mu\nu} \right) U^{-1}.$$

Expand $U \simeq 1 - i\tau_k \theta^k$, use (5.9), multiply both sides by τ_j and take the trace, and use (5.65) and the antisymmetry of f_{jkl} to show that

$$F^j_{\mu\nu} \to F^j_{\mu\nu} + f_{jkl}\theta^k F^l_{\mu\nu}.$$

For $U(1)$ the second term is zero; for a non-Abelian theory the second term implies that $F^j_{\mu\nu}$ transforms as the adjoint representation [see eqns. (5.14) and (5.8)]. However, $F^j_{\mu\nu}F^{\mu\nu}_j$ is gauge invariant, by an analogous proof.

7.3 (a) Let a gauge transform be defined by $\psi'(x) = G(x)\psi(x)$. Then, from (7.19)

$$\psi'(x_1) = U'_\gamma(x_0, x_1)\psi'(x_0) \qquad G(x_1)\psi(x_1) = U'_\gamma(x_0, x_1)G(x_0)\psi(x_0)$$

$$G(x_1)U_\gamma(x_0, x_1)\psi(x_0) = U'_\gamma(x_0, x_1)G(x_0)\psi(x_0),$$

and since $\psi(x_0)$ is arbitrary we obtain (7.20) on dropping $\psi(x_0)$ and multiplying from the right by $G^{-1}(x_0)$. For a *closed path*, $x_0 = x_1$ and (7.20) becomes a *similarity transform* [see eq. (5.5)]; the trace of a matrix is unchanged by a similarity transform so $\mathrm{Tr}\, U_\gamma(x, x)$ is gauge invariant.   (b) Multiply by τ_k, take the trace, and use (5.65).

7.4 From the definitions (7.7) and the generators $\sigma_i/2$ and $\lambda_i/2$ [see (1.78) and (5.63)]

$$A^{SU(2)}_\mu = \frac{1}{2} \begin{pmatrix} A^3_\mu & A^1_\mu - iA^2_\mu \\ A^1_\mu + iA^2_\mu & -A^3_\mu \end{pmatrix}$$

$$A^{SU(3)}_\mu = \frac{1}{2} \begin{pmatrix} A^3_\mu + A^8_\mu/\sqrt{3} & A^1_\mu - iA^2_\mu & A^4_\mu - iA^5_\mu \\ A^1_\mu + iA^2_\mu & -A^3_\mu + A^8_\mu/\sqrt{3} & A^6_\mu - iA^7_\mu \\ A^4_\mu + iA^5_\mu & A^6_\mu + iA^7_\mu & -\frac{2}{\sqrt{3}}A^8_\mu \end{pmatrix}.$$

7.5 The axial gauge-fixing condition [(7.22b) and (7.25)] is $f_j = A_3^j = 0$. Under an infinitesimal transformation (7.9)

$$f_j\left(A_\mu'\right) = A_3^j(x) + \frac{1}{g}\partial_3\theta^j + f_{jkl}\theta^k(x)A_3^l = \frac{1}{g}\partial_3\theta^j,$$

and the response matrix (7.27) is

$$\left(\frac{\delta f}{\delta\theta}\right)_{jk} \equiv \frac{\delta f_j}{\delta\theta_k} = \frac{1}{g}\partial_3\delta_{jk}.$$

This is independent of the gauge fields A_μ so $\text{Det}(\delta f/\delta\theta)$ may be removed from the integral in (7.26) and absorbed in a normalization; it plays no role in the field theory dynamics. For an Abelian theory the gauge transformation is (7.9), less the final term [i.e., (2.120)]. Again, for gauge fixing conditions of the form (7.25) the response matrix (7.27) is independent of the gauge field A_μ, and the determinant may be removed from the functional integral (7.26). For example, in Lorentz gauge $f = \partial^\mu A_\mu = 0$, and from (2.120),

$$\text{Det}\left(\delta f/\delta\chi\right) = \text{Det}\left(-\partial^\mu\partial_\mu\right),$$

which is independent of the gauge field. More generally, for non-Abelian theories in nonaxial gauges the Faddeev–Popov determinant will depend on D^μ instead of ∂^μ. Thus it depends on the gauge fields A^μ and cannot be factored out of the functional integral (see Cheng and Li, 1984, §9.1).

CHAPTER 8

8.1 (a) The requirement $U\left|0\right\rangle = \left|0\right\rangle$ means

$$e^{i(L_1\theta_1+L_2\theta_2+\ldots+L_N\theta_N)}\left|0\right\rangle = \left|0\right\rangle,$$

where L_i denotes the group generators. By expanding the exponential, this can be satisfied for arbitrary θ_i only if $L_i\left|0\right\rangle = 0$ $(i = 1, 2, \ldots, N)$. Thus the condition $U\left|0\right\rangle \neq \left|0\right\rangle$ (spontaneous symmetry breaking) implies that at least one generator fails to annihilate the vacuum.   (b) Use the first derivative of V to find the extrema, and the sign of the second derivative to ensure that the extremum is a minimum. Substitute (8.6) in (8.1) and use (8.4) to obtain (8.7). Compare the quadratic term in ϕ' with (2.41) to obtain $m_{\phi'}$.

8.2 The potential is

$$V = \tfrac{1}{2}\mu^2\left(\phi_1^2 + \phi_2^2 + \phi_3^2\right) + \tfrac{1}{4}\lambda\left(\phi_1^2 + \phi_2^2 + \phi_3^2\right)^2.$$

Use first and second derivatives to show that the potential has a minimum either when $\phi_1 = \phi_2 = \phi_3 = 0$ with $\mu^2 > 0$ (Wigner mode), or when

$$|\phi| = \sqrt{\phi_1^2 + \phi_2^2 + \phi_3^2} = \sqrt{\frac{-\mu^2}{\lambda}} \equiv v$$

with $\mu^2 < 0$ (Goldstone mode). Thus the degenerate vacua lie on the surface of a sphere of radius v in the isospace. Choose the vacuum on the 3-axis:

$$\langle \phi \rangle_0 = v\hat{\epsilon}_3,$$

where $\hat{\epsilon}_3$ is a unit vector in isospin space. The ground state is no longer $SO(3)$ invariant, but it is still invariant under $SO(2)$ rotations about the 3-axis. Introduce the shifted field χ through

$$\phi_3(x) = \chi(x) + v,$$

and rewrite the Lagrangian in terms of the physical fields ϕ_1, ϕ_2, and χ. Identify the masses from the quadratic Lagrangian terms: a massive scalar with $m_\chi = (2\lambda v^2)^{1/2}$ and two Goldstone bosons with $m_{\phi_1} = m_{\phi_2} = 0$.

8.3 See §12.3, Exercises 8.2 and 12.4, and Taylor (1976), §5.2 and §5.5.

CHAPTER 9

9.1 For example, the family (ν_e, e_L) transforms under the gauge group $SU(2)_w \times U(1)_y$. The simplest assumption is that this constitutes a weak $SU(2)$ doublet, so $t = \frac{1}{2}$ and $t_3 = \pm\frac{1}{2}$. Using $Q_\nu = 0$ and $Q_e = -1$, eq. (9.2), and requiring that $y_\nu = y_e$ gives the first two lines of Table 9.1. As another example, the right-handed fermions do not enter the charged-current weak interactions so e_R must be an $SU(2)_w$ singlet: $t = t_3 = 0$. Then from (9.2), $y = -2$ for $Q = -1$.

9.2 The charges are given by (6.14) and (1.118) inserted in (2.193):

$$T_+(t) = \frac{1}{2} \int d^3x \, \nu^\dagger (1 - \gamma_5)e \qquad T_-(t) = (T_+(t))^\dagger \qquad Q_{em} = \int d^3x \, e^\dagger e.$$

Then from (2.89) we find

$$[T_+(t), T_-(t)] \neq 2Q_{em},$$

so the set of generators $\{T_\pm, Q_{em}\}$ is not closed under commutation [compare (5.42)]. Cheng and Li (1984), §11.2, describe how to close the algebra, either by adding a fourth gauge boson [this gives the standard $SU(2) \times U(1)$ electroweak model] or by introducing new (heavy—to explain why they are not seen) fermions in the multiplets to modify the currents so that they close under $SU(2)$. This latter approach was suggested by Georgi and Glashow (1972) and requires no neutral weak currents; it was made less attractive by the discovery of such currents.

9.3   An explicit Dirac mass term is of the form $m\bar{\psi}\psi$ [see (2.46)]. From Exercise 6.2e we have

$$\bar{\psi}\psi = \bar{L}R + \bar{R}L,$$

where L and R are left- and right-handed components. But in the electroweak theory L is an $SU(2)$ doublet while R is a singlet [see eqns. (9.1a–d)]. Therefore $\bar{\psi}\psi$ cannot be invariant under $SU(2)$. However, (9.14) is invariant since the gauge doublets $\bar{L}$ and ϕ (and $\phi^\dagger$ and L) couple to an $SU(2)$ singlet. Explicitly, (9.14) is

$$\mathcal{L}_{\text{f–s}} = -G_{\text{e}} \left[\bar{e}_{\text{R}} \left(\phi^- \; \bar{\phi}^0 \right) \begin{pmatrix} \nu_{\text{L}} \\ e_{\text{L}} \end{pmatrix} + (\bar{\nu}_{\text{L}} \; \bar{e}_{\text{L}}) \begin{pmatrix} \phi^+ \\ \phi^0 \end{pmatrix} e_{\text{R}} \right].$$

Multiply out and use

$$\begin{pmatrix} \phi^+ \\ \phi^0 \end{pmatrix} = \frac{1}{\sqrt{2}} \begin{pmatrix} 0 \\ v + \eta \end{pmatrix}$$

in unitary gauge [see eq. (9.20)] to obtain (9.21).

9.4   (a) Comparing with the Feynman rules in Fig. 9.1 to determine C_{V} and C_{A}, we have

$$\Gamma = \frac{Gm_W^3}{6\sqrt{2}\pi} \cong 242 \text{ MeV},$$

where (6.12) and $m_W = 82$ GeV were used.   (b)–(c) The total width of the W may be estimated by allowing it to decay to all available leptons and hadrons, subject to basic conservation laws. The partial width for each decay mode follows from the formula for Γ in part (a) and the vertices given in Fig. 9.1. Assuming weak universality, three generations of leptons, and three generations of quarks each with three colors, the total width is the sum of these partial widths:[‡]

$$\Gamma_{\text{total}} \simeq (3 + 3 \times 3)\Gamma(W \rightarrow e\nu) \simeq \frac{\sqrt{2}Gm_W^3}{\pi} \simeq 2.9 \text{ GeV},$$

where the result of part (a) was used. Notice that we have ignored fermion

---

[‡]*Note added in proof:* recent lower limits on the mass of the top quark force some modification of the assumptions used in this exercise and the next.

masses, and that the Cabibbo mixing only redistributes the strength and has been omitted in estimating the total width. From parts (a) and (c) we have the branching ratio

$$\frac{\Gamma(W \to l\bar{\nu}_l)}{\Gamma(W \to \text{everything})} \simeq \frac{1}{12}$$

for $l = e, \mu, \tau$.

9.5 (a) Compare the formula for Γ given in Exercise 9.4a with the Feynman rule for the ffZ^0 vertex given in Fig. 9.1 and show that

$$C_V^2 + C_A^2 = 2C_L(f)^2 + 2C_R(f)^2 \qquad g_X = \frac{g}{\cos \theta_w}.$$

(b) Insert the values of $C_{L,R}(f)$ given in Fig. 9.1 into the equation derived in part (a). For example,

$$\Gamma_{Z\nu\bar{\nu}} = \frac{g^2}{96\pi \cos^2 \theta_w} m_Z = \frac{Gm_Z^3}{12\sqrt{2}\pi} \simeq 175 \text{ MeV},$$

where we have taken $m_Z \simeq 93$ GeV. From this result we may write

$$\Gamma_{Zf\bar{f}} = 4N_c \left[C_L(f)^2 + C_R(f)^2 \right] \Gamma_{Z\nu\bar{\nu}},$$

where N_c is a color factor ($N_c = 1$ for leptons, 3 for quarks). For example,

$$\Gamma_{Ze^+e} = \left[4(\sin^2 \theta_w - \tfrac{1}{2})^2 + 4\sin^4 \theta_w \right] \Gamma_{Z\nu\bar{\nu}},$$

$$\Gamma_{Zu\bar{u}} = 3 \left(1 - \tfrac{8}{3} \sin^2 \theta_w + \tfrac{32}{9} \sin^4 \theta_w \right) \Gamma_{Z\nu\bar{\nu}}$$

$$\Gamma_{Zd\bar{d}} = 3 \left(1 - \tfrac{4}{3} \sin^2 \theta_w + \tfrac{8}{9} \sin^4 \theta_w \right) \Gamma_{Z\nu\bar{\nu}}.$$

Assuming universality and summing over leptons and quarks, the total width is

$$\Gamma_Z^{\text{total}} = 3 \left[2(1 + D_q) - 4(1 + D_q) \sin^2 \theta_w + 8(1 + \tfrac{5}{9} D_q) \sin^4 \theta_w \right] \Gamma_{Z\nu\bar{\nu}},$$

where D_q is the number of color triplet, weak isospin quark doublets energetically accessible to the Z^0 decay. For $D_q = 3$ we have $\Gamma_Z^{\text{total}} \simeq 2.9$ MeV; for $D_q = 2$ instead, $\Gamma_Z^{\text{total}} \simeq 2.2$ MeV (see the note added in proof to the Exercise 9.4 solution).

9.6 The quark neutral currents are of the form

$$J_N \simeq \left[(\bar{u}\,\bar{d}_c)_L \gamma^\mu \tau_3 \begin{pmatrix} u \\ d_c \end{pmatrix}_L + (\bar{c}\,\bar{s}_c)_L \gamma^\mu \tau_3 \begin{pmatrix} c \\ s_c \end{pmatrix}_L \right] Z_\mu^0$$

where τ_3 is a Pauli matrix. Multiplying out the first term and using (6.36) gives terms diagonal in flavor plus the flavor-changing terms in (9.40). However, the second term in conjunction with (9.41) gives terms that exactly cancel the flavor-changing pieces of the first term, so Z^0 only couples diagonally to flavor in this current. This is the GIM mechanism in lowest order. A simple discussion of higher order flavor-changing terms and the constraint they place on the mass of the charmed quark may be found in Aitchison and Hey (1982), §6.4.

CHAPTER 10

10.1 (a) For spinless point particles in the Breit frame, angular momentum conservation allows the absorption of longitudinal ($L_z = 0$) photons, but not transverse ($L_z = \pm 1$) photons (see Fig. 10.13). By similar reasoning, $S = \frac{1}{2}$ point particles can absorb transverse photons, but not longitudinal ones: helicity is conserved in electromagnetic interactions with massless leptons (Exercise 6.2e), so the z-component of the parton angular momentum must change by ± 1 unit in the interaction. (b) See Close (1979), §16.3.3.1; Perkins (1987), §5.7 and Table 5.9. Data may be found in Aguilar-Benitez et al. (1988), "Meson Summary Table".

10.2 The 4-momentum transfer is (see Fig. 10.7)

$$
\begin{aligned}
q^2 &= (k - k')^2 = (E - E')^2 - (\mathbf{k} - \mathbf{k}')^2 \\
&= E^2 - 2EE' + E'^2 - \mathbf{k}^2 + 2\mathbf{k} \cdot \mathbf{k}' - \mathbf{k}'^2 \\
&= 2m^2 + 2|\mathbf{k}||\mathbf{k}'|\cos\theta - 2EE' \\
&\simeq 2|\mathbf{k}||\mathbf{k}'|(\cos\theta - 1),
\end{aligned}
$$

where θ is the scattering angle, $m^2 = E^2 - \mathbf{k}^2 = E'^2 - \mathbf{k}'^2$, and $m \to 0$ so $EE' \to |\mathbf{k}||\mathbf{k}'|$ in the last step. Therefore, $q^2 \leqslant 0$ (spacelike).

10.3 (a) See Cheng and Li (1984), §7.1; Leader and Predazzi (1982), §14.3. (b) We wish to produce a spin-averaged cross section (10.7) that is Lorentz and gauge invariant. To construct a rank-2 Lorentz tensor we have available $g^{\mu\nu}$, $\epsilon_{\mu\nu\alpha\beta}$, γ^μ, the 4-momentum transfer q^μ, and the proton 4-momentum p^μ. We are spin averaging so γ^μ will not contribute and the most general combination of terms giving a rank-2 tensor is

$$
\begin{aligned}
W^{\mu\nu} =\, &A_1 g^{\mu\nu} + A_2 p^\mu p^\nu + A_3(p^\mu q^\nu - p^\nu q^\mu) \\
&+ A_4 q^\mu q^\nu + A_5(p^\mu q^\nu + p^\nu q^\mu) + A_6 \epsilon^{\mu\nu\alpha\beta} p_\alpha q_\beta,
\end{aligned}
$$

where the A_n may be functions of Lorentz scalars. However, the contraction in (10.7) is with a symmetric lepton tensor [see (3.102)], so the antisymmetric terms in $W^{\mu\nu}$ may be omitted: $A_3 = A_6 = 0$. [For neutrino scattering the structure of the leptonic current allows a parity-violating term and $A_6 \neq 0$; this gives an additional structure function in (10.20) and (10.21) relative to

(10.15).] Now current conservation, $\partial_\mu j^\mu = 0$, applied to (10.9) requires

$$q_\mu W^{\mu\nu} = q_\nu W^{\mu\nu} = 0 \qquad \text{(gauge invariance)}.$$

[see the alternative form of (10.9) in part (a) of this exercise]. Impose this condition, demand that the coefficients of p^ν and q^ν vanish separately, and use the definitions $W_1 = -A_1$ and $W_2 = A_2 M^2$ to obtain (10.11).

10.4 (a) In the lab frame $p = (M, 0, 0, 0)$, and

$$\nu = \frac{p \cdot q}{M} = q_0 = E - E' \qquad q = (E - E', \mathbf{k} - \mathbf{k}').$$

Likewise,

$$y = \frac{p \cdot q}{p \cdot k} = \frac{q_0}{k_0} = \frac{E - E'}{E} = \frac{\nu}{E}.$$

(b) From (10.13), $x = -q^2 / 2M\nu$. But $q^2 \leqslant 0$ (Exercise 10.2) and $\nu \geqslant 0$ (Exercise 10.4a), so $x \geqslant 0$. From (10.13) and (10.10c),

$$x = \frac{-q^2}{2M\nu} = 1 + \frac{M^2 - W^2}{2M\nu},$$

and since $W^2 \geqslant M^2$ we have $x \leqslant 1$. Combining with the previous result gives $0 \leqslant x \leqslant 1$. From part (a),

$$y = 1 - \frac{E'}{E};$$

therefore, $0 \leqslant y \leqslant 1$.

10.5 (a) For elastic scattering $W^2 = M^2$, so (10.10c) implies that $2M\nu = Q^2$, and from (10.13), $x = 1$. For resonance production $W^2 = (M')^2$, where $(M')^2$ is the (mass)^2 of the resonance; hence,

$$2M\nu = Q^2 + (M'^2 - M^2).$$

For real photons $Q^2 = 0$ (photoproduction on the horizontal axis of Fig. 10.8).
(b) Assume the infinite momentum frame so that masses and transverse momenta of partons may be neglected, and let ξp be the 4-momentum carried by a struck parton of effective mass m (p is the proton 4-momentum). If the parton absorbs 4-momentum q from the virtual photon, its 4-momentum becomes $\xi p + q$ and $E^2 = \mathbf{p}^2 + m^2$ requires

$$(\xi p + q)^2 = m^2 \simeq 0.$$

Expand assuming $|\xi^2 p^2| = \xi^2 M^2 \ll q^2$, which gives $\xi \simeq -q^2 / 2p \cdot q$. The scalar product $p \cdot q$ is invariant, so we can evaluate it in any frame; in the lab frame $p = (M, 0, 0, 0)$ and

$$\xi \simeq \frac{-q^2}{2M q_0} = \frac{Q^2}{2M\nu} = x$$

(see Exercise 10.4a). Therefore, with these assumptions (see the reservations in §10.4 about transforming results from the infinite momentum frame to the lab frame) the Bjorken scaling variable x may be interpreted as the fraction of the proton 4-momentum carried by the struck parton (see Fig. 10.10).

10.6 (a) Equation (10.23) follows immediately from (10.19) and (10.22). Use (10.22) to obtain (10.25) and (10.26). To prove (10.27), use (10.25)–(10.26) and assume

$$q_{\mathrm{v}} = u_{\mathrm{v}} + d_{\mathrm{v}} \qquad 2q_{\mathrm{s}} \simeq 2\bar{q}_{\mathrm{s}} \simeq u_{\mathrm{s}} + d_{\mathrm{s}} + \bar{u}_{\mathrm{s}} + \bar{d}_{\mathrm{s}}$$

to show that $F_2^{\nu} = x(q_{\mathrm{v}} + 2q_{\mathrm{s}})$ and $xF_3^{\nu} = xq_{\mathrm{v}}$. (b) Evaluate $F_2^{\mathrm{en}}/F_2^{\mathrm{ep}}$ using (10.19a,b), neglecting strange quarks. The upper limit follows from assuming $d + \bar{d} \gg u + \bar{u}$; the lower limit from assuming $d + \bar{d} \ll u + \bar{u}$.

10.7 Define $\beta_0 = (33 - 2N_{\mathrm{f}})/12\pi$ and use $\log\left[\exp(\beta_0\alpha)^{-1}\right] = 1/(\beta_0\alpha)$ to write (10.40) as

$$\alpha(Q^2) = \frac{12\pi}{(33 - 2N_{\mathrm{f}})\log\left[\frac{Q^2}{\mu^2}\exp\left(1/\beta_0\alpha(\mu^2)\right)\right]}.$$

Eq. (10.41) follows if we define $\Lambda^2 = \mu^2\exp\left[-1/\beta_0\alpha(\mu^2)\right]$.

CHAPTER 11

11.1 The Dirac charge conjugation operator in the Pauli–Dirac representation is $C = i\gamma^2\gamma^0$ (see Exercise 1.8c); if ψ annihilates a particle, the charge conjugate field

$$\psi^{\mathrm{c}} = C\bar{\psi}^{\mathrm{T}} = C\gamma^0\psi^*$$

annihilates an antiparticle. We may write [see eq. (6.17)]

$$\begin{aligned}\psi_{\mathrm{L}}^{\mathrm{c}} &= \tfrac{1}{2}(1 - \gamma_5)\psi^{\mathrm{c}} = \tfrac{1}{2}(1 - \gamma_5)C\bar{\psi}^{\mathrm{T}} \\ &= C\tfrac{1}{2}\left[\bar{\psi}(1 - \gamma_5)\right]^{\mathrm{T}} = C\left(\bar{\psi}_{\mathrm{R}}\right)^{\mathrm{T}}.\end{aligned}$$

Thus the charge conjugate of a right-handed field is left-handed (see also Exercise 6.2).

11.2 The essential point is that the coupling constants g and g' of the $SU(2)_{\mathrm{w}} \times U(1)_y$ subgroup of $SU(5)$ are not independent [in contrast to the GSW theory, where the $SU(2) \times U(1)_y$ group is not embedded in a larger group]. We may obtain a relationship between g and g' by analyzing the $SU(5)$ charge operator Q. The charge is an additive quantum number so the corresponding operator is a linear combination of diagonal $SU(5)$ generators [compare eq. (5.70) for $SU(3)$]. There are $N - 1 = 4$ independent diagonal generators for $SU(5)$ (Cartan subalgebra; see §5.3.2), but two of them are associated with the $SU(3)_c$ subgroup; since the color field does not couple to electrical charge, we may write

$$Q = t_3 + \frac{y}{2} = t_3 + Kt_0,$$

where t_3 and t_0 are diagonal generators associated with the $SU(2)$ and $U(1)$ subgroups of $SU(5)$, respectively, and y is the weak hypercharge. The constant K relates the normalization of the operators y and t_0, and may be obtained by examining the charges of an $SU(5)$ representation, say the fundamental multiplet **5**. The weak hypercharges of the $SU(5)$ fundamental [see (11.6)]

$$\mathbf{5} = \left(d^R, d^B, d^G, e^+, \bar{\nu}_e\right)_R$$

(where the subscript R denotes right-handed and the superscripts R, B, and G indicate colors) are (Table 9.1 and its extension to antiparticles)

$$y = \left(-\tfrac{2}{3}, -\tfrac{2}{3}, -\tfrac{2}{3}, +1, +1\right).$$

The $SU(5)$ generator $t_0 = \lambda_0/2$ of the $U(1)$ subgroup is given by [see Cheng and Li (1984), eq. (14.2)]

$$\lambda_0 = \sqrt{\tfrac{1}{15}} \, \mathrm{Diag}\,(2,2,2,-3,-3),$$

where the standard $SU(N)$ normalization (5.65) is used. Comparing the eigenvalues of y and λ_0 for the **5** and requiring $y = K\lambda_0$ gives $K = -\sqrt{5/3}$. Therefore, comparing the interaction terms in the GSW and $SU(5)$ theories, the GSW $U(1)$ coupling g' and the $SU(5)$ subgroup $U(1)$ coupling g_1 are related by $g'^2 = \tfrac{3}{5}g_1^2$. [The non-Abelian couplings do not need such renormalization since they are sufficiently constrained by the commutator algebra to uniquely specify the normalization required for them to be a generator of $SU(5)$—see (5.9) and (5.65).] Utilizing (9.27b), and that *at the unification mass* $g = g_2 = g_1$ [where g_2 is the $SU(2)_w$ subgroup coupling constant], we obtain

$$\sin^2 \theta_w = \frac{g'^2}{g^2 + g'^2} = \frac{(3/5)g_1^2}{g_1^2 + (3/5)g_1^2} = \frac{3}{8}.$$

11.3 See Georgi and Glashow (1974). The group must be at least of rank 4 to allow an $SU(3) \times SU(2) \times U(1)$ subgroup, and it must be simple so that there is only one coupling constant. There are many possibilities, but only $SU(5)$ survives the requirements of low-energy phenomenology that the unifying group have complex representations, and that it allow both fractionally and integrally charged particles. It can accommodate the known particles in anomaly-free representations (see Exercise 11.4), which satisfies the final criterion.

11.4 See Cheng and Li (1984), §14.1. Choose $T^a = T^b = T^c = Q$, where Q

is the charge operator, and evaluate for the $\bar{\mathbf{5}}$ and $\mathbf{10}$ representations of eq. (11.6):

$$A\left(\bar{\mathbf{5}}\right) = \operatorname{Tr} Q^3\left(\bar{\mathbf{5}}\right) = 3\left(\tfrac{1}{3}\right)^3 + 0^3 + (-1)^3 = -\tfrac{8}{9}$$
$$A\left(\mathbf{10}\right) = \operatorname{Tr} Q^3\left(\mathbf{10}\right) = 3\left(\tfrac{2}{3}\right)^3 + 3\left(-\tfrac{1}{3}\right)^3 + 3\left(-\tfrac{2}{3}\right)^3 + (1)^3 = +\tfrac{8}{9}.$$

CHAPTER 12

12.1 See Lee (1981), §17.3; Baym (1982c).

12.2 The simpler ones are worked in Ch. 5 (see Exercise 5.12). The more complicated ones require the full set of rules given in §5.3.4. For example, utilizing (5.79) and (5.84)

$$qq\bar{q} = \mathbf{3} \otimes \mathbf{3} \otimes \bar{\mathbf{3}} = \left(\ \square\square \ \oplus\ \begin{array}{c}\square\\\square\end{array}\right) \otimes \begin{array}{c}\boxed{1}\\\boxed{2}\end{array}$$

$$= \square\ \oplus\ \boxed{\ \ \ \boxed{1}\ }_{\boxed{2}}\ \oplus\ \begin{array}{c}\boxed{\ \ \boxed{1}}\\\boxed{2}\end{array}\ \oplus\ \square$$

$$= \mathbf{3} \oplus \mathbf{15} + \bar{\mathbf{6}} \oplus \mathbf{3},$$

where we have used

$$\begin{array}{c}\square\square\\\square\end{array} = \square$$

for $SU(3)$ dimensionalities. (A diagram $\boxed{\ \ \ \boxed{1}\boxed{2}}$ would be excluded by rule 1c; a diagram with four vertical boxes by rule 1b, etc.). Multiplying again by $\bar{q}$ we find

$$qq\bar{q}\bar{q} = \mathbf{3} \otimes \mathbf{3} \otimes \bar{\mathbf{3}} \otimes \bar{\mathbf{3}}$$
$$= \mathbf{1} \oplus \mathbf{1} \oplus \mathbf{8} \oplus \mathbf{8} \oplus \mathbf{8} \oplus \mathbf{8} \oplus \mathbf{10} \oplus \overline{\mathbf{10}} \oplus \mathbf{27},$$

so $qq\bar{q}\bar{q}$ contains two $SU(3)$ singlets. Multiplying the results in eq. (12.8) [e.g., $qqqq = (qqq)q = (\mathbf{1} \oplus \mathbf{8} \oplus \mathbf{8} \oplus \mathbf{10}) \otimes \mathbf{3}$] we find that $qqqqqq$ is the simplest quarks-only configuration beyond qqq that can have a color singlet.

12.3 (a) Write the charges (12.15), displaying explicit matrix indices; then calculate the commutators directly. For example (let $t^a \equiv \tau^a/2$)

$$[Q_a(t), Q_b(t)] = \int d^3x\, d^3y\, t^a_{ij} t^b_{kl} [q^\dagger_i(x) q_j(x), q^\dagger_k(y) q_l(y)].$$

Expand the commutator using

$$[AB, CD] = [A, C]DB + C[A, D]B + A[B, C]D + AC[B, D]$$

and replace the commutators with anticommutators using $[A, B] = 2AB -$

$\{A, B\}$. Then use (2.89), that summation indices are dummy indices, and that for matrices a and b the elements of the commutator matrix are $[a, b]_{ij} = a_{ik}b_{kj} - b_{ik}a_{kj}$, to obtain

$$[Q_a(t), Q_b(t)] = \int d^3x \, q^\dagger(x)[t^a, t^b]q(x) = i\epsilon_{abc}Q_c(t),$$

where (5.30) and (12.15) have been used. By an analogous procedure we find $[Q_{a5}, Q_{b5}] = i\epsilon_{abc}Q_c$ and $[Q_a, Q_{b5}] = i\epsilon_{abc}Q_{c5}$. With these commutators and (12.16) it is then a simple exercise to obtain the algebra (12.17). (b) The Dirac parity operation is $Pq(\mathbf{x}, t)P^{-1} = \gamma_0 q(-\mathbf{x}, t)$, up to a phase. Applying this to (12.15) gives

$$PQ_i(t)P^{-1} = Q_i(t) \qquad PQ_{i5}(t)P^{-1} = -Q_{i5}(t),$$

where $\{\gamma_5, \gamma_\mu\} = 0$ has been used. [Notice that this last result applied to (12.29) implies opposite parities for $|\psi_+\rangle$ and $|\psi_-\rangle$.] Then from (12.16),

$$PQ_\pm P^{-1} = Q_\mp.$$

12.4 (a) Substitute the first-order expansions (12.21), neglect terms of order α^2, and use standard γ-matrix identities. (b) The conserved currents associated with the isospin and axial symmetries of the Lagrangian density follow from the Noether equation (2.192). For the isospin currents V_μ^a the elements of the representation matrices t_{ij}^a are $(\tau^a/2)_{ij}$ for the nucleon $SU(2)$ fundamental doublet, and $-i\epsilon_{aij}$ for the pion isotriplet (adjoint representation, see Exercise 5.7), resulting in a conserved vector current

$$V_\mu^a = \overline{N}\gamma_\mu \frac{\tau^a}{2} N + \epsilon_{abc}\pi^b\left(\partial_\mu \pi^c\right).$$

For the axial current the first two infinitesimal transformations in (12.21b) may be written

$$\begin{pmatrix} \pi_a' \\ \sigma' \end{pmatrix} = \left[\begin{pmatrix} 1 & 0 \\ 0 & 1 \end{pmatrix} + \begin{pmatrix} 0 & -\alpha_a \\ \alpha_a & 0 \end{pmatrix} \right] \begin{pmatrix} \pi_a \\ \sigma \end{pmatrix}.$$

Comparing with the canonical form $R \simeq 1 + i\alpha_a t^a$, we identify the infinitesimal generator

$$t^a = \begin{pmatrix} 0 & i \\ -i & 0 \end{pmatrix}.$$

From (12.21b), the infinitesimal generators for the nucleons are $t^a = \frac{1}{2}\tau^a\gamma^5$. Using these matrices and $\mathcal{L} = \mathcal{L}_0$ in (2.192), we obtain the conserved axial vector current

$$A_\mu^a = -(\partial_\mu \sigma)\pi_a + (\partial_\mu \pi_a)\sigma + \overline{N}\frac{\tau^a}{2}\gamma_\mu \gamma^5 N.$$

The second term of (12.20a) conserves isospin but breaks chiral invariance; the 4-divergence of A_μ^a comes entirely from it (§2.8.3). It can be computed by making an infinitesimal axial transformation depending on the parameters α_i, and then using $\partial(\delta\mathcal{L})/\partial\alpha_i = \partial_\mu A_i^\mu$, where $\delta\mathcal{L}$ is the change in the Lagrangian density brought about by the transformation. For the linear σ-model [see (12.21) and (12.22)]

$$\partial_\mu A_i^\mu = \frac{\partial(\epsilon\alpha^i\pi^i)}{\partial\alpha_i} = \epsilon\pi^i.$$

Comparing with (12.18) we identify $\epsilon = m_\pi^2 f_\pi$, under the PCAC assumption.

12.5 See Itzykson and Zuber (1980), §6–2–1; Cheng and Li (1984), §6.4. Let E = number of external lines, I = number of of internal lines, and V = number of vertices in a diagram. The number of independent loops (independent internal momenta) is

$$L = I - (V - 1),$$

where the term -1 comes from overall momentum conservation for the graph. The quantity L may be related to powers of $\hbar$ by keeping track of the $\hbar$'s introduced in the quantization (which are suppressed in our $\hbar = 1$ notation). These are (1) a factor of $\hbar$ for the free momentum space propagators [e.g. (3.62)], which comes from a factor of $\hbar$ in the canonical quantization [e.g., (2.61)]; (2) a factor $1/\hbar$ for each vertex, coming from a corresponding factor in the time evolution operators [see (3.21)]. Thus the total number of powers of $\hbar$ is

$$P = E + I - V = L + E - 1,$$

and if $L \gg E$ we obtain $P \simeq L$.

12.6 Assume the minimum of (12.23) to be on the σ axis ($\pi = 0$) and require $\partial V/\partial\sigma = 0$ to give

$$\lambda f(f^2 - v^2) = \epsilon$$

for the minimum f. Assume ϵ small and expand to give

$$f(\epsilon) = v + \frac{\epsilon}{2\lambda v^2} + \mathcal{O}\left(\epsilon^2\right).$$

Use this and $\sigma = \sigma' + f$ in (12.20); relate the coefficient of $\overline{N}N$ to the fermion mass and the coefficients of $\pi \cdot \pi$ and σ^2 to the meson masses [see (2.41) and (2.46)]; this yields (12.35).

12.7 The vector and axial currents are derived in Exercise 12.4; the corresponding charges are given by (2.193); the commutators of the fermion terms were already calculated in Exercise 12.3, so we need to calculate the commutators only for the terms involving π and σ. This can be done by analogy with Exercise 12.3a [remember that the structure constants $-i\epsilon_{abc}$ are matrix elements of the generators of the $SU(2)$ adjoint (pions, $T = 1$) representation, and that boson fields obey the commutators (2.32)]. By this procedure one finds that the charges of the linear σ-model satisfy the same commutators as those found in Exercise 12.3, which implies the chiral algebra (12.17).

12.8 See Georgi (1983), Ch. XVI. The sum of the three terms $M^0 + M' + M''$ is the QCD mass term (10.36). For free flavor $SU(3)$ quarks we may write $\mathcal{L} = \mathcal{L}_0 - \mathcal{L}'$, where $\mathcal{L}_0 = i\bar{q}\partial\!\!\!/q$ and

$$\mathcal{L}' = m_u \bar{u}u + m_d \bar{d}d + m_s \bar{s}s \qquad q = \begin{pmatrix} u \\ d \\ s \end{pmatrix} .$$

Under infinitesimal $SU(3)$ flavor transformations

$$q_i \to q_i + i\alpha_a(\lambda_a/2)_{ij} q_j$$

[where λ_a is given by (5.63)], $\mathcal{L}_0$ is invariant but $\mathcal{L}'$ is not unless $m_u = m_d = m_s$; $\mathcal{L}_0$ is also invariant under infinitesimal axial transformations

$$q_i \to q_i + i\alpha_a(\lambda_a/2)_{ij}\gamma_5 q_j,$$

but $\mathcal{L}'$ is not. Therefore, in a generalization of Exercise 12.4, $\mathcal{L}$ is invariant under chiral $SU(3)_L \times SU(3)_R$ transformations provided $m_u = m_d = m_s = 0$. By similar consideration for $SU(2)$ transformations on the u and d fields (neglecting strange quarks), we find that $SU(2)_L \times SU(2)_R$ symmetry requires $m_u = m_d = 0$, while $SU(2)$ isospin requires $m_u = m_d$. Thus if $m_u \neq m_d \neq m_s$, only the charge symmetry remains. For a discussion of the physical implications for the hadronic spectrum, see Cheng and Li (1984), Ch. 5.

12.9 (a) Hermitian conjugate both sides of (12.45), multiply from the right by γ_0, and use $(\gamma^0)^2 = 1$ and $\gamma_0 \gamma_\mu^\dagger \gamma_0 = \gamma_\mu$ to show that

$$\bar{q} = -i\bar{q}n_\mu\gamma^\mu.$$

This and (12.45) imply that $\bar{q}q = -\bar{q}q$, which means that on the surface

$$i\bar{q}n^\mu\gamma_\mu q = \bar{q}q = 0,$$

which establishes (12.47). (b) Under an infinitesimal chiral transformation (12.21b) of the fermion fields, the Dirac terms are invariant if fermion masses are neglected, but the surface term of (12.43) is not [see Exercise 12.4a]. We have

$$\mathcal{L} \to \mathcal{L} + \delta\mathcal{L} \qquad \delta\mathcal{L} = i\bar{q}(x)\alpha_a\frac{\tau^a}{2}\gamma_5 q(x)\Delta_s.$$

Then the divergence (on S) of the axial current is (see Exercise 12.4b)

$$\partial_\mu A_a^\mu = \frac{\partial(\delta\mathcal{L})}{\partial\alpha_a} = \frac{i}{2}\bar{q}(x)\tau^a\gamma_5 q(x)\Delta_s,$$

which is (12.49).

12.10 Show, using $[x_i, p_j] = i\delta_{ij}$, that for the Hamiltonian $H = \boldsymbol{\alpha} \cdot \mathbf{p} + \beta m$,

$$[H, \boldsymbol{\Sigma}] = 2i(\boldsymbol{\alpha} \times \mathbf{p}) \neq 0 \qquad [H, \mathbf{L}] = -i(\boldsymbol{\alpha} \times \mathbf{p}) \neq 0$$

$$[H, \mathbf{J}] = [H, \mathbf{L} + \tfrac{1}{2}\boldsymbol{\Sigma}] = 0 \qquad \boldsymbol{\Sigma} \equiv \begin{pmatrix} \boldsymbol{\sigma} & 0 \\ 0 & \boldsymbol{\sigma} \end{pmatrix} \qquad \mathbf{L} \equiv \mathbf{r} \times \mathbf{p}.$$

12.11 (a) The quark content of Σ^0 and Λ is uds, so the ud pairs carry all the isospin. The spins are $J_\Lambda = J_{\Sigma^0} = \tfrac{1}{2}$; the isospins are $T_\Lambda = 0$ and $T_{\Sigma^0} = 1$; we assume that for the ground states the orbital angular momentum is $L = 0$. Any pair of quarks in a qqq color singlet state must have an antisymmetric color wavefunction (see Exercise 12.14), so the non-color part of the wavefunction of any pair in the singlet qqq is symmetric. For the ud in Σ^0 we have

$$(\text{isospin}) \times (\text{orbital}) = (T{=}1, \text{symmetric}) \times (L{=}0, \text{symmetric});$$

hence the spin must be symmetric for ud ($S = 1$ triplet, spins parallel). This one unit of spin couples with $J = \tfrac{1}{2}$ for the s-quark to give $J_{\Sigma^0} = \tfrac{1}{2}$. For Λ the isospin wavefunction for ud is antisymmetric ($T = 0$), so the spins must be antiparallel ($S = 0$, antisymmetric) for ud, implying that the total spin $J = \tfrac{1}{2}$ of the Λ is carried by the strange quark s (Close, 1982). (b) For a meson (color **1**) composed of a color **3** quark and a color $\bar{\mathbf{3}}$ antiquark,

$$\langle \mathbf{F}_1 \cdot \mathbf{F}_2 \rangle = \tfrac{1}{2} \left[\langle \mathbf{F}^2 \rangle_1 - \langle \mathbf{F}_1^2 \rangle_3 - \langle \mathbf{F}_2^2 \rangle_{\bar{3}} \right]$$
$$= \tfrac{1}{2} \left(0 - \tfrac{4}{3} - \tfrac{4}{3} \right) = -\tfrac{4}{3},$$

where Table 5.4 has been used for the Casimirs. From Exercise 12.14, each qq in a baryon must be in a $\bar{\mathbf{3}}$ and

$$\langle \mathbf{F}_1 \cdot \mathbf{F}_2 \rangle = \tfrac{1}{2} \left[\langle \mathbf{F}^2 \rangle_{\bar{3}} - \langle \mathbf{F}_1^2 \rangle_3 - \langle \mathbf{F}_2^2 \rangle_3 \right] = -\tfrac{2}{3}.$$

12.12 Use the normalized proton wavefunction found in Exercise 5.14b to show that $\mu_p = \tfrac{4}{3}\mu_u - \tfrac{1}{3}\mu_d$. For example, the first term contributes

$$\sum_{i=1}^{3} \frac{2}{\sqrt{18}} \cdot \frac{2}{\sqrt{18}} \langle u_\uparrow u_\uparrow d_\downarrow | \mu_i \sigma_3^i | u_\uparrow u_\uparrow d_\downarrow \rangle = \frac{8\mu_u - 4\mu_d}{18},$$

where, for instance

$$\mu_3 \sigma_3^{(3)} | u_\uparrow u_\uparrow d_\downarrow \rangle = \mu_d(-1) | u_\uparrow u_\uparrow d_\downarrow \rangle.$$

Construct an $SU(3)$ flavor octet wavefunction for the neutron as in Exercise 5.14b (the proton and neutron wavefunctions are related by the interchange $u \leftrightarrow d$), and show that

$$\mu_n = -\tfrac{1}{3}\mu_u + \tfrac{4}{3}\mu_d.$$

Finally use $\mu_u = -2\mu_d$ (if $m_u = m_d$) to show that $\mu_p/\mu_n = -3/2$.

12.13 Use the same procedures as in Exercises 5.12 and 12.2 to find the irreducible representation content of the direct products. For example, in Exercise 5.12d we proved that

$$\mathbf{8} \otimes \mathbf{8} = \mathbf{1} \oplus \mathbf{8} \oplus \mathbf{8} \oplus \mathbf{10} \oplus \overline{\mathbf{10}} \oplus \mathbf{27},$$

so if the gluons G transform as a color $\mathbf{8}$, GG has a color singlet in its decomposition.

12.14 Two-quark states:

$$q \otimes q = \mathbf{3} \otimes \mathbf{3} = \underset{\text{antisymm.}}{\boxed{}} \oplus \underset{\text{symm.}}{\boxed{}} = \overline{\mathbf{3}} \oplus \mathbf{6}.$$

To determine the corresponding wavefunctions we must construct all possible *standard arrangements of Young tableaux*. These correspond to Young diagrams for which no row is longer than the row above it, with a set of positive integers placed in the boxes subject to the following restrictions: (1) The numbers do not decrease from left to right in a row. (2) The numbers increase from top to bottom in a column. (3) If N states are available to a particle, the number j in a box satisfies $1 \leqslant j \leqslant N$. For quarks $N = 3$ (color states), and the standard arrangements for $\mathbf{3} \otimes \mathbf{3} = \overline{\mathbf{3}} \oplus \mathbf{6}$ are

symmetric : $\boxed{1\,1}$ $\boxed{1\,2}$ $\boxed{1\,3}$ $\boxed{2\,2}$ $\boxed{2\,3}$ $\boxed{3\,3}$

antisymmetric : $\begin{smallmatrix}\boxed{1}\\\boxed{2}\end{smallmatrix}$ $\begin{smallmatrix}\boxed{1}\\\boxed{3}\end{smallmatrix}$ $\begin{smallmatrix}\boxed{2}\\\boxed{3}\end{smallmatrix}$

In these diagrams the boxes represent particles and the numbers represent states (for this example there are two particles with three states available). A row of boxes is an injunction to symmetrize in the particle labels and a column of boxes implies antisymmetrization (Lichtenberg, 1978, §4.3); for example,

$$\boxed{1\,2} \to \frac{1}{\sqrt{2}}(\psi_1\psi_2 + \psi_2\psi_1),$$

where $1/\sqrt{2}$ is a normalization factor. After replacing the state numbers with color labels ($1 = R$, $2 = B$, $3 = G$) the corresponding color wavefunctions are

Symmetric: RR, $RB + BR$, $RG + GR$, BB, $GB + BG$, GG,

Antisymmetric: $RB - BR$, $RG - GR$, $BG - GB$,

where we have omitted normalization factors. Now $\overline{\mathbf{3}} \otimes \mathbf{3} = \mathbf{8} \oplus \mathbf{1}$ and $\mathbf{6} \otimes \mathbf{3} = \mathbf{10} \oplus \mathbf{8}$. Hence, to produce a qqq color singlet the qq in the antisymmetric $\overline{\mathbf{3}}$ must combine with the remaining q (in a $\mathbf{3}$).

12.15 The string of total length d_0 has energy density σ and the endpoints move at light velocity. Thus the local velocity at any point in the string is

$v/c = 2r/d_0$, where r is the distance from the center. The total mass is (neglect quark masses)

$$M = 2 \int_0^{d_0/2} \frac{\sigma dr}{\sqrt{1 - \frac{v^2}{c^2}}} = \frac{\sigma d_0 \pi}{2}.$$

The total angular momentum is (neglect quark spins)

$$J = 2 \int_0^{d_0/2} \frac{\sigma r v dr}{\sqrt{1 - \frac{v^2}{c^2}}} = \frac{\sigma d_0^2 \pi}{8}.$$

Solving these two relations for d_0^2 and equating gives

$$J = \alpha' M^2 \qquad \alpha' = \frac{1}{2\pi\sigma}.$$

The constant term of (12.83) could arise from the neglected contributions of the quarks.

12.16 First show from (12.87) and (12.88) that

$$\mathbf{A}_\mu = f_\pi^2 (\hat{\boldsymbol{\pi}} \partial_\mu \theta + \cos\theta \sin\theta \, \partial_\mu \hat{\boldsymbol{\pi}});$$

then show that $f_\pi D_\mu \Phi$ gives the same result (Brown, 1982a).

12.17 The $\eta_c(2980)$ is much broader (about 10 MeV compared with 68 keV for J/ψ). A plausible explanation is that the η_c can decay by two-gluon OZI suppressed diagrams (because $C = +1$), rather than by the three-gluon graphs required for $C = -1$ (Fig. 12.18). This is analogous to two-photon decay for the $^1S_0$, $C = +1$ positronium state (*para;* $\tau = 1.25 \times 10^{-10}$s), compared with the three-photon decay of the $^3S_1$, $C = -1$ positronium state (*ortho;* $\tau = 1.4 \times 10^{-7}$s). Perkins (1987) gives a simple introduction to the physics of positronium and to the parallels with $q\bar{q}$ systems.

CHAPTER 13

13.1 For an Abelian gauge group (13.2) reduces to

$$U(A, B) = \exp\left[ig \int_A^B A_\mu(x) \, dx^\mu \right].$$

Performing a gauge transformation $A_\mu(x) \to A_\mu(x) - \partial_\mu \phi(x)$ gives

$$U(A, B) \to \exp\left[ig \int_A^B A_\mu(x) \, dx^\mu - ig \int_A^B \partial_\mu \phi(x) \, dx^\mu \right]$$

$$= e^{ig\phi(A)} U(A, B) e^{-ig\phi(B)},$$

which is the transformation law for a field with opposite charges at A and B [see (2.144)].

13.2 This exercise requires expansion of group elements U, some algebra, and attention to the required orders to be retained at each algebraic step (see the discussion in Wilson, 1977, Appendix A). Evaluate (13.15) and (13.16) using (13.16b) written in the form (see Fig. 13.2)

$$U_p = U_{n\mu}\left(U^{\dagger}_{n+\nu,\mu}U_{n+\mu,\nu} + \left[U_{n+\mu,\nu}, U^{\dagger}_{n+\nu,\mu}\right]\right)U^{\dagger}_{n\nu}.$$

Expand the U's [see (13.6)] to show that the commutator is

$$\left[U_{n+\mu,\nu}, U^{\dagger}_{n+\nu,\mu}\right] \simeq ig^2 a^2 f_{abc} A^a_{n+\mu,\nu} A^b_{n+\nu,\mu} T_c,$$

where f_{abc} is a structure constant and T_c a group generator. Insert this and expand all remaining U's to show that

$$U_p \simeq 1 + iga^2(\partial_\nu A^a_\mu - \partial_\mu A^a_\nu)T_a + ig^2 a^2 f_{abc} A^a_\nu A^b_\mu T_c,$$

where we have used relations such as

$$U_{n\mu}U^{\dagger}_{n+\nu,\mu} \simeq 1 + iga\left(-A^a_{n\mu} + A^a_{n+\nu,\mu}\right)T_a \underset{a\to 0}{\simeq} 1 + iga^2 T_a \partial_\nu A^a_\mu.$$

Thus comparing with (7.10b), $U_p \simeq 1 - ia^2 g F^a_{\mu\nu} T_a$. But since U_p is an element of the $SU(3)$ group we also have

$$U_p = e^{-i\alpha_a T_a} \simeq 1 - i\alpha_a T_a.$$

Therefore $\alpha_a = a^2 g F^a_{\mu\nu}$ and

$$U_p \simeq 1 - iga^2 F^a_{\mu\nu} T_a - \tfrac{1}{2}g^2 a^4 (F^a_{\mu\nu} T_a)^2 + \cdots .$$

The generators of $SU(N)$ are traceless, so

$$\text{Tr}\, U_p \simeq \text{constant} - \tfrac{1}{2}g^2 a^4 F^a_{\mu\nu} F^{\mu\nu}_b \,\text{Tr}\,(T_a T_b) + \cdots .$$

Insert in (13.16a) and (13.15), use (5.65), and take the continuum limit using $a^4 \sum_p \longrightarrow \int d^4 x$ to obtain

$$S \simeq \int d^4 x \, F^a_{\mu\nu} F^{\mu\nu}_a.$$

13.3 (a) See Kogut (1979), §V.C; Creutz (1985), Ch. 9. (b) From (13.19), (13.23), and (13.15)–(13.16)

$$\langle W(c)\rangle \simeq \int [dU]\, \text{Tr}\, U_c(x,x)\exp\left[-\frac{2}{g^2}\sum_p \text{Tr}\,(\text{Re}\,U_p)\right],$$

where $[dU] \equiv \prod dU_{ij}$ and the normalization has been ignored. For strong coupling g^2 is large and we may expand the exponential

$$\langle W(c) \rangle \simeq \int [dU]\, \mathrm{Tr}\, U_c(x, x)$$

$$\times \left[1 - \frac{2}{g^2} \sum_p \mathrm{Tr}\, U_p + \frac{1}{2!} \left(\frac{2}{g^2} \right)^2 \sum_{pp'} \mathrm{Tr}\, U_p\, \mathrm{Tr}\, U_{p'} + \ldots \right]$$

where $U_p \equiv \mathrm{Re}\, U_p$. The invariant $SU(3)$ group integration has the properties

$$\int [dU]U_{ij} = 0 \qquad \int [dU]U_{ij}U_{kl}^{\dagger} = \tfrac{1}{3}\delta_{il}\delta_{jk}.$$

Hence the lowest order nonvanishing contributions to $\langle W(c) \rangle$ come whenever the closed contour c is tiled exactly once with the plaquettes U_p brought down from the exponential by the expansion. Otherwise, there will be at least one unpaired link for which $\int [dU]U_{ij} = 0$. For example, given a 3×3 Wilson loop the first nonvanishing contribution in the strong coupling expansion comes from the following tiling with plaquettes (Creutz, 1985, Ch. 10; Kogut, 1983, §III):

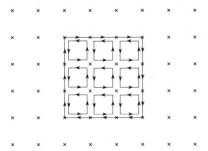

Notice that every directed link is paired with a link oriented in the opposite direction. If N_c is the minimum number of plaquettes required to tile the area enclosed by the contour c, then in leading nonvanishing order

$$\langle W(c) \rangle \simeq \left(\frac{1}{g^2} \right)^{N_c} \simeq (g^2)^{-A(c)/a^2} = e^{-\log(g^2)A(c)/a^2},$$

where the area of the contour is $A(c) = a^2 N_c$. Thus the strong coupling Wilson loop exhibits area law behavior; comparing with (13.29), the string tension is $\sigma \simeq \log(g^2)/a^2$. From the strong coupling expansion and the $SU(3)$ group integration it is clear that as $\beta \to 0$ we have

$$\langle W(c) \rangle \simeq \int [dU]\, \mathrm{Tr}\, U_c(x, x) = 0$$

(see Creutz, 1985, Ch. 10; Cheng and Li, 1984, §10.5). (c) We require that at fixed R

$$a\frac{d}{da}V(R,a,g) = \left(a\frac{\partial}{\partial a} + a\frac{\partial g}{\partial a}\frac{\partial}{\partial g}\right)V(R,a,g)$$

$$= \left(a\frac{\partial}{\partial a} - B(g)\frac{\partial}{\partial g}\right)V(R,a,g) = 0,$$

where (13.37) was used. Insertion of the specific form given for $V(R,a,g)$ yields (13.38). (d) Use $Q \simeq 1/a$ in (13.37); apply the result to (10.40) to obtain (13.38); apply to (10.39) to obtain

$$-\frac{B(g)}{g} = -\frac{4}{3}\frac{g^2}{16\pi^2} + \cdots$$

for QED with one family of fermions.

13.4 The field $\phi(x,t)$ is a scalar; a Lorentz boost may be implemented by the replacement

$$(x - x_0) \rightarrow (x - x_0 - \beta t)/\sqrt{1 - \beta^2},$$

where $\beta = v/c$. Verify that (13.59) with this replacement is also a solution of the kink equations of motion (13.52) and (13.57), except that the kink width is Lorentz contracted,

$$W = \frac{1}{m} \rightarrow \frac{1}{m}\sqrt{1 - \beta^2},$$

and the kink energy

$$E = \frac{2\sqrt{2}m^3}{3\lambda} \longrightarrow \frac{1}{\sqrt{1 - \beta^2}}\left(\frac{2\sqrt{2}m^3}{3\lambda}\right)$$

(see Rajaraman, 1982, §2.3).

13.5 (a) This is the proof of Derrick's theorem, which may be found in many places. For example, see Lee (1981), §7.4; Coleman (1977), §2.4. (b) From (13.54), (13.53), and (13.55)

$$E = \int 2U\,dx = \int_{\phi_1}^{\phi_2}\sqrt{2U}\,d\phi$$

for the kink. Substitution of (13.57) and the limits (13.58) gives (13.61). From (13.65), the static Sine–Gordon solutions also satisfy (13.55). Using (13.66), (13.55), (13.67), and (13.68) leads to (13.70b).

13.6 From (13.52) we may obtain (13.56), and

$$x = \pm\int_0^{\phi(x)}\frac{d\phi}{\phi\left(\phi^2 - 1\right)},$$

where we have set $x_0 = 0$ and $\phi(x_0) = 0$. Doing the integral and solving for $\phi(x)$ gives

$$\phi = \pm \left(1 - e^{\pm 2x}\right)^{-1/2}.$$

These are topological solitons with

$$\left(\phi(-\infty), \phi(+\infty)\right) = (1, 0) \quad (0, 1) \quad (-1, 0) \quad (0, -1),$$

a localized energy density (13.53), and finite energy (13.54).

13.7 (a) Use the parameterization $B_\mu = (g/i)\tau_a A_\mu^a$ and the identity

$$\partial_\mu \left(U U^{-1}\right) = 0 = \left(\partial_\mu U\right) U^{-1} + U \partial_\mu \left(U^{-1}\right)$$

in the first line of (7.8). (b) Use (13.81) and (7.10a) with $G_{\mu\nu} \equiv (g/i)\tau_a F_{\mu\nu}^a$, and the result from part (a) to prove (13.83). (c) The total energy is conserved: after $t \to -i\tau$ we have

$$-E = \frac{1}{2}\left(\frac{dx}{d\tau}\right)^2 - V(x).$$

Solve for a zero-energy Euclidian classical path connecting the (zero-energy) ground states at $x = \pm x_0$:

$$\tau = \pm \int_0^x \frac{dx'}{\sqrt{2V(x)}} = \pm \frac{1}{x_0\sqrt{2}} \tanh^{-1}\left(\frac{-x}{x_0}\right).$$

Invert this to find the desired formula for $x(\tau)$. Show that the Euclidian action is [see (4.38)]

$$S_{\rm E} \equiv -i S_{t \to -i\tau} = \frac{1}{2}\int \left[\left(\frac{dx}{d\tau}\right)^2 + V\right] d\tau = 2 \int V d\tau = \frac{4}{3}\sqrt{2}\, x_0^3.$$

The semiclassical tunneling amplitude is

$$T \simeq e^{-S_{\rm E}} \simeq \exp\left(-\tfrac{4}{3}\sqrt{2}x_0^3\right).$$

13.8 First invoke $\epsilon_{\mu\nu\rho\sigma}\epsilon_{\mu\nu\kappa\lambda} = 2\left(\delta_{\rho\kappa}\delta_{\sigma\lambda} - \delta_{\rho\lambda}\delta_{\sigma\kappa}\right)$ and (13.88) to prove that

$$\tilde{G}_{\mu\nu}\tilde{G}_{\mu\nu} = G_{\mu\nu}G_{\mu\nu}.$$

Use this and the inequality given in the hint to show that

$$-\int d^4x \, {\rm Tr}\left(G_{\mu\nu}G_{\mu\nu}\right) \geqslant \mp \int d^4x \, {\rm Tr}\left(\tilde{G}_{\mu\nu}G_{\mu\nu}\right),$$

and (13.87) and (13.82) to show that

$$S \geqslant \frac{8\pi^2}{g^2}|Q|.$$

Within a given homotopic sector (13.82) and (13.87) show that the lower bound $S = \left(8\pi^2/g^2\right)|Q|$ occurs when

$$G_{\mu\nu} = \pm\tilde{G}_{\mu\nu};$$

hence, by Hamilton's principle [eq. (2.1)] selfdual or anti-selfdual fields extremize the action and are solutions of the Yang–Mills equations within each Q-sector. See Rajaraman (1982), §4.3; Ryder (1985), §10.5.

13.9 The Euclidian dual of the dual is

$$\tilde{\tilde{G}}_{\mu\nu} = \tfrac{1}{4}\epsilon_{\mu\nu\rho\sigma}\epsilon_{\rho\sigma\alpha\beta}G_{\alpha\beta}.$$

In the Euclidian metric $\epsilon_{\mu\nu\rho\sigma}\epsilon_{\rho\sigma\alpha\beta} = 2(\delta_{\mu\alpha}\delta_{\nu\beta} - \delta_{\mu\beta}\delta_{\nu\alpha})$, which leads to the symbolic expression

$$\tilde{\tilde{G}} = \varepsilon^2 G = G,$$

with the eigenvalues of the dualizing operator $\varepsilon = \pm 1$; thus $\tilde{G} = \pm G$. On the other hand, for a Minkowski metric [see (2.131)] one finds

$$\tilde{\tilde{G}} = -G \qquad \varepsilon = \pm i \qquad \tilde{G} \neq \pm G.$$

See Rajaraman (1982), §4.3 and Ryder (1985), §10.5.

13.10 See Balachandran (1986), §II.2; Meissner and Zahed (1986), §§5–6. This is a particular example of introducing additional fields (effectively, through the Skyrme term) to stabilize the soliton and circumvent Derrick's theorem (Exercise 13.5a). The lower bound was originally obtained by Skyrme (1961), but it also occurs in magnetic monopole physics where it is called the Bogomol'ny bound.

CHAPTER 14

14.1 Equation (15.10) with $g_i = 3$ gives the equilibrium number of pions n_π in a unit volume. Then

$$T = \left(\frac{\pi^2 n_\pi}{3\zeta(3)}\right)^{1/3} \simeq (3n_\pi)^{1/3},$$

and if the pions are spheres of radius R, the temperature at which a unit volume is filled with pions [compare (14.3)] is $T_0 \simeq (9/4\pi)^{1/3}R^{-1}$. For realistic pion radii this gives $T_0 \simeq 200$–300 MeV.

14.2 For (14.12) equate (14.8) and (14.10); for (14.13) use (14.9), (14.11), and (14.12).

14.3 See Exercise 14.4 and Bernard (1974); the key is a comparison of the Feynman propagator with the partition function.

14.4 Here we follow Cleymans, Gavai, and Suhonen (1986), §3.2; see also Kapusta (1989), Ch. 2. From (14.25) and the assumption $\mu = 0$,

$$Z = \int dx \left\langle x | e^{-\beta H} | x \right\rangle \simeq \int dx \left\langle x | (e^{-\varepsilon H})^n | x \right\rangle,$$

where $\beta = 1/T$ has been divided into n small segments, $\beta = n\varepsilon$, and we employ orthonormal position eigenstates $|x\rangle$. Using (4.4a) $n-1$ times gives

$$Z = \int dx \prod_{i=1}^{n-1} dx_i \left[\langle x | e^{-\varepsilon H} | x_1 \rangle \langle x_1 | e^{-\varepsilon H} | x_2 \rangle \dots \langle x_{n-1} | e^{-\varepsilon H} | x \rangle \right]$$

$$= \int \prod_{i=1}^{n} dx_i \prod_{i=1}^{n} \langle x_{i-1} | e^{-\varepsilon H} | x_i \rangle,$$

with a periodicity condition $x_0 = x_n$. Evaluate the matrix elements using orthonormal momentum eigenstates satisfying $\int dp \, |p\rangle \langle p| = 2\pi$ [eq. (4.4b)], $\langle p \mid x \rangle = \exp(-ipx)$ [eq. (4.2)], and $H = \frac{1}{2}p^2 + \frac{1}{2}\omega^2 x^2$ (unit mass oscillator):

$$\langle x_{i-1} | e^{-\varepsilon H} | x_i \rangle = \frac{1}{4\pi^2} \int dp\, dp' \, \langle x_{i-1} \mid p \rangle \langle p | e^{-\varepsilon p^2/2} | p' \rangle \langle p' | e^{-\varepsilon \omega^2 x^2/2} | x_i \rangle$$

$$= \frac{1}{4\pi^2} \int dp\, dp' \, e^{ipx_{i-1}} e^{-\frac{1}{2}\varepsilon p'^2} \langle p \mid p' \rangle e^{-\frac{1}{2}\varepsilon \omega^2 x_i^2} \langle p' \mid x_i \rangle$$

$$= \frac{1}{4\pi^2} e^{-\frac{1}{2}\varepsilon \omega^2 x_i^2} \int dp \, e^{ip(x_{i-1}-x_i)} e^{-\frac{1}{2}\varepsilon p^2}$$

$$= \text{constants} \times \exp\left[-\frac{1}{2} \left(\frac{(x_{i-1} - x_i)^2}{\varepsilon} + \varepsilon \omega^2 x_i^2 \right) \right],$$

where we have used

$$\langle p \mid p' \rangle = \delta(p - p') \int_{-\infty}^{+\infty} dp \, e^{-a^2 p^2 \pm bp} = \frac{\sqrt{\pi}}{a} e^{b^2/4a^2}.$$

Insert into the expression for Z and take the limit $\varepsilon \to 0$ with $\varepsilon \equiv dt$ and $x_{i-1} - x_i \equiv dx_i$ to obtain $Z \simeq \sum_{\text{paths}} \exp(-S)$ with

$$S = \int_0^\beta \left[\frac{1}{2} \left(\frac{dx}{dt} \right)^2 + \frac{1}{2} \omega^2 x^2 \right] dt,$$

and with the periodic boundary conditions $x(0) = x(\beta)$.

14.5 We have already obtained the essential result in Exercise 6.5b. The 4-fermion interaction term in the Lagrangian density must be of the form

$$\mathcal{L}' = g\bar{\psi}\psi\bar{\psi}\psi.$$

From Appendix A,

$$[\mathcal{L}] = [M]^4 \qquad [\psi] = [\bar{\psi}] = [M]^{3/2} \qquad [g] = [M]^{-2}.$$

But coupling constants of negative mass dimension signal non-renormalizable theories (§6.3.3).

14.6 See Matsui and Satz (1986). The dominant mechanism would likely be hard parton–parton interactions making $c\bar{c}$ pairs. In a confining regime the interaction between c and $\bar{c}$ could then sometimes produce a J/ψ. In a deconfining regime, if the temperature is sufficiently high that the screening radius is less than the binding radius of a J/ψ (see §14.3 and Fig. 12.20), the c and $\bar{c}$ are more likely to separate until cooling of the plasma allows formation of open-charm mesons ($c\bar{u}, u\bar{c}, ...$); thus J/ψ production is suppressed. See Matsui and Satz (1986) and Gyulassy (1988) for a discussion of whether this effect might be obscured by other factors.

CHAPTER 15

15.1 (a) A lower bound is given by the photons in the 2.7 K microwave background. From (15.14) and (15.15), $S \simeq (RT)^3$. Taking

$$T = 2.7 \text{ K} = 2.25 \times 10^{-13} \text{ GeV} \qquad R \simeq 3 \times 10^9 \text{ LY} = 1.4 \times 10^{41} \text{ GeV}^{-1}$$

[see (A.4) and (A.11)], we estimate

$$S \geqslant (RT)^3 \simeq 3 \times 10^{85},$$

in dimensionless units. (b) From (15.2) assuming $\Lambda = 0$, the critical density (corresponding to $k = 0$; see Fig. 15.1) is $\rho_c = 3H^2/8\pi G$. Then from (15.2) and (15.15)

$$\frac{\Omega - 1}{\Omega} = \frac{3k}{8\pi G\rho R^2} = \frac{3ks^{2/3}}{8\pi G\rho S^{2/3}},$$

where $\Omega \equiv \rho/\rho_c$. Thus,

$$S = \left(\frac{3k}{8\pi G}\frac{\Omega}{\rho(\Omega - 1)}\right)^{3/2} s = \left(\frac{k}{H^2(\Omega - 1)}\right)^{3/2} s.$$

where $\Omega/\rho = 1/\rho_c = 8\pi G/3H^2$ was used. With the numbers given in the exercise and $|k| = 1$, we find $S \gtrsim 10^{87}$. The large value of S found here or in part (a) is viewed as unnatural because S is fixed by initial conditions in the standard cosmological models, and the standard models give no reason to expect other than an initial S of order unity (Blau and Guth, 1987). From the results obtained above,

$$\frac{\rho - \rho_c}{\rho} = \frac{\Omega - 1}{\Omega} \simeq \rho^{-1} R^{-2},$$

which is $t^{2/3}$ for matter dominated and t for radiation dominated when (15.5) and (15.6) are used.

15.2 (a) If the last two terms are neglected, (15.2) yields

$$t \simeq \sqrt{\frac{3}{8\pi G}} \int_0^{R(t)} \frac{dR}{R} \rho^{-1/2}.$$

Utilizing (15.5) leads to $R \propto t^{1/2}$ for relativistic matter and $R \propto t^{2/3}$ for non-relativistic matter [eqns. (15.6)]. (b) For a matter or radiation dominated cosmology, the horizon $l_H(t)$ is the distance that light could have traveled since the initial singularity. In $c = 1$ units:

$$l_H(t) = R(t) \int_0^t \frac{dt'}{R(t')}.$$

Use of the results from part (a) gives $l_H(t) \simeq 2t = 1/H$ for a radiation dominated universe and $l_H(t) \simeq 3t = 2/H$ for one that is matter dominated. (c) From (15.2) without the last two terms, and (15.3),

$$\frac{\ddot{R}}{R} = -\frac{4\pi G}{3}(\rho + 3p).$$

Thus $\dot{R}$ decreases with time unless $\rho + 3p \leqslant 0$. For a standard cosmology the horizon may be approximated as the inverse Hubble parameter, $1/H = R/\dot{R}$ [see part(a)]. Thus

$$\frac{R}{1/H} \simeq \dot{R},$$

and if $\dot{R}$ decreases with time the ratio of the distance scale to the horizon also decreases with time $[l_H(t)/R(t) \simeq t^{1/2}$ (radiation dominated) or $t^{1/3}$ (matter dominated), from (15.6)]: in a standard cosmology the horizon grows more rapidly than the universe expands, and objects not presently in causal contact have never been in causal contact. With inflation we will see that $\rho + 3p$ becomes negative and objects can expand outside of causal contact during inflation, and then come back into causal contact during standard big bang evolution after inflation ends (§15.4.1).

15.3 Assume the universe to have been matter dominated for most of its history, so $R(t) \simeq t^{2/3}$. Then the coordinate distance between μ-wave source and observer is

$$r \simeq \int_{t_r}^{t_0} \frac{dt}{t^{2/3}} = 3\left(t_0^{1/3} - t_r^{1/3}\right),$$

where t_0 is the present time and t_r is the time of hydrogen atom recombination. The coordinate horizon distance at t_r is (see Exercise 15.2b)

$$l_H = \int_0^{t_r} t^{-2/3} dt = 3t_r^{1/3},$$

and the number of horizons separating the sources in opposite directions is

$$n_H \simeq \frac{2r}{l_H} \simeq 2\left[\left(\frac{t_0}{t_r}\right)^{1/3} - 1\right] \simeq 90,$$

where we have used $t_0 = 10^{10}$ yr and $t_r = 10^5$ yr (see Blau and Guth, 1987).

15.4 From (15.31) and (15.32),

$$\frac{dR}{dt} \simeq R\sqrt{\frac{8\pi G\left(\rho_0 + \rho'\right)}{3}}.$$

If $T \gg T_c$ we expect $\rho' \gg \rho_0$ [see (15.30)], so

$$\frac{dR}{dt} \simeq RT^2 \simeq R^{-1},$$

which is satisfied if $R \simeq t^{1/2}$. On the other hand, if $T \ll T_c$ we obtain from (15.30) and (15.31)

$$dR/dt \simeq R/l,$$

with l given by (15.35); this is satisfied by $R \simeq \exp(t/l)$.

15.5 Neutrinos are kept in equilibrium by reactions such as

$$\nu\bar{\nu} \leftrightarrow e^+ e^- + \dots$$

The cross sections σ are weak so they must depend on G_F^2, where $G_F \simeq 1.17 \times 10^{-5}$ GeV^{-2} is the Fermi constant [eq. (6.12)]. Since σ has dimension $[M]^{-2}$ the constant G_F^2 must be multiplied by something of dimension $[M]^2$; the only possibility is the square of the temperature, so $\sigma \simeq G_F^2 T^2$ (see §6.3.1). The two-body reaction rate is (§15.1.2)

$$\Gamma \simeq n\sigma v \simeq \left(T^3\right)\left(G_F^2 T^2\right)(1) \simeq G_F^2 T^5,$$

where (15.10) has been used for the relativistic ($v = 1$) neutrinos and all constants have been dropped. The ratio of Γ to the expansion rate H is

$$\frac{\Gamma}{H} \simeq \frac{G_F^2 T^5}{(T^2/M_P)} = G_F^2 M_P T^3 \simeq \left(\frac{T}{1\,\text{MeV}}\right)^3,$$

where we have used $H \simeq T^2/M_P$ from (15.19), and eq. (15.20). Thus the weak interactions freeze out at $T \simeq 1$ MeV, which occurs about 1 sec after the big bang (see Fig. 14.3 and Kolb, 1987, §1.4).

References

ABERS AND LEE (1973)   E. S. Abers and B. W. Lee, *Phys. Rep.* **C9**, 1 (1973).

ABBOTT AND PI (1986)   L. F. Abbott and S.-Y. Pi (eds.), *Inflationary Cosmology* (World Scientific, 1986).

ADKINS AND NAPPI (1984)   G. S. Adkins and C. R. Nappi, *Nucl. Phys.* **B233**, 109 (1984).

ADKINS, NAPPI, AND WITTEN (1983)   G. S. Adkins, C. R. Nappi, and E. Witten, *Nucl. Phys.* **B228**, 552 (1983).

AGUILAR-BENITEZ ET AL. (1986)   M. Aguilar-Benitez et al., "Review of Particle Properties", *Phys. Lett.* **170B**, 1 (1986).

AGUILAR-BENITEZ ET AL. (1988)   M. Aguilar-Benitez et al., "Review of Particle Properties", *Phys. Lett.* **204B**, 1 (1988).

AITCHISON (1972)   I. J. R. Aitchison, *Relativistic Quantum Mechanics* (MacMillan, 1972).

AITCHISON (1982)   I. J. R. Aitchison, *An Informal Introduction to Gauge Field Theories* (Cambridge University Press, 1982).

AITCHISON AND HEY (1982)   I. J. R. Aitchison and A. J. G. Hey, *Gauge Theories in Particle Physics* (Adam Hilger Ltd, 1982).

ALBRECHT AND STEINHARDT (1982)   A. Albrecht and P. J. Steinhardt, *Phys. Rev. Lett.* **48**, 1220 (1982).

ALTARELLI (1986)   G. Altarelli, in *Proc. XXIII Int. Conf. on High Energy Physics, Vol. 1* (ed. by S. C. Loken; World Scientific, 1986).

ALTARELLI AND PARISI (1977)   G. Altarelli and G. Parisi, *Nucl. Phys.* **B126**, 298 (1977).

ANDERSON (1963)   P. W. Anderson, *Phys. Rev.* **130**, 439 (1963).

ARNISON ET AL. (1983)   G. Arnison et al., *Phys. Lett.* **122B**, 103 (1983).

AUGUSTIN ET AL. (1974)   J.-E. Augustin et al., *Phys. Rev. Lett.* **33**, 1406 (1974).

AUBERT ET AL. (1974)   J. J. Aubert et al., *Phys. Rev. Lett.* **33**, 1404 (1974).

BAGNAIA AND ELLIS (1988)   P. Bagnaia and S. D. Ellis, *Annu. Rev. Nucl. Part. Sci.* **38**, 659 (1988).

BALACHANDRAN (1986)   A. P. Balachandran, in *High Energy Physics 1985, Vol. 1* (ed. by M. J. Bowick and F. Gürsey; World Scientific, 1986).

BALACHANDRAN ET AL. (1982)   A. P. Balachandran, V. P. Nair, S. G. Rajeev, and A. Stern, *Phys. Rev. Lett.* **49**, 1124 (1982).

BALACHANDRAN ET AL. (1983)   A. P. Balachandran, V. P. Nair, S. G. Rajeev, and A. Stern, *Phys. Rev.* **D27**, 1153 (1983).

BANDER (1981)   M. Bander, *Phys. Rep.* **75**, 205 (1981).

BANNER ET AL. (1983)   M. Banner et al., *Phys. Lett.* **122B**, 476 (1983).

BARBOUR ET AL. (1986)   I. Barbour, N.-E. Behilil, E. Dagotto, F. Karsch, A. Moreo, M. Stone, and H. W. Wyld, *Nucl. Phys.* **B275**, 296 (1986).

BARDEEN ET AL. (1978)   W. A. Bardeen, A. J. Buras, D. W. Duke, and T. Muta, *Phys. Rev.* **D18**, 3998 (1978).

BARTEL ET AL. (1982)   W. Bartel et al., *Phys. Lett.* **108B**, 140 (1982).

BAYM (1969)   G. Baym, *Lectures on Quantum Mechanics* (Benjamin, 1969).

BAYM (1981)   G. Baym, in *Statistical Mechanics of Quarks and Hadrons* (ed. by H. Satz; North–Holland, 1981).

BAYM (1982a)   G. Baym, in *Quarks and Nuclear Forces* (ed. by D. Fries and B. Zeitnitz; Springer–Verlag, 1982).

BAYM (1982b)   G. Baym, in *Quark Matter Formation and Heavy Ion Collisions* (ed. by M. Jacob and H. Satz; World Scientific, 1982).

BAYM (1982c)   G. Baym, *Prog. Part. Nucl. Phys.* **8**, 73 (1982).

BAYM (1984)   G. Baym, *Nucl. Phys.* **A418**, 433c, 525c (1984).

BÉG AND SIRLIN (1974)   M. A. B. Bég and A. Sirlin, *Annu. Rev. Nucl. Part. Sci.* **24**, 379 (1974).

BELAVIN ET AL. (1975)   A. A. Belavin, A. M. Polyakov, A. S. Schwartz, and Y. S. Tyupkin, *Phys. Lett.* **59B**, 85 (1975).

BEREZIN (1966)   F. A. Berezin, *The Method of Second Quantization* (Academic Press, 1966).

BERGER AND COESTER (1987)   E. L. Berger and F. Coester, *Annu. Rev. Nucl. Part. Sci.* **37**, 463 (1987).

BERNARD (1974)   C. W. Bernard, *Phys. Rev.* **D9**, 3312 (1974).

BHADURI (1988)   R. K. Bhaduri, *Models of the Nucleon: From Quarks to Soliton* (Addison–Wesley, 1988).

BJORKEN (1969)   J. D. Bjorken, *Phys. Rev.* **179**, 1547 (1969).

BJORKEN (1983)   J. D. Bjorken, *Phys. Rev.* **D27**, 140 (1983).

BJORKEN AND DRELL (1964)   J. D. Bjorken and S. D. Drell, *Relativistic Quantum Mechanics* (McGraw–Hill, 1964).

BJORKEN AND DRELL (1965)   J. D. Bjorken and S. D. Drell, *Relativistic Quantum Fields* (McGraw–Hill, 1965).

BJORKEN AND PASCHOS (1969)   J. D. Bjorken and E. A. Paschos, *Phys. Rev.* **185**, 1975 (1969).

BLAU AND GUTH (1987)   S. K. Blau and A. H. Guth, in *300 Years of Gravitation* (ed. by S. W. Hawking and W. Israel; Cambridge University Press, 1987).

BLUDMAN AND KLEIN (1962)   S. Bludman and A. Klein, *Phys. Rev.* **131**, 2364 (1962).

BOEHM AND VOGEL (1984)   F. Boehm and P. Vogel, *Annu. Rev. Nucl. Part. Sci.* **34**, 125 (1984).

BOGOLIUBOV AND SHIRKOV (1983)   N. N. Bogoliubov and D. V. Shirkov, *Quantum Fields* (Benjamin/Cummings, 1983).

BOHR AND MOTTELSON (1975)   A. Bohr and B. R. Mottelson, *Nuclear Structure, Vol. II* (Benjamin, 1975).

BOTT (1956)   R. Bott, *Bull. Soc. Math. France* **84**, 251 (1956).

BRANDENBERGER (1985)   R. H. Brandenberger, *Rev. Mod. Phys.* **57**, 1 (1985).

BRINK AND SATCHLER (1968)   D. M. Brink and G. R. Satchler, *Angular Momentum* (Clarendon Press, 1968).

BRODSKY AND MONIZ (1986)   S. Brodsky and E. Moniz (eds.), *Workshop on Nuclear Chromodynamics* (World Scientific, 1986).

BROWN (1982a)   G. E. Brown, *Prog. Part. Nucl. Phys.* **8**, 147 (1982).

BROWN (1982b)   G. E. Brown, *Nucl. Phys.* **A374**, 63c (1982).

BROWN ET AL. (1984)   G. E. Brown, A. D. Jackson, M. Rho, and V. Vento, *Phys. Lett.* **140B**, 285 (1984).

BROWN AND RHO (1979)   G. E. Brown and M. Rho, *Phys. Lett.* **82B**, 177 (1979).

BROWN AND RHO (1983)   G. E. Brown and M. Rho, *Phys. Today,* February (1983).

BROWN AND RHO (1986)   G. E. Brown and M. Rho, *Comm. Nucl. Part. Phys.* **15**, 245 (1986).

BURAS (1980)   A. J. Buras, *Rev. Mod. Phys.* **52**, 199 (1980).

BURNETT AND SHARPE (1990)   T. H. Burnett and S. R. Sharpe, *Annu. Rev. Nucl. Part. Sci.* **40**, 328 (1990).

BUSSIERE ET AL. (1988)   A. Bussiere et al., *Zeit. Phys.* **C38**, 117 (1988).

BUSZA (1984)   W. Busza, *Nucl. Phys.* **A418**, 635c (1984).

BUSZA AND LEDOUX (1988)   W. Busza and R. Ledoux, *Annu. Rev. Nucl. Part. Sci.* **38**, 119 (1988).

CABIBBO (1963)   N. Cabibbo, *Phys. Rev. Lett.* **10**, 531 (1963).

CALLAN AND GROSS (1969)   C. G. Callan and D. J. Gross, *Phys. Rev. Lett.* **22**, 156 (1969).

CALLAN, DASHEN, AND GROSS (1976)   C. G. Callan, R. F. Dashen, and D. J. Gross, *Phys. Lett.* **63B**, 334 (1976).

CALLAN, DASHEN, AND GROSS (1979)   C. G. Callan, R. F. Dashen, and D. J. Gross, *Phys. Rev.* **D19**, 1826 (1979).

CAMPBELL (1978)   D. K. Campbell, in *Nuclear Physics with Heavy Ions and Mesons, Vol. 2* (ed. by R. Balian, M. Rho, and G. Ripka; North–Holland, 1978).

CAPSTICK AND ISGUR (1986)   S. Capstick and N. Isgur, *Phys. Rev.* **D34**, 2809 (1986).

CAPSTICK ET AL. (1986)   S. Capstick, S. Godfrey, N. Isgur, and J. Paton, *Phys. Lett.* **175B**, 457 (1986).

CASHER (1979)   A. Casher, *Phys. Lett.* **83B**, 395 (1979).

CELMASTER AND GONSALVES (1979)   W. Celmaster and R. J. Gonsalves, *Phys. Rev.* **D20**, 1420 (1979).

CHAICHIAN AND NELIPA (1984)   M. Chaichian and N. F. Nelipa, *Introduction to Gauge Field Theories* (Springer–Verlag, 1984).

CHANOWITZ (1988)   M. S. Chanowitz, *Annu. Rev. Nucl. Part. Sci.* **38**, 323 (1988).

CHEN (1989)   J.-Q. Chen, *Group Representation Theory for Physicists* (World Scientific, 1989).

CHENG AND LI (1984)   T.-P. Cheng and L.-F. Li, *Gauge Theory of Elementary Particle Physics* (Clarendon Press, 1984).

CHENG AND O'NEILL (1979)   D. C. Cheng and G. K. O'Neill, *Elementary Particle Physics: An Introduction* (Addison–Wesley, 1979).

CHODOS AND THORN (1975)   A. Chodos and C. B. Thorn, *Phys. Rev.* **D12**, 2733 (1975).

CHRISTENSON ET AL. (1964)   J. H. Christenson et al., *Phys. Rev. Lett.* **13**, 138 (1964).

CLEYMANS, GAVAI, AND SUHONEN (1986)   J. Cleymans, R. V. Gavai, and E. Suhonen, *Phys. Rep.* **130**, 217 (1986).

CLOSE (1979)   F. E. Close, *An Introduction to Quarks and Partons* (Academic Press, 1979).

CLOSE (1982a)   F. E. Close, in *Quarks and Nuclear Forces* (ed. by D. Fries and B. Zeitnitz; Springer–Verlag, 1982).

CLOSE (1982b)   F. Close, *Phys. Scripta* **25**, 86 (1982).

COLEMAN (1975)   S. Coleman, in *Laws of Hadronic Matter* (ed. by A. Zichichi; Academic Press, 1975). Reprinted as Ch. 5 of Coleman (1985).

COLEMAN (1977)   S. Coleman, in *New Phenomena in Subnuclear Physics* (ed. by A. Zichichi; Plenum Press, 1977). Reprinted as Ch. 6 of Coleman (1985).

COLEMAN (1985)   S. Coleman, *Aspects of Symmetry* (Cambridge University Press, 1985).

COLEMAN AND WEINBERG (1973)   S. Coleman and E. Weinberg, *Phys. Rev.* **D7**, 1888 (1973).

COLLINS (1984)   J. C. Collins, *Renormalization* (Cambridge University Press, 1984).

COLLINS, MARTIN, AND SQUIRES (1989)   P. D. B. Collins, A. D. Martin, and E. J. Squires, *Particle Physics and Cosmology* (Wiley Interscience, 1989).

COLLINS AND PERRY (1975)   J. C. Collins and M. J. Perry, *Phys. Rev. Lett.* **34**, 1353 (1975).

COLLINS AND SOPER (1987)   J. C. Collins and D. E. Soper, *Annu. Rev. Nucl. Part. Sci.* **37**, 383 (1987).

CREUTZ (1979)   M. Creutz, *Phys. Rev. Lett.* **43**, 553 (1979).

CREUTZ (1980a)   M. Creutz, *Phys. Rev.* **D21**, 2308 (1980).

CREUTZ (1980b)   M. Creutz, *Phys. Rev. Lett.* **45**, 313 (1980).

CREUTZ (1985)   M. Creutz, *Quarks, Gluons, and Lattices* (Cambridge University Press, 1985).

CREUTZ, JACOBS, AND REBBI (1979)   M. Creutz, L. Jacobs, and C. Rebbi, *Phys. Rev.* **D20**, 1915 (1979).

CVITANOVIC (1983) P. Cvitanovic, *Field Theory*, NORDITA Lecture Notes (Copenhagen, 1983).

DALITZ (1982)   R. H. Dalitz, *Prog. Part. Nucl. Phys.* **8**, 7 (1982).

DASHEN (1969)   R. Dashen, *Phys. Rev.* **183**, 1245 (1969).

DAVIS, GOLDHABER, AND NIETO (1975)   L. Davis, A. S. Goldhaber, and M. M. Nieto, *Phys. Rev. Lett.* **35**, 1402 (1975).

DeGrand et al. (1975)   T. DeGrand, R. L. Jaffe, K. Johnson, and J. Kiskis, *Phys. Rev.* **D12**, 2060 (1975).

Derrick (1964)   G. H. Derrick, *J. Math. Phys.* **5**, 1252 (1964).

De Rújula, Georgi, and Glashow (1975)   A. De Rújula, H. Georgi, and S. L. Glashow, *Phys. Rev.* **D12**, 147 (1975).

DeTar and Donoghue (1983)   C. E. DeTar and J. F. Donoghue, *Annu. Rev. Nucl. Part. Sci.* **33**, 235 (1983).

Dine, Fischler, and Srednicki (1981)   M. Dine, W. Fischler, and M. Srednicki, *Phys. Lett.* **104B**, 199 (1981).

Dirac (1958)   P. A. M. Dirac, *The Principles of Quantum Mechanics* (Oxford University Press, 1958).

Dolan and Jackiw (1974)   L. Dolan and R. Jackiw, *Phys. Rev.* **D9**, 3320 (1974).

Dothan and Biedenharn (1987)   Y. Dothan and L. C. Biedenharn, *Comm. Nucl. Part. Phys.* **17**, 63 (1987).

Drees and Montgomery (1983)   J. Drees and H. E. Montgomery, *Annu. Rev. Nucl. Part. Sci.* **33**, 383 (1983).

Duke and Owens (1985)   D. W. Duke and J. F. Owens (eds.), *Advances in Lattice Gauge Theory* (World Scientific, 1985).

Duke and Roberts (1985)   D. W. Duke and R. G. Roberts, *Phys. Rep.* **120**, 275 (1985).

Dylla and King (1973)   H. F. Dylla and J. G. King, *Phys. Rev.* **A7**, 1224 (1973).

Elitzur (1975)   S. Elitzur, *Phys. Rev.* **D12**, 3978 (1975).

Elliott and Dawber (1979)   J. P. Elliott and P. G. Dawber, *Symmetry in Physics, Vols. 1 and 2* (Oxford University Press, 1979).

Ellis et al. (1977)   J. Ellis, M. K. Gaillard, D. V. Nanopoulos, and S. Rudaz, *Nucl. Phys.* **B131**, 285 (1977).

Englert and Brout (1964)   F. Englert and R. Brout, *Phys. Rev. Lett.* **13**, 321 (1964).

Faddeev (1976)   L. D. Faddeev, in *Methods in Field Theory* (ed. by R. Balian and J. Zinn-Justin; North–Holland, 1976).

Faddeev and Popov (1967)   L. D. Faddeev and V. N. Popov, *Phys. Lett.* **25B**, 29 (1967).

Fetter and Walecka (1971)   A. L. Fetter and J. D. Walecka, *Quantum Theory of Many-Particle Systems* (McGraw–Hill, 1971).

Feynman (1949)   R. P. Feynman, *Phys. Rev.* **76**, 749, 769 (1949).

Feynman (1961)   R. P. Feynman, *Quantum Electrodynamics* (Benjamin, 1961).

Feynman (1972)   R. P. Feynman, *Photon–Hadron Interactions* (Benjamin, 1972).

Feynman and Gell-Mann (1958)   R. P. Feynman and M. Gell-Mann, *Phys. Rev.* **109**, 193 (1958).

Feynman and Hibbs (1965)   R. P. Feynman and A. R. Hibbs, *Quantum Mechanics and Path Integrals* (McGraw–Hill, 1965).

Fisher (1967)   M. E. Fisher, *Rep. Prog. Phys.* **30**, 615 (1967).

FISHER (1982)   M. E. Fisher, lecture notes presented at "Advanced Course on Critical Phenomena," Merensky Inst. of Physics, Stellenbosch, S. Africa (January, 1982).

FLÜGGE (1982)   G. Flügge, in *Quarks and Nuclear Forces* (ed. by D. Fries and B. Zeitnitz; Springer–Verlag, 1982).

FRANZINI AND LEE-FRANZINI (1983)   P. Franzini and J. Lee-Franzini, *Annu. Rev. Nucl. Part. Sci.* **33**, 1 (1983).

FUCITO ET AL. (1981)   R. Fucito, E. Maninari, G. Parisi, and C. Rebbi, *Nucl. Phys.* **B180**, 369 (1981).

FUKUGITA (1988)   M. Fukugita, *Nucl. Phys. (Proc. Suppl.)* **B4**, 105 (1988).

FUKUGITA (1989)   M. Fukugita, *Nucl. Phys. (Proc. Suppl.)* **B9**, 291 (1989).

GAILLARD, LEE, AND ROSNER (1975)   M. K. Gaillard, B. W. Lee, and J. L. Rosner, *Rev. Mod. Phys.* **47**, 277 (1975).

GASIOROWICZ (1966)   S. Gasiorowicz, *Elementary Particle Physics* (Wiley, 1966).

GASIOROWICZ AND ROSNER (1981)   S. Gasiorowicz and J. L. Rosner, *Am. J. Phys.* **49**, 954 (1981).

GELL-MANN (1962)   M. Gell-Mann, *Phys. Rev.* **125**, 1067 (1962).

GELL-MANN (1964)   M. Gell-Mann, *Physics* **1**, 63 (1964).

GELL-MANN AND LEVY (1960)   M. Gell-Mann and M. Levy, *Nuovo Cim.* **16**, 705 (1960).

GEORGI (1982)   H. Georgi, *Lie Algebras in Particle Physics* (Benjamin/Cummings, 1982).

GEORGI AND GLASHOW (1972)   H. Georgi and S. L. Glashow, *Phys. Rev. Lett.* **28**, 1494 (1972).

GEORGI AND GLASHOW (1974)   H. Georgi and S. L. Glashow, *Phys. Rev. Lett.* **32**, 438 (1974).

GEORGI, QUINN, AND WEINBERG (1974)   H. Georgi, H. R. Quinn, and S. Weinberg, *Phys. Rev. Lett.* **33**, 451 (1974).

GERVAIS AND JACOB (1983)   J.-L. Gervais and M. Jacob, (eds.), *Non-Linear and Collective Phenomena in Quantum Physics* (World Scientific, 1983).

GILMORE (1974)   R. Gilmore, *Lie Groups, Lie Algebras, and Some of Their Applications* (Wiley, 1974).

GLASHOW (1961)   S. L. Glashow, *Nucl. Phys.* **22**, 579 (1961).

GLASHOW (1979)   S. L. Glashow, *Physica* **96A**, 27 (1979).

GLASHOW, ILIOPOULOS, AND MAIANI (1970)   S. L. Glashow, J. Iliopoulos, and L. Maiani, *Phys. Rev.* **D2**, 1285 (1970).

GODFREY AND ISGUR (1985)   S. Godfrey and N. Isgur, *Phys. Rev.* **D32**, 189 (1985).

GOLDSTEIN (1981)   H. Goldstein, *Classical Mechanics* (Addison–Wesley, 1981).

GOLDSTONE (1961)   J. Goldstone, *Nuovo Cim.* **19**, 154 (1961).

GOLDSTONE, SALAM, AND WEINBERG (1962)   J. Goldstone, A. Salam, and S. Weinberg, *Phys. Rev.* **127**, 965 (1962).

GOTTLIEB ET AL. (1989)   S. Gottlieb, W. Liu, R. L. Renken, R. L. Sugar, and D. Toussaint, *Phys. Rev.* **D40**, 2389 (1989).

GREEN (1985)   M. B. Green, *Nature* **314**, 409 (1985).

GREINER AND MÜLLER (1989)   W. Greiner and B. Müller, *Quantum Mechanics: Symmetries* (Springer–Verlag, 1989).

GROSS (1979)   D. J. Gross, *Phys. Rep.* **49**, 143 (1979).

GROSS AND WILCZEK (1973)   D. J. Gross and F. Wilczek, *Phys. Rev. Lett.* **30**, 1343 (1973).

GURALNIK, HAGEN, AND KIBBLE (1964)   G. S. Guralnik, C. R. Hagen, and T. W. B. Kibble, *Phys. Rev. Lett.* **13**, 585 (1964).

GURALNIK, HAGEN, AND KIBBLE (1968)   G. S. Guralnik, C. R. Hagen, and T. W. B. Kibble, *Adv. Particle Phys.* **2**, 567 (1968).

GUTH (1981)   A. H. Guth, *Phys. Rev.* **D23**, 347 (1981).

GYULASSY (1988)   M. Gyulassy, *Zeit. Phys.* **C38**, 361 (1988).

HALZEN AND MARTIN (1984)   F. Halzen and A. D. Martin, *Quarks and Leptons: An Introductory Course in Modern Particle Physics* (Wiley, 1984).

HAMERMESH (1962)   M. Hamermesh, *Group Theory and Its Application to Physical Problems* (Addison–Wesley, 1962).

HARRIS (1975)   E. G. Harris, *Introduction to Modern Theoretical Physics, Vols. 1 and 2* (Wiley, 1975).

HASENFRATZ (1987)   P. Hasenfratz, in *Proc. XXIII Int. Conf. High Energy Phys., Vol. 1* (ed. by S. C. Loken; World Scientific, 1987).

HASENFRATZ AND HASENFRATZ (1980)   A. Hasenfratz and P. Hasenfratz, *Phys. Lett.* **93B**, 165 (1980).

HASENFRATZ AND HASENFRATZ (1985)   A. Hasenfratz and P. Hasenfratz, *Annu. Rev. Nucl. Part. Sci.* **35**, 559 (1985).

HELLER (1982)   L. Heller, in *Quarks and Nuclear Forces* (ed. by D. Fries and B. Zeitnitz; Springer–Verlag, 1982).

HERB ET AL. (1977)   S. W. Herb et al., *Phys. Rev. Lett.* **39**, 252 (1977).

HERNÁNDEZ ET AL. (1990)   J. J. Hernández et al., "Review of Particle Properties", *Phys. Lett.* **239B**, 1 (1990).

HIGGS (1964)   P. W. Higgs, Phys. Rev. Lett. **12**, 132 (1964); **13**, 508 (1964).

HINCHCLIFFE (1986)   I. Hinchcliffe, *Annu. Rev. Nucl. Part. Sci.* **36**, 505 (1986).

HUANG (1982)   K. Huang, *Quarks, Leptons, and Gauge Fields* (World Scientific, 1982).

IIZUKA (1966)   J. Iizuka, *Prog. Theor. Phys. (Suppl.)* **37–8**, 21 (1966).

ISGUR (1980)   N. Isgur, (ed.), *Proc. of Baryon '80*, Univ. of Toronto, Toronto (1980).

ISGUR AND KARL (1978)   N. Isgur and G. Karl, *Phys. Rev.* **D18**, 4187 (1978).

ISGUR AND KARL (1979a)   N. Isgur and G. Karl, *Phys. Rev.* **D19**, 2653 (1979).

ISGUR AND KARL (1979b)   N. Isgur and G. Karl, *Phys. Rev.* **D20**, 1191 (1979).

ISGUR AND KARL (1983)   N. Isgur and G. Karl, *Phys. Today*, November (1983).

ITZYKSON AND NAUENBERG (1966)   C. Itzykson and M. Nauenberg, *Rev. Mod. Phys.* **38**, 95 (1966).

ITZYKSON AND ZUBER (1980)   C. Itzykson and J.-B. Zuber, *Quantum Field Theory* (McGraw–Hill, 1980).

JACKIW (1986)   R. Jackiw, in *High Energy Physics 1985, Vol. 1* (ed. by M. Bowick and F. Gürsey; World Scientific, 1986).

JACKIW AND REBBI (1976)   R. Jackiw and C. Rebbi, *Phys. Rev. Lett.* **37**, 172 (1976).

JACKSON (1975)   J. D. Jackson, *Classical Electrodynamics* (Wiley, 1975).

JACKSON AND RHO (1983)   A. D. Jackson and M. Rho, *Phys. Rev. Lett.* **51**, 751 (1983).

JAUCH AND ROHRLICH (1955)   J. M. Jauch and F. Rohrlich, *The Theory of Photons and Electrons* (Addison–Wesley, 1955).

JOHNSON (1978)   K. Johnson, *Phys. Lett.* **78B**, 259 (1978).

JOHNSON AND THORN (1976)   K. Johnson and C. B. Thorn, *Phys. Rev.* **D13**, 1934 (1976).

JONES (1984)   W. V. Jones, *Nucl. Phys.* **A418**, 139c (1984).

JONES ET AL. (1987)   W. V. Jones, Y. Takahashi, B. Wosiek, and O. Miyamura, *Annu. Rev. Nucl. Part. Sci.* **37**, 71 (1987).

KADANOFF (1976)   L. P. Kadanoff, in *Phase Transitions and Critical Phenomena, Vol. 5A* (ed. by C. Domb and M. S. Green; Academic Press, 1976).

KAJANTIE (1984)   K. Kajantie, *Nucl. Phys.* **A418**, 41c (1984).

KAJANTIE AND MCLERRAN (1987)   K. Kajantie and L. McLerran, *Annu. Rev. Nucl. Part. Sci.* **37**, 293 (1987).

KAPUSTA (1989)   J. I. Kapusta, *Finite-Temperature Field Theory* (Cambridge University Press, 1989).

KAPUSTA AND STROTTMAN (1981)   J. Kapusta and D. Strottman, *Phys. Lett.* **106B**, 33 (1981).

KARSCH (1988)   F. Karsch, *Zeit. Phys.* **C38**, 147 (1988).

KARSCH ET AL. (1987)   F. Karsch, J. B. Kogut, D. K. Sinclair, and H. W. Wyld, *Phys. Lett.* **188B**, 353 (1987).

KAYSER (1985)   B. Kayser, *Comm. Nucl. Part. Phys.* **14**, 69 (1985).

KIM (1979)   J. E. Kim, *Phys. Rev. Lett.* **43**, 103 (1979).

KITTEL (1976)   C. Kittel, *Introduction to Solid State Physics* (Wiley, 1976).

KOBAYASHI AND MASKAWA (1973)   M. Kobayashi and T. Maskawa, *Prog. Theor. Phys.* **49**, 652 (1973).

KOGUT (1979)   J. B. Kogut, *Rev. Mod. Phys.* **51**, 659 (1979).

KOGUT (1980)   J. B. Kogut, *Phys. Rep.* **67**, 67 (1980).

KOGUT ET AL. (1982)   J. Kogut, M. Stone, H. W. Wyld, J. Shigemitsu, S. H. Shenker, and D. K. Sinclair, *Phys. Rev. Lett.* **48**, 1140 (1982).

KOGUT (1983)   J. B. Kogut, *Rev. Mod. Phys.* **55**, 775 (1983).

KOGUT (1984)   J. B. Kogut, *Nucl. Phys.* **A418**, 381c (1984).

KOGUT, PEARSON, AND SHIGEMITSU (1981)   J. B. Kogut, R. P. Pearson, and J. Shigemitsu, *Phys. Lett.* **98B**, 63 (1981).

KOGUT AND SINCLAIR (1989)   J. B. Kogut and D. K. Sinclair, *Phys. Rev.* **D39**, 636 (1989).

KOGUT AND SUSSKIND (1975)   J. Kogut and L. Susskind, *Phys. Rev.* **D11**, 395 (1975).

KOLB (1987)   E. W. Kolb, in *From the Planck Scale to the Weak Scale: Toward a Theory of the Universe* (ed. by H. E. Haber; World Scientific, 1987).

KOLB AND TURNER (1983)   E. W. Kolb and M. S. Turner, *Annu. Rev. Nucl. Part. Sci.* **33**, 645 (1983).

KOLB AND TURNER (1990)   E. W. Kolb and M. S. Turner, *The Early Universe* (Addison–Wesley, 1990).

KOVACS, SINCLAIR, AND KOGUT (1987)   E. V. E. Kovacs, D. K. Sinclair, and J. B. Kogut, *Phys. Rev. Lett.* **58**, 751 (1987).

KÖPKE AND WERMES (1989)   L. Köpke and N. Wermes, *Phys. Rep.* **174**, 67 (1989).

KUO AND PANTALEONE (1989)   T. K. Kuo and J. Pantaleone, *Rev. Mod. Phys.* **61**, 937 (1989).

KWONG, ROSNER, AND QUIGG (1987)   W. Kwong, J. L. Rosner, and C. Quigg, *Annu. Rev. Nucl. Part. Sci.* **37**, 325 (1987).

LANDAU AND LIFSHITZ (1958)   L. D. Landau and E. M. Lifshitz, *Statistical Physics* (Addison–Wesley, 1958).

LANDAU AND LIFSHITZ (1971)   L. D. Landau and E. M. Lifshitz, *The Classical Theory of Fields* (Addison–Wesley, 1971).

LANGACKER (1981)   P. Langacker, *Phys. Rep.* **72**, 185 (1981).

LAUTRUP AND NAUENBERG (1980)   B. Lautrup and M. Nauenberg, *Phys. Lett.* **95B**, 63 (1980).

LEADER AND PREDAZZI (1982)   E. Leader and E. Predazzi, *An Introduction to Gauge Theories and the 'New Physics'* (Cambridge University Press, 1982).

LEE (1972)   B. W. Lee, *Chiral Dynamics* (Gordon and Breach, 1972).

LEE (1975)   T. D. Lee, *Rev. Mod. Phys.* **47**, 267 (1975).

LEE (1981)   T. D. Lee, *Particle Physics and Introduction to Field Theory* (Harwood, 1981).

LEE AND YANG (1956)   T. D. Lee and C. N. Yang, *Phys. Rev.* **104**, 254 (1956).

LEE AND ZINN-JUSTIN (1972)   B. W. Lee and J. Zinn-Justin, *Phys. Rev.* **D5**, 3121, 3137, 3155 (1972); **D7**, 1049 (1973).

LEIBBRANDT (1975)   G. Leibbrandt, *Rev. Mod. Phys.* **47**, 849 (1975).

LEPAGE AND BRODSKY (1980)   G. P. Lepage and S. J. Brodsky, *Phys. Rev.* **D22**, 2157 (1980).

LICHTENBERG (1978)   D. B. Lichtenberg, *Unitary Symmetry and Elementary Particles* (Academic Press, 1978).

LINDE (1982)   A. D. Linde, *Phys. Lett.* **108B**, 389 (1982).

LINDE (1983)   A. D. Linde, *Phys. Lett.* **129B**, 177 (1983).

LIPKIN (1966)   H. J. Lipkin, *Lie Groups for Pedestrians* (North–Holland, 1966).

LIU (1987)   K.-F. Liu (ed.), *Chiral Solitons* (World Scientific, 1987).

MA (1973)   S.-K. Ma, *Rev. Mod. Phys.* **45**, 589 (1973).

MA (1976)   S.-K. Ma, *Modern Theory of Critical Phenomena* (Benjamin, 1976).

MAHAN (1981)   G. D. Mahan, *Many-Particle Physics* (Plenum Press, 1981).

MANDL (1959)   F. Mandl, *Introduction to Quantum Field Theory* (Interscience, 1959).

MARCIANO (1983)   W. J. Marciano, in *Proc. 1983 Int. Symp. Lepton and Photon Interactions* (ed. by D. Cassel and D. Kreinick; Cornell University Press, 1983).

MARCIANO AND PARSA (1986)   W. J. Marciano and Z. Parsa, *Annu. Rev. Nucl. Part. Sci.* **36**, 171 (1986).

MASSMANN, RING, AND RASMUSSEN (1975)   H. Massmann, P. Ring, and J. O. Rasmussen, *Phys. Lett.* **57B**, 417 (1975); H. Massmann, Ph. D. Thesis, U. C. Berkeley (1975).

MATSUI AND SATZ (1986)   T. Matsui and H. Satz, *Phys. Lett.* **178B**, 416 (1986).

MATTUCK (1976)   R. D. Mattuck, *A Guide to Feynman Diagrams in the Many-Body Problem* (McGraw–Hill, 1976).

McLERRAN (1982)   L. McLerran, in *Quark Matter Formation and Heavy Ion Collisions* (ed. by M. Jacob and H. Satz; World Scientific, 1982).

McLERRAN (1984)   L. McLerran, *Nucl. Phys.* **A418**, 401c (1984).

McLERRAN (1986)   L. McLerran, *Rev. Mod. Phys.* **58**, 1021 (1986).

McLERRAN (1987)   L. McLerran, in *Proc. XXIII Int. Conf. on High Energy Physics* (ed. by S. C. Loken; World Scientific, 1987).

McLERRAN AND SVETITSKY (1981)   L. McLerran and B. Svetitsky, *Phys. Rev.* **D24**, 450 (1981).

McVOY (1965)   K. W. McVoy, *Rev. Mod. Phys.* **37**, 84 (1965).

MEISSNER AND ZAHED (1986)   U.-G. Meissner and I. Zahed, *Adv. Nucl. Phys.* **17**, 143 (1986).

MESSIAH (1958)   A. Messiah, *Quantum Mechanics, Vols. 1 and 2* (Wiley, 1958).

METROPOLIS ET AL. (1953)   N. Metropolis, A. W. Rosenbluth, M. N. Rosenbluth, A. H. Teller, and E. Teller, *J. Chem. Phys.* **21**, 1087 (1953).

MIAKE ET AL. (1988)   Y. Miake et al., *Zeit. Phys.* **C38**, 135 (1988).

MILLER (1984)   G. S. Miller, *Int. Rev. Nucl. Phys.* **1**, 190 (1984).

MILLER AND GEORGE (1972)   W. H. Miller and T. F. George, *J. Chem. Phys.* **56**, 5668 (1972); T. F. George and W. H. Miller, *J. Chem. Phys.* **57**, 2458 (1972).

MILTON (1983)   K. A. Milton, *Phys. Rev.* **D27**, 439 (1983).

MISNER, THORNE, AND WHEELER (1973)   C. W. Misner, K. S. Thorne, and J. A. Wheeler, *Gravitation* (W. H. Freeman, 1973).

MORSE (1969)   P. M. Morse, *Thermal Physics* (Benjamin/Cummings, 1969).

MUELLER (1981)   A. Mueller, *Phys. Rep.* **73**, 237 (1981).

MUIRHEAD (1965)   H. Muirhead, *The Physics of Elementary Particles* (Pergamon Press, 1965).

MUTA (1987)   T. Muta, *Foundations of Quantum Chromodynamics* (World Scientific, 1987).

MYHRER (1984)   F. Myhrer, *Int. Rev. Nucl. Phys.* **1**, 326 (1984).

MYHRER AND WROLDSEN (1988)   F. Myhrer and J. Wroldsen, *Rev. Mod. Phys.* **60**, 629 (1988).

NAMBU (1960)   Y. Nambu, *Phys. Rev. Lett.* **4**, 380 (1960).

NAMBU AND JONA-LASINIO (1961)   Y. Nambu and G. Jona-Lasinio, *Phys. Rev.* **122**, 345; **124**, 246 (1961).

NASH AND SEN (1983)   C. Nash and S. Sen, *Topology and Geometry for Physicists* (Academic Press, 1983).

NEGELE AND ORLAND (1988)   J. W. Negele and H. Orland, *Quantum Many-Particle Systems* (Addison–Wesley, 1988).

NIELSEN (1981)   N. K. Nielsen, *Am. J. Phys.* **49**, 1171 (1981).

NIELSEN AND OLESEN (1973)   H. B. Nielsen and P. Oleson, *Nucl. Phys.* **B61**, 45 (1973).

NIEMI AND SEMENOFF (1986)   A. J. Niemi and G. W. Semenoff, *Phys. Rep.* **135**, 99 (1986).

OKA AND YAZAKI (1984)   M. Oka and K. Yazaki, *Int. Rev. Nucl. Phys.* **1**, 489 (1984).

OKUBO (1963)   S. Okubo, *Phys. Lett.* **5**, 163 (1963).

OKUN (1982)   L. B. Okun, *Leptons and Quarks* (North–Holland, 1982).

OLINTO (1988)   A. V. Olinto, *Zeit. Phys.* **C38**, 303 (1988).

OLIVE (1990)   K. A. Olive, *Phys. Rep.* **190**, 307 (1990).

OLIVE AND SCHRAMM (1985)   K. A. Olive and D. N. Schramm, *Comm. Nucl. Part. Phys.* **15**, 69 (1985).

O'RAIFEARTAIGH (1986)   L. O'Raifeartaigh, *Group Structure of Gauge Theories* (Cambridge University Press, 1986).

PAGELS (1975)   H. Pagels, *Phys. Rep.* **16**, 219 (1975).

PARK ET AL. (1985)   H. S. Park et al., *Phys. Rev. Lett.* **54**, 22 (1985).

PECCEI AND QUINN (1977)   R. D. Peccei and H. R. Quinn, *Phys. Rev.* **D16**, 1791 (1977).

PERKINS (1984)   D. H. Perkins, *Annu. Rev. Nucl. Part. Sci.* **34**, 1 (1984).

PERKINS (1987)   D. H. Perkins, *Introduction to High Energy Physics* (Addison–Wesley, 1987).

PI (1985)   S.-Y. Pi, *Comm. Nucl. Part. Phys.* **14**, 273 (1985).

PISARSKI (1982)   R. D. Pisarski, *Phys. Lett.* **110B**, 155 (1982).

PLUNIEN, MÜLLER, AND GREINER (1986)   G. Plunien, B. Müller, and W. Greiner, *Phys. Rep.* **134**, 87 (1986).

POKORSKI (1987)   S. Pokorski, *Gauge Field Theories* (Cambridge University Press, 1987).

POLITZER (1973)   H. D. Politzer, *Phys. Rev. Lett.* **30**, 1346 (1973).

POLONYI AND SZLACHANYI (1982)   J. Polonyi and K. Szlachanyi, *Phys. Lett.* **110B**, 395 (1982).

POLONYI AND WYLD (1984)   J. Polonyi and H. W. Wyld, *Nucl. Phys.* **A418**, 491c (1984).

POLYAKOV (1974)   A. M. Polyakov, *JETP Lett.* **20**, 194 (1974).

POLYAKOV (1975)   A. M. Polyakov, *Phys. Lett.* **59B**, 82 (1975).

POLYAKOV (1977)   A. M. Polyakov, *Nucl. Phys.* **B120**, 429 (1977).

PRESKILL (1984)   J. Preskill, *Annu. Rev. Nucl. Part. Sci.* **34**, 461 (1984).

PRIMACK, SECKEL, AND SADOULET (1987)   J. R. Primack, D. Seckel, and B. Sadoulet, *Annu. Rev. Nucl. Part. Sci.* **38**, 751 (1988).

QUIGG (1983)   C. Quigg, *Gauge Theories of the Strong, Weak, and Electromagnetic Interactions* (Benjamin/Cummings, 1983).

RAJARAMAN (1982)   R. Rajaraman, *Solitons and Instantons* (North–Holland, 1982).

RAMOND (1981)   P. Ramond, *Field Theory: A Modern Primer* (Benjamin, 1981).

REBBI (1983)   C. Rebbi (ed.), *Lattice Gauge Theories and Monte Carlo Simulations* (World Scientific, 1983).

RENTON (1990)   P. Renton, *Electroweak Interactions* (Cambridge University Press, 1990).

RHO (1982)   M. Rho, *Prog. Part. Nucl. Phys.* **8**, 103 (1982).

RHO (1984)   M. Rho, *Annu. Rev. Nucl. Part. Sci.* **34**, 531 (1984).

RHO (1985)   M. Rho, *Nucl. Phys.* **A434**, 639c (1985).

RHO, GOLDHABER, AND BROWN (1983)   M. Rho, A. S. Goldhaber, and G. E. Brown, *Phys. Rev. Lett.* **51**, 747 (1983).

RING AND SCHUCK (1980)   P. Ring and P. Schuck, *The Nuclear Many-Body Problem* (Springer–Verlag, 1980).

RITH (1988)   K. Rith, *Zeit. Phys.* **C38**, 317 (1988).

RIVERS (1987)   R. J. Rivers, *Path Integral Methods in Quantum Field Theory* (Cambridge University Press, 1987).

ROBERTSON AND KNAPP (1988)   R. G. H. Robertson and D. A. Knapp, *Annu. Rev. Nucl. Part. Sci.* **38**, 185 (1988).

ROMAN (1969)   P. Roman, *Introduction to Quantum Field Theory* (Wiley, 1969).

ROSS (1984)   G. G. Ross, *Grand Unified Theories* (Benjamin/Cummings, 1984).

RYDER (1985)   L. H. Ryder, *Quantum Field Theory* (Cambridge University Press, 1985).

SAKHAROV (1967)   A. D. Sakharov, *JETP Lett.* **5**, 24 (1967).

SAKURAI (1967)   J. J. Sakurai, *Advanced Quantum Mechanics* (Addison–Wesley, 1967).

SALAM (1968)   A. Salam, in *Elementary Particle Theory* (ed. by N. Svartholm, Almqvist and Wiksell; Stockholm, 1968).

SASAKI (1983)   M. Sasaki, in *Proc. Workshop on Grand Unified Theories and the Early Universe* (ed. by M. Fukugita and M. Yoshimura), KEK 83–13, Tsukuba, Japan (1983).

SATZ (1983)   H. Satz, *Nucl. Phys.* **A400**, 541c (1983).

SATZ (1984)   H. Satz, *Nucl. Phys.* **A418**, 447c (1984).

SATZ (1985a)   H. Satz, *Annu. Rev. Nucl. Part. Sci.* **35**, 245 (1985).

SATZ (1985b)   H. Satz, in *Advances in Lattice Gauge Theory* (ed. by D. W. Duke and J. F. Owens; World Scientific, 1985).

SCHENSTED (1976)   I. V. Schensted, *A Course on the Application of Group Theory to Quantum Mechanics* (NEO Press, 1976).

SCHIFF (1968)   L. I. Schiff, *Quantum Mechanics* (McGraw–Hill, 1968).

SCHRAMM AND OLIVE (1984)   D. N. Schramm and K. A. Olive, *Nucl. Phys.* **A418**, 289c (1984).

SCHWARZ (1982)   J. H. Schwarz, *Phys. Rep.* **89**, 223 (1982).

SCHWEBER (1961)   S. S. Schweber, *An Introduction to Relativistic Quantum Field Theory* (Row–Peterson, 1961).

SCHWINGER (1969)   J. Schwinger, *Particles and Sources* (Gordon and Breach, 1969).

SHURYAK (1980)   E. V. Shuryak, *Phys. Rep.* **61**, 71 (1980).

SHURYAK (1981)   E. V. Shuryak, *Phys. Lett.* **107B**, 103 (1981).

SHURYAK (1984)   E. V. Shuryak, *Phys. Rep.* **115**, 151 (1984).

SHURYAK (1988)   E. V. Shuryak, *The QCD Vacuum, Hadrons, and the Superdense Matter* (World Scientific, 1988).

SIKIVIE (1982)   P. Sikivie, *Phys. Rev. Lett.* **48**, 1156 (1982).

SKYRME (1961)   T. H. R. Skyrme, *Proc. Roy. Soc. London* **A260**, 127 (1961).

SKYRME (1962)   T. H. R. Skyrme, *Nucl. Phys.* **31**, 556 (1962).

SLANSKY (1981)   R. Slansky, *Phys. Rep.* **79**, 1 (1981).

SÖDING AND WOLF (1981)   P. Söding and G. Wolf, *Annu. Rev. Nucl. Part. Sci.* **31**, 231 (1981).

SOHNIUS (1985)   M. F. Sohnius, *Phys. Rep.* **128**, 39 (1985).

STEINHARDT (1984)   P. J. Steinhardt, *Comm. Nucl. Part. Phys.* **12**, 273 (1984).

STEINHARDT (1986)   P. J. Steinhardt, in *High Energy Physics 1985, Vol. 2* (ed. by M. Bowick and F. Gürsey; World Scientific, 1986).

STÜCKELBERG (1941)   E. C. G. Stückelberg, *Helv. Phys. Acta* **14**, 588 (1941).

SUDARSHAN AND MARSHAK (1958)   E. C. G. Sudarshan and R. E. Marshak, *Phys. Rev.* **109**, 1860 (1958).

SVETITSKY (1986)   B. Svetitsky, *Phys. Rep.* **132**, 1 (1986).

SVETITSKY AND YAFFE (1982)   B. Svetitsky and L. G. Yaffe, *Nucl. Phys.* **B210**, 423 (1982).

TAYLOR (1976)   J. C. Taylor, *Gauge Theories of Weak Interactions* (Cambridge University Press, 1976).

TAYLER (1981)   R. J. Tayler, *Prog. Part. Nucl. Phys.* **6**, 5 (1981).

THOMAS (1983)   A. W. Thomas, *"Towards a Description of Nuclear Physics at the Quark Level"* (Th. 3668–CERN). Lectures presented at International School of Nucl. Phys., Erice (1983).

THOMAS (1984)   A. W. Thomas, *Adv. Nucl. Phys.* **13**, 1 (1984).

'T HOOFT (1971)   G. 't Hooft, *Nucl. Phys.* **B33**, 173 (1971); **B35**, 167 (1971).

'T HOOFT (1974a)   G. 't Hooft, *Nucl. Phys.* **B79**, 276 (1974).

'T HOOFT (1974b)   G. 't Hooft, *Nucl. Phys.* **B72**, 461 (1974).

'T HOOFT (1978)   G. 't Hooft, in *Particles and Fields* (ed. by D. H. Boal and A. N. Kamal; Plenum Press, 1978).

'T HOOFT AND VELTMAN (1972)   G. 't Hooft and M. Veltman, *Nucl. Phys.* **B44**, 189 (1972).

TOUSSAINT (1987)   D. Toussaint, in *Proc. Salt Lake City Meeting* (ed. by C. DeTar and J. Ball; World Scientific, 1987).

TOUSSAINT ET AL. (1979)   D. Toussaint, S. B. Treiman, F. Wilczek, and A. Zee, *Phys. Rev.* **D19**, 1036 (1979).

TRILLING (1981)   G. H. Trilling, *Phys. Rep.* **75**, 57 (1981).

TURNER (1990)   M. S. Turner, *Phys. Rep.* **197**, 67 (1990).

TURNER AND SCHRAMM (1979)   M. S. Turner and D. N. Schramm, *Phys. Today*, September (1979).

UKAWA (1989)   A. Ukawa, *Nucl. Phys.* **A498**, 227c (1989).

VAN HOVE AND POKORSKI (1975)   L. Van Hove and S. Pokorski, *Nucl. Phys.* **B86**, 243 (1975).

VENTO ET AL. (1980)   V. Vento, M. Rho, E. M. Nyman, J. H. Jun, and G. E. Brown, *Nucl. Phys.* **A345**, 413 (1980).

VENTO, BAYM, AND JACKSON (1981)   V. Vento, G. Baym, and A. D. Jackson, *Phys. Lett.* **102B**, 97 (1981).

VEPSTAS AND JACKSON (1990)   L. Vepstas and A. D. Jackson, *Phys. Rep.* **187**, 111 (1990).

WEGNER (1971)   F. J. Wegner, *J. Math. Phys.* **12**, 2259 (1971).

WEINBERG (1967)   S. Weinberg, *Phys. Rev. Lett.* **19**, 1264 (1967).

WEINBERG (1978)   S. Weinberg, *Phys. Rev. Lett.* **40**, 223 (1978).

WEINBERG (1979)   S. Weinberg, *Phys. Rev. Lett.* **42**, 850 (1979).

WEISE (1989)   W. Weise, *Nucl. Phys.* **A497**, 7c (1989).

WILCZEK (1978)   F. Wilczek, *Phys. Rev. Lett.* **40**, 279 (1978).

WILCZEK (1982)   F. Wilczek, *Annu. Rev. Nucl. Part. Sci.* **32**, 177 (1982).

WILETS (1987)   L. Wilets, in *Chiral Solitons* (ed. by K.-F. Liu; World Scientific, 1987).

WILSON (1974)   K. G. Wilson, *Phys. Rev.* **D10**, 2445 (1974).

WILSON (1976)   K. G. Wilson, *Phys. Rep.* **23**, 331 (1976).

WILSON (1977)   K. G. Wilson, in *New Phenomena in Subnuclear Physics* (ed. by A. Zichichi; Plenum Press, 1977).

WILSON (1979)   K. G. Wilson, *Sci. Am.* **241**, 158 (1979).

WILSON (1983a)   K. G. Wilson, *Rev. Mod. Phys.* **55**, 583 (1983).

WILSON (1983b)   K. G. Wilson, in *Proc. 1983 Int. Symp. on Lepton and Photon Interactions at High Energy* (ed. by D. G. Cassell and D. L. Kreinick; Cornell University Press, 1983).

WILSON AND KOGUT (1974)   K. G. Wilson and J. Kogut, *Phys. Rep.* **12**, 75 (1974).

WITTEN (1979)   E. Witten, *Nucl. Phys.* **B160**, 57 (1979).

WITTEN (1983)   E. Witten, *Nucl. Phys.* **B223**, 422, 433 (1983).

WONG (1986)   C. W. Wong, *Phys. Rep.* **136**, 1 (1986).

WU (1984)   S. L. Wu, *Phys. Rep.* **107**, 59 (1984).

WU ET AL. (1957)   C. S. Wu, E. Ambler, R. W. Hayward, D. D. Hoppes, and R. P. Hudson, *Phys. Rev.* **105**, 1413 (1957).

WYBOURNE (1974)   B. G. Wybourne, *Classical Groups for Physicists* (Wiley, 1974).

YAFFE AND SVETITSKY (1982)   L. G. Yaffe and B. Svetitsky, *Phys. Rev.* **D26**, 963 (1982).

YANG AND MILLS (1954)   C. N. Yang and R. L. Mills, *Phys. Rev.* **96**, 191 (1954).

ZAHED AND BROWN (1986)   I. Zahed and G. E. Brown, *Phys. Rep.* **142**, 1 (1986).

ZIMAN (1969)   J. M. Ziman, *Elements of Advanced Quantum Theory* (Cambridge University Press, 1969).

ZWEIG (1964)   G. Zweig, CERN reports 8182/TH401 and 8419/TH412 (1964).

Index

593